DATE DUE

JUL 2 0 2004			
6-15			
FEB 16 2011			

Statistical Analysis of Categorical Data

WILEY SERIES IN PROBABILITY AND STATISTICS
APPLIED PROBABILITY AND STATISTICS SECTION

Established by WALTER A. SHEWHART and SAMUEL S. WILKS

A complete list of the titles in this series appears at the end of this volume.

Statistical Analysis of Categorical Data

CHRIS J. LLOYD

A Wiley-Interscience Publication
JOHN WILEY & SONS, INC.
New York • Chichester • Weinheim • Brisbane • Singapore • Toronto

This book is printed on acid-free paper. ♾

Published simultaneously in Canada.

Library of Congress Cataloging-in-Publication Data:

Lloyd, Christopher J.
Statistical analysis of categorical data / Christopher J. Lloyd.
p. cm. — (Wiley series in probability and statistics. Applied probability and statistics)
"A Wiley-Interscience publication."
Includes bibliographical references (p. –) and index.
ISBN 0-471-29008-4 (acid-free paper)
1. Multivariate analysis. I. Title. II. Series.
QA278.L564 1999
519.5′35–DC21 98-39058
CIP

Printed in the United States of America.

10 9 8 7 6 5 4 3 2

To E. J. Williams
A Scholar and a Gentleman

Preface

A large amount of the data a statistician sees is categorical in nature: In biomedicine, ecology, social sciences, and marketing, to name but a few, categorical data dominates. Statisticians who do not have a grounding in the statistics of categorical data will quickly find themselves in deep water, deep enough that no chi-square test of independence will save them.

Detecting and analyzing statistical patterns in counts involves more than "a few tweaks of the knob" of analysis of variance. Once we leave the realm of normal least squares, statistics becomes less straightforward but more interesting. Fortunately, the past 30 years has seen much progress in clarifying the principles and practice of modeling count data. As a result of this rapid development, most statistics and biostatistics departments now offer courses in categorical data analysis.

This book is intended to be suitable for graduate statistics students in the U.S. system or for Coursework Masters students in the British or Australian systems. It is assumed that students have a background in basic inference and probability theory, linear algebra, and calculus. While students may well follow the methods without this background, I would question the wisdom of doing so. A summary of the chapter contents is as follows.

- Chapter 1: Review of statistical theory, including some advanced topics like the Score statistic and transformed central limit theorem.
- Chapter 2: Distribution theory of Poisson and Multinomial variables and the connections between these distributions.
- Chapter 3: Analysis of 2×2, $2 \times k$, and $2 \times 2 \times k$ contingency tables.
- Chapter 4: Logistic regression; Dose-response models; diagnostics; overdispersion; weighted least squares.
- Chapter 5: Smoothing of binary data using kernel methods; Further applications of smoothing to residual analysis and goodness-of-fit.
- Chapter 6: Log-linear models and their connection with Multinomial logit models; ordinal data; incomplete tables; multiple record systems.

- Chapter 7: Conditional methods covering most basic data forms covered in Chapters 4 and 6, and illustrated using Statxact and Logxact.

The core material of any categorical data analysis course is logistic regression (Chapters 3 and 4) and log-linear models (Chapter 6). The approach to both is regression based and I have tried to appeal as much as possible to intuition that students would have gained in undergraduate linear models courses. The essential theoretical background is covered in Chapters 1 and 2. Chapter 1 is a wide ranging review of statistical theory, with an emphasis on count data. This need not be covered in lectures, however, I believe that good students especially benefit from having ready access to basic theory and from relating this theory to applications.

Several modern topics, that are rapidly becoming essential knowledge for a statistician, have contributed to the book being rather long but also unique in its coverage. Ordinal data methods, at least those that fit easily into the categorical data methodology, are covered in Section 6.4. Chapter 5 covers smoothing of binomial data, especially the use of smoothing in residual analysis. Chapter 7 is on conditional inference which does not appear in any graduate program I have seen, perhaps for lack of a suitable textbook. However, the proliferation of computer packages for conditional inference makes neglect of this material harder and harder to justify.

How would an instructor use this book? A more elementary course could be based on Chapters 2 and 3, the first two sections of Chapter 4, and the first three sections of Chapter 6. I have covered this material in 24 one-hour lectures. A more advanced course or second course could cover Chapters 5 and 7, as well as the material on ordinal data, incomplete tables, and multiple record systems in Chapter 6.

The theory is illustrated on a wealth of real data sets, in many cases with annotated `GLIM` output. Where `GLIM` is inadequate I use the "gold standard" packages appropriate to the method. Brief manuals for `GLIM`, `MINITAB`, *Splus*, `LogXact`, and `StatXacT` are provided in a *TeX*.dvi file at the Wiley website http://www.wiley.com. This site also contains certain *Splus* functions and data sets used throughout the book, a set of solutions to selected exercises and a Fortran program for illustrating conditional inference on 2×2 tables.

This book has its genesis in a Masters course taught at La Trobe University between 1991 and 1995. Over those years, the lecture notes evolved into something that might form the foundation of a textbook. I got consistent feedback from ex-students that the material covered was more useful in the workplace than perhaps any other course they had studied, so I decided to embark on this long labor of love while visiting the University of Waterloo in 1993.

My thanks are due to several people: First, to Lina for proofreading the final version for typos. All of you who know me will appreciate just how many typos I can pack into 500 pages. To my sons Ali and Oli for preventing me from becoming a one-dimensional academic. To Steve Quigley of John Wiley

& Sons who was supportive from the start. Finally, the La Trobe staff were ideal colleagues and I learned a great deal during my years there.

CHRIS LLOYD

Hong Kong
July 1998

Contents

1. The Tools of Statistical Inference **1**

1.1 Statistical Models, 2
1.2 Fitting Statistical Models, 9
1.3 Large Sample Estimation Theory, 19
1.4 Testing One Parameter, 32
1.5 Testing Several Parameters, 40
1.6 Transforming Data and Parameters, 51
1.7 Exercises, 59

2. Distribution Theory for Count Data **68**

2.1 Categorical Data, 68
2.2 Multinomial Models, 69
2.3 Poisson Models, 80
2.4 The Multinomial-Poisson Connection, 84
2.5 Goodness-of-Fit, 87
2.6 Confidence Intervals, 101
2.7 Exercises, 107

3. Binary Contingency Tables **119**

3.1 Binomial Data, 120
3.2 Comparing Two Binomial Samples, 124
3.3 Comparing Several Binomial Samples, 131
3.4 Analysis of Several 2×2 Tables, 143
3.5 Inference Using Weighted Least Squares, 150
3.6 Some Cautions and Qualifications, 153
3.7 Exercises, 163

4. Binomial Regression Models **175**

4.1 Linear Binomial Models, 175
4.2 Diagnosis of Model Inadequacy, 198
4.3 Modeling Relative Risk, 219
4.4 Modeling Over-Dispersion, 226
4.5 Approximation by Normal GLM's, 236
4.6 Inference on Generalized Linear Models, 248
4.7 Exercises, 252

5. Smoothing Binomial Data **265**

5.1 Smoothing a Conditional Mean, 265
5.2 Smoothing Binary Data, 272
5.3 Estimating Attributable Response, 283
5.4 Smoothing Residuals, 287
5.5 Goodness-of-fit Against a Smooth, 294
5.6 Exercises, 296

6. Poisson Regression Models **298**

6.1 Log-Linear Models, 298
6.2 Multinomial Logit Models, 311
6.3 Modeling Independence, 326
6.4 Some Models for Ordinal Data, 338
6.5 Models for Incomplete Tables, 352
6.6 Multiple Record Systems, 367
6.7 Exercises, 384

7. Conditional Inference **395**

7.1 Conditional Inferences for 2×2 Tables, 396
7.2 Conditional Inferences for $2 \times k$ Tables, 407
7.3 Conditional Inferences on Linear Models, 413
7.4 Conditional Inferences for $r \times c$ Tables, 418
7.5 Conditional Inferences for $2 \times 2 \times k$ Tables, 422
7.6 Conditional Inference for Three-Way Tables, 427
7.7 Conditional Logistic Regression, 431
7.8 Conditional Goodness-of-Fit Tests, 435
7.9 Saddlepoint Approximation*, 437

Index **463**

CHAPTER 1

The Tools of Statistical Inference

Statistical analysis begins with a set of data and a set of questions about the world. The statistical analyst has the task of trying to understand the data and answer the questions. This involves looking for patterns in the data, some of which may be easy to see and others not so easy. How important are these patterns? Are they even real or just things we found from looking so hard? Some of these patterns could be reflections of the underlying mechanism that generated the data and provide partial answers to our questions. But as in any scientific investigation, we can never see the underlying mechanism directly, and we will never really know the truth. The statistician's job then is to discover the answers that the data seem to point to, and then to assign evidential weight to these answers in a systematic and quantitative manner.

Usually we know, or think we know, something about the mechanism that generated the data. Background about how the data were collected, as well as expert knowledge about the physical or biological processes underlying the data, mean that we can often exclude some descriptions of the data right from the start. Our background knowledge is summarized in a *statistical model* and we examine the data through the filter of this assumed model. Devising useful statistical models is not easy to do and is even less easy to teach, but a wide range of models will be studied in this book and their field of application should become clear from the many detailed examples provided. Armed with the model, the data are examined and combined with the model, a process called *fitting*. This leads to the more purely technical topics of statistical estimators and tests, and much of this chapter provides the essential theory of estimation and testing.

It is imperative to good statistical practice that the uncertainty of the results be evaluated. This rests on the very simple philosophy that a statement about whose accuracy we know nothing is worse than useless. First, uncertainty arises from the usually random nature of the data themselves; we must always consider the possibility that the data look the way they do by mere chance. Thus, estimators need a standard error or an associated confidence interval; tests should ideally be accompanied by an estimate of their power. A second source of uncertainty is misspecification of the model. While quantifying such uncertainty is often difficult, results based on several alternative well-fitting models may produce quite different results. Presenting results based on only one of them would then convey a false sense of accuracy. As a matter of principle, a single best statistical analysis should never be the aim of the analyst.

In this chapter some central statistical ideas are reviewed, with particular emphasis on categorical data. Most students will have already encountered notions of estimator, test, P-value, likelihood, etc., but important topics often ignored at the undergraduate level are covered and many detailed examples are given. The student seeking a deeper knowledge of statistical theory should find many of his/her questions answered. Of course there are many excellent alternative texts on statistical theory available and this chapter is not designed to compete with them. It does afford an opportunity to focus attention on categorical data, to establish some notation, and to provide a ready reference for specific results later, as required. More difficult or detailed sections are marked (*) and could be omitted the first time you read this book.

Section 1.1 covers the basic elements of a statistical model, including how hypotheses about the data can be expressed in terms of the parameters of the model. In Section 1.2, maximum likelihood estimation is described and then illustrated in situations of varying complexity. We consider the asymptotic theory of estimation in Section 1.3 and go through some very detailed calculations to see how accurately the approximations perform in practice. Asymptotic theory of testing is presented in Sections 1.4 and 1.5 and the classical trinity of Wald, Score, and Likelihood ratio statistics are compared and contrasted both in theory and through detailed calculations. Section 1.6 covers the more technical issues of approximating the distributions of transformed variables.

1.1 STATISTICAL MODELS

All serious statistical analyses begin with a statistical model which connects the known data with unknown quantities of interest. It provides a link between what we know and what we would like to know, without which no inference would be possible. In this section the main elements of statistical models are described.

1.1.1 The Elements of a Statistical Model

This book concerns categorical data and so we will take the data to be a set of counts $Y_1, \ldots, Y_k$ of k different categories of events. A *statistical model* is a set of assumptions about the joint distribution of the data. This set of assumptions naturally divides into two components.

The first component of a statistical model is an assumption that the distribution of the data, commonly called the *error distribution*, comes from some specific set, often a *parametric family*, of distributions. If conditions had been different then the distribution of the data may have been different, but still a member of this family. The role of the error distribution is to describe the *random* variations of the data about any systematic features or patterns. For continuous data, with which this book is not concerned, the normal distribution is often assumed. For categorical data the most common error distributions are Poisson and Multinomial. The mean values of the data will be denoted by $e_i = \mathrm{E}(Y_i)$ where expectation is with respect to the error distribution. The error distribution describes the random variation of the data Y_i about their mean values e_i.

The second component of a statistical model, called the *systematic* component, is a statement about the underlying pattern of the data. Commonly, this is a statement about the mean values of the Y_i, called a *regression function*. For instance, we might assume $\mathrm{E}(Y_i)$ is a linear function of some measured covariate x_i. The actual data are regarded as having been generated by unexplained random variation about this regression function. Much of statistical analysis is concerned with examining which regression functions are suggested by the data. Some regression functions will be suggested by interest or logic, such as "the mean cure rates for the two cancer drugs are equal." Sometimes we may not have any preconceptions or prior hypotheses about the pattern of the data. In either case, one goal of statistical analysis is to discover simple regression functions which summarize the main patterns of the data without sacrificing important features.

A further part of the statistical analyst's job will be to check that the error distribution does apply to the data and to give some quantitative statements about how seriously the results could be compromised if in fact it does not apply. Commonly, however, the error distribution will not be under suspicion either because the logic of the data themselves suggest the family (which is the case with the data set below) or because the underlying family has already been subjected to statistical scrutiny; it is the exact shape of the systematic component that is at issue. In this chapter at least, we will be more concerned with fitting models that specify different restrictions on the parameters, *the underlying family being assumed true*. The term *fitting* is statistical jargon for estimating all unknown parameters.

1.1.2 Data on Serious Road Accidents

Australia is a rather odd multicultural nation down at the bottom of Asia where I was born and bred and in the early 80's, automobile accident rates were very high (by international standards) and there was much interest in finding the causes. The data set in Table 1.1 will be used throughout this chapter to illustrate the statistical techniques presented. It lists information on the 84 motor accidents in the state of Victoria during the first week of 1985 which were serious enough to require hospitalization of one or more people. These 84 accidents were categorized according to two criteria; first, whether or not a fatality was involved; second, whether the driver of the impacting vehicle was under or over 21 years of age. Age of the driver is often cited as a factor in road fatalities and age-specific legislation has since been imposed both in Australia and in other countries.

Table 1.1. Serious Accident Data—Victoria, 1985

Over 21		Under 21	
Fatal	Nonfatal	Fatal	Nonfatal
11	62	4	7

What is a reasonable error distribution for the serious accident data? Random events in time are usually modeled by the Poisson distribution so it seems natural to take the four data values to be four independent Poisson random variables with possibly different means.

How confident can we be in the Poisson assumption? For the number of events in a fixed amount of time to be Poisson distributed, it is sufficient for events to occur *singly* and *independently*. It is not necessary that the underlying accident rate be constant over time or across the population of drivers.

Certainly accidents occur singly by definition—if several vehicles are involved then this is still called a single accident. It seems equally plausible that an accident at one time and location would not affect the chance of an accident at other times and locations except perhaps in the immediate locality of an accident where traffic might be slowed and diverted around the accident scene. Thus, from background knowledge about these data, we may be quite confident that the four counts listed in Table 1.1 are realizations of four independent Poisson variables.

1.1.3 Parameters of a Statistical Model

In the previous section we decided to treat the four accident counts as four independent Poisson random variables. Let us denote the means parameters of these Poisson distributions by $\lambda_1, \lambda_2, \lambda_3, \lambda_4$, these being the unknown parameters. For this model the number of data values and the number of parameters both equal 4.

A statistical model with as many parameters as data values is called *full* or *saturated*. The full model is appropriate if absolutely nothing is known about likely values of the data. We learn very little about the data by assuming the full model and a simpler model is usually one aim of the statistical analysis. We will denote the full model for the accident data by $\mathcal{M}_1$.

At the other extreme, suppose we specify four values for $\lambda_1, \lambda_2, \lambda_3, \lambda_4$ in advance of seeing the data. The data values Y_1, Y_2, Y_3, Y_4 are then four random variables with known distributions. Observing the data tells us nothing extra about these distributions and we could just as well generate four Poisson observations from the computer as from the slower and more tragic mechanism of road accidents. When everything about the expected counts of accidents of different types is assumed known, there is no need for data at all.

Useful statistical models will fall between these extremes of complete knowledge and complete ignorance. They should allow for some lack of knowledge about the paramters and use the data themselves to try to reduce the level of ignorance.

Issues of interest concerning a data set can usually be expressed as restrictions on parameters. There may also exist prior information about these parameters that may again be expressed as restrictions on, or relations between, them. Each set of restrictions defines a different statistical model. A statistical model can be expressed in many different ways and it will often be convenient to express a model in terms of a smaller number of parameters rather than a larger number of parameters and a set of restrictions. It is also desirable for the small number of parameters to each have a *physical meaning*, although this is not always easy to achieve.

Example: Some Alternative Parameters Consider the full model for the accident data described above. Each parameter $\lambda_1, \ldots, \lambda_4$ is the expected number of accidents of a certain type and so each parameter does have a direct physical meaning. Let us consider an alternative set of parameters that could also physically describe the situation. The mean weekly accident rate for drivers over 21 is $\lambda_1 + \lambda_2$ which we denote θ_1. Of these, we would expect λ_1 to be fatal. Thus, the probability of an accident being fatal if the driver is over 21 is $\lambda_1/(\lambda_1 + \lambda_2)$. Call this p_1. Similarly, for younger drivers $\theta_2 = \lambda_3 + \lambda_4$ is the weekly accident rate and $p_2 = \lambda_3/(\lambda_3 + \lambda_4)$ the probability the accident is fatal. The four parameters $(\theta_1, p_1, \theta_2, p_2)$ all have clear physical interpretations of more interest, it might be argued, than the original mean parameters $\lambda_1, \lambda_2, \lambda_3, \lambda_4$. After a little algebra the mean values are

$$e_1 = p_1\theta_1, \qquad e_2 = (1 - p_1)\theta_1, \qquad e_3 = p_2\theta_2, \qquad e_4 = (1 - p_2)\theta_2. \qquad (1.1)$$

We might further reparametrize this model by letting $\rho = \theta_2/\theta_1$ represent the ratio of expected accident counts for younger compared to older drivers. If this parameter were greater than one then, on average, more younger drivers than older drivers are involved in accidents. We have four parameters p_1, p_2, θ_1, ρ related to the λ_i by the relations

$$\lambda_1 = p_1\theta_1, \lambda_2 = (1 - p_1)\theta_1, \qquad \lambda_3 = \rho p_2\theta_1, \qquad \lambda_4 = \rho(1 - p_2)\theta_1.$$

Remember: Reparametrizing a model does not really change it, and should not alter how well it fits, nor any prediction or test result. All of the above models are the same (full) model expressed in a different way.

Example: Some Alternative Models What models might be of interest for the accident data? Some models reflect previous knowledge. For the sake of argument let us suppose that it was known beforehand that 23.3% of serious accidents can be expected to be fatal. This might be based on previous data although it would be important to consider whether previously collected data was relevant to 1985. Now since the expected proportion of fatal accidents is

$$p = \frac{\lambda_1 + \lambda_3}{\lambda_1 + \lambda_2 + \lambda_3 + \lambda_4}$$

the assumption of 23.3% fatal accidents corresponds to a restriction

$$0.767\lambda_1 - 0.233\lambda_2 + 0.767\lambda_3 - 0.233\lambda_4 = 0 \tag{1.2}$$

on the original mean parameters. There are really only three unknown parameters in the model now, since by knowing three one can automatically calculate the fourth. We can bring this out more clearly by again reparametrizing this model, which will be denoted $\mathcal{M}_2$. Let us rename the mean number of fatalities $\phi_1 = \lambda_1 + \lambda_3$ and nonfatalities $\phi_2 = \lambda_2 + \lambda_4$ in which case the model restriction above becomes $\phi_1/(\phi_1 + \phi_2) = 0.233$ which is equivalent to $\phi_2 = 3.292\phi_1$. The expected counts in the four categories of the accident data can now be expressed

$$e_1 = \lambda_1, \qquad e_2 = \lambda_2, \qquad e_3 = \phi_1 - \lambda_1, \qquad e_4 = 3.292\phi_1 - \lambda_2. \tag{1.3}$$

The three parameters λ_1, λ_2 and ϕ_1 all have clear physical meanings. A fourth parameter is no longer needed because $p = 0.233$ is *assumed.*

Models are also generated by hypotheses of interest. Do both older and younger drivers have the same 23.3% fatality rate when they are involved in an accident? In terms of the parameters p_1, p_2, θ_1, θ_2 we are interested in the simple restriction p_1 0.233 $= p_2$. In terms of the mean parameters we are imposing

$$0.767\lambda_1 - 0.233\lambda_2 = 0 \qquad 0.767\lambda_3 - 0.233\lambda_4 = 0. \tag{1.4}$$

This model, which will be denoted $\mathcal{M}_3$, has essentially two parameters. The model may be expressed in terms of the mean parameters plus the above restrictions. It may be expressed entirely in terms of the parameters θ_1 and θ_2 giving expected values

$$e_1 = 0.233\theta_1, \qquad e_2 = 0.767\theta_1, \qquad e_3 = 0.233\theta_2, \qquad e_4 = 0.767\theta_2 \tag{1.5}$$

or in terms of λ_1 and λ_3 (see Table 1.2).

Example: An Independence Model Let us now abandon the assumption that 23.3% of serious accidents can be expected to be fatal and concentrate on the issue of driver age. Do younger and older drivers have the same expected proportion of fatal accidents amongst serious accidents, i.e., are p_1 and p_2 equal?

Table 1.2. Parametric Models for the Serious Accident Data

	Over 21		Under 21	
	Fatal	Nonfatal	Fatal	Nonfatal
$\mathcal{M}_1$	λ_1	λ_2	λ_3	λ_4
	λ_1	λ_2	$\phi_1 - \lambda_1$	$\phi_2 - \lambda_4$
	$p_1\theta_1$	$(1-p_1)\theta_1$	$p_2\theta_2$	$(1-p_2)\theta_2$
$\mathcal{M}_2$	λ_1	λ_2	$\phi_1 - \lambda_1$	$3.292\phi_1 - \lambda_2$
$\mathcal{M}_3$	λ_1	$3.292\lambda_1$	λ_3	$3.292\lambda_3$
	$0.233\theta_1$	$0.767\theta_1$	$0233\theta_2$	$0.767\theta_2$
$\mathcal{M}_4$	λ_1	λ_2	λ_3	$\lambda_2\lambda_3/\lambda_1$
	$p\theta_1$	$(1-p)\theta_1$	$p\theta_2$	$(1-p)\theta_2$

We denote this model $\mathcal{M}_4$. In words, the severity of the accident is independent of age-group. Under this hypothesis the expected counts for the four categories are $p\theta_1$, $(1-p)\theta_1$, $p\theta_2$, $(1-p)\theta_2$. In terms of the original mean parameters our assumption that older and younger drivers have the same expected proportion of fatal accidents corresponds to the restriction

$$\frac{\lambda_1}{\lambda_1 + \lambda_2} = \frac{\lambda_3}{\lambda_3 + \lambda_4} \Leftrightarrow \lambda_1\lambda_4 = \lambda_2\lambda_3. \tag{1.6}$$

Thus $\mathcal{M}_4$ imposes a nonlinear restriction on the original mean parameters of the full model. In terms of the parameters p_1, p_2, θ_1, θ_2 however, the restriction is linear being simply $p_1 = p_2 = p$. However we choose to express this model, there are essentially three unknown parameters.

The parametrizations of the different statistical models for the accident data are summarized in Table 1.2. This by no means exhausts the possibilities for this data set.

1.1.4 Statistical Sub-Models

We say that one model is a *sub-model* of another if it is a special case of the other model. Models $\mathcal{M}_2$, $\mathcal{M}_3$, $\mathcal{M}_4$ are all special cases of the full model $\mathcal{M}_1$. Of course every possible model will be a special case of, and therefore a sub-model of, the saturated or full model. In addition, $\mathcal{M}_3$ is a sub-model of $\mathcal{M}_2$ since $\mathcal{M}_3$ assumes that the proportion of fatal accidents is 23.3% not only overall but separately for both older and younger drivers. Similarly, $\mathcal{M}_3$ is a sub-model of $\mathcal{M}_4$ since $\mathcal{M}_4$ assumes equal fatality rates for the two age groups while $\mathcal{M}_3$ further specifies the common value 0.233.

A sequence of models $M_1, M_2, M_3, \ldots$ where each model is a sub-model of the previous one is said to be *nested*. Two examples of a nested sequence of models for the accident data are $\mathcal{M}_1$, $\mathcal{M}_2$, $\mathcal{M}_3$ and $\mathcal{M}_1$, $\mathcal{M}_4$, $\mathcal{M}_3$. Neither model $\mathcal{M}_2$ nor $\mathcal{M}_4$ is a sub-model of the other. Both impose a single restriction on the four parameters of the full model and so both involve essentially three unknown parameters. The assumptions that the overall fatality rate is 23.3% (model 2) and that the fatality rates for the two age-groups are equal (model 4) are quite distinct assumptions. Neither model logically implies or is a special case of the other. The intersection of the two nonnested models $\mathcal{M}_2$ and $\mathcal{M}_4$ is actually $\mathcal{M}_3$ and the logical relation between these four models is represented in Figure 1.1.

A significant part of statistical analysis and theory concerns comparing a statistical model with sub-models. When one has a nested sequence of models one will often begin by assuming the largest model and systematically test whether the next sub-model is an acceptable simplification. Alternative sequences of nested models might lead us to different models. For instance, we might be led to $\mathcal{M}_2$ in the sequence $\mathcal{M}_1$, $\mathcal{M}_2$, $\mathcal{M}_3$ and to $\mathcal{M}_4$ in the sequence $\mathcal{M}_1$, $\mathcal{M}_4$, $\mathcal{M}_3$.

An obvious next step would be to compare $\mathcal{M}_2$ and $\mathcal{M}_4$. Unfortunately, so long as one takes the common frequentist approach to statistics, there are difficulties in comparing these two models because neither is a sub-model of the other. First, it is unclear which hypothesis should be the null which makes interpretation difficult. Second, the hypotheses cannot be reduced to a test about the parameter of either model which invalidates the asymptotic distributional results described in Section 1.5. For the reader already familiar with generalized linear models, comparing two alternative link functions is a comparison of two nonnested models.

Nonnested models can be formally tested by embedding them in a larger model, though the issue of which is the null remains. Other approaches include Stone's (1954) cross-validation and the celebrated criterion of Akaike (1973).

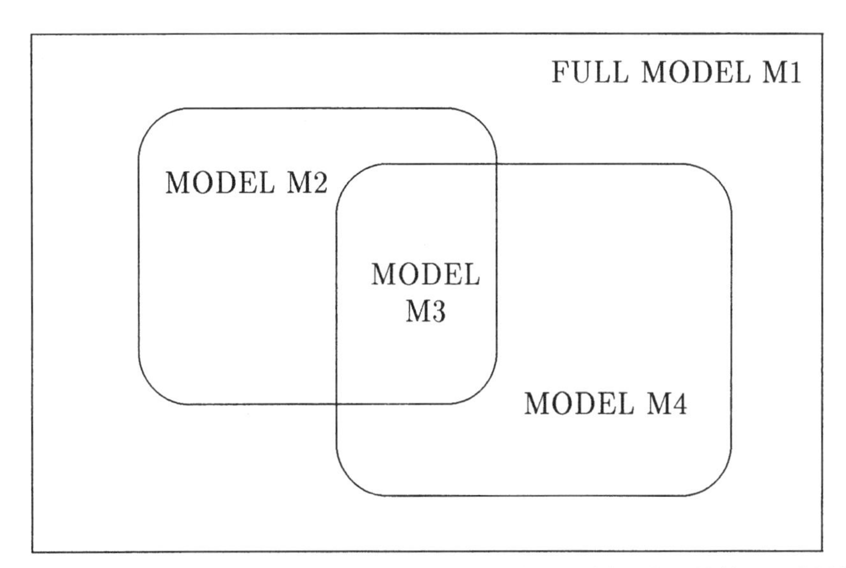

Figure 1.1. Logical relationship of four models for the serious accident data. Neither model M2 nor M4 is a sub-model of the other.

SUMMARY:

A statistical model comprises: a family of probability distributions which describes the uncertainty in the data (the random component) and a set of relations between the mean values of the data expressible in terms of a set of parameters and describing the underlying trend in the data (the systematic component). A particular model can be parametrized in many ways without affecting its essential properties. The full model has as many parameters as observations and always fits the data perfectly. A model is a sub-model if it is a special case of another. There are difficulties in comparing nonnested models.

1.2 FITTING STATISTICAL MODELS

Fitting is statistical jargon for estimating the unknown parameters of a model. We study here the estimation method known as maximum likelihood, which is historically the most important and also the most commonly used for categorical data. The general problem is that we have a set of data $Y_1, \ldots, Y_k$ and a model which specifies the distributions of these random variables in terms of a set of p unknown parameters $\theta = (\theta_1, \ldots, \theta_p)$. The aim is to estimate θ.

The mean value of Y_i depends on θ and once θ is estimated we can estimate this mean value. We use the notation

$$e_i(\theta) = \mathrm{E}(Y_i; \theta), \qquad \hat{e}_i = e_i(\hat{\theta}),$$

where the $\hat{e}_i$ are called *fitted* values. Fitted values estimate what the data values would be on average if we could observe them repeatedly under the same conditions. If one imagines the data as being made up of a systematic component or trend with random error added then the fitted values estimate the trend without the obscuring influence of the random error. Fitted values automatically satisfy the systematic component of the model while being as close as possible to the data values. They are a *smoothed* version of the data.

1.2.1 Maximum Likelihood

The method of *maximum likelihood* (ML) was developed by R. A. Fisher (1922, 1925) and largely replaced more ad hoc methods (such as least squares and method of moments) as the standard estimation method. The strength of ML is its inherent logic, its extremely wide scope, and its high efficiency under wide conditions (see Section 1.3). The main weakness of ML is that the entire distribution of the data must be modeled. This involves extra assumptions whose failure might have adverse effects on estimation precision.

In general, the ML estimator of an unknown parameter θ based on a set of data $Y_1, Y_2, \ldots, Y_k$ is obtained by first writing down the *likelihood function* for θ and then maximizing this function with respect to θ. For discrete data, the likelihood function is the probability of the observed data values

$$L(\theta; y_1, \ldots, y_k) = \Pr(Y_1 = y_1, \ldots, Y_k = y_k; \theta) \tag{1.7}$$

and considered as a function of θ; it is simply "the probability of the observed data values occurring." The maximum likelihood estimate is the value of θ maximizing this function, and so $L(\theta)$ may be multiplied by any function of the data without any effect. The *logic* of ML estimation is compelling and almost universally applicable; one finds those parameter values for which the data as a whole is most likely. The data is then most consistent with the model in the sense that it is more likely supposing these parameter values than any others.

For the accident data, the probability of the entire data is

$$\Pr(y_1, y_2, y_3, y_4) = \left\{e^{-\lambda_1}\frac{\lambda_1^{y_1}}{y_1!}\right\}\left\{e^{-\lambda_2}\frac{\lambda_2^{y_2}}{y_2!}\right\}\left\{e^{-\lambda_3}\frac{\lambda_3^{y_3}}{y_3!}\right\}\left\{e^{-\lambda_4}\frac{\lambda_4^{y_4}}{y_4!}\right\}.$$

Treated as a function of the parameters $(\lambda_1, \lambda_2, \lambda_3, \lambda_4)$, this is the likelihood function. In practice, it is often easier to work with the natural logarithm of the likelihood function denoted $\ell(\theta)$. Since *log* is a 1-1 function, ℓ is maximized at the same value of θ as is L. Since the likelihood may be multiplied by any constant, it follows that any constant may be *added* to the log-likelihood and so additive terms that are not functions of the parameters may be dropped. For the accident data the log-likelihood function is

$$\begin{aligned}\ell(\lambda_1, \lambda_2, \lambda_3, \lambda_4; \mathbf{y}) = {} & -\lambda_1 - \lambda_2 - \lambda_3 - \lambda_4 \\ & + 11\log\lambda_1 + 62\log\lambda_2 + 4\log\lambda_3 + 7\log\lambda_4\end{aligned} \tag{1.8}$$

the factorial terms $y_i!$ having been dropped. The different models give rise to different likelihoods by substituting known expressions for the four parameters into the above likelihood.

Example: ML Estimation for Full Model Under the full model, the above log-likelihood separates into four functions each depending on exactly one of the λ_i. Maximizing each separately, it is simple to show that L is maximized when $\lambda_1 = 11$, $\lambda_2 = 62$, $\lambda_3 = 4$, and $\lambda_4 = 7$. It makes no difference to the results if we parametrize in terms of $p_1, p_2, \theta_1, \theta_2$. To see this, substitute the relations [see Eq. (1.1)] into the above likelihood to obtain the reparametrized likelihood

$$\begin{aligned}\ell(p_1, p_2, \theta_1, \theta_2; \mathbf{y}) = {} & -\theta_1 - \theta_2 + 73\log\theta_1 + 11\log\theta_2 + 11\log p_1 \\ & + 62\log(1 - p_1) + 4\log p_2 + 7\log(1 - p_2).\end{aligned} \tag{1.9}$$

Again, this breaks into four functions each of which is easily maximized separately. The result is $\hat{\theta}_1 = 73$, $\hat{\theta}_2 = 11$, $\hat{p}_1 = 11/73$, $\hat{p}_2 = 4/11$. From these it follows that the estimates of the λ_i are again the data values themselves.

Example: ML Estimation for $\mathcal{M}_3$ This model supposes that $p_1 = p_2 = 0.233$ are known values. To obtain the log-likelihood we just drop terms in Eq. (1.9) that involve p_1 and p_2 obtaining

$$\ell(\theta_1, \theta_2; \mathbf{y}) = -\theta_1 - \theta_2 + 73 \log \theta_1 + 11 \log \theta_2.$$

Again, this is a sum of two functions each of which can be separately maximized and we obtain $\hat{\theta}_1 = 73$, $\hat{\theta}_2 = 11$. The fitted values are

$$\hat{e}_1 = 0.233\hat{\theta}_1 = 17.009, \qquad \hat{e}_2 = .767\hat{\theta}_1 = 55.991,$$
$$\hat{e}_3 = .233\hat{\theta}_2 = 2.563, \qquad \hat{e}_4 = 0.767\hat{\theta}_2 = 8.437.$$

Example: ML Estimation for $\mathcal{M}_2$ Under this model the expected proportion of fatalities is 23.3% overall. Let us choose the parametrization $\theta = (\lambda_1, \lambda_2, \phi_1)$ given in Table 1.2. Substituting these relations into the full log-likelihood gives

$$\begin{aligned}\ell(\lambda_1, \lambda_2, \phi_1) &= -\sum_{i=1}^{4} \lambda_i + \sum_{i=1}^{4} y_i \log \lambda_i \\ &= -4.292\phi_1 + y_1 \log \lambda_1 + y_2 \log \lambda_2 \\ &\quad + y_3 \log(\phi_1 - \lambda_1) + y_4 \log(3.292\phi_1 - \lambda_2). \qquad (1.10)\end{aligned}$$

This is a smooth function of the parameters but does not separate into simple functions of each parameter as was the case for the previous two models. Algorithms for maximizing this likelihood are not obvious.

To maximize a differentiable function of three variables we can compute partial derivatives and equate them to zero. This gives three equations:

$$\begin{aligned}\frac{\partial \ell}{\partial \lambda_1} &= \frac{y_1}{\lambda_1} - \frac{y_3}{\phi_1 - \lambda_1} = 0 \\ \frac{\partial \ell}{\partial \lambda_2} &= \frac{y_2}{\lambda_3} - \frac{y_4}{3.292\phi_1 - \lambda_2} = 0 \\ \frac{\partial \ell}{\partial \phi_1} &= -4.292 + \frac{y_3}{\phi_1 - \lambda_1} + \frac{3.292 y_4}{3.292\phi_1 - \lambda_2} = 0. \qquad (1.11)\end{aligned}$$

There are many algorithms for solving these three equations. One way is as follows. For known values of λ_1, λ_2, the third equation can be rearranged into a quadratic. This is easily solved for ϕ_1 giving a first guess at the fitted values $\hat{e}_i$. The first two equations are equivalent to

$$\lambda_1 = e_3 \frac{y_1}{y_3}, \qquad \lambda_2 = e_4 \frac{y_2}{y_4}$$

which gives us two new values for (λ_1, λ_2). We then solve the quadratic again and iterate until convergence.

While this is a rather ad hoc algorithm, for this particular model it is successful and for the given data values ends up giving $\hat{\lambda}_1 = 14.353$, $\hat{\lambda}_2 = 57.892$ and $\hat{\phi}_1 = 19.572$. Remembering that the model $\mathcal{M}_2$ has a systematic component given by the relations $e_1 = \lambda_1$, $e_2 = \lambda_2$, $e_3 = \phi_1 - \lambda_1$, $e_4 = 3.292\phi_1 - \lambda_2$ we obtain the four fitted values in Table 1.3. Notice that the estimated value of ϕ_1 is precisely 23.3% of the data total 84, though it is not obvious from the above three equations that this should be the case. *Remember:* The fitted values always follow the regression function.

1.2.2 Sufficient Statistics

Sufficiency is a central concept in both the theory and practice of inference. A statistic S is sufficient for a parameter θ if it contains all the information about that parameter. It is then sufficient to know the value of S and other aspects of the data are unnecessary. The mathematical expression of this idea is as follows. Suppose the data Y is modeled by a family of probability distributions $p_Y(y;\theta)$ indexed by a set of parameters θ. Let S be a function of the data. Then S is sufficient for θ if the distribution of Y given S no longer depends on θ.

In an obvious notation, this corresponds to the factorization

$$p_Y(y;\theta) = p_S(s;\theta)p_{Y|S=s}(y;s).$$

Only the first factor on the right-hand side is relevant to θ. Once $S = s$ is observed, all other aspects of the data have a completely known distribution

Table 1.3. Fitted Values for the Serious Accident Data

	Over 21		Under 21	
Model	Fatal	Nonfatal	Fatal	Nonfatal
$\mathcal{M}_1$	11.000	62.000	4.000	7.000
$\mathcal{M}_2$	14.353	57.892	5.219	6.536
$\mathcal{M}_3$	17.009	55.991	2.563	8.437

and observing them is equivalent to drawing a random number from the known distribution $p_{Y|S=s}$. In practice, conditional distributions can be quite difficult to derive and the definition of sufficiency is *not* usually used to identify sufficient statistics in a given problem. Rather, the following connection with likelihood is used.

(Neyman's) Factorization Theorem. A statistic $S(Y)$ is sufficient for the parameter θ in the model $\mathcal{M} = \{p_Y(y;\theta) : \theta \in \Theta\}$ if and only if, for some functions m_1 and m_2, the log-likelihood can be written in the form

$$\ell(\theta; y) = m_1(s;\theta) + m_2(y).$$

Proofs abound in standard textbooks [see Azzalini (1996, Theorem 2.3.7)]. To use the result in practice recall that the log-liklihood function is only defined up to an additive function of the data. Hence, the above theorem says that S is sufficient if and only if $\ell(\theta; y)$ depends on the data y only through s. We simply ask "What statistics would I have to know in order to draw the log-likelihood function?" The answer will not be unique. For instance, the entire data y is always sufficient by this test. We would like a smaller set of statistics to be sufficient if possible, thus reducing the dimension of the problem. The smallest such set is called *minimal* sufficient. The mathematical definition requires a minimal sufficient statistic to be a function of every other sufficient statistic. But in practice we just ask "What is the smallest set of statistics I would have to know in order to draw the log-likelihood function?"

Minimal sufficient statistics (MSS) are also not unique since any one-to-one transform is also minimal sufficient. This agrees with our intuition that the amount of information in a statistic is not affected by taking the logarithm or any other reversible transform. It also means we can take whatever representation of the MSS is most convenient, for instance, the one whose distribution is easiest to derive. In Chapter 7, it is especially important to be able to write down a likelihood function and immediately recognize the minimal sufficient statistics for various parameters. A brief account of sufficiency is found in Cox and Hinkley (1974, Section 2.2). Our discussion ends with three examples.

Example 1: Binary Trials Suppose we have n independent binomial trials with probability of success p_i on trial i. The probability function of the ith binary observation Y_i is

$$p(y; p_i) = p_i^y (1 - p_i)^{1-y} \qquad y = 0, 1$$

and so the log-likelihood function is

$$\ell(p_1, \ldots, p_n) = \sum_{i=1}^{n} \log\{p_i^{y_i}(1 - p_i)^{1-y_i}\}$$
$$= \sum_{i=1}^{n} \left\{y_i \log\left(\frac{p_i}{1 - p_i}\right) + \log(1 - p_i)\right\}$$

and to know this function of $p_1, \ldots, p_n$ we must know every binary observation $y_1, \ldots, y_n$. There is no reduction of the data to a smaller MSS. However, suppose that the p_i all equal p. Then the log-likelihood function becomes

$$\ell(p) = \log\left(\frac{p}{1 - p}\right) \sum_{i=1}^{n} y_i + n \log(1 - p).$$

To draw this function of p it is sufficient to know the total number of successes $S = \sum Y_i$, which is consequently sufficient for p. Knowing anything less than this we could not draw the function so this sufficient statistic is minimal. It also has a simple binomial distribution so we would probably choose S rather than, say, $\log(S)$ to work with.

Example 2: Random Number of Binary Trials. Continuing example 1, suppose the number of binary trials N was actually a random variable with $Pn(\lambda)$ distribution. Then the distribution of the data Y factors into (1) the conditional binomial distribution of Y given N, times (2) the Poisson distribution of N. Taking the logarithm we obtain the log-likelihood

$$\ell(p, \lambda) = \log\left(\frac{p}{1 - p}\right) \sum_{i=1}^{n} y_i + n \log(1 - p) - \lambda + n \log \lambda.$$

Since N is now part of the data, the minimal sufficient statistic for the parameter $\theta = (p, \lambda)$ is $(\sum Y_i, N)$, or equivalently

$$S = \left\{\sum_{i=1}^{n} Y_i, \sum_{i=1}^{n} (1 - Y_i)\right\},$$

the number of successes and the number of failures. It is a good exercise to reparametrize this likelihood, by letting $\lambda_1 = \lambda p$ and $\lambda_2 = \lambda(1 - p)$. This gives the log-likelihood

$$\ell(\lambda_1, \lambda_2) = s_1 \log \lambda_1 + s_2 \log \lambda_2 - \lambda_2 - \lambda_2. \tag{1.12}$$

The MSS is still S_1, S_2 and reparametrization generally has no effect on the MSS. Note that Eq. (1.12) is actually the log-likelihood for two independent Poisson random variables S_1, S_2 and in fact these statistics do have Poisson distributions with parameters λ_1 and λ_2. The fact that binomial variables become Poisson when the number of trials N is Poisson is of great practical significance to modeling count data. This is discussed further in Section 2.4.

Example 3: Serious Accident Data For the model $\mathcal{M}_2$ we wrote down the log-likelihood function

$$\ell(\lambda_1, \lambda_2, \phi_1) = -4.292\phi_1 + y_1 \log \lambda_1 + y_2 \log \lambda_2 + y_3 \log(\phi_1 - \lambda_1) + y_4 \log(3.292\phi_1 - \lambda_2) \qquad (1.13)$$

and so the whole data Y_1, Y_2, Y_3, Y_4 are sufficient for the parameters (λ_1, λ_2, ϕ_1). Suppose that λ_1 were known. Then the log-likelihood is a function of the of the parameters (λ_2, ϕ_1) and the $y_1 \log \lambda_1$ term disappears. Thus Y_2, Y_3, Y_4 is sufficient for (λ_2, ϕ_1).

From the definition of sufficiency, it follows that the distribution of the data given this sufficient statistic is free of λ_2 and ϕ_1. This is easy to check. The distribution of the data Y_1, Y_2, Y_3, Y_4 given Y_2, Y_3, Y_4 is just the distribution of Y_1 given Y_2, Y_3, Y_4. Since the counts are independent this is just the unconditional distribution of Y_1, which is $Pn(\lambda_1)$ and indeed does not depend on λ_2 or ϕ_1.

1.2.3 Algorithms for Maximum Likelihood*

The mathematical problem of maximizing the log-likelihood function $\ell(\theta)$ is a very general one, and different methods will be appropriate depending on the specific function. Nevertheless, some general statements can be made and there are some general algorithms.

In an earlier section we maximized the likelihood function for (λ_1, λ_2, ϕ_1) under $\mathcal{M}_2$ using an ad hoc algorithm. There is a much easier approach. Let $\gamma_1 = \lambda_1/\phi_1$ and $\gamma_2 = \lambda_2/\phi_1$. Then the log-likelihood function Eq. (1.13) becomes

$$\ell(\theta_1, \theta_2, \phi_1) = -4.292\phi_1 + (y_1 + y_2 + y_3 + y_4) \log \phi_1 + y_1 \log \gamma_1 + y_2 \log \gamma_2 + y_3 \log(1 - \gamma_1) + y_4 \log(3.292 - \gamma_2)$$

and this function does separate into separate functions of ϕ_1, γ_1, and γ_2. Differentiating with respect to ϕ_1 and equating to zero we have

$$\hat{\phi}_1 = 0.233(Y_1 + Y_2 + Y_3 + Y_4).$$

*This section, and subsequent ones marked with an asterisk, are more difficult and more detailed and could be omitted the first time this book is read.

Differentiating with respect to the γ_i, equating to zero, and multiplying by $\hat{\phi}_1$ gives

$$\hat{\lambda}_1 = \frac{0.233Y_1(Y_1 + Y_2 + Y_3 + Y_4)}{Y_1 + Y_3},$$

$$\hat{\lambda}_1 = \frac{0.233Y_2(Y_1 + Y_2 + Y_3 + Y_4)}{Y_2 + Y_4}.$$

We see in this example that by suitable reparamatrization, the estimators can be written down in closed form and no special algorithm is required. Unfortunately, computers do not share even our own meager intelligence and usually cannot recognize when a log-likelihood function simplifies. Statistical packages use general algorithms. Indeed, it is very easy to make a computer package work for many minutes computing estimates that are obvious from inspection of the likelihood.

Our problem is to maximize an objective function ℓ whose maximum we suppose is at a stationary point, i.e., where the partial derivatives are all zero. We will denote the vector of partial derivates by U. For instance, Eq. (1.11) gives the components of a three-dimensional derivative vector. We want to solve $U(\theta) = 0$. There are many algorithms for finding the root $\hat{\theta}$ of a nonlinear function U. For functions of a scalar variable θ write

$$U(\hat{\theta}) \approx U(\theta) + (\hat{\theta} - \theta)U'(\theta)$$

with small error if θ is close to $\hat{\theta}$. Rearranging gives

$$\hat{\theta} = \theta - \frac{U(\theta)}{U'(\theta)} \tag{1.14}$$

which suggests an iterative scheme for finding $\hat{\theta}$; begin with an initial guess θ and then generate the next guess by the right-hand side of Eq. (1.14). For a p dimensional parameter θ, U has p components. The previous iterative scheme generalizes to

$$\hat{\theta} = \theta + \mathcal{J}(\theta)^{-1}U(\theta) \tag{1.15}$$

where $\mathcal{J}$ is minus the $p \times p$ matrix of partial derivatives of U with respect to θ, called the observed information matrix. The scheme [see Eq. (1.15)] is called the Newton–Raphson algorithm, being the multidimensional extension of the Newton algorithm [see Eq. (1.14)]. A closely related method replaces the matrix $\mathcal{J}$ by its expectation and this algorithm is known as *Fisher scoring*.

When the Newton algorithm converges, it does so very quickly, the convergence rate being *quadratic*. Convergence is faster the closer $U(\theta)$ is to linear, or equivalently the closer $l(\theta)$ is to quadratic. For nonlinear U, the scheme will converge provided that the initial guess θ is close enough to $\hat{\theta}$ for U to be monotonic and convex in the neighborhood of θ and $\hat{\theta}$. Similar comments apply to the Newton–Raphson algorithm.

1.2.4 Errors of Estimation*

Estimates are not perfect and almost always differ from the true value to some extent. Since we never know the true value we never observe the actual error of any given estimate. The classical approach to describing estimation error is to examine how close the estimator is to the true value, *on average*.

If $\hat{\theta}$ is an estimator for the parameter θ then two simple quantities describing its reliability are its bias and standard deviation defined as

$$\text{bias}(\hat{\theta}) = \text{E}(\hat{\theta}) - \theta, \text{s.d.}(\hat{\theta}) = \sqrt{\text{Var}(\hat{\theta})}.$$

Bias measures the tendency of the estimator to be too large or small and zero bias means that *on average* the estimate gives the true value. Standard deviation measures how variable the estimator would be if used again and again. For a given amount of data, it is typically possible to reduce bias at the expense of standard deviation and vice versa. If the amount of data is increased, then both will be reduced. The standard deviation often depends on the unknown parameter θ, which may be estimated. The term *standard error* is used for an estimate of the standard deviation of an estimator.

There is an important and often overlooked source of estimation error and this is an erroneous model; it is important in practice to consider to what extent the model assumptions influence the estimation of a particular parameter that might be of interest. Bias, standard error, and the effect of the model should all be considered in evaluating the accuracy of an estimate.

Example: Serious Accident Data Consider estimating the probability $p_2 = \lambda_3/(\lambda_3 + \lambda_4)$, of an accident involving a younger driver being fatal. Under the full model the estimate is

$$\hat{p}_2 = \frac{Y_3}{Y_3 + Y_4} = \frac{4}{11}.$$

The distribution of this estimator is quite simple, at least if we treat the denominator of 11 accidents as fixed. Since each may be fatal or nonfatal, Y_3 has the conditional binomial distribution with parameters $(y_3 + y_4, p_2)$. Therefore the conditional mean and variance is

$$\text{E}_c(\hat{p}_2) = p_2, \qquad \text{Var}_c(\hat{p}_2) = \frac{p_2(1 - p_2)}{y_3 + y_4}$$

and so unconditionally

$$\mathrm{E}(\hat{p}_2) = p_2, \qquad \mathrm{Var}(\hat{p}_2) = p_2(1 - p_2)\mathrm{E}\left(\frac{1}{Y_3 + Y_4}\right).$$

The estimator is unbiased for p_2. The variance can be estimated by replacing p_2 by $\hat{p}_2$ and replacing $\mathrm{E}\{(Y_3 + Y_4)^{-1}\}$ by $(y_3 + y_4)^{-1}$. This gives the standard error

$$\mathrm{se}(\hat{p}_2) = \sqrt{\frac{\frac{4}{11}\left(1 - \frac{4}{11}\right)}{4 + 7}} = 0.145.$$

Typically, estimators have a more complicated distribution than binomial. For instance, under model $\mathcal{M}_2$ we estimate p_2 by

$$\hat{p}_2 = \frac{\hat{\lambda}_3}{\hat{\lambda}_3 + \hat{\lambda}_4} = \frac{5.219}{5.219 + 6.536} = 0.4438.$$

We might anticipate that this estimator is less variable than the full model estimator since we have used extra information, namely that the overall proportion of fatalities is 23.3%. This means that the full weight of the data is used in estimating λ_3 and λ_4, not just Y_3 and Y_4. However, the distribution of this estimator is very complicated because both $\hat{\lambda}_3$ and $\hat{\lambda}_4$ are complicated functions of the data. Nevertheless, the variance can be approximated using techniques described in Section 1.6 and we obtain the standard error 0.130 in this case. Thus, the standard error *is* less than that of the simple estimator but not much less.

Example: The Effect of the Model Consider model $\mathcal{M}_2$ applied to the serious accident data. The fitted values 14.353, 57.892, 5.219, 6.536 given in Table 1.3 agree reasonably well with the data values 11, 62, 4, 7 and in fact this model does pass most statistical tests of adequacy of fit. However, the overall fit might not be of primary importance. We might be more interested in p_1, p_2, the fatality probabilities for older and younger drivers, respectively. Under the full model we have direct estimates $\hat{p}_1 = 11/73 = 0.1507$, $\hat{p}_2 = 4/11 = 0.3636$ from the raw data, whereas under model $\mathcal{M}_2$ we have $\hat{p}_1 = 0.1986$, $\hat{p}_2 = 0.4438$.

Under $\mathcal{M}_2$ both estimates of the fatality probabilities are higher than the corresponding proportions from the raw data. This is because the model $\mathcal{M}_2$ *assumes* an overall probability of fataility of 23.3%, whereas for the data only $15/84 = 17.8\%$ of accidents were fatal. The estimation procedure under $\mathcal{M}_2$ automatically inflates the fitted values $\hat{e}_1$ and $\hat{e}_3$, and essentially refuses to believe the low figure of 17.8%. Indeed, by assuming $p > 0.233$ we can obtain even higher estimates of p_1 and p_2. This will produce better estimates provided that the model $\mathcal{M}_2$ is true but if it is false then the estimates will all be *mistakenly* inflated.

This issue runs to the heart of statistical modeling; if one fits a model then statistical accuracy is typically enhanced assuming the model is true. On the other hand, if the model is false then falsely imposing it introduces all kinds of biases into the analysis. Thus, it is important to always check whether the data are consistent with the model and to anticipate the kinds of biases involved if the model is wrong, as we have just done. The important point is that accurate estimation of a *particular* parameter of interest may not be automatically achieved by finding a model which seems to fit quite well. The particular estimate may be quite sensitive to the model, and this may not be revealed by a large standard error.

Which then is the best estimate? The estimate 0.4438 is based on an assumption that 23.3% of accidents are expected to be fatal which results in much higher estimates of the probability of fatality than indicated by the raw data. The improvement in standard error is modest (0.145 to 0.130) and so little is gained in terms of statistical variability of the estimators; both are quite inaccurate. Unless one was very confident indeed in the 23.3% assumption, it would be wiser to use the simpler and safer estimate 0.3636.

SUMMARY:

The most common method of parameter estimation is maximum likelihood. Various algorithms exist for finding ML estimates, the most general of which are Newton–Raphson and Fisher scoring. Both achieve rapid convergence when they converse and for many models convergence is assured. Errors of estimation due to randomness of the data are summarized by standard error and bias. Errors due to misspecification of the model may have quite severe effects on some parameter estimates but very little on others.

1.3 LARGE SAMPLE ESTIMATION THEORY

In the classical approach to statistics, an estimator $\hat{\theta}$ is judged according to its distribution and how close this distribution is to the unknown parameter θ being estimated. Bias and standard error give some idea of estimation accuracy but to answer questions such as "How likely is it that $\hat{\theta}$ is within 0.1 of θ" the distribution is needed. Occasionally, an analytic expression for the distribution can be written down, but more often it cannot.

Example: Serious Accident Data, Model $\mathcal{M}_2$ Recall model $\mathcal{M}_2$ which assumes that the probability p of an accident being fatal is 0.233. The four mean values under this model were λ_1, λ_2, $\phi_1 - \lambda_1$, $3.292\phi_1 - \lambda_2$. Earlier we found closed formulas for the estimators of $(\lambda_1, \lambda_2, \phi_1)$, namely

$$\hat{\lambda}_1 = \frac{0.233Y_1(Y_1 + Y_2 + Y_3 + Y_4)}{1 + Y_3/Y_1}, \qquad \hat{\lambda}_2 = \frac{0.767(Y_1 + Y_2 + Y_3 + Y_4)}{1 + Y_4/Y_2}. \tag{1.16}$$

and $\hat{\phi}_1 = 0.233(Y_1+Y_2+Y_3+Y_4)$. Suppose we were interested in the accuracy of $\hat{\phi}_1 = 19.572$ which estimates the mean weekly fatality rate. The random variable $Y_1 + Y_2 + Y_3 + Y_4$ is a sum of independent Poisson variables and so is itself Poisson with mean $\lambda_1 + \lambda_2 + \lambda_3 + \lambda_4$ which equals $\phi_1/0.233$ under the model $\mathcal{M}_2$. We can use this distribution to make direct statements about how close $\hat{\phi}_1$ is to any hypothesized value. For instance, suppose that the true weekly fatality rate were 25. What is the probability we would obtain an estimate as low as the observed value 19.572? If $\phi_1 = 25$ then $Y_1+Y_2+Y_3+Y_4$ has Poisson distribution with mean $25/0.233 = 107.30$. Since $\hat{\phi} \leq 19.572$ exactly when $Y_1 + Y_2 + Y_3 + Y + 4 \leq 84$

$$\Pr(\hat{\phi}_1 \leq 19.572) = \Pr(Pn(107.30) \leq 84) = 0.0116.$$

It is thus very *unlikely* that $\hat{\phi}_1$ would be as low as 19.572 if the true value is 25. Figure 1.2 displays the full distribution of $\hat{\phi}_1$, obtained by plotting the Pn(107.3) distribution and rescaling the x-axis by 0.233.

There are two reasons why we were able to complete the calculation above. First, $\hat{\phi}_1$ had a distribution that is *simple*. Second, this distribution depended

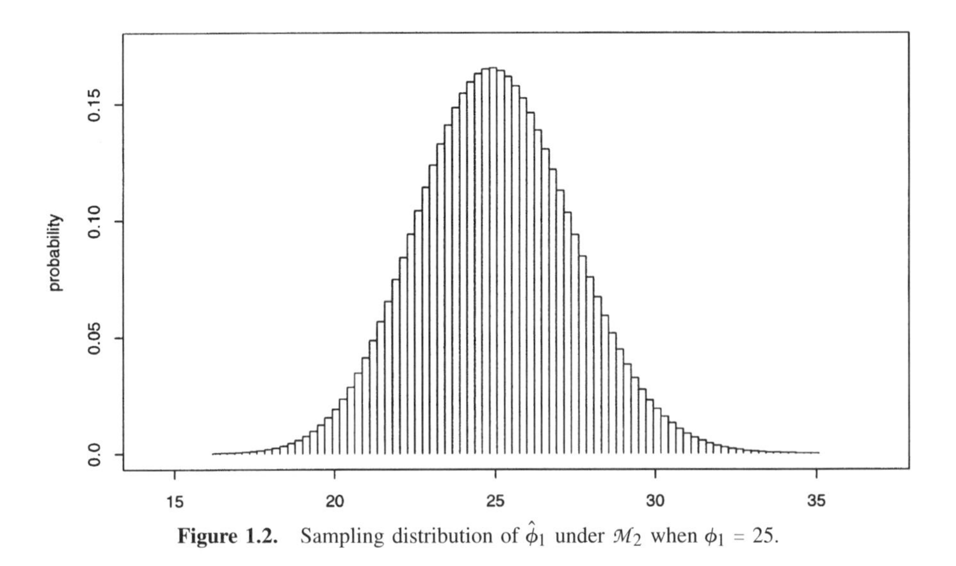

Figure 1.2. Sampling distribution of $\hat{\phi}_1$ under $\mathcal{M}_2$ when $\phi_1 = 25$.

only on ϕ_1, not on the other parameters. In practice such a friendly state of affairs is rare. Consider for instance $\hat{\lambda}_1$ whose formula is given above. The distribution is that of a nonlinear combination of Poisson variables and virtually impossible to derive exactly. Even if we could derive it, the distribution does not depend on λ_1 only. It also depends on λ_2 and ϕ_1. In this context, λ_2 and ϕ_1 are called *nuisance parameters* as they are not of direct interest for inference on λ_1 and yet their unknown values have a bearing on, and complicate, inference on λ_1. Similarly λ_1 and ϕ_1 are nuisance parameters for inference on λ_2.

The rest of this section is concerned with showing how the joint distribution of the ML estimator can be approximated by a certain multivariate normal distribution and in examining in detail its application to the serious accident data. We give informal derivations of the basic asymptotic results for the special case of an independent set of discrete data. The main results hold in considerably more generality and basic conditions will be indicated where appropriate. It is assumed that the reader is familiar with the Law of Large Numbers and Central Limit Theorem for sums of independent random variables.

1.3.1 Likelihood, Score, and Information

Let $X = (X_1, \ldots, X_n)$ be a set of independent random variables with distribution depending on a vector of parameters $\theta = (\theta_1, \theta_2, \ldots, \theta_p)$ whose true value is known to lie within some specified space Ω. It is convenient to work with the logarithm of the likelihood function rather than the likelihood function itself since this is typically a sum of independent terms and is maximized at exactly the same value as is the ordinary likelihood. The log-likelihood function

$$\ell(\theta) = \log L(\theta) \sum_{i=1}^{n} \log p_i(X_i; \theta) \qquad \theta \in \Omega$$

where p_i is the probability function of the ith random variable, X_i. The derivative of l with respect to θ_j is denoted

$$U_j = \sum_{i=1}^{n} \frac{\partial \log p_i(X_i; \theta)}{\partial \theta_j} \qquad j = 1, \ldots, p$$

and in all but exceptional circumstances U_j has mean zero. The p-dimensional derivative vector taken as a whole is called the *score function*. Since it is usual for ℓ to be a smooth function of its parameters, and for the maximum to be achieved at an interior point of Ω, the likelihood is maximized by simultaneously equating all elements of the score function to their zero expectation. This system of equations is known as the *likelihood equations*.

We next introduce the concept of *information* about a parameter θ. The *observed* information matrix is a symmetric matrix $\mathcal{J}(\theta)$ with (j,k)th component

$$\mathcal{J}_{jk} = -\frac{\partial U_j}{\partial \theta_k} = -\frac{\partial^2 l}{\partial \theta_j \partial \theta_k}(X;\theta).$$

Since $\hat{\theta}$ is a *maximum* of the log-likelihood function, $\mathcal{J}(\theta)$ is also nonnegative definite when evaluated at $\hat{\theta}$.

The observed information matrix depends on the data and so is itself a random variable. Since ℓ is a sum of n independent terms, one from each data point X_i, the same is true of each entry of the random matrix $\mathcal{J}$. Thus, provided that the variances of these terms are bounded, the Law of Large Numbers implies that the information matrix will converge element by element to the matrix of expectations

$$I_{ij} = -\mathrm{E}\left(\frac{\partial U_j}{\partial \theta_k}\right)$$

known as the *expected information* matrix. It is left as an exercise to show that provided one may move the mixed derivative $\partial^2/\partial\theta_j\partial\theta_k$ across a summation sign that

$$\mathrm{Cov}\{U_j(\theta), U_k(\theta)\} = -\mathrm{E}\left(\frac{\partial U_j(\theta)}{\partial \theta_k}\right) \tag{1.17}$$

so that expected information may be alternatively viewed as the variance matrix of the vector score function U. Thus $I(\theta)$ is a symmetric nonnegative definite matrix for any value θ. Reparametrizing the model will generally change the information matrix [see Eq. (1.22)] and so we normally talk of the information *about* θ.

1.3.2 The Asymptotics of ML Estimation

The random vector $U(\theta)$ is a sum of n independent mean zero random variables with variance matrix $I(\theta)$. The ordinary central limit theorem thus implies that

$$U \xrightarrow{d} \mathcal{N}_p(0, I(\theta)) \tag{1.18}$$

where $\xrightarrow{d}$ denotes convergence in distribution as the sample size n grows unbounded. An alternative version of Eq. (1.18) replaces the expected information I by the observed information $\mathcal{J}$; this is not surprising since these two matrices differ asymptotically by a negligible amount. Note, however, that $\mathcal{J}$ is usually easier to calculate since it does not require any expectations.

In order to show that the ML estimator $\hat{\theta}$ also has an asymptotically multivariate normal distribution, we show that it is approximately linearly related to the score function. A one step Taylor expansion of U_j with respect to the p parameters $\theta_1, \ldots, \theta_p$ about the ML estimate $\hat{\theta}$ is

$$U_j(\hat{\theta}) = U_j(\theta) + \sum_{k=1}^{p} \frac{\partial U_j(\theta)}{\partial \theta_k} (\hat{\theta}_k - \theta_k) + R \tag{1.19}$$

where the remainder term R is of smaller asymptotic size than the leading terms of the formula. The left-hand side is zero by definition of $\hat{\theta}$ and so on rearranging and converting to matrix notation we have

$$U(\theta) \approx \mathcal{J}(\theta)(\hat{\theta} - \theta) \approx I(\theta)(\hat{\theta} - \theta).$$

Isolating $\hat{\theta}$ and noting that $U(\theta)$ is asymptotically normal with variance $I(\theta)$, this gives us two versions of a central limit theorem for $\hat{\theta}$,

$$\hat{\theta} \xrightarrow{d} \mathcal{N}_p\{\theta, I^{-1}(\theta)\} \qquad \hat{\theta} \xrightarrow{d} \mathcal{N}_p\{\theta, \mathcal{J}^{-1}(\theta)\}, \tag{1.20}$$

where the inverse of either the expected or observed information may be used as the asymptotic variance of $\hat{\theta}$. One may also estimate θ by $\hat{\theta}$ in these information matrices. The issue of whether expected or observed information gives the better approximation to the variance of $\hat{\theta}$ is discussed a little later.

The framework of n independent random variables is unnecessarily restrictive and the limiting results hold in much greater generality. Roughly what is required is that the information about the parameter grows unboundedly. One criterion, relevant to normality of Poisson data, is that the smallest eigenvalue of the information matrix diverges, or equivalently that the smallest expected value diverges. This is enough to show that the score function is asymptotically normal. Provided then that the ML estimator is consistent [see Wald (1949)] the argument from Eq. (1.19) follows through as above. In the classical independent data framework the smallest eigenvalue criterion is equivalent to letting the sample size n diverge.

1.3.3 Information and Asymptotic Efficiency

Let θ be a vector parameter and $\hat{\theta}$ be a p-dimensional statistic with expectation $g(\theta)$ and variance matrix V. For instance, V_{jj} is the variance of the jth component $\hat{\theta}_j$ of $\hat{\theta}$. Let I be the expected information matrix and G the $p \times p$ matrix of partial derivatives of the map g. Then under mild conditions that allow certain derivatives to be moved across certain integrals

$$V - GI^{-1}G^T \geq 0 \tag{1.21}$$

where ≥ 0 denotes that the matrix at left is nonnegative definite. This result is usually credited to Rao (1946) and Cramer (1946) and known as the Cramer–Rao bound, although it was essentially given by Fisher (1925).

Interpretation of Diagonal Elements For the special case that $\hat{\theta}$ is an unbiased estimate of θ, the matrix G is the identity and we conclude that $V - I^{-1}$ is nonnegative definite. Nonnegative definite matrices have nonnegative values along the diagonal so

$$V_{jj} = \text{Var}(\hat{\theta}_j) \geq (I^{-1})_{jj}.$$

The expected information matrix I thus determines a theoretical lower bound on the variance of unbiased estimators $\hat{\theta}_j$ and the larger this information the lower this bound will be. Where information is high/low, accurate/inaccurate estimation of the parameters is possible. This is the main justification for calling I information.

Large Sample Efficiency of MLE We have been assuming that the data is a set of independent variables which causes the log-likelihood, score function, and observed information to be expressed as sums of independent variables. In this case, I is a sum of positive terms, one for each data value, and so expected information always increases (and never decreases) roughly in proportion to the amount of data n. Observed information, being close to expected information, will also tend to steadily increase. We have also seen that the ML estimator of θ converges in distribution to normal with variance matrix I^{-1}. Drawing these ideas together, and supposing that n is large enough for these approximations to be accurate, we are able to make the following statements about ML estimation: *The accuracy of the ML estimator increases as more data is collected. For a given sample size, the accuracy of the ML estimator is determined by the amount of information. The ML estimator is asymptotically as accurate as possible since it attains the Cramer–Rao bound* [*see Eq. (1.21)*].

Reparametrization The score and information are with respect to a parameter θ. Suppose we reparametrize the model in terms of parameters $\omega = g(\theta)$ where g is a one-to-one transformation. Let G be the derivative matrix of this transformation, i.e., $G_{ij} = \partial\omega_i/\partial\theta_j$. Then using the chain rule for differentiation, the score function for the parameter ω is related to the score function for θ by

$$U_\theta = G^T U_\omega.$$

Since the information matrix is the variance matrix of the score function this implies

$$I_\theta = G^T I_\omega G. \tag{1.22}$$

Taking the inverse of both sides and pre- and post-multiplying by the inverse of G and G^T we find that

$$I_\omega^{-1} = G I_\theta^{-1} G^T$$

which would be the asymptotic variance matrix of $\hat{\Omega} = g(\hat{\theta})$. A more general expression for the asymptotic variance of a noninvertible function $g(\hat{\theta})$ is given in Section 1.6.

Interpretation of Off-Diagonal Elements The Cramer–Rao bound also affords an important interpretation of the off-diagonal elements and gives a direct measure of the effect of one model parameter on another. To illustrate: suppose θ has dimension two and the first component is to be estimated. What does the bound say about estimating θ_1 when θ_2 is also unknown? The expected information matrix is

$$I(\theta) = \begin{pmatrix} I_{11} & I_{12} \\ I_{12} & I_{22} \end{pmatrix}$$

whose inverse is

$$I^{-1}(\theta) = \frac{1}{I_{11} I_{22} - I_{12}^2} \begin{pmatrix} I_{22} & -I_{12} \\ -I_{21} & I_{11} \end{pmatrix}.$$

By inequality [see Eq. (1.21)] any unbiased estimate of θ_1 will have variance at least as large as the upper left entry of this matrix, i.e.,

$$\mathrm{Var}(\hat{\theta}_1) \geq I_{22}/(I_{11} I_{22} - I_{12}^2). \tag{1.23}$$

In contrast if θ_2 were known then the expected information would simply be the 1×1 matrix I_{11} and the lower bound on the variance of unbiased estimates of θ_1 would be $1/I_{11}$. The lower bound in Eq. (1.23) is always larger than this and its reciprocal is often called the expected information about θ_1 when θ_2 is unknown and written

$$I_{1/2}(\theta) = I_{11} - I_{12}^2/I_{22}.$$

The relative loss of information about one parameter from not knowing the other measures the *nonorthogonality* of θ_1 and θ_2 and is denoted ρ^2 since

$$\frac{I_{11} - I_{1/2}}{I_{11}} = \frac{I_{12}^2}{I_{11} I_{22}} = \rho^2(U_1, U_2), \tag{1.24}$$

being the squared correlation of the components U_1, U_2 of the score function. Large nonorthogonalities (close to 1) mean assumptions about one parameter have a large effect on inference about the other parameter. Specifically, the asymptotic variance of one parameter estimate will decrease by the factor $(1-\rho^2)$ if the other parameter is assumed known. The change in the *value* of one estimate can also be predicted from ρ. First, write the first order approximation

$$0 = \frac{\partial l}{\partial \theta_1}(\hat{\theta}_0) = \frac{\partial l}{\partial \theta_1}(\theta) - \mathcal{I}_{11}(\hat{\theta}_{10} - \theta_1) - \mathcal{I}_{12}(\theta_{20} - \theta_2)$$

where $\hat{\theta}_0 = (\hat{\theta}_{10}, \theta_{20})$ and $\hat{\theta}_{10}$ is the ML estimate of θ_1 when θ_2 is assumed to equal θ_{20}. Rearrange this to isolate $\hat{\theta}_{10}$ and differentiate with respect to the assumed value θ_{20} to obtain

$$\frac{\partial \hat{\theta}_{10}}{\partial \theta_{20}} = -\frac{\mathcal{I}_{12}}{\mathcal{I}_{11}}$$

giving the rate of change of the ML estimator of θ_1 as θ_2 is moved away from its ML estimate $\hat{\theta}_2$. Rescaling this derivative by dividing $\hat{\theta}_i$ by their standard errors $\mathcal{I}_{ii}^{-1/2}$ the rate of change simply becomes $-\rho$. In other words, for every fraction of a standard error we increase θ_2 from its ML estimate, the ML estimate of θ_1 will decrease by roughly ρ fractions of a standard error.

Nonorthogonality is an important diagnostic for data analysis. In the course of model building we will perhaps set several parameters to zero after suitable tests. If these parameters are highly nonorthogonal to the parameter of primary interest then we have identified two mathematical effects. First, the value of the estimate will change and second, the standard error will reduce. If we were wrong to set the other parameters to zero then we will have a worse estimate about which we are erroneously more confident. The orthogonality matrix can be obtained from a regression package by inverting the covariance matrix of the parameter estimates and then converting to a squared correlation matrix.

1.3.4 Application to Poisson Models*

The limit theorems follow from the normality of the score function in Eq. (1.18). For binomial data, which are common in this book, the preceding theory can be applied directly—the sample size n is the number of binary trials. But when the data is Poisson, as is commonly assumed for count data, the normal limit usually applies as the means of the Poisson variables diverge, not as the sample size diverges. It would take too much space to examine Poisson models in general, and the reader is referred to Haberman (1974). However, it is important to give an idea of how the large n results can be cajoled into results for Poisson variables with large means.

It is easy to make the required link between the sample size and the mean of a Poisson variable. Suppose Y has mean $\lambda = \alpha T$, for instance, if α is the accident rate and T is the number of days over which counts take place. Then Y can be written

$$Y = Y_1 + Y_2 + \cdots + Y_T$$

where Y_i is the number of accidents on day i with $Pn(\alpha)$ distribution. From the ordinary central limit theorem, Y will be approximately normal as the number of days T diverges, i.e., as the mean $\lambda = \alpha T$ diverges.

Moving to the multiparameter case, consider the full model $\mathcal{M}_1$, with each mean parametrized as $\lambda_j = T\alpha_j$. The score function with respect to $\alpha = (\alpha_1, \alpha_2, \alpha_3, \alpha_4)$ is

$$U_j(\alpha) = -T + \frac{Y_j}{\alpha_j}$$

and this score function is asymptotically normal if, and only if, Y_j is asymptotically normal; Y_j converges to normal as the "sample size" T diverges. This is equivalent to all the expected values λ_j diverging. We can also express this as a condition on the information matrix. The observed information about the parameters α_i is

$$\mathcal{I}_{jk}(\alpha) = -\frac{\partial U_j}{\partial \alpha_k} = \begin{cases} Y_j/\alpha_j^2 & j = k \\ 0 & j \neq k \end{cases}$$

while the expected information is $I(\alpha) = T \operatorname{diag}(1/\alpha_i)$, the expected value of the observed information. The smallest eigenvalues of these matrices diverge iff T diverges. Thus, the normal limit holds if (i) the sample size T diverges, (ii) the expected values $\mathrm{E}(Y_j)$ all diverge, or (iii) the information, as measured by the smallest eigenvalue, diverges. For more complicated Poisson models the same conclusions can be reached, but precise conditions differ from model to model. There are circumstances when the normal limit holds even when all the expected values do not diverge [see Haberman (1977)].

Example: Serious Accident Data, Model $\mathcal{M}_2$ For the $(\lambda_1, \lambda_2, \phi_1)$ parametrization, the score function $U = (U_1, U_2, U_3)^T$ was given earlier as

$$\frac{\partial \ell}{\partial \lambda_1} = \frac{Y_1}{\lambda_1} - \frac{Y_3}{\lambda_3}$$

$$\frac{\partial \ell}{\partial \lambda_2} = \frac{Y_2}{\lambda_2} - \frac{Y_4}{\lambda_4}$$

$$\frac{\partial \ell}{\partial \phi_1} = -4.292 + \frac{Y_3}{\lambda_3} + \frac{3.292 Y_4}{\lambda_4}.$$

The observed information matrix $\mathcal{J}$ is obtained by differentiating each component of this vector U with respect to each of the parameters λ_1, λ_2, ϕ_1 and the expected information matrix I is obtained either by finding the expectation of each element of $\mathcal{J}$ or by calculating the variance matrix of U from the independent Poisson distributions of the Y_i. The observed and expected information matrices are respectively

$$\begin{pmatrix} \frac{Y_1}{\lambda_1^2} + \frac{Y_3}{\lambda_3^2} & 0 & -\frac{Y_3}{\lambda_3^2} \\ 0 & \frac{Y_2}{\lambda_2^2} + \frac{Y_4}{\lambda_4^2} & -\frac{3.292 Y_4}{\lambda_4^2} \\ -\frac{Y_3}{\lambda_3^2} & -\frac{3.292 Y_4}{\lambda_4^2} & \frac{Y_3}{\lambda_3^2} + \frac{3.292^2 Y_4}{\lambda_4^2} \end{pmatrix},$$

$$\begin{pmatrix} \frac{1}{\lambda_1} + \frac{1}{\lambda_3} & 0 & -\frac{1}{\lambda_3} \\ 0 & \frac{1}{\lambda_2} + \frac{1}{\lambda_4} & -\frac{3.292}{\lambda_4} \\ -\frac{1}{\lambda_3} & -\frac{3.292}{\lambda_4} & \frac{1}{\lambda_3} + \frac{3.292^2}{\lambda_4} \end{pmatrix}$$

These matrices are distinct, even when estimates are substituted for the parameters, since under $\mathcal{M}_2$ the ML estimates are *not* just the data values. The inverses of $\hat{I}$ and $\hat{\mathcal{J}}$, which approximate the variance matrix of the three estimators, are shown in Table 1.4.

One use of information is in assessing the sensitivity of the model to the various parameters. Using the estimated expected information $\hat{I}$ and converting to a correlation matrix gives the estimated orthogonality matrix

$$\hat{R} = \begin{pmatrix} 1.000 & 0.000 & -0.279 \\ 0.000 & 1.000 & -0.897 \\ -0.297 & -0.897 & 1.000 \end{pmatrix}$$

Table 1.4. Approximations to the Variance Matrix.[a]

Actual[b]	6.45	9.91	3.36
Obssserved	7.45	9.89	3.34
Expected	6.28	9.89	3.35
Actual[b]	—	46.01	13.54
Observed	—	45.38	13.49
Expected	—	45.77	13.49
Actual[b]	—	—	4.58
Observed	—	—	4.56
Expected	—	—	4.56

[a]Each figure estimates the variance matrix of $(\hat{\lambda}_1, \hat{\lambda}_2, \hat{\phi}_1)$ assuming the parameter values estimated under model $\mathcal{M}_2$.
[b]From 100,000 simulations.

Since $\hat{\rho}_{12} = 0$, estimation of λ_1 is substantially unaffected by what we assume about λ_2. On the other hand, since $\hat{\rho}_{23} = -0.897$, assumptions about λ_2 will have a very large impact on estimates of ϕ_1 and assuming that λ_2 is completely unknown inflates the variance by a factor $1/(1 - \hat{\rho}^2_{23}) = 5.11$.

The second and more common use of information is in estimating the asymptotic variance matrix of estimators. According to the limit results of Section 1.3.2, the estimator $(\hat{\lambda}_1, \hat{\lambda}_2, \hat{\phi}_1)$ should have an approximate three dimensional normal distribution with covariance matrix equal to the inverse of the matrix I or $\mathcal{J}$ given above. For instance, $\hat{\lambda}_1$ should be approximately $\mathcal{N}(\lambda_1, V)$ where V is approximated by 7.45 using observed information, and 6.28 using expected. Figure 1.3 shows the results of 10,000 simulations of $\hat{\lambda}_1$ assuming the

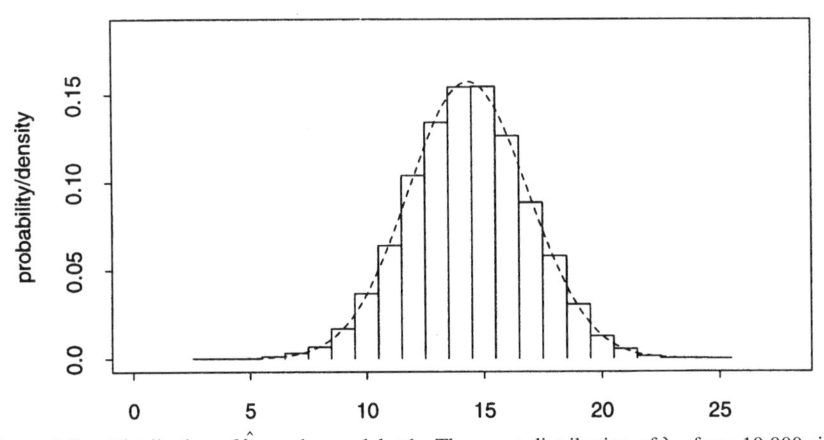

Figure 1.3. Distribution of $\hat{\lambda}_1$ under model $\mathcal{M}_2$. The exact distribution of λ_1 from 10,000 simulations using estimated parameter values compared to the normal approximation using expected information.

parameter values $(\lambda_1, \lambda_2, \phi_1)$ equal their ML estimates (14.353, 57.892, 19.572). Superimposed is the $\mathcal{N}(14.353, 6.28)$ distribution. Apparently, the distribution is extremely close to normal in shape. The true variance of these simulations was actually 6.45, closer to the expected information than the observed.

1.3.5 Observed Versus Expected Information*

Either the observed or expected information can be used in the CLT. Which, if any, is better? There are certainly cases where using expected information gives the variance of the ML estimator exactly. For example, for the full model $\mathcal{M}_1$, the score function for $(\lambda_1, \lambda_2, \lambda_3, \lambda_4)$ has components

$$U_j = -1 + Y_j/\lambda_j.$$

The ML estimator of λ_j is Y_j and the variance matrix of these four estimators is diag(λ_j). The expected information matrix is $I = \text{diag}(1/\lambda_j)$ which upon inversion gives the exact variance matrix. The observed information, $\mathcal{J} = \text{diag}(Y_j/\lambda_j^2)$ is random and only gives the correct variance on average. We might expect that using I will give better approximations to probabilities than using $\mathcal{J}$.

What about models where neither I nor $\mathcal{J}$ gives the exact variance? The observed and expected information matrices for $(\hat{\lambda}_1, \hat{\lambda}_2, \hat{\phi}_1)$ under $\mathcal{M}_2$ were shown in the last section. After substituting the estimated parameter values and finding the inverse of the matrices, this gave the values in Table 1.4. The main point of disagreement is in the (1,1) term, where according to observed information the variance of $\hat{\lambda}_1$ should be 7.45 but according to expected information it should be 6.28. Which gives the better approximation to the variance of $\hat{\lambda}_1$, expected or observed information based methods?

Since the ML estimates (14.353, 57.892, 19.572) were substituted in both matrices, it makes sense to ask "What is the true variance of $\hat{\lambda}_1$ when model $\mathcal{M}_2$ is true with these parameter values?" To answer this, four Poisson variables with means (14.353, 57.892, 5.219, 6.536) were, these being fitted values under $\mathcal{M}_2$. For each simulated set of four counts, Eq. (1.16) was used to calculate $\hat{\lambda}_1$, $\hat{\lambda}_2$, $\hat{\phi}_1$. This was repeated 100,000 times giving a very accurate estimate of the true variances and covariances of the estimators, also displayed in Table 1.4. It looks like a win for expected information, since the true variance of $\hat{\lambda}_1$ was 6.45.

Both the previous examples point toward superiority of expected information but both examples are misleading. In the first example, the expected information gives the exact variance, a rare circumstance. In the second example, the superior performance of expected information is just good luck—changing the data slightly can make observed information appear better. In fact it can be shown theoretically that neither method is better in general. As estimators of $\mathrm{Var}(\hat{\theta})$, both $\hat{\imath}^{-1}$ and $\hat{\jmath}^{-1}$ have the same asymptotic mean squared error.

However, there are strong arguments that observed information leads to a better estimator of $\mathrm{Var}(\hat{\theta}|A)$ where A is called an *ancillary* statistic. This is a statistic whose distribution does not depend on, or hardly depends on, θ but whose value has a large influence on the shape of the likelihood function. In this case, it has been argued that the value of A should be treated as fixed and inference evaluated conditional upon its observed value.

What is the ancillary statistic here? Since we are *assuming* that 23.3% of the accidents are fatal, the statistic $Y_1 + Y_3$ should be around 23.3% as large as the total $T = Y_1 + Y_2 + Y_3 + Y_4$. In fact conditional on this total $Y_1 + Y_3 \stackrel{d}{=} Bi(T, 0.233)$ which depends on no unknown parameters at all. The value of this statistic might just as well be provided by a computer as by road accidents and it tells us nothing about the parameter λ_1. In fact the statistic

$$A = \frac{2.365(Y_1 + Y_3 - 0.233T)}{\sqrt{T}}$$

has very close to a standard normal distribution conditional on T and also unconditionally. For the given data $a = -1.18$, which is not improbably low. What is the distribution of $\hat{\lambda}_1$ conditional on this value of a?

Let us look at the variance of λ_1 conditional on A being within a small range of values. The real line was divided up into 25 disjoint intervals such that the probability of a standard normal variable being within each interval was 0.04. From the 100,000 simulations of $\hat{\lambda}_1$, the values of a were also computed. For each interval, we computed the sample variance of the subset of simulated values of $\hat{\lambda}_1$ whose corresponding value of A is within that interval. I have plotted this conditional variance against the center of the interval in Figure 1.4.

Since there is still some simulation variation, the plot has been smoothed. At $a = -1.18$ the extrapolated value of the conditional variance is 7.40. To within simulation error, which here is about 0.03, this is the true variance of $\hat{\lambda}_1$ conditional on A being close to -1.18. The observed information estimate, 7.45, is a much better approximation to this than the expected information estimate, 6.28, which estimates the variance *averaged* across other values of A which might have, but in fact did not, happen.

In Chapter 4 we describe a wide class of models called generalized linear models, in particular, models that have so-called canonical link. For these models, the expected and observed information matrices are identical. The choice does not arise and in such models there is no underlying ancillary on which to condition. This perhaps explains why little attention has been given to these issues in texts on data analysis. However, in general the matrices can differ substantially, and observed information would appear to be the better choice. As a minor point, observed information is also easier to calculate.

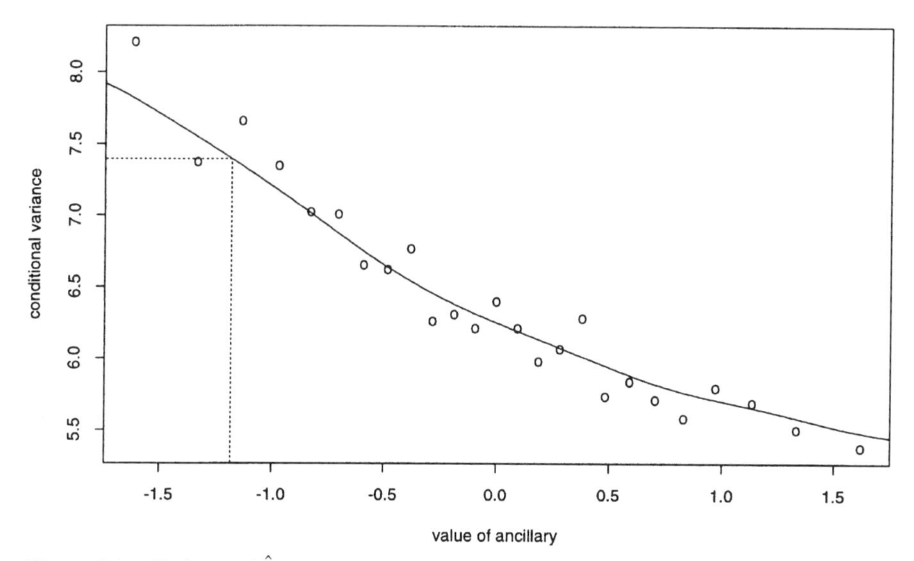

Figure 1.4. Variance of $\hat{\lambda}_1$ conditional on A = a. Each point gives the sample variance of the simulated values of $\hat{\lambda}_1$ for which A lies in the interval whose center is plotted on the x-axis. A smooth curve has been superimposed to allow extrapolation at $a = -1.118$.

SUMMARY:

There is a lower limit on the variance of unbiased estimators that depends on the inverse of the information matrix. The ML estimator asymptotically achieves this bound. Off-diagonal elements of the information matrix measure the sensitivity of the model's fit to each parameter. Under mild conditions, ML estimators have asymptotically multivariate normal distributions with mean equal to the true parameter vector. The variance matrix may be approximated by the observed or the expected information, and when these differ the former is preferred.

1.4 TESTING ONE PARAMETER

A model typically involves several unknown parameters, collected together in a parameter vector θ. Many hypotheses of interest are hypotheses about a single quantity. For instance "the mean number of accidents per week is 10." Even an hypothesis like $\theta_1 = 0.6\theta_3$ is an hypothesis about the *single* parameter θ_1/θ_3 although at first glance it appears to involve two parameters. In general, we have a model with parameters $\theta = (\theta_1, \ldots, \theta_p)$ and an hypothesis of the form

$$\mathcal{H}_0 : \omega = \omega_0$$

where $\omega = g(\theta_1, \ldots, \theta_p)$ is a parameter of interest to us, perhaps being a complicated nonlinear function g of the elementary parameters θ_i, and ω_0 is a specified value, often but not always zero. We denote the ML estimator of ω by $\hat{\Omega}$ and the observed value of this $\hat{\omega}$. Note that $\hat{\omega}$ can be computed by substituting estimates of the θ_i into g, i.e., $\hat{\omega} = g(\hat{\theta}_1, \ldots, \hat{\theta}_p)$.

In this section we will consider tests based on $\hat{\Omega}$ although there are certainly other test statistics available (see Section 1.5). We introduce the concept of the P-value, discuss some of the problems in its computation, and give detailed numerical illustrations using the accident data.

1.4.1 Evidence and the Use of P-Values

Testing based on P-values is a natural extension of scientific hypothesis falsification. In the so-called hard sciences, an hypothesis will be proposed and an experiment designed to disprove it. Such falsifying experiments aim to produce data sets that are *logically inconsistent* with the hypothesis. When data are subject to random variation it will not be possible to conclude that any hypothesis is false with logical certainty. The data may well be an exceedingly improbable anomaly pointing to the hypothesis being false when it is in fact true. The best that we can do then is to say that the results, while not in logical contradiction with, are exceedingly unlikely under the hypothesis. This suggests that the hypothesis is false and that some other hypothesis under which the results are not so improbable is true. In the words of Fisher (1956, p. 43).

> Either an intrinsically improbable event has occured or the offered hypothesis is not correct.

and we choose the second of these exhaustive possibilities. The notion of finding hypotheses which make the data more rather than less probable is at the heart of maximum likelihood. In the extreme case that the observed data has zero probability, and only in this case, may we conclude with certainty that the hypothesis is false.

How to measure the "improbability of the data?" An obvious consideration is to look at how far $\hat{\omega}$ is from the hypothesized value ω_0. Larger differences indicate more strongly that ω_0 is not the true value. How improbable is it that $\hat{\omega}$ would be as extreme as the value observed, assuming $\mathcal{H}_0$ is in fact true? The smaller is this probability, or *P-value*, the more doubt falls on the hypothesis $\mathcal{H}_0$.

To be specific, for testing the null hypothesis $\mathcal{H}_0 : \omega = \omega_0$ against the one sided alternative $\mathcal{H}_1 : \omega > \omega_0$ the P-value is

$$\text{P} - \text{value} = \Pr(\hat{\Omega} \geq \hat{\omega}; \mathcal{H}_0). \tag{1.25}$$

For testing against $\mathcal{H}_1 : \omega < \omega_0$ the P-value would be the lower tail probability from this same *null distribution*.

Measuring evidence with P-values does pose some logical problems and we will discuss some of the important issues in Section 1.5.6. We have been stressing P-values rather than a strict *accept–reject* approach to evaluating hypotheses. The P-value can indeed be used to make a formal test of size α by simply rejecting $\mathcal{H}_0$ if and only if it is smaller than α. Many elementary textbooks couch hypothesis tests in the form "Reject $\mathcal{H}_0$ if $\hat{\omega}$ exceeds some critical value obtained from the sampling distribution of $\hat{\Omega}$." The main advantage of quoting a P-value instead is that the decision of the test size may be left to others; the statistician may simply quote the weight of evidence that the data throws *against* the null hypothesis.

1.4.2 Computation of P-Values

The P-value is computed from the null distribution of $\hat{\Omega}$. This might prove impossible if (i) the distribution is too difficult to derive, (ii) the distribution depends not only on ω but on other nuisance parameters. In practice, the first problem is dealt with by finding a normal approximation to the sampling distribution of $\hat{\omega}$. The second problem is dealt with by replacing nuisance parameters by maximum likelihood estimates, either in the exact distribution of $\hat{\Omega}$ if this is available, or in the normal approximation.

Estimated P-Values Write the parameter θ in the form (ω, ψ) where ω is being tested and ψ is a vector of nuisance parameters. Suppose the distribution of $\hat{\Omega}$ depends on θ both through ω and through ψ. Let $\hat{\psi}$ be an estimator of ψ. Then the *estimated* P-value is defined to be

$$\hat{\text{P}} - \text{value} = \Pr(\hat{\Omega} \geq \hat{\omega}; \omega = \omega_0, \psi = \hat{\psi}). \tag{1.26}$$

Since the probability is computed assuming $\omega = \omega_0$, it would seem best to let the estimator $\hat{\psi}$ be the Ml estimator of ψ, under the restriction $\omega = \omega_0$. This is obtained by substituting $\omega = \omega_0$ into the likelihood and maximizing with respect to ψ only. In practice, most computer packages report estimated P-value, using the *unrestricted* ML estimator to estimate ψ. Asymptotically it makes no difference if $\mathcal{H}_0$ is true since both estimators converge to the true value ψ and will hardly differ. When $\mathcal{H}_0$ is untrue, however, the tests defined by these two ways of estimating the P-value may have different powers.

While the estimated P-value can at least be calculated it is no longer the probability of obtaining results as extreme as those observed. Rather, it is an estimate of this probability and the danger clearly exists that our estimated P-value will be significantly in error.

Example: Testing λ_1 Under Model $\mathcal{M}_2$ Suppose that on previous evidence we believed that the weekly fatality rate for older drivers is 10 and we want to know if the rate has increased this year. The hypotheses to be tested are $\mathcal{H}_0 : \lambda_1$

$= 10$ against $\mathcal{H}_1 : \lambda_1 > 10$. To make things less trivial, we test these hypotheses assuming $\mathcal{M}_2$ which assumed the probability of a serious accident being fatal is 0.233. The parameters of this model are $(\lambda_1, \lambda_2, \phi_1)$ and the parameter of interest is here $\omega = \lambda_1$. The ML estimate was 14.353 computed from the formula

$$\hat{\lambda}_1 = \frac{0.233Y_1(Y_1 + Y_2 + Y_3 + Y_4)}{Y_1 + Y_3}.$$

This statistic is a complicated function of the data and it is impossible to write down a formula for its distribution. However, we can solve this problem by simulating the distribution. The more important difficulty is that the distribution of $\hat{\lambda}_1$ depends on the nuisance parameters $\psi = (\lambda_2, \phi_1)$. So there is simply no way to compute the P-value

$$\Pr(\hat{\lambda}_1 \geq 14.353; \lambda_1 = 10)$$

until we specify values for λ_2 and ϕ_1.

The ML estimate of ψ is $(\hat{\lambda}_2, \hat{\phi}_1) = (57.892, 19.572)$ and we could substitute these values for (λ_2, ϕ_1). However, it is preferable to compute ML estimates assuming $\lambda_1 = 10$. To find this we substitute $\lambda_1 = 10$ into the log-likelihood (see Eq. (1.10)] and then maximize with respect to λ_2 and ϕ_1 by solving the score equations

$$\frac{\partial l}{\partial \lambda_2} = \frac{y_2}{\lambda_2} - \frac{y_4}{3.292\phi_1 - \lambda_2} = 0$$

$$\frac{\partial l}{\partial \phi_1} = -4.292 + \frac{y_3}{\phi_1 - 10} + \frac{3.292y_4}{3.292\phi_1 - \lambda_2} = 0.$$

These can be solved using the Newton–Raphson algorithm and we find that $\tilde{\psi} = (53.692, 18.152)$. Substituting these for the nuisance parameters defines our estimated P-value.

Because the distribution of $\hat{\lambda}_1$ is so complicated we will simulate it. For this we need the means $\lambda_1, \lambda_2, \lambda_3, \lambda_4$ for our four Poisson distributed data values. Using the equations for the fitted values under $\mathcal{M}_2$ and the values $\lambda_1 = 10$, $\lambda_2 = 53.692$, $\phi_1 = 19.152$ gives the four fitted values in the second line of Table 1.5. We can now perform a simulation to approximate

$$\Pr(\hat{\lambda}_1 \geq 14.353; \lambda_1 = 10, \lambda_2 = 53.692, \lambda_3 = 8.152, \lambda_4 = 6.062). \tag{1.27}$$

We generate four Poisson random variables with these four means. We compute $\hat{\lambda}_1$ from these four simulated data values. This is repeated many times. It was found that for 3952 of 100,000 simulations $\hat{\lambda}_1$ was 14.353 or larger. Hence

Table 1.5. Expected and Fitted Values for Testing $\mathcal{H}_0 : \lambda_1 = 10$

	Over 21		Under 21	
	Fatal	Nonfatal	Fatal	Nonfatal
$\mathcal{H}_0$	10	λ_2	$\phi_1 - 10$	$3.292\phi_1 - \lambda_2$
Fitted ($\mathcal{H}_0$)	10.000	53.692	8.152	6.062
Fitted ($\mathcal{M}_2$)	14.353	57.892	5.219	6.536
Data	11	62	4	7

our estimated P-value is $3952/100{,}000 = 0.0395$. This probability may not be exactly equal to Eq. (1.27) but with so many simulations the error is very small, around 0.0006 as measured by standard deviation.

Since the P-value turned out to be quite small, there does seem to be considerable evidence that the expected weekly number of fatal accidents involving younger drivers exceeds 10. This is a surprising result bearing in mind there were only 11 such accidents from the raw data! However, throughout this analysis we have been assuming that the proportion of fatal accidents is $p = 0.233$ (model $\mathcal{M}_2$). With this assumption, the 11 accidents are not taken at face value. Rather, the data collectively point to an average rate of 14.353. Clearly, if we dropped the assumption of $\mathcal{M}_2$ then the ML estimate of $\hat{\lambda}_1$ becomes 11 and there would be very little evidence that the fatal accident rate for older drivers exceeds 10.

Normal Approximations to P-Values It is a pain to conduct a simulation every time we want a P-value. The Central Limit Theorems stated in Eq. (1.20) are fundamental tools for approximating the distributions of ML estimators and play a major practical role in calculating P-values. The involvement of any nuisance parameters in the distribution of $\hat{\Omega}$ is simplified to a formula for the approximate variance

$$\text{Var}(\hat{\theta}) \simeq \mathcal{J}(\omega, \psi)^{-1}$$

and the top left component of this matrix, which could depend on ω and ψ, is the approximate variance of $\hat{\Omega}$. Estimating ψ by the ML or restricted ML estimator then allows simple and direct computation of the P-value from the approximate normal distribution.

Example: (Continued) The observed information matrix for the three parameters $\lambda_1, \lambda_2, \phi_1$ under model $\mathcal{M}_2$ was given on p. 28. Since we want the *null* distribution we substitute the fitted values in the second line of Table 1.5 obtained from taking $\lambda_1 = 10$ and $(\lambda_2, \phi_1) = (53.692, 19.152)$. Taking the inverse of the resulting matrix gives the asymptotic variance matrix

$$\hat{I}_0^{-1} = \begin{pmatrix} 5.774 & 6.892 & 2.330 \\ 6.892 & 42.451 & 12.510 \\ 2.330 & 12.510 & 4.229 \end{pmatrix}$$

of the ML estimator $(\hat{\lambda}_1, \hat{\lambda}_2, \hat{\phi}_1)^T$. We are interested here only in the marginal distribution of $\hat{\lambda}_1$ which is approximated by $\mathcal{N}(10, 5.774)$. This gives the approximate value 0.035 for the probability of $\hat{\lambda}_1$ being as large as its observed value 14.353 in good agreement with the simulation value 0.0395.

Conditional P-Values A completely alternative method of dealing with the problem of nuisance parameters is to instead calculate a conditional P-value. Much more is said about this in Chapter 7. The method requires us to identify a statistic S sufficient for the nuisance parameter ψ. It then follows that the distribution of the data conditional on S depends only on ω, not on ψ. In particular we can calculate the *conditional* P-value,

$$\Pr(\hat{\Omega} \geq \hat{\omega} \,|\, S; \omega_0). \tag{1.28}$$

There are several important issues. Is the conditional P-value as meaningful a measure of evidence as the original P-value? Why should we calculate the probability treating the observed value of S as fixed? Some reasons for conditioning have already been given in Section 1.3.5 but do not apply directly here. Computation of conditional distributions is also sometimes much more difficult than unconditional ones. Finally, a minimal sufficient statistic for the nuisance parameters may be the whole data. In this case, the conditional P-value is equal to 1 and the method fails.

Example: (Continued) To apply the conditional method we need a sufficient statistic for $\psi = (\lambda_2, \phi_1)$. From the null means in the first line of Table 1.5 the log-likelihood under the null model is

$$\begin{aligned} \ell(\lambda_2, \phi_1) = {} & -4.292\phi_1 + y_2 \log \lambda_2 + y_3 \log(\phi_1 - 10) \\ & + y_4 \log(3.292\phi_1 - \lambda_2) \end{aligned}$$

and so the sufficient statistic $S = (Y_2, Y_3, Y_4)$ and the conditional P-value is

$$\Pr(\hat{\lambda}_1 \geq 14.353 \,|\, Y_2 = 62, Y_3 = 4, Y_4 = 7, \lambda_1 = 10).$$

Now recall that

$$\hat{\lambda}_1 = \frac{0.233 Y_1 (Y_1 + Y_2 + Y_3 + Y_4)}{Y_1 + Y_3}$$

which is a monotone increasing function of Y_1. If follows that, conditional on Y_2, Y_3, Y_4, the set $\{\hat{\lambda}_1 \geq 14.353\}$ is just the set $\{Y_1 \geq 11\}$. Futhermore, Y_1 is independent of Y_2, Y_3, Y_4 and so the conditional probability is the same as the unconditional probability. Thus, our so-called conditional P-value is

$$\Pr(Y_1 \geq 11; \lambda_1 = 10) = 0.4169,$$

a very different value to 0.035. Conditional P-values are not necessarily close to unconditional P-values even asymptotically. Notice that this P-value is exactly the same as the ordinary P-value we would calculate assuming the full model. Conditioning on S has essentially relaxed $\mathcal{M}_2$ to the full model in this example. Again the reader is referred to Chapter 7 for further discussion.

1.4.3 Example: The Goodness-of-Fit of $\mathcal{M}_2$*

Since the assumption that $\mathcal{M}_2$ is true is so critical to the above test result it would be judicious to test the adequacy of this assumption. In general, *goodness-of-fit tests* refer to a test of a particular model against the *full* model. The rationale for this is that if we really wish to test the validity of a model we should assume as little as possible and let the data speak entirely for themselves; the full model assumes least of all possible parametric models. This particular goodness-of-fit test reduces to a test of a single parameter because of the fact that $\mathcal{M}_2$ is specified by the *single* parametric restriction $p = 0.233$ or equivalently

$$0.767\lambda_1 - 0.233\lambda_2 + 0.767\lambda_3 - 0.233\lambda_4 = 0.$$

Calling the left-hand side ω we wish to test whether or not $\omega = 0$ against the alternative that $\omega \neq 0$; goodness-of-fit tests are by their nature always two-sided. The ML estimator of ω under $\mathcal{M}_1$ is

$$\hat{\Omega} = 0.767Y_1 - 0.233Y_2 + 0.767Y_3 - 0.233Y_4 \tag{1.29}$$

and the observed value is $\hat{\omega} = -4.587$. We need to find the probability of obtaining an estimate as large as this *in absolute value* if in fact ω is zero.

The fitted values under the null hypothesis $\mathcal{M}_2$ are listed in line 3 of Table 1.5. From 100,000 simulations using these parameter values the absolute value of $\hat{\Omega}$ was as large as 4.587 on 22,746 occasions. The P-value is 0.227 which provides very little evidence against $\mathcal{M}_2$. The Central Limit Theorem [see Eq. (1.20)] here leads to simple normal approximations to the Poisson distributions of (Y_1, Y_2, Y_3, Y_4). The estimator $\hat{\Omega}$ is a simple linear combination of these with null mean zero and null variance

$$0.767^2(14.353) + 0.233^2(57.892) + 0.767^2(5.219) + 0.233^2(6.536) = 14.928$$

and gives the approximate (estimated) P-value $\Pr(|\hat{\Omega}| \geq 4.587) = 0.236$ in good agreement with the simulation.

It is pertinent to note that even though there is little evidence against the model $\mathcal{M}_2$, assuming that it is true has a large effect on inferences about λ_1. This introduces extra uncertainty into our inference which will not be accounted for by the usual standard error estimate. The sensitivity of inferences on λ_1 to assumptions about ω can be traced to the high nonorthogonality of these parameters.

1.4.4 One-Sided and Two-Sided Tests*

When ω represents the effect of some treatment, one-sided tests of ω are quite common because of the prior expectation that the treatment effect will be in a particular direction. What are the relative merits of one- and two-sided tests and how much difference does it make anyway?

For testing $\mathcal{H}_0 : \omega = \omega_0$ against $\mathcal{H}_1 : \omega > \omega_0$, the P-value is the probability of the estimator $\hat{\Omega}$ being as large as its observed value $\hat{\omega}$. When the alternative hypothesis is $\mathcal{H}_1 : \omega < \omega_0$ we would naturally calculate the probability of $\hat{\Omega}$ being as small as its observed value. For the two-sided alternative $\mathcal{H}_1 : \omega \neq \omega_0$ one naturally rejects H_0 if $\hat{\Omega}$ is too much larger or smaller than ω_0 so the two-sided P-value is

$$P_{2s} = \Pr(|\hat{\Omega} - \omega_0| > |\hat{\omega} - \omega_0|).$$

If $\hat{\Omega}$ has a symmetric distribution then it is easy to show that the two-sided P-value is twice the one-sided P-value. Conversely, to convert a two-sided P-value to a one-sided P-value

$$P_{1s} = \begin{cases} 0.5P_{2s} & \hat{\omega} \in \mathcal{H}_1 \\ 1 - 0.5P_{2s} & \hat{\omega} \notin \mathcal{H}_1 \end{cases}.$$

When $\hat{\Omega}$ is approximated by the (symmetric) normal distribution, these assertions are approximately true.

The choice between a one- and two-sided test is but one example of the sensitivity of statistical tests to the choice of hypotheses. The one-sided test is certainly more powerful than the two-sided test because it is tailored to be sensitive in the particular direction of interest. This is both useful and dangerous—dangerous because if one decides to test the specific one-sided alternative *after* looking at the data, one has essentially decided to halve the P-value. This is certainly *not* statistically valid and makes a mockery of the notion of a P-value.

In principle, all hypotheses and their alternatives should be stated in advance of looking at the data. Where one decides to test hypotheses that are generated by the data then the P-value cannot be interpreted in the usual way. Data driven hypotheses may well be of interest but great care should be taken in assigning statistical weight to tests of these hypotheses.

SUMMARY:

Weight of evidence against an hypothesis is often measured by the P-value. This is the probability of the ML estimate being as extreme as its observed value. When the distribution of the estimator depends on nuisance parameters the P-value cannot be calculated. In this case, the P-value may be estimated, approximated, or computed conditionally. One-sided tests can be obtained from two-sided tests by roughly halving the P-value but the decision to do a one-sided test should be made before seeing the data.

1.5 TESTING SEVERAL PARAMETERS

Many hypotheses cannot be expressed in terms of a single parameter. Hypotheses which say something about essentially more than one parameter require more general techniques. In the most general case, consider testing an hypothesis $\mathcal{H}_0$ which specifies exactly $r \leq p$ restrictions on the parameters $\theta_1, \ldots, \theta_p$. Such an hypothesis can be expressed in the form

$$\mathcal{H}_0 : g(\theta) = \omega_0$$

where g is a function from $\Re^p$ to $\Re^r$ and ω_0 is the hypothesized value of the parameter vector $\omega = g(\theta)$. The components $\omega_1, \ldots, \omega_r$ are interest parameters and the null hypothesis specifies the values of all these parameters. We concentrate on testing the alternative hypothesis $\mathcal{H}_1 : \omega \neq \omega_0$ as one-sided tests of several parameters are quite rare in practice and are rather complicated. Let $\hat{\omega} = g(\hat{\theta})$ be the ML estimator of ω with jth component $\hat{\omega}_j$. We could certainly test each ω_j separately using the estimate $\hat{\omega}_j$ and methods of the previous section, but we want to test them *simultaneously*.

Example: Goodness-of-Fit of $\mathcal{M}_3$ Model $\mathcal{M}_3$ assumed that the probability of an accident being fatal was 0.233, both for younger and older drivers. Testing the goodness-of-fit of this model is a test of *two* simultaneous assumptions, that $p_1 = 0.233$ and $p_2 = 0.233$. In terms of the original parameters λ_i we are testing the null hypothesis that both $\omega_1 = 0.767\lambda_1 - 0.233\lambda_2$ and $\omega_2 = 0.767\lambda_3 - 0.233\lambda_4$ are zero. The ML estimators of ω_1, ω_2 are

$$\hat{\Omega}_1 = 0.767Y_1 - .233Y_2, \qquad \hat{\Omega}_2 = .767Y_3 - .233Y_4$$

Estimating the λ_i by their ML estimates under $\mathcal{H}_0 = \mathcal{M}_3$, namely (17.009, 55.991, 2.563, 8.437) we obtain the variances 13.015, 1.962. We could easily test $\omega_1 = 0$ by computing $\Pr(|\hat{\Omega}_1| \geq 6.009)$ from the approximate $\mathcal{N}(0, 13.015)$ distribution and similarly test $\omega_2 = 0$ by computing $\Pr(|\hat{\Omega}_2| \geq 1.437)$ from the approximate $\mathcal{N}(0, 1.962)$.

There are many ways to combine the single tests into a simultaneous test. In this section we describe three common, and in practice similar, methods.

1.5.1 Wald's Maximum Likelihood Statistic

It is helpful in what follows to write the parameter θ as (ω, ψ) where ψ represents all the other $p - r$ parameters besides the interest parameter ω. Let $\hat{\psi}_0$ denote the restricted ML estimator of ψ assuming $\omega = \omega_0$ and let $\hat{\theta}_0 = (\omega_0, \hat{\psi}_0)$.

The ML estimator $\hat{\theta}$ has an approximately multivariate normal distribution with mean θ and $p \times p$ variance matrix given by the inverse of the information matrix I_{-1}. Let $V(\omega, \psi)$ denote the $r \times r$ sub-matrix in the upper left-hand corner. Then $\hat{\Omega}$ is approximately multivariate normal with mean ω and variance matrix $V(\omega, \psi)$. Under $\mathcal{H}_0$ we may estimate this variance matrix by $\hat{V}_0 = V(\hat{\theta}_0)$. The quadratic form

$$W = (\hat{\Omega} - \omega_0)^T \hat{V}_0^{-1} (\hat{\Omega} - \omega_0) \tag{1.30}$$

is called the Wald or ML statistic and was proposed by Wald (1941). Under the null hypothesis, the distribution of W is approximately χ^2 with degrees of freedom equal to the number of parameters under test, r. In the simple case where ω is scalar, and so testing $\mathcal{H}_0$ is a test of a single parameter, W is just the square of the usual z-statistic $(\hat{\Omega} - \omega_0)/\text{s.e.}(\hat{\Omega})$.

In practice, computer packages more commonly use $\hat{V} = V(\hat{\theta})$ rather than $\hat{V}_0 = V(\hat{\theta}_0)$ for the variance matrix of $\hat{\Omega}$. In this case a computational advantage of the Wald statistic is that only the model $\mathcal{H}_1$ need be fitted. A logical disadvantage of the Wald statistic is that a different value of the test statistic is obtained if the problem is reparametrized, i.e., if we express the null hypothesis in terms of parameter $\omega' = h(\omega)$ where h is an invertible transform. This is actually an advantage since by judicious choice of h the χ^2 distributional approximation can be improved.

Example: Goodness-of-Fit of $\mathcal{M}_3$ (Continued) The two ML estimators $\hat{\Omega}_1 = 0.767Y_1 - 0.233Y_2$ and $\hat{\Omega}_2 = 0.767Y_3 - 0.233Y_4$ are independent since the Y_i are independent. Thus, the covariance is zero and the variance matrix $\hat{V}_0$ of $(\hat{\Omega}_1, \hat{\Omega}_2)^T$ is diagonal with entires 13.015, 1.962. Hence

$$W = (\hat{\Omega}_1 \quad \hat{\Omega}_2)^T \begin{pmatrix} 13.015 & 0 \\ 0 & 1.962 \end{pmatrix}^{-1} \begin{pmatrix} \hat{\Omega}_1 \\ \hat{\Omega}_2 \end{pmatrix}$$

$$= \frac{\Omega_1^2}{13.015} + \frac{\hat{\Omega}_2^2}{1.962}$$

with observed value $1.6636^2 + 1.0249^2 = 3.818$. If we accept that the estimators $\hat{\Omega}_1$ and $\hat{\Omega}_2$ are approximately normal then it is obvious that W has the χ_2^2 distribution since it is a total of two squared standard normal variables. The P-value from observing $\omega = 3.818$ is $\Pr(\chi_2^2 \geq 3.818) = 0.148$. Note that if the estimators had not been independent then the quadratic form W would have included a term in $\hat{\Omega}_1 \hat{\Omega}_2$.

1.5.2 Rao's Score Statistic

The score function $U(\theta)$ is a vector of length p. In Section 1.3.1 we pointed out that under mild conditions, $U(\theta)$ has mean zero and variance equal to the expected information matrix and that furthermore the distribution is approximately multivariate normal. The efficient score statistic, more commonly known as the *score* statistic, is the standardized quadratic form

$$S = U(\hat{\theta}_0)^T \hat{I}_0^{-1} U(\hat{\theta}_0) \tag{1.31}$$

where $\hat{I}_0 = I(\hat{\theta}_0)$ is the information matrix with $\hat{\theta}_0$ substituted for θ.

Proposed by Rao (1947), the score statistic judges the hypothesis $\mathcal{H}_0$ by how far $U(\hat{\theta}_0)$ is from its null mean of zero. The null distribution of S is approximately χ^2 with r degrees of freedom and it tends to be larger when $\mathcal{H}_0$ is false. A computational advantage of S is that only the model $\mathcal{H}_0$ need be fitted. The value of the score statistic is also unaffected if the null hypothesis is reexpressed in terms of the parameter $\omega' = h(\omega)$.

Why does S have the χ_r^2 distribution? Since $U(\hat{\theta}_0)$ is a vector of length p, it looks like it should have p degrees of freedom, not r. To see why not, it is again helpful to write θ in the form (ω, ψ). The score function then breaks into an r-vector U_ω and a second vector U_ψ. Now $\hat{\theta}_0$ is obtained by equating only U_ψ to zero. Thus, the full score vector has its last $p - r$ components equal to zero. It is then easy to see that

$$S = (U_\omega(\hat{\theta}_0)^T 0)\hat{I}_0^{-1} \begin{pmatrix} U_\omega(\hat{\theta}_0) \\ 0 \end{pmatrix} = U_\omega(\hat{\theta}_0)^T \hat{V}_0 U_\omega(\hat{\theta}_0)$$

and so S is an r-dimensional quadratic form just like the Wald statistic.

Example: Score Test for Poisson Data For the serious accident data we had four independent Poisson data values. The score function under the full model has components $U_j = -1 + Y_j/\lambda_j$. The information matrix is diagonal with entries $1/\lambda_j$. Looking at the definition of S we see that

$$U^T I^{-1} U = \sum_{j=1}^{4} U_j^2 / I_{jj} = \sum_{j=1}^{4} \lambda_j \left(-1 + \frac{Y_j}{\lambda_j} \right)^2.$$

The score statistic is obtained by substituting $\hat{\theta}_0$ for θ everywhere it appears. In other words we replace λ_j by the fitted values $\hat{e}_j$ under $\mathcal{H}_0$. A little algebra and S becomes

$$S = \sum_{j=1}^{4} \frac{(Y_i - \hat{e}_j)^2}{\hat{e}_j}. \tag{1.32}$$

The expression (1.32) is easy to understand intuitively since each term in the sum compares a data point Y_j to its fitted value $\hat{e}_j$ and the greater the disagreement the larger S will be. This statistic is called the Pearson χ^2 statistic and is discussed in detail in Section 2.5.3.

Let us compare S for testing the goodness-of-fit of $\mathcal{M}_3$ so that we can make a comparison with the earlier result using W. The fitted values under this model were 17.009, 55.991, 2.563, 8.437. On substituting these and the data values 11, 62, 4, 7 into the above equation the answer is 3.517. The corresponding P-value is $\Pr(\chi_2^2 > 3.517) = 0.172$. The value of W for testing the same hypotheses was 3.818 with P-value 0.148.

1.5.3 The Likelihood-Ratio Statistic

The likelihood function $L(\theta)$ measures the likelihood of a particular value θ in light of the data. Under the hypothesis $\mathcal{H}_0$, the likelihood function takes maximum value $L(\hat{\theta}_0)$ and under the more general hypothesis $\mathcal{H}_1$ its maximum value is $L(\hat{\theta}) > L(\hat{\theta}_0)$. The ratio is always greater than 1 and measures how much more probable are the data assuming $\mathcal{H}_1$ than $\mathcal{H}_0$. A sufficiently large ratio casts doubt on the hypothesis $\mathcal{H}_0$. The statistic

$$LR = 2\{\ell(\hat{\theta}) - \ell(\hat{\theta}_0)\} \tag{1.33}$$

is often called the *generalized* likelihood ratio statistic as it generalizes the simple likelihood ratio for testing two simple hypotheses as proposed by Neyman and Pearson (1928, 1933). The null distribution of LR is again χ^2 with r degrees of freedom and it tends to be larger when $\mathcal{H}_0$ is false. A computational disadvantage of the LR statistic is that both the null and alternative models must be fitted to compute the LR statistic. Like the Score statistic, the LR statistic is unaffected by change of parametrization.

Example: (Continued) The general log-likelihood function for four Poisson variables is

$$\ell(\lambda) = \sum_{i=1}^{4} (-\lambda_i + Y_i \log \lambda_i).$$

Under the null hypothesis, ML estimates of the λ_j are the fitted values under that hypothesis. Under the alternative (full) model ML estimates are the data values. Hence, a general formula for the LR statistic for the accident data is

$$LR = 2\sum_{i=1}^{4} (\hat{e}_i - Y_i) + 2\sum_{j=1}^{4} Y_i \log\left(\frac{Y_i}{\hat{e}_i}\right).$$

For both hypotheses $\mathcal{H}_0 = \mathcal{M}_3$ and $\mathcal{H}_1 = \mathcal{M}_1$ the fitted values total 84 and so the first term of this expression is zero and we obtain

$$\begin{aligned} LR = 2\sum_{i=1}^{4} Y_i \log\left(\frac{Y_i}{\hat{e}_i}\right) &= -9.587 + 12.641 \\ &+ 3.561 - 2.614 = 3.999. \end{aligned}$$

The approximate P-value associated with this observed value is $\Pr(\chi_2^2 > 3.999) = 0.135$. Upon replacing 4 with k, the expression at left applies quite generally to testing goodness-of-fit for models of independent Poisson data. Further details are given in Chapter 2.

Profile Likelihood The *profile* log-likelihood function is defined as

$$\hat{\ell}(\omega_0) = \ell(\omega_0, \hat{\psi}_0)$$

where $\hat{\psi}_0$ is the ML estimator of the nuisance parameters ψ assuming $\omega = \omega_0$. It is the ordinary log-likelihood function partially maximized over the nuisance parameters ψ and when there are no nuisance parameters—and only in this case—does it reduce to the ordinary log-likelihood.

When ω is scalar a plot of $\hat{\ell}(\omega)$ against ω graphically summarizes our state of knowledge about the parameter; large/small values of the profile indicate plausible/implausible values of the parameter and the function has its maximum value at $\hat{\omega}$. None of the standard packages calculate the profile likelihood as standard output.

In terms of profile likelihood, the LR statistic is

$$LR(\omega_0) = 2\{\hat{\ell}(\hat{\omega}) - \hat{\ell}(\omega_0)\} \tag{1.34}$$

which is a positive function of ω_0 with minimum value zero at $\hat{\omega}$. This statistic is output by the major packages. By varying ω_0 one can therefore construct the profile likelihood function; see examples in later chapters.

1.5.4 Relation Between the Statistics*

The three statistics W, S, and LR all have approximate χ^2_r distribution but give distinct answers in most problems. In the Score and Wald statistics the expected information may be replaced by observed information without invalidating the approximate χ^2 result. In this section we investigate the relationship between the three statistics.

For testing a scalar parameter ω with no nuisance parameter in the model, the geometrical relationship is shown in Figure 1.5. In each case the function plotted is a hypothetical score function $U(\omega)$ and the statistics LR, S, and W are displayed as shaded areas. In representing W and S the information $\mathcal{I}(\omega_0)$ has been used but in the fourth display the statistic W^* uses $\mathcal{I}(\hat{\omega})$. It is left to the reader to be convinced that the statistics can indeed be represented by these areas.

How close these areas are depends on how close to linear is $U(\omega)$ between $\hat{\omega}$ and ω_0. The score function is linear *exactly* for some models, for instance, linear models for normally distributed data. In this case all three statistics are identical. This virtually never holds for count data models. The score function is linear *approximately* if $\hat{\omega}$ happens to be close to ω_0. In this case all the statistics are small and agree that there is no evidence against the null. The score function is linear *asymptotically* between $\hat{\omega}$ and ω_0, provided ω_0 is the true value. For large samples, the three statistics are asymptotically equivalent under the null hypothesis.

Disagreement between the three statistics typically occurs when there *is* evidence against $\mathcal{H}_0$ and each of the three statistics measures this evidence differently. Often it will make little practical difference which is used: when there is no evidence against $\mathcal{H}_0$ the statistics agree; when there is considerable evidence against $\mathcal{H}_0$ the statistics disagree about the precise amount of evidence but all point strongly against $\mathcal{H}_0$.

These statements continue to be good when θ and ω are vectors. This can be shown by using the (ω, ψ) parametrization and developing Taylor expansions of the different statistics. Some details are given in Cox and Hinkley (1974, Chapter 9) and general results in Serfling (1980).

1.5.5 Reason for Preferring LR*

When $\mathcal{H}_0$ is false $\hat{\Omega}$ tends to differ from ω_0, $U(\hat{\theta}_0)$ tends to differ from zero, and $L(\hat{\theta})$ tends to be much larger than $L(\hat{\theta}_0)$. Thus, all three statistics tend to

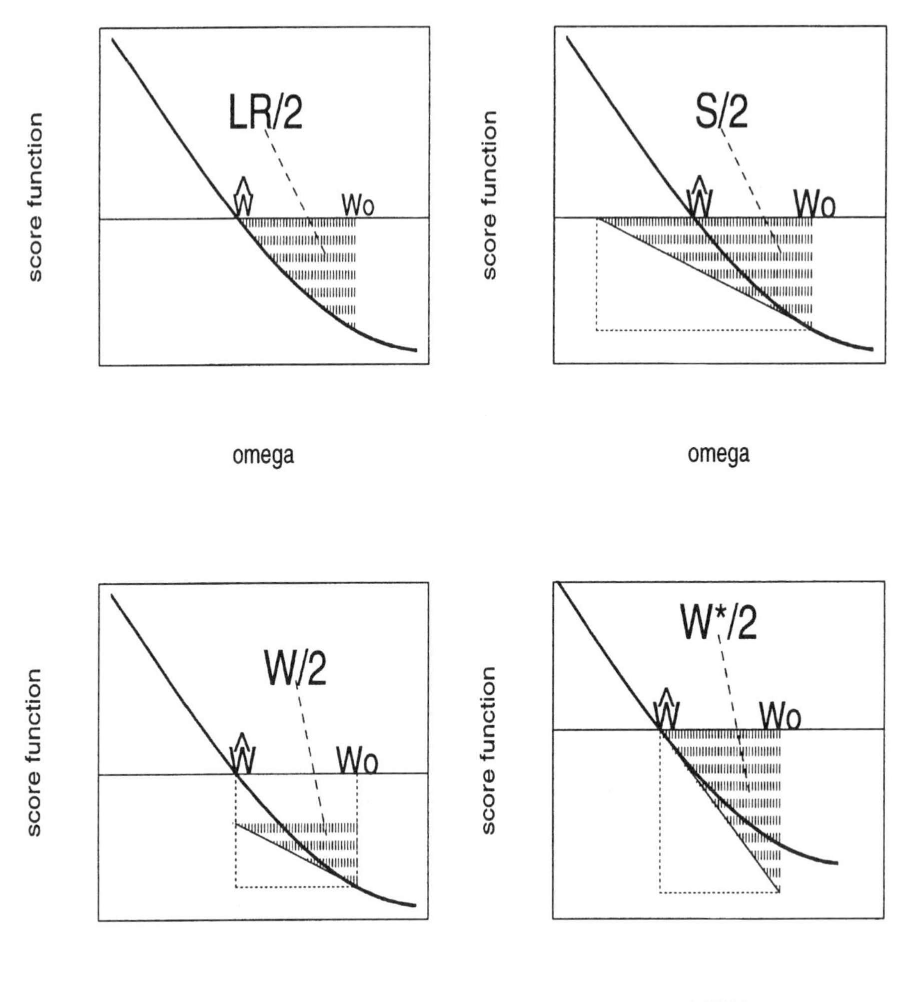

Figure 1.5. Geometrical relationship of three statistics. The ML, Score, and LR statistics can all be represented as linear approximations to the area between the score function and the axis. Only when the score function is linear do all these statistics agree exactly.

be larger than χ_r^2 under the alternative hypothesis and so in each case $\mathcal{H}_0$ is rejected if the test statistic T is too large.

If one of these tests had better power we might prefer it over the others. Let $T(Y, \omega_0)$ represent any one of these three statistics. If the true value of ω is ω_0 we saw that the common distribution is asymptotically χ_r^2 whose mean value is r and whose variance is $2r$. More generally if the true value $\omega = \omega_0 + \delta/\sqrt{n}$ then it can be shown that

$$T(Y, \omega_0) \xrightarrow{d} \chi_r^2(c^2)$$

known as noncentral χ^2 [see Cox and Hinkley (1974, p. 318) and Wald (1943)]. The mean of this distribution is $r+c^2$ and the variance is $2r+4c^2$. For the Wald statistic

$$c^2(\omega, \omega_0) = (\omega - \omega_0)^T V^{-1}(\omega - \omega_0)$$

where V is the variance matrix of $\hat{\Omega}$.

This result describes exactly how T tends to be larger when $\mathcal{H}_0$ is false. The greater the noncentrality parameter the larger T tends to be and the greater will be the power for the particular alternative ω. Apparently, power will tend to be greater for more remote alternatives and for models where the variance matrix V is "small." The important point is that for alternatives a distance $\delta/\sqrt{n}$ away from ω_0, the three statistics have the same power. For more remote alternatives the distributions are not approximated by noncentral χ^2 but decades of theoretical work has failed to point clearly to the superiority of any one of them.

In the absence of compelling theoretical comparisons of the three tests, the LR test is to be preferred. First, it is based on the sensible criterion that we will reject $\mathcal{H}_0$ if the most supported parameter value under $\mathcal{H}_1$ makes the data sufficiently more likely than does the most supported parameter value under $\mathcal{H}_0$. Thus, the earlier value 3.999 for the LR statistic means that the data can be made $\exp(3.999/2) = 7.385$ times more likely under $\mathcal{H}_1$ than under $\mathcal{H}_0$. Second, likelihood based inferences respect the boundaries of the parameter space such as [0, 1] for a probability, whereas the Wald and Score statistics may not. Third, when the likelihood surface is flat both Wald and Score statistics become numerically unstable, whereas the maximized likelihood function is not.

Realistic conditions do exist under which LR tests break down but under these conditions the alternative Wald or Score statistics also break down. Appropriate conditional, marginal, or otherwise adjusted likelihoods are required in such circumstances. An example especially relevant to categorical data is given in Section 3.6.4.

1.5.6 Interpreting P-Values*

Since P-values play a major role in inference it is important not to interpret them naively or wrongly. Some of the criticisms leveled at P-values are worth considering, if for no other reason than to better understand what they are and are not.

First, evidence against the null hypothesis is not necessarily evidence in favor of the alternative hypothesis. A simple example is testing the null value $p = 0.5$ against the alternative $p = 0.75$ from observing $x = 60$ successes from $n = 100$ binomial trials. The P-value is $\Pr(X \geq 60) = 0.0284$ so we would reject the null value $p = 0.5$ in favor of 0.75 at level 5%; but there were 60% successes, which is closer to 0.5 than to 0.75. Indeed, the likelihood ratio is about 30 to 1, in favor of $p = 0.5$. If we reverse the hypotheses and make $p = 0.75$ the null value then the P-value is even smaller, namely $\Pr(X \leq 60; p = 0.75) = 0.0007$. The example demonstrates that data which points against the null hypothesis need not point towards the alternative; the data may be hostile to both hypotheses.

Such examples are an argument against hypothesis testing rather than P-values though it does reinforce the important point that P-values do not compare two hypotheses but measure evidence against one of them in the direction of the other.

Second, some statisticians argue against P-values as a measure of evidence at all. For instance, Edwards (1972) argued on axiomatic grounds that $L(\hat{\theta})/L(\hat{\theta}_0)$ summarizes all the relevant information concerning the hypotheses that the data have to offer, for instance a maximized LR statistic of 10.75 tells us that the data are $\exp(10.75/2) = 215.9$ times less probable under $\mathcal{H}_0$ than under $\mathcal{H}_1$. This is called the *pure likelihood* approach to inference.

The number 215.9 is large but how reliable is it? Depending on the nature of the data and the hypotheses, a given value of the LR might be incredibly significant or quite within the realm of chance. We calculate $\Pr(LR > 10.75|\mathcal{H}_0)$ or the P-value to quantify how much weight to place on the number 215.9. Suppose that testing $\mathcal{H}_0$ against $\mathcal{H}_1$ was a test of $r = 8$ parameters. Then using the χ^2_8 distribution we obtain the P-value 0.216. This should temper our enthusiasm about rejecting $\mathcal{H}_0$. However impressive the number 215.9 and however impeccable the logical properties of LR as a measure of evidence, there is a very real sense in which the data provide hardly any *reliable* evidence against $\mathcal{H}_0$.

Bayes factors are similar to likelihood ratios except that the likelihoods are averaged with respect to an assumed distribution for the parameters. While it is claimed that this ratio better reflects the relative evidence for the hypotheses in the presence of prior information the question again arises—what is the probability of obtaining a Bayes factor as extreme as the one obtained? This leads us back to the P-value but based on the Bayes, rather than on a standard, test statistic. It seems unavoidable that no matter what measure of evidence we use, no matter how impeccable its logical foundations, we will still want to know whether what it tells us is to be believed or whether it might not be misleading us by bad luck.

The story is not finished yet, however. The fact that the evidence against $\mathcal{H}_0$ is so unreliable ought not to make us report that $\mathcal{H}_0$ may be safely assumed to be true. Lack of evidence against an hypothesis is not the same as support; if our experimental technique is so coarse that it almost never uncovers evidence under any circumstances then our failure to uncover evidence is itself little evidence one way or the other. The frequentist measure of experimental quality is power, discussed in elementary text books. Performing a low power test almost always gives a nonsignificant P-value and so leaves us little wiser. Only for a high power test is a nonsignificant P-value supportive of the null hypothesis.

1.5.7 Distribution of the P-Value

A test statistic $T(Y, \omega_0)$ whose distribution is known under $\mathcal{H}_0 : \omega = \omega_0$ generates the P-value

$$P(t, \omega_0) = \Pr(T \geq t; \mathcal{H}_0). \tag{1.35}$$

Since T is a random variable, the P-value $P(T, \omega_0)$ is also a random variable that takes values between 0 and 1. What is its distribution?

Suppose we observe the value t giving the P-value p. Since $P(t, \omega_0)$ is a monotonic *decreasing* function of t

$$p = \Pr(T \geq t; \mathcal{H}_0) = \Pr(P(T, \omega_0) \leq p; \mathcal{H}_0) \quad p \in [0, 1]. \tag{1.36}$$

Saying that $\Pr(P(T) \leq p) = p$ is identical to saying that the null distribution of $P(T)$ is *uniform* on [0,1]. Suppose we obtained a P-value of 0.044. The chance of our obtaining a value as small as this if $\mathcal{H}_0$ is actually true is exactly 0.044. It is because of this property that we can use P-values directly for formal testing. The decision rule is to reject $\mathcal{H}_0$ if the P-value is less than α, which falsely rejects $\mathcal{H}_0$ with probability α as required.

There is a slight complication to this argument when dealing with categorical data because the distribution of $P(T)$ will then be discrete; there will be a countable, though perhaps very large, set of possible values that we can get for the P-value. The relation Eq. (1.36) then only holds for these values of p, *not* for all possible values in [0, 1]. This hardly makes any difference to the above argument except that the probability that $P(T)$ is α or less will not be exactly α unless we are lucky enough that α is *exactly* a possible value of P. Instead, if p^* is the next possible value of P smaller than α then

$$\Pr(P(T) \leq \alpha) = \Pr(P(T) \leq p^*) = p^* \leq \alpha$$

and so we arrive at the conclusion that the test

$$\text{Reject } \mathcal{H}_0 \text{ if } P(T, \omega_0) \leq \alpha \tag{1.37}$$

falsely rejects $\mathcal{H}_0$ with probability *not more than* α. Such a test is called *conservative*.

1.5.8 Confidence Sets from Tests

A confidence set for a vector parameter ω is a set $R_\alpha(Y)$ calculated from the data $Y_1, \ldots, Y_k$ with the property that

$$\Pr(\omega \in R_\alpha(Y)) = 1 - \alpha \tag{1.38}$$

for a specified value α. The set $R_\alpha(Y)$ is interpreted as a range of plausible values of the parameter vector ω and the plausibility is measured by the confidence coefficient $1 - \alpha$. The larger this value is the more confident we may be that the parameter ω does lie within the set $R_\alpha(Y)$ but the cost will be that the set $R_\alpha(Y)$ is larger. Conversely, if we make the set $R_\alpha(Y)$ smaller we will generally be less confident that it contains the parameter ω.

When ω is a single parameter then the region $R_\alpha(Y)$ will be a subset of the real line and in most circumstances it will be an interval with a lower endpoint $L_\alpha(Y)$ and upper endpoint $U_\alpha(Y)$. The confidence set property above then becomes

$$\Pr(L_\alpha(Y) \leq \omega \leq U_\alpha(Y)) = 1 - \alpha \tag{1.39}$$

and $[L_\alpha(Y), U_\alpha(Y)]$ is called a $(1 - \alpha)\%$ confidence *interval* for ω.

There is an intimate relationship between confidence sets and tests. To see this, replace the $\leq$ by a $>$ sign in Eq. (1.36) implying that

$$\Pr(P(T, \omega) > p) = 1 - p. \tag{1.40}$$

This immediately implies that

$$R_\alpha(Y) = \{\omega : P(T, \omega) > \alpha\} \tag{1.41}$$

is a $1 - \alpha$ confidence set for ω. To construct a $(1 - \alpha)\%$ confidence set we collect together all parameter values ω that are accepted by a size α test. Conversely, the confidence set automatically allows us to test any specified value of ω at level α by simply checking to see if this value is in the set.

Since $P(t, \omega)$ is monotonically decreasing in t, we can also express the confidence set as $\{\omega : T(Y, \omega) < q_\alpha\}$ where q_α is the upper $1 - \alpha$ quantile of the distribution of T. Plotting $T(y, \omega)$ against ω allows us to directly read off the confidence set.

There are some difficulties with the above argument, however. First, for categorical data we saw that $\Pr(P(T) \leq \alpha) \leq \alpha$ rather than equal to α. This implies that $R_\alpha(Y)$ will have coverage at least $1 - \alpha$ and so is said to be *conservative*. A more important difficulty is that the distribution of $T(Y, \omega)$ must be known to find the critical value t_α and then the region $R_\alpha(Y)$. If the distribution of T depends on nuisance parameters or is very complicated, an approximation to its distribution might be used. This gives an approximation to q_α and an approximate interval. Different choices of test statistic give different approximate intervals.

Examples of confidence intervals are not given here. Section 2.6 further develops those ideas in the context of categorical data and gives numerous illustrations.

SUMMARY:

When an hypothesis involves $r > 1$ parameters, one must decide how much weight to give to each parameter in the test. Three automatic test statistics for doing this are the Wald, Score, and LR statistics. These statistics will be numerically close when they are small and less close when they are large. They have the same approximate distribution which is (noncentral) χ_r^2. Small P-values cannot be interpreted as evidence for the alternative hypothesis but only as evidence against the null in the direction of the alternative. A confidence interval for ω is constructed by collecting together all parameter values accepted at a specified level. Conversely, a test may be performed by observing whether or not the null value of the parameter is within a confidence interval. Discreteness leads to conservative confidence intervals. When the P-value depends on nuisance parameters and these are estimated, the generated confidence interval will only be approximate.

1.6 TRANSFORMING DATA AND PARAMETERS

The main result of Section 1.3 was that the ML estimator $\hat{\theta}$ has an approximate multivariate normal distribution with mean θ and variance equal to the inverse of the information matrix. In this section we show that the limiting distribution of almost any combination of the components of the ML estimator θ is also normally distributed. Thus, if we are interested in the square of θ_1 or the ratio of θ_1 to θ_2, one can approximate the distribution of the ML estimators of these by a certain normal distribution. We also study the effect of reparametrizing a model on estimates and hypothesis tests.

1.6.1 The O_p Calculus

We first introduce a special notation convenient in the algebra of asymptotics. This notation is a generalization of the Bachmann–Landau O-notation used in asymptotics of mathematical functions [see Landau (1927), De Bruijn (1970)].

Let X_n and Z_n be two sequences of random variables. The sequence X_n is $O_p(Z_n)$ if

$$Z_n^{-1} X_n \xrightarrow{d} F \tag{1.42}$$

where $\xrightarrow{d}$ denotes convergence in distribution and F is a distribution which is neither concentrated at zero nor infinity. One often translates this as "X_n is of probable order Z_n." Often Z_n is a sequence of numbers. For instance, $X_n = O_p(n^{-1/2})$ means that the random variables eventually decrease at rate $n^{-1/2}$ as n increases. The sequence X_n is $o_p(Z_n)$ if

$$Z_n^{-1} X_n \xrightarrow{d} 0 \tag{1.43}$$

and is often translated "X_n is of probable order smaller than Z_n." When X_n and Z_n are nonrandom we drop the subscript p.

The same notation may be applied to vector or matrix valued sequences of random variables X_n. For instance, the observed information matrix $\mathcal{J}$ is typically $O_p(I)$ which means that $\mathcal{J}I^{-1}$ converges in probability to a bounded full rank random matrix, the identity matrix in most cases. Similarly, both $\mathcal{J}$ and I are $O_p(n)$ which means that $n^{-1}I$ and $n^{-1}\mathcal{J}$ converge in probability to a bounded full rank matrix. The calculus of the O_p notation is given in an exercise.

As an illustration of the efficiency of this notation recall the result [see Eq. (1.20)] which stated that $\hat{\theta}$ was approximately normally distributed with mean θ and variance matrix $V = I^{-1}$. If $V^{1/2}$ is a symmetric nonnegative definite matrix such that $V^{1/2}V^{1/2} = V$ and Z is multivariate standard normal then

$$\theta + V^{1/2}Z \stackrel{d}{=} \mathcal{N}_p(\theta, V)$$

and so it follows that $\hat{\theta}$ has the same distribution as $\theta + V^{1/2}Z$. The difference between the true distribution of $\hat{\theta}$ and this normal approximation can be accounted for by adding a random variable $O_p(\hat{\theta} - \theta)^2$ which can be seen by carefully going through the Taylor expansion argument used to derive Eq. (1.20) but doing a *second order* expansion. In the new notation we write

$$\hat{\theta} = \theta + V^{1/2}Z + O_p(\hat{\theta} - \theta)^2. \tag{1.44}$$

This equation is entirely equivalent to the central limit result but is more convenient theoretically since it separates the distribution of $\hat{\theta}$ into the normal component $\theta + V^{1/2}Z$ and the O_p error term which is small, provided $(\hat{\theta} - \theta)^2$ is small. Since $V = O(n^{-1})$, Eq. (1.44) also immediately implies that $\hat{\theta} = \theta + O_p(n^{-1/2})$ or in other words, the difference between the estimator of θ and the true value decreases like $n^{-1/2}$. Consequently the error term in Eq. (1.44) is $O_p(n^{-1})$. On taking expectation we obtain

$$\mathrm{E}(\hat{\theta}) = \theta + O(n^{-1}) \tag{1.45}$$

since Z has mean zero and since $O_p(n^{-1})$ has mean $O(n^{-1})$. Thus, the ML estimator $\hat{\theta}$ of θ has bias of order $1/n$ despite the fact that $\hat{\theta}$ differs from θ by $O_p(n^{-1/2})$.

1.6.2 Transforming the Central Limit Theorem

In this section we show that if $\hat{\theta}$ has approximate normal distribution then so does $\hat{\Omega} = g(\hat{\theta})$ for any reasonable function g. We also find the mean vector and variance matrix of this approximating distribution.

To find the approximate distribution of $\hat{\Omega} = g(\hat{\theta})$ suppose that g is differentiable and let $G(\theta)$ denote the Jacobian. Then a first-order Taylor expansion of $g(\hat{\theta})$ about θ is

$$\hat{\Omega} = g(\theta) + G(\theta)(\hat{\theta} - \theta) + O_p(n^{-1}). \tag{1.46}$$

The error term is $O_p(n^{-1})$ because it involves quadratic forms in $\hat{\theta} - \theta$ which we decided above is $O_p(n^{-1/2})$. Substituting Eq. (1.44) gives

$$\begin{aligned}\hat{\Omega} &= \omega + G(V^{1/2}Z + O_p(n^{-1})) + O_p(n^{-1}) \\ &= \omega + GV^{1/2}Z + O_p(n^{-1})\end{aligned} \tag{1.47}$$

which says that $\hat{\Omega}$ is approximately a linear transform of the multivariate normal variable Z. The mean is ω and the asymptotic variance

$$\mathrm{Var}(\hat{\Omega}) = (GV^{1/2})(GV^{1/2})^T = GVG^T \tag{1.48}$$

and so recalling that $V(\theta) = I^{-1}(\theta)$ we conclude that

$$\hat{\Omega} \xrightarrow{d} \mathcal{N}_r(\omega, G\hat{I}^{-1}G^T). \tag{1.49}$$

This is a generalization of Eq. (1.20) and reduces to it when g is the identity map. As usual we may replace $I(\theta)$ by $\mathcal{J}(\theta)$ if desired and estimate θ in either of these matrices without invalidating the limit (see Exercise 1.14). Equation (1.48) is often used to obtain the matrix V_0 used in the Wald statistic.

The expressions for the mean and variance in equations like (1.49) are *not* the limiting mean or variance of $g(\hat{\theta})$. It is quite possible that the mean and variance of $g(\theta)$ will not exist and yet its distribution will be approximated by a distribution which does have finite mean and variance, namely the normal distribution. The terminology *asymptotic mean and variance* is used for the mean and variance of a distribution to which a statistic converges. In some cases this may also be the limiting mean and variance or even the exact mean and variance but in many cases it will be neither. The method [see Eq. (1.46)] of approximating a complicated statistic by a linear function of a well-understood statistic is called the δ-method.

Example 1 Consider the hypothesis that amongst fatal accidents, the proportion of older drivers is $\pi = 0.6$. The observed proportion was $11/15$. To make the calculation less trivial, assume model $\mathcal{M}_2$ so that there are $p = 3$ unknown parameters $\lambda_1, \lambda_2, \phi_1$. In terms of these parameters $\pi = \lambda_1/\phi_1$, and one way of writing the null hypothesis is $\mathcal{H}_0 : \lambda_1 - 0.6\phi_1 = 0$. We estimate

$$\omega = g(\lambda_1, \lambda_2, \phi_1) = \lambda_1 - 0.6\phi_1 = 0$$

by $\hat{\Omega} = \hat{\lambda}_1 - 0.6\hat{\phi}_1$ with observed value of 2.610. The Jacobian matrix of the transform g is

$$G = \begin{pmatrix} \dfrac{\partial \omega}{\partial \lambda_1} & \dfrac{\partial \omega}{\partial \lambda_2} & \dfrac{\partial \omega}{\partial \phi_1} \end{pmatrix} = (1.0 \quad 0.0 \quad -0.6).$$

The matrix $\hat{I}^{-1}$ was given in Section 1.4 and so Eq. (1.49) says that $\hat{\Omega}$ is normal with mean ω and variance

$$(1.0 \quad 0.0 \quad -0.6)\begin{pmatrix} 628 & 9.89 & 3.35 \\ 9.89 & 45.77 & 13.49 \\ 3.35 & 13.49 & 4.56 \end{pmatrix}\begin{pmatrix} 1.0 \\ 0.0 \\ -0.6 \end{pmatrix} = 3.906. \tag{1.50}$$

The value of the Wald statistic is therefore $2.610^2/3.906 = 1.745$ which is not particularly large for a χ_1^2 variable. We thus easily accept $\mathcal{H}_0$.

An alternative way of expressing the same hypothesis is $\mathcal{H}_0 : \lambda_1/\phi_1 = 0.6$ which is a *nonlinear* combination of the parameters. The estimate of

$$g(\lambda_1, \lambda_2, \phi_1) = \lambda_1/\phi_1$$

is $\hat{\Omega} = \hat{\lambda}_1/\hat{\phi}_1$ with observed value $14.353/19.572 = 0.733$ which is greater than 0.6. The Jacobian of the transform g is

$$G(\hat{\theta}_0) = \begin{pmatrix} \dfrac{1}{\hat{\phi}_1} & 0 & -\dfrac{\hat{\lambda}_1}{\hat{\phi}_1^2} \end{pmatrix} = (0.0543 \quad 0 \quad 0.0326)$$

and we find that $G\hat{I}_0^{-1}G^T = 0.0130$. Thus, Eq. (1.49) says that approximately $\hat{\Omega} \overset{d}{=} \mathcal{N}(0.6,\ 0.0130)$ under $\mathcal{H}_0$ and the value of the Wald statistic is $(0.733 - 0.6)^2/0.013 = 1.361$. The P-value associated with this is 0.122 while the P-value associated with the earlier value 1.745 is 0.093.

The statistics $\hat{\lambda}_1 - 0.6\hat{\phi}_1$ and $\hat{\lambda}_1/\hat{\phi}_1$ are distinct test statistics and give different P-values for the *same data*. Thus, they would have different power properties. A simulation revealed that the normal approximation to $\hat{\lambda}_1/\hat{\phi}_1$ is barely adequate, the true P-value being 0.132, whereas the true P-value based on $\hat{\lambda}_1 - 0.6\hat{\phi}_1$ is 0.095. A statistic having an almost normal distribution is merely convenient in practice. However since more refined calculations are seldom done, nonnormal statistics are avoided. Thus, $\hat{\lambda}_1 - 0.6\hat{\phi}_1$ would be preferred here although LR is also an attractive option.

Example 2: How Accurate is $\hat{p}_2$? Let us return to an issue raised in Section 1.2 where we estimated

$$\hat{p}_2 = \frac{5.219}{5.219 + 6.536} = 0.4438$$

under model $\mathcal{M}_2$ and considered what the variance of this estimator might be. In the present notation we want to calculate the variance of

$$\hat{p}_2 = g(\hat{\lambda}_1, \hat{\lambda}_2, \hat{\phi}_1) = \frac{\hat{\lambda}_3}{\hat{\lambda}_3 + \hat{\lambda}_4} = \frac{\hat{\phi}_1 - \hat{\lambda}_1}{4.292\hat{\phi}_1 - \hat{\lambda}_1 - \hat{\lambda}_2}$$

expressed in terms of the parameters λ_1, λ_1, ϕ_1 of $\mathcal{M}_2$. Using the quotient rule for differentiation the derivative matrix of this transformation is

$$G = \left(-\frac{\lambda_4}{(\lambda_3 + \lambda_4)^2} \quad \frac{\lambda_3}{(\lambda_3 + \lambda_4)^2} \quad \frac{3.292\lambda_1 - \lambda_2}{(\lambda_3 + \lambda_4)^2} \right)$$
$$= (-0.0473 \quad 0.0378 \quad -0.0770)$$

after substituting the parameter estimates under $\mathcal{M}_2$ listed in Table 1.3. Using the estimated expected information matrix under $\mathcal{M}_2$ the estimated variance of $\hat{p}_2$ is found to be

$$(-0.0473 \quad 0.0378 \quad -0.0770) \begin{pmatrix} 6.28 & 9.89 & 3.35 \\ 9.89 & 45.77 & 13.49 \\ 3.35 & 13.495 & 4.56 \end{pmatrix} \begin{pmatrix} -0.0473 \\ 0.0378 \\ -0.0770 \end{pmatrix}$$

giving 0.00171 and standard error $\sqrt{0.00171} = 0.130$.

A histogram of the distribution of $\hat{p}_2$ is shown in Figure 1.6 from 10,000 simulations under $\mathcal{M}_2$ with the ML parameter estimates used. The overall shape of the distribution is close to normal, however, there are "spikes" at zero and one that visually deviate from this pattern. Looking at the formula for the ML estimators in Eq. (1.16) it follows that $\hat{p}_3$ equals zero exactly when $Y_3 = 0$ and equals one exactly when $Y_4 = 0$. These events happen with respective probabilities $\exp\{-5.219\} = 0.00541$ and $\exp\{-6.536\} = 0.00145$ and so we expect about 54 and 14 occurrences, respectively, in 10,000 simulations. These spikes become smaller *exponentially* as the expected values λ_3, λ_4 increase, but for the value supposed here the normal approximation to extreme tail probabilities will be very poor. This is typically the case with normal approximations anyway—poor accuracy in the extreme tails—however the present example graphically demonstrates this. The standard error of $\hat{p}_2$ from the simulations was

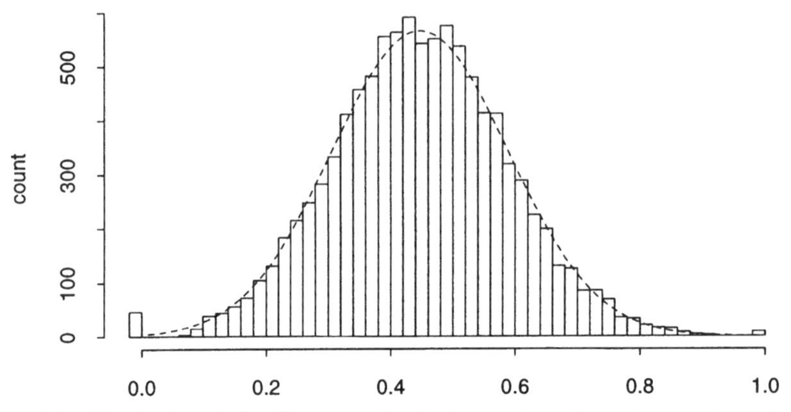

Figure 1.6. Distribution of $\hat{p}_2$. The exact distribution under estimated parameter values is approximated from 10,000 simulations. The normal curve was variance estimated from the expected information matrix. The spike at left is exponentially small.

0.141 which is larger than our estimate 0.130, largely because of the mass at zero.

SUMMARY:

The ML estimator of virtually any parameter at all is approximately normally distributed when the information is large. For a given sample size, the approximation will be good for some parameters and poor for others. There is a simple formula for finding the asymptotic mean and variance of a transformed parameter estimate. The maximized likelihood, LR statistic, and score statistic are unaffected by reparametrizing a model while the Wald statistic is not. This is a source of practical strength but theoretical weakness for the Wald statistic.

FURTHER READING

Section 1.1: Statistical Models

1.1.4 *For nonnested hypotheses the LR is not chi-square. Cox (1961, 1962) suggested embedding the models in a super model so that classical chi-square tests can be applied. Properties were investigated by Chambers and Cox (1967), Jackson (1968) and ideas further developed by Atkinson (1969, 1970). The upshot of the research is that such tests often have low power and that their null distributions converge to standard distributions extremely slowly. Bayesian analysis allows the direct comparison of unrelated hypotheses via their averaged likelihoods, known as the Bayes factor. A variation by Aitkin (1990, 1991) averages the likelihood with respect to* *the posterior distribution of the parameters. This method is controversial but seems to improve the frequency performance at the cost of coherence.*

Section 1.2: Fitting Statistical Models

1.2.1 *The essentials of maximum likelihood theory were laid down by Fisher (1922, 1925, 1934) in three ground-breaking papers. He derived the property of asymptotic efficiency and related it to sufficiency as well as suggesting conditioning to recover information when the ML estimator is insufficient. Asymptotic properties of ML estimation were revealed some time later. Consistency and efficiency were discussed in Wilks (1938), Cramer (1946), and very generally in Wald (1949). When the dimension of the parameter grows unbounded then the classical arguments for consistency fail. The first concrete examples of inconsistency of ML were given by Neyman and Scott (1948) who also provided consistent estimates for these cases by appropriate conditioning; see also Kiefer and Wolfowitz (1956), Bahadur (1958), and Rao (1961). General results for multinomial data were given in Rao (1957, 1958).*

1.2.3 *There are other algorithms used for maximizing likelihood functions with special characteristics. For instance, the EM algorithm is useful for likelihoods based on data sets with missing or truncated observations. The iterative proportional fitting algorithm, developed by Deming and Stephan (1940), is a simple method of fitting log-linear models to contingency tables. This is a less general algorithm than Newton–Raphson with advantages in computational complexity. However, its rate of convergence may be slow.*

Section 1.3: Large Sample Theory of Estimation

1.3.3 *Generalizations of the Cramer–Rao inequality were given by Bhattacharya (1946) based on regressing the score function on higher derivatives of the log-likelihood often resulting in tighter (i.e., higher) lower bounds when nuisance parameters are present. Variance bounds in nonregular cases were obtained in Chapman and Robbins (1951). An alternative approach is to concentrate on the estimating equation rather than the estimator, pioneered in Godambe (1960) and unified in Godambe and Heyde (1985).*

1.3.5 *The term ancillary statistic was first coined by Fisher (1934) where he made several highly intuitive but unsubstantiated claims about their use such as "the function of an ancillary statistic is analogous to providing a true, in place of an approximate, weight (i.e., standard error) for ... the estimate." The lucid discussion of Cox (1958) focused attention on the deficiencies of Fisher's theory. Questions of uniqueness and existence were addressed by Basu(1959, 1964) and those of choice between alternatives by Cox (1971). A survey of the subject may be found in the entry ANCILLARY STATISTICS in the Encyclopaedia of Statistical Science, written by this author.*

Efron and Hinkley (1978) first clarified the sense in which observed information may give better estimates of conditional variance than expected information. They obtained unequivocal results for location families and more general results for regular likelihoods provided the likelihood is close to normal. The situation has *been greatly clarified by the work of Barndorff–Nielsen (1983). The highly accurate p^* formula for the conditional distribution of the ML estimator is essentially the observed likelihood function reflected about its maximum. This extends the work of Fisher (1934) and Pitman (1938) for location families. Observed properties of the likelihood function, rather than average properties, determine the conditional distribution of the ML estimator.*

Section 1.4: Hypotheses Concerning a Single Parameter

1.4.1 *A general discussion of measures of evidence and the shortcomings of the P-value is given in Schafer (1982), the discussions of Pratt (1987) and Good (1987), the article by Berger and Sellke (1987), as well as Good (1957) and Good and Crook (1974). This is recommended reading for all students who are interested in the logical basis for the many modeling decisions they make in the course of a data analysis.*

Section 1.5: Hypotheses Concerning Many Parameters

1.5.1, 1.5.2, 1.5.3 *The chi-square distribution of the generalized LR statistic was demonstrated by Wilks (1938). Wald (1943) demonstrated the asymptotic equivalence of Wald and LR. The Score statistic of Rao (1947) was further developed by Bartlett (1953 a, b, 1955). Optimality issues were further developed in Neyman (1959). For count data there is evidence that the chi-square approximation to the null distribution is poorer for the LR statistic than for the score statistic; see Larntz (1978), Koehler and Larntz (1980), and Koehler (1986).*

Refined asymptotic expansions of the power functions for the different test statistics have been studied by Peers (1971) and Harris and Peers (1980) but none of the three emerges as uniformly superior. The only general context in which uniformly most powerful tests exist are the one-sided test of linear functions of canonical parameters in exponential families and in these cases the LR and Score tests coincide and are optimal; see Cox and Hinkley (1974, p. 107). For one-sided tests of scalar parameters the Score test maximizes the local power and this sometimes translates into more power among unbiased two-sided tests. The second-order powers of the three tests are identical and the third-order power loss depends on the model through the statistical curvature of Efron (1975) which allows general model-independent comparison of the three tests. Detailed computations by Amari (1985), using concepts of differential geometry and a special kind of averaging, show that the Score statistic has third-order power deficiency uniformly worse than the Wald statistic but the LR has smaller third-order loss than the Wald for most of the parameter space; see Amari (1985, Section 6.3). Mukerjee (1992) compared the LR and Score statistics. In essence he found that the power of the Score test is more sensitive to the alternative than the LR but on average is slightly (third-order) higher. The models, alternatives, and range of values of n for which these results translate into superior actual power is not clear.

1.5.6 *Blyth and Staudte (1995) have suggested measuring the evidence for $\mathcal{H}_1$ compared to $\mathcal{H}_0$ by $LR/(1+\kappa LR)$ where κ can be adjusted rather like a type I error rate. This approach leads to what they call confidence profiles. Berger and Sellke (1987,* *p. 112) have argued that P-values "... can be highly misleading measures of the evidence ... against a null hypothesis" and general results are given to show that for testing two-sided hypotheses P-values exaggerate the evidence against the null compared to Bayesian methods. This is only an argument against P-values if one accepts the Bayesian measure. Moreover, Casella and Berger (1987) demonstrate that this conflict does not occur for testing one-sided hypotheses.*

Section 1.6: Transformations of Data and Parameters

1.6.2 *Invariance was first applied to location and scale problems in Hotelling (1931) and Pitman (1938). Further developments aimed at making inferential procedures respect any basic symmetry in the structure of a statistical problem. Together with sufficiency, conditionality, and likelihood principles, invariance was studied by Birnbaum (1969) as a guiding principle of statistics in its own right. Requiring invariance can sometimes conflict with other compelling criteria such as minimum expected loss in a decision theory framework. Artificial examples have also demonstrated that different classes of tests and estimators can be obtained by applying invariance before or after sufficiency reduction. Invariance arguments are strongest when the underlying symmetry generating them is real rather than mathematical. General discussions of invariant tests may be found in Lehmann (1959), Zacks (1971), and of the invariance principle and its relation to other unifying principles in Chapters 2 and 5 of Cox and Hinkley (1974).*

1.7 EXERCISES

1.1. A question of interest that arises from the serious accident data set is "*do younger drivers have the same probability as older drivers of being involved in a fatal accident given that they are involved in a serious accident?*." This is model $\mathcal{M}_4$ of Section 1.1.3. Substitute the parametric relations in the last line of Table 1.2 into the general likelihood function [see Eq. (1.8)] on p. 10, and then fit this model using maximum likelihood.

1.2. Recall the parametrization of the full model for the accident data given in Eq. (1.1) and that model $\mathcal{M}_4$ imposes the restriction $p_1 = p_2 = p$. Substitute these relations into Eq. (1.8) to obtain the likelihood for the parameters θ_1, θ_2, p and estimate them by ML. Show that the fitted values are the same as in Exercise 1.1.

1.3. On p. 22 it was stated that the information matrix may also be calculated by finding the variance matrix of the score function. For the case of a single parameter

$$\frac{\partial^2 \ell}{\partial \theta^2} = \sum_{i=1}^{k} \frac{p_i'' p_i - (p_i')^2}{p_i^2},$$

where p' is the derivative of $p_i(X_i, \theta)$ with respect to θ. By writing an expression for $I = -\mathrm{E}(\partial^2 l/\partial\theta^2)$ and assuming that any derivatives may be taken across the integral, show that $E(U') = -E(U^2)$ and then, that $I = \mathrm{Var}(U)$.

1.4. Consider the score function under $\mathcal{M}_2$ given by Eq. (1.11). Show that this score function has mean, the zero vector. Confirm the calculation of the expected information matrix on p. 28 first, by using expected minus derivatives and second, by using the variance.

1.5. On p. 28 expressions were given for the information matrices about λ_1, λ_2, ϕ_1 under $\mathcal{M}_2$. Substituting ML estimates,

a. compare the observed and expected information matrices;

b. evaluate the nonorthogonality of the parameters using Eq. (1.24);

c. predict the variance of $\hat{\lambda}_2$ when ϕ_1 is known and if ϕ_1 were known to equal its estimate.

1.6 Refer to Section 1.4.3 where several tests of a single parametric hypothesis were performed for the serious accident data.

a. Test the hypothesis $\lambda_1 = 9$ against $\lambda_1 > 9$ but do not assume the model $\mathcal{M}_2$ but rather the full model $\mathcal{M}_1$. The calculation is easy since under the full model $\hat{\lambda}_1 = Y_1$.

b. Retest the hypotheses assuming that model $\mathcal{M}_2$ holds. Compare the P-values. What do you think is the main reason for the difference?

1.7. Express the hypothesis "the expected proportion of fatal accidents among serious accidents for older drivers is 20%" in terms of a linear hypothesis about the parameters λ_i with integer coefficients. Test the hypothesis using the Wald statistic. Give the estimated P-value (i) using the usual standard error, (ii) using the standard error based on *null* expected values, and (iii) a simulation using these null expected values.

1.8 In Section 1.4 we tested the hypothesis $\mathcal{H}_0 : \lambda_1 = 10$ against the alternative $\lambda_1 > 10$ and obtained the P-value 0.035 assuming $\mathcal{M}_2$.

a. Use the Wald, Score, and LR tests to test $\mathcal{H}_0 : \lambda = 9$ against the two-sided alternative.

b. Discuss why they differ so much in this example, by referring to which aspects of the data each statistic depends on.

1.9. The following Splus commands simulate an observation of $\hat{\Omega}$ defined by Eq. (1.29) under the fitted model $\mathcal{M}_2$ described on p. 38.

```
y_rpois(4,c(14.353,57.892,5.219,6.536))
omegahat_.767*y[1]-.233*y[2]+.767*y[3]-.233*y[4]
```

Repeat this simulation as many times as possible and confirm the P-value obtained at the end of Section 1.4.3.

1.10. On p. 15, the log-likelihood for the model $\mathcal{M}_2$ was expressed in terms of parameters $\gamma_1 = \lambda_1/\phi_1$, $\gamma_2 = \lambda_2/\phi_1$, and ϕ_1.

a. Suppose we are interested in testing γ_1. What would be the sufficient statistic for the nuisance parameters (γ_2, ϕ_1)?

b. Show that the conditional P-value is based on the distribution of Y_1 given $Y_1 + Y_3$. By considering the $y_1 + y_3$ fatal accidents as binary trials, what is this conditional distribution?

1.11 Imagine a statistican decides on the following policy. "Clients want significant results so in reporting the significance of parameter estimates I will always give a P-value for a one-sided alternative in the direction of the estimate." Explain how the statistician will reject the null hypothesis about twice as often as desired.

1.12. A single observation $x = 3$ is taken on a known Poisson random variable with mean $\mu_0 = 1$ under $\mathcal{H}_0$ and mean $\mu_1 = 10$ under $\mathcal{H}_1$.

a. Show that the LR is 8.103 in favor of $\mathcal{H}_0$. How strong is this evidence in favor of $\mathcal{H}_0$? In particular, what is the chance of obtaining an LR this $\mathcal{H}_0$-favorable, if $\mathcal{H}_1$ is actually true?

b. Compute the P-value taking $\mathcal{H}_0$ as the null hypothesis and also taking $\mathcal{H}_1$ as the null hypothesis. Show that the ratio of these P-values is 7.769 in favor of $\mathcal{H}_0$. Thus, the claim (see Further Reading), that taking tail regions $T \geq t$ of the distribution of the test statistic will exaggerate the evidence is not generally true.

1.13. Recall the definition of O_p and o_p given on p. 51. Show that

a. For any constant c, $cO_p(n^{-\alpha}) = O_p(n^{-\alpha})$;

b. For any constant c, $co_p(n^{-\alpha}) = o_p(n^{-\alpha})$;

c. $O_p(n^{-\alpha})O_p(n^{-\beta}) = O_p(n^{-(\alpha+\beta)})$;

d. $O_p(1)/n^{\alpha} = O_p(n^{-\alpha})$;

e. $O_p(n^{-\alpha}) + O_p(n^{-\beta}) = O_p(n^{-\gamma})$, where $\gamma = min(\alpha, \beta)$.

1.14. Recall the O_p calculus studied in the previous exercise.

a. Recall that $V = I^{-1}$ and substitute the relation $\mathcal{J} = I + O_p(n^{-1/2})$ into Eq. (1.47). Then conclude that expected information may be replaced by observed information in Eq. (1.49).

b. Substitute $\hat{\theta} = \theta + O_p(n^{-1/2})$ into Eq. (1.47) and use the O_p calculus to conclude that expected information I can be replaced by estimated expected information $\hat{I}$ in Eq. (1.49).

1.15. The derivation of the transformation Eq. (1.49) began with the ML estimate but could equally have begun with any normally distributed variable. The more general result is that if X is any approximately multivariate normal variable with mean μ and variance matrix $\vee$ and if $Y = g(X)$, then Y is approximately multivariate normal with mean $g(\mu)$ and variance matrix $G \vee G^T$.

a. Consider the ordinary central limit result that the sample mean $\overline{X} \xrightarrow{d} \mathcal{N}(\mu, \sigma^2/n)$. Using the O_p notation, show that for any differentiable function g

$$g(\overline{X}) \xrightarrow{d} \mathcal{N}(g(\mu), (g'(\mu)\sigma)^2/n).$$

b. For a particular set of data on a positive random variable the sample mean is 2.64 and the sample variance 3.75. Calculate the student-t statistic for testing $\mu = 2$ and compare its value to an asymptotically equivalent statistic based on $g(\overline{X}) = \overline{X}^2$ instead of $\overline{X}$.

c. Calculate the student-t statistics based on $g(\overline{X}) = 1/\overline{X}$ and $\log \overline{X}$.

d. Assuming you had the raw data, which of the four t-statistics would you use?

1.16. The asymptotic mean and variance of a sequence of random variables is the mean and variance of its limiting distribution. This differs from its limiting mean and variance which in turn differs from the actual mean and variance. For each of the following distributional statements decide if the mean and variance of the limiting distribution is exact, limiting, or merely asymptotic. We take $X \stackrel{d}{=} Bi(n, p)$ and $Y \stackrel{d}{=} Pn(\mu)$.

a. $X \xrightarrow{d} \mathcal{N}(np, np(1-p))$ for large n.

b. $Y^2 \xrightarrow{d} \mathcal{N}(\mu(\mu+1), 4\mu^2)$ for large μ.

c. $Y^{-1} \xrightarrow{d} \mathcal{N}(\mu^{-1}, \mu^{-3})$ for large μ.

d. $Y \xrightarrow{d} \mathcal{N}(\mu, \mu)$ for large μ.

e. $\log(X/(n-X)) \xrightarrow{d} \mathcal{N}(\log[p/(1-p)], 1/[np(1-p)])$ for large n.

1.17. Recall the test of the hypothesis that $\pi = 60\%$ of fatal accidents involve older drivers performed in detail on p. 54. We will repeat this test now without assuming $\mathcal{M}_2$, i.e., under the full model.

a. Compare the estimators of π under the full model with the estimate under $\mathcal{M}_2$. Try to explain why assuming $\mathcal{M}_2$ has relatively little effect on the estimator.

b. Express $\pi = 0.6$ as a linear hypothesis about the λ_i and carry out the Wald test.

1.18. A set of five independent Poisson variables have means parametrized in terms of unknown parameters θ and λ. The data and model are summarized in Table 1.6.

Table 1.6. Data from a Simple Poisson Model

Data	12	54	79	112	14
Model	θ	θ^2	$\theta\lambda$	λ^2	λ

a. Derive the score function and by equating it to zero find ML estimates of θ and λ.

b. Find the observed and expected information matrices $\mathcal{J}$ and I.

c. Give standard errors for $\hat{\theta}$ and $\hat{\lambda}$.

d. Calculate the LR statistic for testing the goodness-of-fit of the model. State the approximate P-value.

e. Calculate the Score statistic for testing the goodness-of-fit of the model. State the approximate P-value.

f. Perform a simulation to investigate the true distributions of the LR and Score goodness-of-fit statistics.

1.19. In late 1959 a pollster interviewed 1000 randomly chosen voters and asked them to identify their preferred presidential candidate. Three months later another sample of 1000 voters was surveyed (see Table 1.7)

Table 1.7. Data From Two Political Polls

October 1959		January 1960	
Nixon	Kennedy	Nixon	Kennedy
532	477	502	498

Let p_1 and p_2 be the proportion of voters in the entire American voting population who favored Nixon at the first date and second date, respectively. Then the number of voters, X_1, on the first poll who favored Nixon has the $Bi(1000, p_1)$ distribution while the number of voters, X_2, who favored Nixon on the second poll has the $Bi(1000, p_2)$ distribution.

a. Write down the likelihood function for p_1 and p_2 based on the data X_1, X_2 and fit the model $\mathcal{H}_0 : p_1 = p_2$ by maximum likelihood. Explain this hypothesis and give the ML estimate of the assumed common value of p_1 and p_2.

b. Calculate the LR statistic for testing the goodness-of-fit of this model.

c. Even though it is not the case, take the four counts given in the table to be Poisson with unknown means λ_1, λ_2, λ_3, λ_4. Express p_1, p_2 in terms of these mean parameters.

d. Show that the hypothesis $p_1 = p_2$ expressed in terms of the Poisson mean parameters is $\mathcal{H}_0 : \mu_4 = \mu_2\mu_3/\mu_1$.

e. Write down the likelihood for the parameters μ_1, μ_2, μ_3, μ_4 under $\mathcal{H}_0$ and fit the model by maximum likelihood. Confirm that the estimate of the assumed common value of p_1 and p_2, the fitted values and the LR statistic for testing goodness-of-fit of this model, are all the same as under the binomial model.

1.20. Suppose there is an unknown number N of animals in a population. The population is sampled so that each animal has an equal chance p of being captured and u_1 animals are observed and marked. These animals are released into the population and allowed to mix and a further sample of animals is captured under the same conditions as the first sample. Of the n_2 animals in this sample, m_2 carry a mark from their previous capture and u_2 are unmarked. Thus the total number of distinct animals seen is $u = u_1 + u_2$ and the total number of captures is $n = u + m_2$. The unknown parameters are N and p with primary interest in estimating N. A typical dataset is given in Table 1.8 for which $n = 11$ and $u = 7$.

Table 1.8. Data from a Simple Animal Recapture Experiment

First Occasion		Second Occasion	
Unmarked (u_1)	Marked	Unmarked (u_2)	Marked (m_2)
6	0	1	4

a. Argue that N cannot be smaller than u or n_2. Then give a lower bound on N from the given data.

b. From the above description of the experiment, argue that u_1 and n_2 have independent $Bi(N, p)$ distributions and that given the value u_1, m_2 has the $Bi(u_1, p)$ distribution.

c. By equating various combinations of u_1, n_2, and m_2 to their expected values, derive the three alternative estimators of N given below and calculate them for the observed data. Is there anything wrong with these estimates?

$$\hat{N}_1 = \frac{u_1^2}{m_2}, \qquad \hat{N}_2 = \frac{u_1 n_2}{m_2}, \qquad \hat{N}_3 = \frac{u_1 n}{2m_2}.$$

d. Write down the likelihood function for N, p from the data u_1, u_2, m_2. Maximize this function with respect to p and substitute the expression for $\hat{p}$ back into the likelihood. Then conclude that the ML estimate of N maximizes

$$\tilde{L}(N) = \frac{N!}{(N-u)!} N^{-n}.$$

For the data $n = 11$, $u = 7$, calculate this function for a range of values of N and then find the ML estimate of N from the given data.

e. Recall Stirling's formula

$$N! \approx \sqrt{2\pi} e^{-N} N^{N+1/2}.$$

By substituting this into the formula for $\tilde{L}(N)$ and taking logarithms show that an approximate formula for $\log \tilde{L}(N)$ is

$$(N - n + \tfrac{1}{2}) \log N - (N - u + \tfrac{1}{2}) \log(N - u).$$

Plot this function against N for $N > u$ and find the same ML estimate as in part d.

1.21. A random variable X has a distribution which is geometric but truncated at 2, i.e., the probability function is

$$\Pr(X = x) = (1 - p)p^{x-2} \quad x = 2, 3, 4, \ldots$$

Exactly 24 independent observations are taken on X with the result of 12 two's, 6 three's, 5 four's, and 1 five.

a. Write down the likelihood function for p and show that the ML estimate of p is 19/41.

b. Find the expected information matrix and calculate the Wald statistic for testing $\mathcal{H}_0 : p = 0.5$.

c. Calculate the LR statistic for testing this same hypothesis.

d. Define $\phi = 1/p$. Show that $\hat{\phi} = 41/19$. Test the hypothesis that $\phi = 2$ using the Wald and LR statistics and compare the results with parts b and c.

e. Perform a simulation of the model under $\mathcal{H}_0$ and compare the adequacy of the normal approximations to $\hat{p}$ and $\hat{\phi}$.

1.22. A random variable X can take four possible values with probabilities given in Table 1.9. One hundred independent observations are taken on X and the numbers in each category are recorded.

Table 1.9. Data from a Simple Multinomial Model

x	1	2	3	4
Frequency	17	19	26	38
Model	$100p_1p_2$	$100p_1(1-p_2)$	$100(1-p_1)p_2$	$100(1-p_1)(1-p_2)$

a. Interpret the probabilities p_1 and p_2.

b. Fit the model. In particular, estimate p_1 and p_2, give the fitted values, and calculate the LR goodness-of-fit statistic.

c. Fit the model $\mathcal{H}_0 : p_1 = p_2 = p$. In particular, estimate p, give the fitted values, and calculate the LR goodness-of-fit statistic.

1.23. A Poisson random variable Y has its mean related to a covariate x by the relation $E(Y) = bx$, the slope parameter b being unknown. A set of data are shown in Table 1.10.

Table 1.10. Data from a Simple Poisson Regression Model

x_i	1.1	2.1	2.7	4.3	5.7
Y_i	12	17	27	32	51

a. Estimate b by maximum likelihood.

b. Give the LR statistic for testing goodness-of-fit.

c. Give the score statistic for testing goodness-of-fit.

1.24. Consider an estimator $\hat{\theta}$ which converges to θ according to some central limit theorem. In the O_p notation of Section 1.6.2 we write this in the form

$$\hat{\theta} = \theta + Z \frac{\sigma}{\sqrt{n}} + Z^2 \frac{b}{n} + o_p(n^{-1}).$$

a. Assuming these quantities exist, show that for large n the variance of $\hat{\theta}$ is σ^2/n and the bias is b/n.

b. Consider the estimator $g(\hat{\theta})$ of $g(\theta)$. Using a two-step Taylor expansion show that the asymptotic bias is

$$bg'(\theta)/n + \sigma^2 g''(\theta)/(2n).$$

c. Let X have the $Bi(n, 0.3)$ distribution. Denote by $\hat{p}$ the usual ML estimator X/n. However, rather than p itself it is desired to estimate $1/(p + 0.1)$. Using the above equation, show that for large n

$$\frac{1}{\hat{p} + 0.1} \overset{d}{=} \mathcal{N}\left(2.5 + \frac{3.281}{n}, \frac{8.203}{n}\right)$$

and confirm this with a simulation for $n = 40$. Suggest a better estimator of $1/(p + 0.1)$ than $1/(\hat{p} + 0.1)$.

d. Let X have the $Pn(\lambda)$ distribution where $\lambda = n\mu$. Using the above equations, show that for large λ

$$X^2 \xrightarrow{d} \mathcal{N}(\lambda(\lambda + 1), 4\lambda^3)$$

and confirm this with a simulation for $\lambda = 20$. Suggest a better estimator of μ^2 than X^2 and give the asymptotic distribution.

1.25. When you calculate a confidence interval with coverage greater than 50%, you often would like to conclude that the complement of this set contains unlikely values of the parameter. This is not always true. Consider a real parameter θ and divide the real line up into three regions I_1, I_2, I_3. Draw a uniform random number u and let

$$I = \begin{cases} I_1 \cup I_2 & u \in (0, 1/3) \\ I_2 \cup I_3 & u \in (1/3, 2/3). \\ I_1 \cup I_3 & u \in (2/3, 1) \end{cases}$$

a. Show that I is a 66.66% confidence interval for θ.

b. Argue that on observing $I_1 \cup I_2$ it is not reasonable to conclude that values of θ in I_3 are unlikely. Is there any sense in which I_3 is unlikely on observing $u < 1/3$?

CHAPTER 2

Distribution Theory for Count Data

In this chapter, we outline the basic distribution theory of multinomial and Poisson models for a single categorical data set and give the specific forms of the statistical inference paradigms studied in Chapter 1. Much of the material is covered in standard undergraduate courses. The notation and statistical treatment however is more general, and it is intended to present as much of the distribution theory as possible for the simplest possible count data, in preparation for the more complex theory presented in later chapters. The reader should have a thorough understanding of basic concepts such as score function, information, likelihood ratio, P-value, null distribution, and power.

2.1 CATEGORICAL DATA

This book concerns the analysis of data which is *categorical* in nature, using ideas and methods which, as much as possible, mirror and extend those used in analyzing normal data. You will already have encountered discrete variables as predictors or factors in normal theory, i.e., grouping labels that are used to explain the data collected on the normal variable Y. Here we suppose that the response variable Y itself is categorical.

There are many common situations where data cannot be considered as the realization of a real continuous random variable. For example, imagine American voters are surveyed about their attitude towards Bill Clinton. Each individual might be asked whether he/she supports or opposes Clinton without any other choice of response. Since there are only two possible answers permitted, this is called a *binary* response. The data might list the aggregate numbers of voters who do and do not support Clinton.

If the gender of each individual was recorded then we might break down the aggregate numbers into male voters and female voters. There are then four possible responses from an individual—male–support, male–oppose, female–support, female–oppose. This is an example of a *categorical* response with four categories. A binary response is a special type of categorical response with only two categories.

Of course, individuals have no choice over their gender which might lead us to treat the support–oppose classification differently to the gender classification. For instance, we might like to think of the data as comprising a set of data for males and a set of data for females, each comprising aggregate counts of supporters and nonsupporters for that gender. We thus have the choice of considering the data to be two separate data sets on a binary response or as a single data set on a four category response. Depending on the model being fitted, one view may be more convenient than the other.

Next, image that the oppose–support classification was actually derived from a finer categorization of voters' attitudes say, strongly oppose, oppose, neutral, support, strongly support. This is an example of an *ordinal* response since there is a natural ordering of the five attitude categories. While ordinal data can be treated as categorical this is usually only at the price of efficiency. The ordering contains information which if utilized enhances the accuracy of our conclusions. If these ordinal data were broken down by sex then we might think of the data as comprising two separate ordinal data sets. Or we might think of it as a single data set on a 10 category response variable which is partly ordinal (in the attitude classification) and partly binary (in the sex classification).

The attitude classification might itself have been condensed from even more refined data. For instance, voters might have been asked to give an approval score out of 100 for Clinton. Such data would be difficult to analyze unless the scale was somehow standardized. For instance, what score would an individual give if he/she strongly approved—80, 90, 95? The score might instead have been derived from a detailed questionnaire concerning 20 political issues about which the voter would express strength of support for Clinton's performance on the same five category attitude scale described above. The score could be calculated by associating these categories with the numbers 1–5 and adding over the 20 issues. This final score would in some sense measure support on a numerical scale. This is an example of a discrete numeric response. Such data will not be addressed by the methods in this book.

2.2 MULTINOMIAL MODELS

Throughout, we adhere to a notation illustrated here for the simplest type of categorical data possible. We consider the so-called *one-sample* problem which, in the case of normal theory, is the problem of inference on the mean μ of a single sample of normal variables.

2.2.1 Categorical Random Variables

A categorical random variable X is a discrete random variable that takes values in one of k categories. For convenience the categories are labeled $1, \ldots, k$ though we could equally use the labels $A, B, C \ldots$. If the random variable X is truly random then we cannot predict its value in advance. Rather an observed value of X has a certain probability of taking each possible value. The entire probabilistic behavior of X is summarized by its probability function

$$p_X(i) = \Pr(X = i) = \pi_i \qquad i = 1, \ldots, k$$

where the π_i are (usually unknown) parameters that add up to one. There are thus really only $k - 1$ of these *fundamental* parameters.

Descriptive Measures Descriptive measures of location, spread, and association are more difficult to define for categorical than for numeric variables. For instance, what is the mean ethnicity of Americans? The mean, median, variance, interquartile range, and covariance are all undefined. The only meaningful measure of location is the *mode* which is the most probable value of X.

Spread can be defined in several ways. A variable X taking k categorical values is most spread when $\Pr(X = i) = 1/k$ for all i and is least spread when $\Pr(X = j) = 1$ for some j. Two measures in common use are

$$V_G(X) = \sum_{i=1}^{k} \pi_i(1 - \pi_i), \qquad V_E(X) = -\sum_{i=1}^{k} \pi_i \log \pi_i.$$

The first, called *Gini concentration*, takes minimum value zero when X is least spread and maximum value $(k-1)/k$ when X is most spread. The second, called *entropy*, takes minimum value zero when X is least spread and maximum value $\log(k)$ when X is most spread [where $0 \log(0)$ is defined to be 0]. These measures are of use in several contexts. First, one might be genuinely interested in seeing how the variability of a categorical variable is affected by various conditions, though more commonly one is interested in the entire distribution. One application is in ecology where the health of an ecosystem might be related to the existence of a sufficient number of different species. In this context entropy is often called the Shannon index of diversity. A second use of spread measures is in defining measures of association of two categorical variables.

Association Measures For categorical variables, the notion of positive or negative association is meaningless but one can still talk about strength of association. If we could better predict the category of X_1 from knowing the category of X_2 then we would say X_1 and X_2 are associated. A natural way to measure improvement in prediction is to measure how much the variability of X_1 decreases when X_2 becomes known, i.e., by

$$\rho^2 = \frac{V(X_1) - E[V(X_1|X_2)]}{V(X_1)}.$$

This equals 1.0 when $V(X_1|X_2)$ is zero for all values of X_2 and only happens when X_1 takes a known value conditional on X_2, i.e.; X_1 is a known function of X_2. For continuous normal variables, ρ^2 is indeed the squared correlation coefficient, hence the notation.

Let us now introduce the notation

$$\Pr(X_1 = i, X_2 = j) = \pi_{ij}, \qquad i = 1, \ldots, k, \qquad j = 1, \ldots, l$$

so that the marginal probability $\Pr(X_1 = i) = \pi_{i\bullet}$ is the sum of π_{ij} over j, etc. Then using Gini concentration for V in ρ^2 gives

$$\tau = \frac{\sum_{i=1}^{k} \sum_{j=1}^{l} \pi_{ij}^2 / \pi_{i\bullet} - \sum_{j=1}^{k} \pi_{\bullet j}^2}{1 - \sum_{j=1}^{l} \pi_{\bullet j}^2}$$

proposed by Goodman and Kruskall (1954) and using entropy for V gives

$$U = -\frac{\sum_{i=1}^{k} \sum_{j=1}^{l} \pi_{ij} \log(\pi_{ij} / \pi_{i\bullet} \pi_{\bullet j})}{\sum_{j=1}^{l} \pi_{\bullet j} \log \pi_{\bullet j}}$$

proposed by Theil (1970) and called the *uncertainty coefficient*. Both indices are invariant to permutation of the classification labels and both measure how much better we can predict X_1 from knowing X_2. However, neither measure is reflexive—different answers are obtained when applied to the transposed array $\{\pi_{ji}\}$.

When X_1 and X_2 are categorical variables defined on the same categories then another measure is

$$\kappa = \frac{\sum_{i=1}^{k} \pi_{ii} - \sum_{i=1}^{k} \pi_{i\bullet} \pi_{\bullet i}}{1 - \sum_{i=1}^{k} \pi_{i\bullet} \pi_{\bullet i}}$$

proposed by Cohen (1960), and will be more powerful at detecting association than τ or U, under such circumstances. This measure is reflexive, and is also invariant to permutation of the (common) row and column labels. One problem with all three association measures is that they tend to be smaller when there are more categories. How large a value determines a *practically* strong association is then unclear although tests of *statistical* significance may still be performed.

Estimates of τ, U, κ are obtained from data by replacing probabilities by observed proportions and have large sample normal distributions. The *Splus* function `associate` at the Wiley website computes these three association measures.

Example 2.1: Marital Status of Danes The information shown in Table 2.1 relates to a survey of 165 Danes who were classified by their age (categorized into 1 of 8 groups) and their marital status. Row percentages have been given in parentheses. For instance, among those 32 individuals surveyed in their forties, 16% were unmarried, 66% married, and 19% divorced.

The Gini concentration V_G and entropy V_E are listed for each age-group, as well as the mode in the last column. Looking at the last three columns only, we can see that the variability steadily increases as age increases, i.e., it becomes more and more difficult to predict someone's marital status from knowing only their age. Only for young persons can their status be predicted well (as single). There are only three categories here, so it is probably better to simply look at the three percentages for each age-group however, for variables with larger number of categories the spread and location can provide a useful summary measure of the data patterns. We will fit a parametric model to these data in Chapter 6.

The variability for the separate age-groups is mostly around 0.7–0.9, not much smaller than the variability ignoring age-groups (around 0.9). We might expect that the association is low. Both $\hat{\tau}$ and $\hat{u}$ give almost identical answers of 0.22 in this example. However, reversing the roles of the row and column factors we obtain $\hat{\tau} = 0.06$, $\hat{u} = 0.11$. It is not appropriate to compute κ as the row and column classification are not the same.

Table 2.1. Marital Status Data

Age	Single	Married	Divorced	Total	$3V_G/2$	$V_E/\log(3)$	Mode
17–21	17(0.94)	1(0.06)	0(0.00)	18	0.16	0.19	S
21–25	16(0.67)	8(0.33)	0(0.00)	24	0.67	0.56	S
25–30	8(0.31)	17(0.65)	1(0.04)	26	0.71	0.70	M
30–40	6(0.19)	22(0.69)	4(0.12)	32	0.73	0.76	M
40–50	5(0.16)	21(0.66)	6(0.19)	32	0.77	0.80	M
50–60	3(0.11)	17(0.61)	8(0.29)	28	0.81	0.82	M
60–70	2(0.12)	8(0.50)	6(0.38)	16	0.89	0.89	M
70+	1(0.11)	3(0.33)	5(0.55)	9	0.85	0.85	D
All ages	58(0.31)	97(0.52)	30(0.16)	185	0.90	0.91	M

Survey of Danes classified by age group and marital status; row percentages given in parentheses.

We will have no more to say in this chapter about measures of spread and association for categorical variables. Rather, our main aim is to make inferences about the parameters π_i not only about their likely values but the likely relations between them and about future likely behavior of the random variable.

2.2.2 Multinomial Distribution

If X is multinomially distributed and if the probability parameters π are known, then we can make precise statements about the future behavior of X limited only by its unavoidably random nature. For example, we can predict how many times X might be seen in a certain category from a specified number of observations and give exact statements about the reliability of our prediction. We would not be able to improve these predictions from observing more data. The probability vector $\pi = (\pi_1, \ldots, \pi_k)$ describes all that we can know about a categorical random variable.

In practice, the probability parameters are at least partially unknown. To infer their values, we need data. A *random sample* is a single sample of n independent observations on X. These n observations will simply be a list of n responses on the k categories. For instance, $n = 10$ independent observations on a categorical variable with $k = 4$, might yield the data set 3221423214. The likelihood for the data set is the probability of the data set treated as a function of the parameters. The probability of obtaining $x = i$ is π_i and so the likelihood will contain a factor π_i exactly once for every time $x = i$ was observed. Denoting

$$Y_i = \#\text{ responses in category } i.$$

the likelihood is

$$L(\pi_1, \ldots, \pi_k; x) = \pi_{x_1}\pi_{x_2}\ldots\pi_{x_n} = \pi_1^{y_1}\pi_2^{y_2}\ldots\pi_k^{y_k}.$$

For instance, in the above example the likelihood would be

$$L(\pi_1, \pi_2, \pi_3, \pi_4; 3221423214) = \pi_3\pi_2\pi_2\pi_1\pi_4\pi_2\pi_3\pi_2\pi_1\pi_4 = \pi_1^2\pi_2^4\pi_3^2\pi_4^2.$$

The statistics $Y_1, \ldots, Y_k$ together define the likelihood and so by the sufficiency principle the data may be summarized by these counts without loss of information. The joint distribution of the sufficient statistics $Y_1, \ldots, Y_k$ is given by the probability function

$$P_M(y_1, \ldots, y_k; n, k, \pi) = \frac{n!}{y_1!y_2!\ldots y_k!}\,\pi_1^{y_1}\pi_2^{y_2}\ldots\pi_k^{y_k} \qquad \sum_{i=1}^{k} y_i = n \tag{2.1}$$

which is called the multinomial distribution with parameters k, $\pi_1, \ldots, \pi_k$ and denoted $M(n, k, \pi)$. Usually k and n are known and it is the probability parameters π_i that are unknown. These fundamental parameters will usually be related or restricted in some way by the statistical model assumed. These relations can be summarized as expressions for the π_i in terms of a smaller number $p \leq k - 1$ of model parameters, denoted $\theta = (\theta_1, \ldots, \theta_p)$.

The factorials in Eq. (2.1) account for all the different *orders* in which we could obtain the total numbers of responses $Y_1, \ldots, Y_k$. There are two connections with the binomial distribution. First, in the special case that $k = 2$, the above formula reduces to the binomial distribution after noting that the π_i sum to 1 and that the y_i sum to n. Second, consider the separate or *marginal* distribution of Y_1. Each observation on X will either be in category 1 (with probability π_1) or not. Thus, Y_1 has the $Bi(n, \pi_1)$ distribution, Y_2 the $Bi(n, \pi_2)$ distribution, etc. The joint distribution of the $Y_1, \ldots, Y_k$ is thus that of k binomial random variables with possibly different probability parameters and constrained to add up to n. The expected values of the Y_i are given by $e_i = n\pi_i$, the variances by $n\pi_i(1 - \pi_i)$, and it can be shown that the covariance of Y_i and Y_j is $-n\pi_i\pi_j$.

A statistical model for the data $Y_1, \ldots, Y_k$ comprises the multinomial distribution together with expressions for the fundamental parameters π_i in terms of a set of model parameters, θ. The joint distribution of the data under the model is then Eq. (2.1) with these expressions $\pi_i(\theta)$ substituted.

2.2.3 Multinomial Likelihood

In this section, expressions for the score function, the information, and the maximized log-likelihood function under the multinomial model are derived. On taking logarithms of Eq. (2.1) we obtain the log-likelihood

$$\ell_M(\theta; y) = c + \sum_{i=1}^{k} y_i \log \pi_i(\theta) \tag{2.2}$$

where c is the logarithm of the factorials and will have no role to play in almost all subsequent results. Denote derivatives of π_i with respect to components of θ by superscripts. For instance, π_2^{13} is the mixed derivative of $\pi_2(\theta)$ with respect to θ_1 and θ_3. Since the π_i sum to one, derivatives such as π_i^j and π_i^{lm} will sum to zero. With this notation, we differentiate Eq. (2.2) to obtain the score function whose jth component is

$$U_{Mj}(\theta) = \frac{\partial \ell_M}{\partial \theta_j} = \sum_{i=1}^{k} y_i \frac{\pi_i^j(\theta)}{\pi_i(\theta)} \tag{2.3}$$

and it is readily checked that the mean of U_{Mj} is zero. In most regular cases, equating the score function to zero gives a unique solution which maximizes the likelihood. Conditions for this will be given in later chapters. The estimate $\hat{\theta}$ determines estimates $\hat{\pi}_i = \pi_i(\hat{\theta})$ of the fundamental parameters π_i which in turn determine estimates of the expected values $e_i = n\pi_i$. These are denoted

$$\hat{e}_i = n\pi_i(\hat{\theta})$$

and are called *fitted values*. The maximized log-likelihood is thus

$$\ell_M(\hat{\theta}) = c + \sum_{i=1}^{k} y_i \log(\hat{e}_i) - n\log(n). \tag{2.4}$$

Expressions for the information follow upon further differentiating U_j. The (l, m)th component of the observed information matrix is

$$\mathcal{I}_{lm} = -\frac{\partial \ell}{\partial \theta_l \theta_m} = \sum_{i=1}^{k} y_i \frac{\pi_i^l \pi_i^m - \pi_i^{lm} \pi_i}{(\pi_i)^2}. \tag{2.5}$$

Using the fact that the π_i^{lm} sum to zero and $\mathrm{E}(Y_i) = n\pi_i$ we find that

$$I_{lm} = E\left(-\frac{\partial \ell}{\partial \theta_l \theta_m}\right) = n \sum_{i=1}^{k} \frac{\pi_i^l \pi_i^m}{\pi_i}. \tag{2.6}$$

This can also be obtained by finding the covariance of U_l and U_m using $\mathrm{Var}(Y_l) = n\pi_l(1 - \pi_l)$ and $\mathrm{Cov}(Y_l, Y_m) = -n\pi_l\pi_m$.

2.2.4 The Full Model

The full model contains as many parameters as data points. While this model is rarely of intrinsic interest in itself we compare more refined models with it in goodness-of-fit tests. In order to find general expressions for Wald, Score, or LR goodness-of-fit statistics we need expressions for the maximized likelihood, score function, and information functions for the full model. The full model has $k - 1$ parameters that we take to be

$$\theta = (\pi_1, \ldots, \pi_{k-1})$$

without loss of generality. Thus

$$\pi_k(\theta) = 1 - \pi_1 - \cdots - \pi_{k-1}$$

and $\pi_i(\theta) = \pi_i$ for $i = 1, \ldots, k - 1$ and so

$$\pi_k^j = -1, \qquad \pi_i^j = \delta_{ij}\ i < k, \qquad \pi_i^{lm} = 0$$

where the Kronecker delta δ_{ij} equals 1 when $i = j$ and zero otherwise. Substituting this into (2.3) gives

$$U_{Mj} = \frac{\partial \ell_M(\pi)}{\partial \pi_j} = \frac{y_j}{\pi_j} - \frac{y_k}{\pi_k}\ j = 1, \ldots, k - 1 \tag{2.7}$$

and so the estimates of the π_i are proportional to the y_i. Since the π_i add to one and the y_i add to n we obtain the estimates $\hat{\pi}_i = y_i/n$ being simply the cell relative frequencies. The fitted values are $\hat{e}_i = n\hat{\pi}_i = y_i$, i.e., the data values themselves. From Eq. (2.4) the maximized log-likelihood is

$$\ell(\hat{\theta}) = c + \sum_{i=1}^{k} y_i \log\ y_i - n \log\ n. \tag{2.8}$$

Substitution into Eqs. (2.5) and (2.6) for the observed and expected information functions gives

$$\mathcal{I}_{lm} = \frac{y_k}{\pi_k^2} + \delta_{lm}\frac{y_l}{\pi_l^2}, \qquad I_{lm} = n\left(\frac{1}{\pi_k} + \frac{\delta_{lm}}{\pi_l}\right). \tag{2.9}$$

The inverse of the expected information matrix has entries $\pi_l(1 - \pi_l)/n$ on the diagonal and entries $-\pi_l\pi_m/n$ off the diagonal. According to asymptotic theory these are the approximate variances and covariances of the estimates $\hat{\pi}_i$. In fact, they are actually the exact variances and covariances which follow easily from the known covariances of the Y_i.

Example 2.2: Distribution of Genotypes One hundred plants are sampled from a large population and classified by *genotype* aa, aA, or AA. The genotype is determined by the two *alleles* each of which may be A or a. Let X_i denote the random variable which records the genotype of the ith sampled plant. This is a categorical random variable on $k = 3$ categories. The fundamental parameters are

$$\pi_1 = \Pr(\text{type aa}), \pi_2 = \Pr(\text{type aA}), \pi_3 = \Pr(\text{type AA})$$

with the relation that $\pi_3 = 1 - \pi_1 - \pi_2$. The data and the expected values are listed in Table 2.2.

Table 2.2. Classification of Plants by Genotype I

Type	aa	aA	AA
Count	18	55	27
Expected	$100\pi_1$	$100\pi_2$	$100\pi_3$

The full model gives estimates $\hat{\pi}_1 = 0.18$, $\hat{\pi}_2 = 0.55$, $\hat{\pi}_3 = 0.27$, and fitted values 18, 55, 27. The maximized log-likelihood is

$$\ell(\hat{\theta}) = c + 18\log(18) + 55\log(55) + 27\log(27) - 100\log(100) = c - 99.099$$

The constant $c = \log(100!/(18!55!27!)) = 94.4678$ and so the maximized (unlogged) likelihood is 0.00437. This figure has no particular significance since *relative* rather than absolute likelihoods form the basis of statistical inference.

The observed and expected information matrices are

$$\mathcal{J} = \begin{pmatrix} \dfrac{18}{(\pi_1)^2} + \dfrac{27}{(\pi_3)^2} & \dfrac{27}{(\pi_3)^2} s \\ \dfrac{27}{(\pi_3)^2} & \dfrac{55}{(\pi_2)^2} + \dfrac{27}{(\pi_3)^2} \end{pmatrix},$$

$$I = 100 \begin{pmatrix} \dfrac{1}{\pi_1} + \dfrac{1}{\pi_3} & \dfrac{1}{\pi_3} \\ \dfrac{1}{\pi_3} & \dfrac{1}{\pi_2} + \dfrac{1}{\pi_3} \end{pmatrix}$$

Substituting estimates of the parameters, these two matrices agree and the common inverse is

$$\begin{pmatrix} 0.1476 & -0.0990 \\ -0.0990 & 0.2475 \end{pmatrix} /100.$$

These values can be obtained directly from the binomial distributions of Y_1 and Y_2. For instance, $\text{Var}(\hat{\pi}_1) = \pi_1(1 - \pi_1)/100$ which on substituting estimates gives 0.001476. Similarly, $\text{Cov}(\hat{\pi}_1, \hat{\pi}_2) = -\pi_1\pi_2/n$ giving -0.000990.

2.2.5 Simple Models

When the fundamental parameters $\pi = (\pi_1, \ldots, \pi_k)$ have known values

$$\pi_0 = (\pi_{10}, \pi_{20}, \ldots, \pi_{k0})$$

there is no further inference about the underlying distribution possible. Everything about the behavior of the categorical variable X and the counts Y_i is known and summarized in the known multinomial distribution Eq. (2.1). A model with all parameters specified and therefore no parameters to estimate is called a *simple* model. Since there are no parameters to estimate there are no score function or information matrices. The "estimates" of the π_i are the known values π_{i0}. The fitted values will be $\hat{e}_i = n\pi_{i0}$ which will typically disagree with the data values. The log-likelihood function at these "estimates" equals

$$\ell(\pi_0) = c - \sum_{i=1}^{k} y_i \log \pi_{i0}.$$

Simple models are rarely of intrinsic interest in themselves except for very artificial and simple experiments.

Example 2.2: (Continued) **Distribution of Genotypes.** According to theory the alleles a and A should be equally distributed across a population provided that characteristics associated with the different genotypes confer no survival advantage or disadvantage. Plants obtain one allele from each parent. If the proportions of types a and A in the population are equal, then each of the two alleles an offspring receives is an independent binomial trial with probability of success 1/2. The offspring of two randomly chosen plants will thus have genotype *aa*, *aA*, *Aa*, or *AA* with equal probability 1/4. Since genotypes *Aa* and *aA* cannot be distinguished theory predicts that the proportions of plants of genotypes *aa*, *aA*, and *AA*, chosen randomly from a large population, will be $\pi_1 = 1/4$, $\pi_2 = 1/2$, and $\pi_3 = 1/4$, respectively. The expected values under this model are 25, 50, and 25 for the three categories. The maximized likelihood is

$$\ell(\pi_0) = c + 18\log(0.25) + 55\log(0.5) + 27\log(0.25) - 100\log\ 100 = c - 100.506.$$

Notice that this maximized log-likelihood is 1.407 smaller than the maximized log-likelihood under the full model. This means that the data 18, 55, 27 are $\exp(1.407) = 4.085$ times less likely assuming the theory than not assuming it. This is not necessarily bad news for the theory since $\ell(\pi_0)$ is necessarily smaller than $\ell(\hat{\pi})$. Could the number 4.085 occur by chance? Section 2.5 studies tests based directly on the sampling distribution of the difference between two maximized log-likelihoods.

2.2.6 Intermediate Models

A model is *intermediate* if it neither specifies all the fundamental parameters nor leaves them all entirely unrestricted, i.e., if it is neither a full nor a simple model.

One goal of data analysis will be to find at least one intermediate model that describes the data adequately. An intermediate model will specify $0 < r < k - 1$ restrictions on the fundamental parameters but it will often be simpler to express the model in terms of $p = k - r - 1$ parameters $\theta_1, \ldots, \theta_p$. The expressions for the fundamental parameters π in terms of the θ_j will then define the model.

A central role of the model is to simplify and explain the data. This is best achieved if all the parameters $\theta_1, \ldots, \theta_p$ have physical interpretations. While this is not always completely achievable a good data analyst will attempt to avoid purely mathematical descriptions of the pattern in a data set.

Example 2.2: (Continued) **Distribution of Genotypes.** If the characteristic associated with the genes confers a survival advantage then we would expect the proportion of A alleles in the population to be larger than $1/2$. Let X' be the number of a alleles carried by an individual (0, 1, or 2). Assuming that parents donate alleles independently, the distribution of X' is $Bi(2, p)$ where p is the proportion of a alleles in the population. Explicitly, the probabilities of the three genotypes are

$$\Pr(aa) = p^2, \Pr(aA) = p(1 - p), \Pr(Aa) = p(1 - p), \Pr(AA) = (1 - p)^2$$

known as Hardy–Weinberg equilibrium. Further, assuming that individuals receive alleles independently of each other the distribution of the frequencies is multinomial with expected values given in Table 2.3.

The single unknown parameter p is to be estimated. This parameter has a direct physical interpretation; it is the proportion of a alleles in the population from which the plants were sampled. Differentiating Eq. (2.3), the score function is

$$U(p) = y_1 \frac{2p}{p^2} + y_2 \frac{2 - 4p}{2p(1 - p)} + y_3 \frac{2p - 2}{(1 - p)^2} = \frac{2y_1 + y_2}{p} - \frac{y_2 + 2y_3}{1 - p}.$$

The solution of equating this to zero is

$$\hat{p} = \frac{2y_1 + y_2}{2(y_1 + y_2 + y_3)} = \frac{2y_1 + y_2}{200}$$

Table 2.3. Classification of Plants by Genotype II

Genotype (X)	aa	aA	AA
a Genes (X')	2	1	0
Count	18	55	27
Expected	$100p^2$	$200p(1 - p)$	$100(1 - p)^2$
Fitted	20.70	49.60	29.70

with observed value $91/200 = 0.455$. This estimate is easy to explain intuitively. Each allele of each of the 100 sampled plants is independent of every other and will be type a with probability p. The data thus comprise 200 Bernoulli trials. Exactly y_1 plants had two a alleles and y_2 plants a single a allele so the total number of a alleles in the sample is $2y_1 + y_2$. The estimate of p is therefore the number of a alleles in the sample from a possible total which is surely a natural estimate of the proportion in the population. The estimated variance of this estimate is based on the $Bi(200, p)$ distribution for $2y_1 + y_2$ and is $0.455 \times 0.545/200$ giving standard error 0.0352. This is identical to the asymptotic variance estimate from the inverse of the expected information. Using Eq. (2.6) this expected information is

$$100 \times \left(\frac{(2p)^2}{p^2} + \frac{(4p-2)^2}{2p(1-p)} + \frac{(2p-2)^2}{(1-p)^2} \right)$$

which after a little algebra reduces to $200/(p(1-p))$. The fitted values are listed in the table. For instance, $\hat{e}_1 = 100 \times (0.455)^2 = 20.70$. The maximized log-likelihood is

$$\ell(\hat{p}) = c + 18\log(20.7) + 55\log(49.6) + 27\log(29.7) - 100\log(100) = c - 99.694.$$

Notice that this value is 0.595 smaller than for the full model but 0.812 larger than for the theoretical (simple) model considered earlier. The data are only 1.813 times less likely assuming the binomial model than not assuming it. They are a further 2.253 times less likely if p is assumed to equal 1/2.

2.3 POISSON MODELS

A distinguishing feature of the multinomial distribution is that the sum of the counts Y_i is fixed and known. Typically, this arises from our having taken a random sample on X of predetermined size n. There are many situations in which the total of the counts is not known in advance. For instance, we might be counting events of different types occurring in time. Even if the amount of time T sampled is fixed in advance, the total of the counts is not known until it is observed, i.e., it is a random variable.

The most common distribution for modeling counts in time is the Poisson distribution. Imagine recording events of k different types that occur singly and independently during a sampling interval $[0, T]$. The probability of an event of type i occurring during the time interval $[t, t+\delta]$ we suppose is $\alpha_i\delta$ for small δ. Let Y_i denote the total number of events of type i that occur. Then it can be shown that $Y_1, \ldots, Y_k$ are sufficient for the parameters $\alpha_1, \ldots, \alpha_k$ and so by the sufficiency principle we may ignore the precise times of the events and simply analyze the total counts. The distribution of each Y_i will be Poisson with mean $\mu_i = \alpha_i T$ and since the Y_i are independent their joint distribution is

$$P_P(y_1, \ldots, y_k; k, \mu) = e^{-\sum \mu_i} \frac{\mu_1^{y_1} \mu_2^{y_2} \ldots \mu_k^{y_k}}{y_1! y_2! \ldots y_k!} \tag{2.10}$$

which it should be noted is rather similar to Eq. (2.1). One difference is that there are k functionally independent variables rather than $k-1$. The *fundamental* parameters $\mu_1, \ldots, \mu_k$ tell us everything about the probabilistic behavior of the counts. These parameters will usually be restricted or related somehow by the statistical model assumed. These relations can be summarized as expressions for the μ_i in terms of a smaller number $p \leq k$ of model parameters, denoted $\theta = (\theta_1, \ldots, \theta_p)$. The statistical model for the data $Y_1, \ldots, Y_k$ comprises the independent Poisson distribution Eq. (2.10) with these expressions for the fundamental parameters.

2.3.1 Poisson Likelihood

We seek expressions for the score function, the information, and the maximized likelihood. On taking logarithms of Eq. (2.10) we obtain

$$\ell_P(\theta; y) = c - \sum_{i=1}^{k} \mu_i(\theta) + \sum_{i=1}^{k} y_i \log \mu_i(\theta). \tag{2.11}$$

Again denote derivatives of $\mu(\theta)$ with respect to components of θ by superscript. Thus, μ_2^{14} is the mixed derivative of μ_2 with respect to θ_1 and θ_4. Unlike the multinomial model, the μ_i are not constrained in their total and so the sums of these derivatives need not equal zero. Taking the derivative of Eq. (2.11) with respect to θ_j gives

$$U_{Pj}(\theta) = \frac{\partial \ell_P}{\partial \theta_j} = \sum_{i=1}^{k} \left\{ \frac{y_i}{\mu_i(\theta)} - 1 \right\} \mu_i^j(\theta) \tag{2.12}$$

the jth component of the score function and it is simple to see that the mean of U_{Pj} is zero. In regular cases, equating the score function to zero gives a unique solution which maximizes the likelihood. The ML estimate $\hat{\theta}$ determines fitted values $\hat{e}_i = \mu_i(\hat{\theta})$. The maximized log-likelihood is thus

$$\ell_P(\hat{\theta}) = c + \sum_{i=1}^{k} y_i \log(\hat{e}_i) - \sum_{i=1}^{k} \hat{e}_i. \tag{2.13}$$

Expressions for the information follow upon further differentiating U_j. The *lm*th component of the observed information matrix is

$$\mathcal{I}_{lm} = \sum_{i=1}^{k} y_i \frac{\mu_i^l \mu_i^m}{\mu_i^2} + \mu_i^{lm}\left(\frac{y_i}{\mu_i} - 1\right) \tag{2.14}$$

and on taking expectation the expected information matrix has *lm*th entry

$$I_{lm} = \sum_{i=1}^{k} \frac{\mu_i^l \mu_i^m}{\mu_i} = T \sum_{i=1}^{k} \frac{\alpha_i^l \alpha_i^m}{\alpha_i}. \tag{2.15}$$

The information thus increases proportionally with the sampling time T which, for Poisson sampling, is analogous to the sample size.

2.3.2 Full, Simple, and Intermediate Models

The full Poisson model imposes no restrictions on the fundamental parameters μ_i at all, in symbols $\mu_i = \theta_i$, $i = 1, \ldots, k$. Thus, $\mu_i^j = \delta_{ij}$ and all second derivatives are zero. Thus Eq. (2.12) becomes

$$U_{Pj} = \frac{y_j}{\mu_j} - 1 \tag{2.16}$$

with solution $\mu_j = y_j$ and so the fitted values are the observed data values. The maximized log-likelihood is

$$\ell_P(\hat{\theta}) = c + \sum_{i=1}^{k} y_i \log y_i - \sum_{i=1}^{k} y_i. \tag{2.17}$$

The observed and expected information matrices are diagonal with entries y_i/μ_i^2 and $1/\mu_i$ respectively. The inverse of these matrices are thus diagonal with entries μ_i^2/y_i and μ_i, respectively. As with the full multinomial model, the inverse of the expected information gives the exact covariance matrix of the estimates $\hat{\mu}_i = y_i$, not just asymptotically.

If the μ_i are all assumed to have known values μ_{i0} then nothing about the distribution of the data is left to model and the model is simple. The "fitted" values are the assumed parameter values μ_{i0} and so the log-likelihood

$$\ell_P(\mu_0) = c + \sum_{i=1}^{k} y_i \log \mu_{i0} - \sum_{i=1}^{k} \mu_{i0}.$$

An intermediate model will neither assume values for all the fundamental parameters nor leave them entirely unrestricted but rather will impose $0 < r < k$ restrictions. The maximized likelihood will be less than the maximized likelihood for the full model and the difference forms the basis of a goodness-of-fit test in Section 2.5.

Example 2.3: Seasonality of Births The observed numbers of births at a certain hospital for four consecutive quarterly periods are given in Table 2.4. There is a preponderance of births in the January and October quarters. We will see whether or not the data provide evidence for different birth rates in the four quarters. The total number of births is a random variable with observed value 151. Since births are random events in time, it is reasonable to model the counts in each quarter by independent Poisson variables with mean rates $\mu_1, \mu_2, \mu_3, \mu_4$.

Our null hypothesis is $\mathcal{H}_0 : \mu_1 = \mu_2 = \mu_3 = \mu_4 = \mu$ or in other words "mean birth rates are the same for all four quarters." The model parameter $\theta = \mu$ has dimension $p = 1$ and $\mu_i^1 = 1$ for all i. Thus Eq. (2.12) becomes

$$U_{Pj} = \sum_{i=1}^{k} y_i/\mu - k$$

with solution $\mu = \sum y_i/k = 37.75$. These are also the fitted values and the maximized log-likelihood Eq. (2.13) is

$$\ell_P(\hat{\theta}) = c - 151 + (55 + 29 + 26 + 41)\log(37.75) = c + 397.279.$$

In contrast, the maximized likelihood Eq. (2.17) for the full model is c + 404.022, being 6.743 larger. We find that the data are $\exp(6.734) \approx 850$ times *less* likely assuming equal rates for the four quarters than allowing the rates to differ. Converting this to a statistical test requires us to consider the sampling distribution of the difference in maximized likelihoods.

Notice that we could also have analyzed these data using the multinomial distribution model by pretending that the 151 births were fixed in advance and noting that each of these 151 births could have occurred in any of the four quarters. The null hypothesis then specifies that the probability of any birth being in a particular quarter is 1/4, this being a simple model. This approach actually leads to identical fitted values and tests. The duality between multinomial and Poisson models is studied in the next section.

Table 2.4. Births for Four Quarters

Quarter	January	April	July	October	Total
Births	55	29	26	41	151

2.4 THE MULTINOMIAL-POISSON CONNECTION

The major difference between the multinomial and Poisson models is that the total of the data is fixed and known for the former, but random for the latter. There is a close connection between these two models and this connection has two practical consequences. First, count data may often be modeled in two equivalent ways with identical results. We may thus choose whichever formulation is theoretically or computationally convenient. For instance, the package GLIM does not support the multinomial distribution but multinomial models may be fitted as Poisson models. Second, we may fit a Poisson model even if the data were not generated by a Poisson process—so long as we are interested in the π_i the Poisson assumption will have no effect on our conclusions.

In both the multinomial and Poisson context we have a data set $Y_1, \ldots, Y_k$ with respective means $\mu_1, \ldots, \mu_k$. Let $N = Y_1 + \cdots + Y_k$ be the total number of events observed considered as a random variable whose mean value is $\tau = \mu_1 + \cdots + \mu_k$ and whose observed value is n. According to the definition of conditional probability we may write

$$\Pr(Y_1 = y_1, \ldots, Y_k = y_k) = \Pr(N = n)\Pr(Y_1 = y_1, \ldots, Y_k = y_k | N = n). \quad (2.18)$$

This factorization corresponds to a logical decomposition of the act of observing the full data $Y_1, \ldots, Y_k$ into the following two steps: (1) observe how many events $n = Y_1 + Y_2 + \cdots + Y_k$ in total were recorded, (2) go through these n events and classify them according to type to obtain the counts $Y_1, \ldots, Y_k$ of each. The second step of classification into k categories is exactly the multinomial framework. Thus, the *conditional* distribution of $Y_1, \ldots, Y_k$ is multinomial with parameters n, k, and probability parameters

$$\pi_i = \Pr(\text{an event is of type } i).$$

The conditional means of the Y_i are $n\pi_i$ and the unconditional means are

$$\mu_i = \mathrm{E}(\mathrm{E}(Y_i|N)) = \mathrm{E}(N\pi_i) = \tau\pi_i.$$

Thus, the original parameters naturally divide into parameters

$$\tau = \mu_1 + \cdots + \mu_k, \qquad \pi_i = \mu_i/\tau_i \qquad i = 1, \ldots, k - 1 \quad (2.19)$$

the first relevant to step 1 and the last $k - 1$ to step 2. If $g(n; \tau)$ is any assumed distribution for the first step then the likelihood Eq. (2.18) becomes

$$L(\pi, \tau; y) = g(n; \tau)P_M(y_1, \ldots, y_k; n, k, \pi). \quad (2.20)$$

The multinomial parameters will typically be modeled through a parameter θ

of dimension p. Now as long as the parameter τ is not related to these parameters, maximization of Eq. (2.20) breaks up into two separate problems; the first maximizing $g(n;\tau)$ to find the ML estimator of τ and the second maximizing the multinomial likelihood to find the ML estimators of the π_i as described in Section 2.2.3. Subject to this proviso, the factor $g(n;\tau)$ will not affect likelihood estimation of the π_i at all. Furthermore, since the first factor depends on τ and not θ while the second depends on θ and not τ, it is easy to show that the observed information factorizes into

$$\mathcal{J}(\theta,\tau) = \begin{pmatrix} \mathcal{J}_M(\theta) & 0 \\ 0 & \mathcal{J}_g(\tau) \end{pmatrix}$$

where $\mathcal{J}_M$ is the information matrix Eq. (2.5) from the multinomial model and $\mathcal{J}_g$ is the information corresponding to $g(n;\tau)$. Thus, the parameters τ and θ are orthogonal (see p. 26) and the asymptotic variance matrix of the $\hat{\pi}_i$ is just the inverse of $\mathcal{J}_M$. Thus, not only are the estimates of the π_i themselves unaffected by the choice of g but their standard errors are also invariant. The fitted values under the model Eq. (2.20) are

$$\hat{e}_i = \hat{\tau}\hat{\pi}_i$$

which will be identical to the fitted values $n\hat{\pi}_i$ under the multinomial model provided that $\hat{\tau} = n$. This seems a natural condition since τ is the mean of N and we have a single observation on this random variable. Finally, if the parameter τ *is* related by the model to the θ_i then the ML estimates of the probability parameters will disagree with the conditional multinomial estimates as will the fitted values. For most statistical models this does not occur.

One possible model for N is the $Pn(\tau)$ distribution for which the ML estimate of τ is indeed n. On substituting the term

$$g(n;\tau) = e^{-\tau}\,\frac{\tau^n}{n!}$$

into Eq. (2.20) it can be checked that the joint distribution is exactly the independent Poisson model Eq. (2.10) with $\mu_i = \tau\pi_i$. Conversely, if we assume the independent Poisson model, then the distribution of the $Y_i, \ldots, Y_k$ conditional on their total is multinomial with parameters $\pi_i = \mu_i/\tau$. Simply stated, the independent Poisson distribution arises from a multinomial with Poisson distributed parameter n. The multinomial distribution arises as the distribution of a set of independent Poissons conditional on their total.

We are very often interested in the π_i rather than in τ. The π_i measure mean event rates *relative* to the total and we are usually most interested in the *relative* expected frequencies of the categories. This describes the *shape* of the data. We are less often interested in the absolute values μ_i; the scale of these is determined by processes that determine the total count N. In the case of a genuine Poisson model it would be determined by the sampling time T decided upon by the experimenter. However the total count N is generated, the conditional multinomial model forms the basis for inference on the probability parameters π_i. This is true *regardless of the distribution assumed for* N. We thus often fit Poisson models to count data even when we know that the data were not generated by a Poisson process. We can do this because the assumption has no effect on the main inferences of interest, namely the pattern, rather than the absolute size, of the expected values. If we are interested in the mean parameter μ_i then we will need to take more care in modeling the total count N.

Example 2.2: (Continued) **Distribution of Genotypes.** Imagine that the number of plants sampled is a random variable N. This might be the case if we sampled all the plants in a particular field. The distribution of N would be unknown and it is difficult to imagine what form this distribution might take. Nevertheless, let us *assume* that the distribution is $Pn(\tau)$. Then $\tau = E(N)$ represents the expected number of plants in the sample. The genetic theory makes no predictions about τ which is presumably determined by factors such as the size and fertility of the field and the vigor of the particular plant species. Rather, the genetic theory says something about the *relative* numbers of different genotypes in the field. Once the number of plants in the field is known the number of each type is determined by the multinomial distribution with fundamental parameters

$$\pi_i = \Pr(\text{plant is of genotype } i).$$

The unrealistic assumption that N is Poisson distributed, together with the knowledge that the counts of the different genotypes are multinomial, is equivalent to assuming that the counts $Y_1, \ldots, Y_k$ are independent Poisson variables. The first of these assumptions has no effect on inference about the π_i. An appropriate Poisson model specifies that the four different genotypes will occur in the *proportions* $p^2 : 2p(1-p) : (1-p)^2$ so we may write the model as

$$\mu_1 = p^2\tau, \mu_2 = 2p(1-p)\tau, \mu_3 = (1-p)^2\tau$$

where $\tau = E(N)$. Fitting this model, and comparing the results with the earlier multinomial treatment, is left as an exercise. You should obtain identical fitted values, parameter estimates, and standard errors.

Example 2.3: (Continued) **Seasonality of Births.** As mentioned earlier, the model specifying equal birth rates for the four quarters may be treated as a multinomial on four categories with equal probability $1/4$ for each cell. This is a simple model and the fitted values are $n\pi_{i0} = 37.75$ as was found for the

Poisson formulation. The maximized log-likelihood under the simple and full multinomial model are not the same as the maximized log-likelihoods under the corresponding Poisson models. Comparing Eq. (2.4) with Eq. (2.13), the Poisson models contain an extra degree of freedom, the random variable N, and so the log-likelihoods are always *smaller* by $n \log n - \log n! - n$. However, the *difference* in maximized log-likelihoods for two models is identical whether we take a Poisson or a multinomial formulation—the data is still 850 times less likely assuming equal birth rates than not.

2.5 GOODNESS-OF-FIT

A model once fitted, forms the basis for statements about parameters of interest, their structural relations, and for predictions of future behavior. It is of obvious importance to examine the adequacy of a model and to quantify our confidence in its correctness. A goodness-of-fit test of model $\mathcal{M}$ is a test of

$$\mathcal{H}_0 : \mathcal{M} \text{ is true} \quad \textit{against} \quad \mathcal{H}_1 : \text{full model.}$$

In most cases, goodness-of-fit tests are implicit tests of several structural relations between the fundamental parameters of the full model, whether this model be multinomial or Poisson. Also, in most cases, the null hypothesis is composite. It is worth reiterating the difficulties that multiparameter tests involve.

When a test is essentially a test of $r > 1$ parameters there are many plausible ways to combine the sufficient statistics of these parameters into a single test statistic and typically no most powerful test will exist. For this reason many reasonable test statistics will be available and goodness-of-fit test statistics are a case in point. When the null hypothesis is composite then the null distribution of any test statistic will typically still depend on nuisance parameters and so at least in general we will not be able to use the test statistic except perhaps approximately. We thus inevitably end up with a test whose size is only approximately what we claim and which is not the most powerful at picking up all kinds of departures from $\mathcal{H}_0$.

As a simple example, take two Poisson observations Y_1, Y_2 with means μ_1, μ_2. The hypothesis $\mathcal{H}_0 : \mu_1 = \mu_2$ is a composite null hypothesis since the common value μ is left unspecified. The natural test statistic is $Y_1 - Y_2$, which is distributed as the difference of two independent $Pn(\mu)$ variables, which still depends on the common mean μ. Thus, $Y_1 - Y_2$ cannot form the basis of an exact test. We might estimate μ by the average of Y_1 and Y_2 and construct the approximately standard normal statistic $(Y_1 - Y_2)/\sqrt{2\bar{Y}}$. The exact distribution of this statistic however still depends on μ, but only slightly if μ is large. As another example, consider the simple hypothesis $\mathcal{H}_0 : (\mu_1, \mu_2) = (15, 40)$. Because this is a test of more than one parameter there is no best test statistic, but common sense suggests using an *increasing* combination of $D_1 = Y_1 - 15$

and $D_2 = Y_2 - 40$ such as

$$T = (Y_1 - 15)^2 + (Y_2 - 40)^2 = D_1^2 + D_2^2.$$

While the null distribution of this statistic is known in principle, it is difficult to work with and would probably need approximation in practice. More importantly, T is not most powerful. For instance, if we were primarily interested in departures of μ_1 from 15, the test statistic D_1 would be more powerful, but is of course completely blind to the value of μ_2. The statistic T is a compromise between detecting departures of μ_1 from 15 and of μ_2 from 40. However, it will actually be less sensitive to departures of the latter type because Y_2 has larger variance than Y_1. The LR, Score, and Wald statistics are at least an automatic way of finding a test statistic "fair" to all departures and which is correspondingly quite powerful for most alternatives. The null distributions are also simply approximated by χ^2. For the present example the Score and LR test statistics are

$$S = \frac{D_1^2}{15} + \frac{D_2^2}{40}, \qquad LR = 2Y_1 \log\left(\frac{D_1}{15} + 1\right) + 2Y_2 \log\left(\frac{D_2}{40} + 1\right) - 2D_1 - 2D_2.$$

2.5.1 Likelihood Ratio Statistics

There are three general prescriptions for testing a general hypothesis specifying r restrictions on the parameters θ of a statistical model (see Section 1.4). Each may be applied to testing goodness-of-fit. The LR test statistic is

$$LR = 2(l_1 - l_0)$$

where ℓ_1 is the maximized log-likelihood under the more general hypothesis $\mathcal{H}_1$ and ℓ_0 is the maximized log-likelihood under the restrictive hypothesis $\mathcal{H}_0$ and is necessarily less than ℓ_1. Recall Eq. (2.2) for the log-likelihood under the multinomial model and Eq. (2.8) for the maximized log-likelihood under the full model. This gives the LR test statistic

$$LR_M = 2 \sum_{i=1}^{k} y_i \log\left(\frac{y_i}{\hat{e}_i}\right) \tag{2.21}$$

where the $\hat{e}_i$ are the fitted values $n\pi_i(\hat{\theta}_0)$ under the null hypothesis and the constants c in the log-likelihoods have canceled. This statistic may be used to test the goodness-of-fit of any multinomial model. If the number of parameters in the model is p (and the number of parameters in the full model is $k - 1$) then the degrees of freedom of the approximating chi-square distribution will be $r = k - 1 - p$. For Poisson models, Eqs. (2.11) and (2.17) give the LR test statistic

$$LR_P = 2 \sum_{i=1}^{k} y_i \log\left(\frac{y_i}{\hat{e}_i}\right) - 2 \sum_{i=1}^{k} (y_i - \hat{e}_i) \tag{2.22}$$

where $\hat{e}_i$ are the fitted values $\mu_i(\hat{\theta}_0)$ under the null hypothesis. This statistic may be used to test the goodness-of-fit of any Poisson model. If the number of parameters in the model is p (and the number of parameters in the full model is k) then the degrees of freedom of the approximating chi-square distribution will be $k - p$.

In most cases the fitted values under the Poisson model have the same total as the data and so the second term in Eq. (2.22) vanishes and the two LR goodness-of-fit statistics have identical form under the multinomial and Poisson models. As described in the last section, multinomial models may be fitted as Poisson models without affecting the fitted values. Thus, the calculated value of the LR goodness-of-fit statistic will be unaffected by whether we choose a multinomial or Poisson formulation of the model. Despite appearances, the degrees of freedom for the two formulations do agree; for any multinomial model the equivalent Poisson model always has one extra parameter, τ. The degrees of freedom for a multinomial model with p parameters is $(k - 1) - p$. The equivalent Poisson model has k data values and $p + 1$ parameters and so the degrees of freedom are $k - (p + 1)$.

The LR goodness-of-fit statistic is a measure of how close the vector of expected values is to the vector of observed values, and it will equal zero if, and only if, the two vectors agree exactly. Naturally there are many other ways of measuring the agreement of two vectors giving alternative statistics. However, only the LR statistic measures the relative likelihood of the observed data under the two hypotheses.

Example 2.2: (Continued) **Distribution of Genotypes.** Table 2.5 shows summary statistics for three models for the plant classification data. The maximized log-likelihoods were calculated earlier. The maximized likelihood ratio

Table 2.5. Comparison of Models for Plants/Genotype Data

	Fitted Values			GOF Statistics		
Genotype	aa	aA	AA	LR	df	P-Value
Full	18	55	27	0	0	
$p = 1/2$	25	50	25	4.085	2	0.245
p Free	20.70	49.60	29.70	1.813	1	0.562

relative to the full model (not the LR statistic), the degrees of freedom under a multinomial formulation, and the P-values from the asymptotic chi-square distribution, are also listed.

The LR test can also be used to test $\mathcal{H}_0 : p = 1/2$ model against the general binomial model, rather than against the full model. The ratio of maximized likelihoods is

$$\frac{L(\hat{\theta}_1)}{L(\hat{\theta}_0)} = \frac{L(\hat{\theta}_1)/L(\hat{\theta})}{L(\hat{\theta}_0)/L(\hat{\theta})} = \frac{4.085}{1.813} = 2.253$$

corresponding to an LR statistic $2\log(2.253) = 1.624$. The difference in the dimensions of the two models is 1 and so the P-value is $\Pr(\chi_1^2 > 1.624) = 0.202$. This measures the extent to which the data points against $p = 1/2$ *assuming a binomial* $(2, p)$ *model*. The goodness-of-fit test simultaneously measures this as well as the extent to which the data measures against the binomial model itself and gave the P-value 0.245. It will usually be possible to obtain more powerful tests by confining attention to a particular issue of interest (is $p = 1/2$?) than performing a goodness-of-fit test. Goodness-of-fit tests are important and indeed essential in their own right but they address general rather than specific aspects of the model.

2.5.2 Deviance and Lack-of-Fit

The likelihood ratio goodness-of-fit statistic for the model $\mathcal{M}$ is the likelihood ratio statistic for testing $\mathcal{M}$ against the full model $\mathcal{F}$, namely twice the difference in the maximized log-likelihoods or in symbols

$$D(\mathcal{M}) = 2(\ell_f - \ell_M). \tag{2.23}$$

Numerically it measures the difference between observed and fitted values. As such it measures the deviation of the observed data from model-based predictions and is thus commonly called the *deviance* of the model. In even simpler terms, deviance measures *lack-of-fit*. The concept of deviance allows a slightly different but useful interpretation of the general LR test. Recall that for testing a model $\mathcal{M}_0$ against a larger model $\mathcal{M}_1$ the LR statistic was $2(l_1 - l_0)$. This can be reexpressed as

$$LR(\mathcal{M}_0 | \mathcal{M}_1) = 2(\ell_f - \ell_0) - 2(\ell_f - \ell_1) = D(\mathcal{M}_0) - D(\mathcal{M}_1). \tag{2.24}$$

The deviance of model $\mathcal{M}_1$ will always be less than the deviance of the model $\mathcal{M}_0$ because $\mathcal{M}_1$ is a more general model. The LR statistic for testing whether $\mathcal{M}_0$ is an acceptable simplification of $\mathcal{M}_1$ is simply the increase in deviance that results from assuming $\mathcal{M}_0$. If this increase is not too large then we conclude that the extra assumption is justified. Likelihood ratio tests are for this reason also called *change in deviance* tests.

The equation $\mathcal{D}(\mathcal{M}_0) = \mathcal{D}(\mathcal{M}_1) + LR(\mathcal{M}_0|\mathcal{M}_1)$ says that the deviance of a more restrictive model $\mathcal{M}_0$ will be larger than the deviance of the more general model $\mathcal{M}_1$ by the LR statistic. Imagine a *nested* sequence of models $\mathcal{M}_0 \subset \mathcal{M}_1 \subset \ldots \subset \mathcal{M}_k$. Let us denote the LR statistic for testing $\mathcal{H}_0 = \mathcal{M}_{j-1}$ against $\mathcal{M}_j$ by $LR(\mathcal{M}_{j-1}|\mathcal{M}_j)$. Then substitution of the deviance relation into itself gives the formula

$$\mathcal{D}(\mathcal{M}_0) = LR(\mathcal{M}_0|\mathcal{M}_1) + LR(\mathcal{M}_1|\mathcal{M}_2) + \cdots + LR(\mathcal{M}_k - 1|\mathcal{M}_k) + \mathcal{D}(\mathcal{M}_k).$$

The deviances strictly increase from $\mathcal{D}(\mathcal{M}_k)$ to $\mathcal{D}(\mathcal{M}_0)$ as the model is made more and more restrictive. Each increase is an LR statistic for testing a particular model against the next one. Thus, the overall lack of fit of the most restrictive model $\mathcal{M}_0$ is broken up into a sum of terms each representing the lack of fit of a particular model compared to the next more general model. The partitioning of the deviance into separate and readily interpreted terms enhances its use as a data analytic tool.

Deviance is closely related to the residual sum of squares (RSS) in normal theory linear models. If RSS_0 and RSS_1 are the sums of squares of deviations of observed and expected values under two nested models $\mathcal{M}_0$ and $\mathcal{M}_1$ of dimensions p_0 and p_1, respectively, then it is simple to show (see Aitkin et al. 1989, pp. 87, 121) that the LR test statistic is

$$LR = n \log\left(\frac{RSS_0}{RSS_1}\right)$$

whose distribution according to the general theory is approximately chi-square with degrees of freedom $p_1 - p_0$. Inference for normal linear models could indeed be carried out on this basis however, because of the special location-scale structure of the normal distribution, the distribution of the LR statistic is actually *determined exactly* under $\mathcal{M}_0$. For historical reasons the statistic

$$F = \frac{(RSS_0 - RSS_1)/(p_1 - p_0)}{RSS_1/(n - p_1)}$$

is usually used and compared to the so-called Fisher or F-distribution. This statistic is a monotonic function of the LR statistic and it is entirely a matter of convention which statistic we use.

One important difference between deviance and RSS is that the latter involves an unknown scale parameter σ^2 in its distribution. Only *ratios* of residual sums of squares have a meaningful statistical interpretation and the deviance is actually the logarithm of such a ratio as given above. Nevertheless, both deviance and RSS are measures of the lack-of-fit of a statistical model. One source of confusion in this regard is the package GLIM which treats normal, multinomial, and Poisson models within the single framework of generalized linear models. The main output of a GLIM fitting procedure is the deviance statistic. Unfortunately, for normal models, the residual SS is output under this name while for all other models the LR statistic relative to the full model is output. The residual SS *is not the deviance of the model* regardless of the GLIM nomenclature.

2.5.3 Pearson's Goodness-of-Fit Statistic

For either a multinomial or Poisson formulated categorical data model, the Pearson goodness-of-fit statistic is

$$S = \sum_{i=1}^{k} \frac{(Y_i - \hat{e}_i)^2}{\hat{e}_i}. \tag{2.25}$$

The Pearson statistic is actually the score statistic as defined on p. 43.

The distribution is approximately χ_r^2 where r is the difference in the number of free parameters under the model to be tested and the full model. Some consider this one of the great discoveries of the twentieth century. The approximation applies an n increases for fixed k or, more accurately, as the smallest of the fitted values $\hat{e}_i$ grows unbounded. Under a simple model specifying the π_i, S has first three cumulants

$$\mathrm{E}(S) = k - 1, \mathrm{Var}(S) = 2(k-1) + O(n^{-1}), \mathrm{skew}(S) = 8(k-1) + O(n^{-1})$$

where the $O(n^{-1})$ terms increase in magnitude with k^2 and $\sum \pi_i^{-1}$ [see McCullagh and Nelder (1989, p. 169)]. Thus, the approximation to χ_k^2 will be poor when some of the π_i are very small, when the number of categories k is large, or when the total of the data n is not large enough. Substantially the same conclusions may be reached for nonsimple models.

Each contribution $(Y_i - \hat{e}_i)^2/\hat{e}_i$ is the square of an approximately standard normal variable. Since Y_i is discrete, this approximation is apparently improved by continuity correction. Thus, the corrected Pearson statistic

$$S = \sum_{i=1}^{k} \frac{(|Y_i - \hat{e}_i| - 1/2)^2}{\hat{e}_i} \tag{2.26}$$

is often used. The correction of 1/2 was given by Yates (1934) and is often called Yates' correction.

When are S and LR Close? In Section 1.5.4 we compared LR, Score, and Wald statistics in general. We found that they are asymptotically equal and numerically close when the ML estimators $\hat{\theta}$ and $\hat{\theta}_0$ are close. In the context of goodness-of-fit, this means the data and fitted values are close. We concluded that LR and Score statistics are numerically close when they take small values but may differ when they take large values. Thus, typically there will be little practical difference between them. It is useful to give an explicit expansion of the LR statistics [see Eqs. (2.21) and (2.22)] about the Pearson statistic [see Eq. (2.25)], and to examine the error terms. Consider the expansion

$$\log(1+x) = x - \frac{x^2}{2} + \frac{x^3}{3} + O(x^4)$$

valid for real values of x small in absolute value. Let $\delta_i = y_i - \hat{e}_i$ and $z_i = \delta_i/\sqrt{\hat{e}_i}$ be ordinary and standardized "residuals." Then approximately

$$2y_i \log\left(\frac{y_i}{\hat{e}_i}\right) \approx 2\delta_i + \frac{(y_i - \hat{e}_i)^2}{\hat{e}_i} - \frac{z_i^3}{3\sqrt{\hat{e}_i}} + R_i \tag{2.27}$$

where the R_i are small remainders. Summing these over i and noting that the δ_i sum to zero for multinomial models we find that

$$LR = S - \sum_{i=1}^{k} \frac{z_i^3}{3\sqrt{\hat{e}_i}} + R \tag{2.28}$$

good for both multinomial and Poisson models. The error term will be a sum of terms of order $z_i^4/\hat{e}_i$. It is clear that for the two statistics to agree either (i) the z_i must all be small, i.e., the fitted and observed values are not too different or (ii) the fitted values under the null hypothesis must all be large. It also helps if the z_i are symmetric about zero. This will only be true for fixed k. If k is allowed to increase then even small error terms might accumulate into a more significant discrepancy.

Example 2.3: (Continued) **Seasonality of Births.** Table 2.6 gives contributions to the Pearson statistic, called z_i above, and contributions to the multinomial and Poisson LR statistics, the difference between these last two being the doubled "residuals" $2\delta_i$. Since the fitted and observed values agree in total, the values of these two statistics are identical. However, only the Poisson contributions agree closely with the Pearson contributions.

The Pearson and LR statistics are 13.848 and 13.486 giving P-values 0.0031 and 0.0037, respectively (using χ_3^2). These values are quite close despite there being some large z_i. The reason is that the fitted values (37.75) are all large.

Table 2.6. Goodness-of-Fit Statistics for Birth Data, I

Quarter	January	April	July	October	Total
Observed (y_i)	55	29	26	41	151
Fitted $(\hat{e}_i)$	37.75	37.75	37.75	37.75	151
$LR_{Mi} = 2y_i \log(y_i/\hat{e}_i)$	41.40	−15.29	−19.39	6.77	13.486
$LR_{Pi} = LR_{Mi} - 2\delta_i$	6.90	2.21	4.11	0.27	13.486
$z_i^2 = (y_i - \hat{e}_i)^2/\hat{e}_i$	7.88	2.03	3.66	0.28	13.848
% Difference	−12.9	8.8	12.4	−2.7	−2.6

The agreement between the contributions is much better for smaller than for larger contributions.

As a final illustration, consider the hypothesis that birth rates in the first quarter are twice the rate of the other three. One possible model for this supposes means $(2\mu, \mu, \mu, \mu)$ for the four cells and fitting this Poisson model gives fitted values and contributions to the goodness-of-fit statistics shown in Table 2.7. Despite the large expected frequencies and the smaller values of z_i, the agreement between the two statistics is not quite as good as for the previous model because the z_i are not as symmetric about zero. The P-values associated with the Pearson and LR tests are 0.201 and 0.173, respectively (using χ_3^2). The only point of conflict is the fourth cell where the fitted value is much too low.

There is not much evidence against this model. However, even the P-values 0.201 and 0.173 overestimate the evidence against the double-birth-rate hypothesis. This is because we have assumed an equal rate for the other three quarters and so these P-values also contain evidence against this assumption. We can construct a better (i.e., more specific) test by fitting the model which specifies that the rate in the first quarter is twice the *average* rates of the last three quarters which is equivalent to parametrizing the cells means as

$$\mu_1 = 2\mu, \mu_2 = 3\mu\lambda_1, \mu_3 = 3\mu\lambda_2, \mu_4 = 3\mu(1 - \lambda_1 - \lambda_2).$$

The fitted values turn out to be 60.4, 27.37, 24.54, 38.69 and the LR goodness-of-fit statistic is 0.813 on 1 degree of freedom giving the P-value 0.367. The likelihood equations for this model cannot be solved directly.

Table 2.7. Goodness-of-Fit Statistics for Birth Data, II

Quarter	January	April	July	October	Total
Observed (y_i)	55	29	26	41	151
Fitted $(\hat{e}_i)$	60.4	30.2	30.2	30.2	151
$z_i^2 = (y_i - \hat{e}_i)^2/\hat{e}_i$	0.48	0.05	0.58	3.86	4.997
$LR_{Pi} = LR_{Mi} - 2\delta_i$	0.50	0.05	0.61	3.46	4.629
% Difference	3.1	1.3	5.0	−10.2	−7.0

For a sequence of nested models the Pearson statistic does not divide into separate terms for testing each model compared to the next more general one as was the case for the LR statistic. However, in as much as the Pearson goodness-of-fit statistic is often close to the deviance, the partition may hold approximately (provided the conditions are appropriate for *all* the nested models). In fact it is possible for the Pearson goodness-of-fit statistic to be larger for $\mathcal{H}_1$ than for $\mathcal{H}_0$ particularly for models where the expected values e_i are too small. For this reason a better Pearson analogue of the LR test of $\mathcal{H}_0$ against $\mathcal{H}_1$ is not the difference in Pearson statistics but

$$S(\mathcal{H}_0, \mathcal{H}_1) = \sum_{i=1}^{k} \frac{(\hat{e}_{i1} - \hat{e}_{i0})^2}{\hat{e}_{i0}}$$

where $\hat{e}_{i1}$, $\hat{e}_{i0}$ are fitted values under $\mathcal{H}_1$ and $\mathcal{H}_0$, respectively.

2.5.4 Other Goodness-of-Fit Statistics

Consider the formula for Pearson's chi-square where the data Y_i are Poisson with parameter μ_i. For large μ_i, each Y_i is approximately normal with mean and variance μ_i and so

$$\left(\frac{Y_i - \mu_i}{\sqrt{\mu_i}}\right)^2$$

is approximately χ_1^2. The sum of squares of these will be approximately χ_k^2. When the μ_i are estimated then a degree of freedom is lost for each estimated parameter. The discussion of transformations in Section 1.6 made it clear that smooth transformations do not invalidate the normal limiting distribution. Thus, if g is any smooth function then $g(Y_i)$ is approximately normal with mean $g(\mu_i)$, variance $g'^2(\mu_i)\mu$, and so

$$\left\{\frac{g(Y_i) - g(\mu_i)}{\sqrt{\mu_i} g'(\mu_i)}\right\}^2$$

will be approximately chi-square. It follows that the Pearson chi-square is a special case of the family

$$S_g = \sum_{i=1}^{k} \frac{\{g(Y_i) - g(\hat{e}_i)\}^2}{g'^2(\hat{e}_i)\hat{e}_i} \tag{2.29}$$

all of which have asymptotic χ^2_{k-p} distributions. The distribution will be closer to chi-square if the underlying normal approximation to $g(Y_i)$ can be improved. For instance, the Poisson distribution has positive standardized skewness $1/\sqrt{\mu}$ which could be nonnegligible for moderate μ which suggests tail-reducing transformations such as log or square root. Now

$$\log(Y_i) \xrightarrow{d} \mathcal{N}\left(\log \mu_i, \frac{1}{\mu_i}\right), \qquad \sqrt{Y_i} \xrightarrow{d} \mathcal{N}\left(\sqrt{\mu_i}, \frac{1}{4}\right)$$

which in turn give the goodness-of-fit statistics

$$S_{\log} = \sum_{i=1}^{k} \hat{e}_i \left\{\log\left(\frac{y_i}{\hat{e}_i}\right)\right\}^2, \qquad S_{root} = 4 \sum_{i=1}^{k} y_i + \hat{e}_i - 2\sqrt{y_i \hat{e}_i}.$$

As with the LR and Pearson statistics, for fixed k these statistics will be close whenever the z_i are small or the $\hat{e}_i$ are large.

Another family of statistics, called *power divergence statistics*, was defined by Cressie and Read (1984). For a specified real parameter λ the statistic

$$\chi^2(\lambda) = \frac{2}{\lambda(\lambda+1)} \sum_{i=1}^{k} y_i \left\{\left(\frac{y_i}{\hat{e}_i}\right)^{\lambda} - 1\right\} \tag{2.30}$$

measures the distance of the vector of fitted values $\hat{e}_i$ from the vector of observed values y_i. The family includes several well-known test statistics as special cases; when $\lambda = 0$ it becomes the LR statistic; when $\lambda = 1$ it is the Pearson statistic; when $\lambda = 1/2$ it is the Freeman–Tukey statistic; when $\lambda = -1$ it is the modified LR statistic, where the roles of y_i and $\hat{e}_i$ are reversed, as proposed by Kullback (1959); when $\lambda = -2$ it is the Neyman or modified Pearson statistic, as proposed by Neyman (1949). For all finite values of λ, the divergence statistic is asymptotically chi-square for fixed k.

For extreme values of λ the divergence statistic becomes highly sensitive to a small number of cells and Cressie and Read (1984) suggested the interval $[0, 3/2]$ for λ which excludes the modified LR and modified Pearson statistics mentioned above. On the basis of different measures of power, they recommended

$$\chi^2(2/3) = \frac{9}{5} \sum_{i=1}^{k} y_i \left\{\left(\frac{y_i}{\hat{e}_i}\right)^{2/3} - 1\right\} \tag{2.31}$$

as a test statistic with both high power and accurate size and claimed good operational properties when the fitted values are all larger than 1 and when

Table 2.8. Goodness-of-Fit Statistics for Birth Data, III

Quarter	January	April	July	October	Total	P-value
Log-transform	0.48	0.05	0.58	3.83	4.945	0.176
Square-root	0.51	0.05	0.63	3.29	4.478	0.214
Modified Pearson	0.53	0.05	0.68	2.84	4.103	0.250
Modified LR	−11.31	−2.45	−9.04	−18.46	4.341	0.227
CR ($\lambda = 2/3$)	−5.99	−1.39	−4.45	16.86	4.853	0.183

$n \geq 10$. This is a compromise between the Pearson and LR statistics. We refer to this statistic as the CR statistic.

Example 2.3: (Continued) **Seasonality of Births.** For the birth data under the model specifying means 2μ, μ, μ, μ, contributions to various chi-square test statistics are given in Table 2.8. These may be further compared with the contributions in Table 2.7. The contributions to the divergence statistic are not necessarily interpretable as standardized residuals (except when $\lambda = \pm 1$). A plot of $\chi^2(\lambda)$ versus λ displayed in Figure 2.1 reveals a function minimized (with minimum value around 3.355) for large negative λ and increasing without limit as λ increases. As noted above however, extreme values of λ invalidate the null chi-square distribution and should not be used.

2.5.5 Approximate Null and Nonnull Distributions*

Accuracy of Approximate Null Distribution When the null hypothesis is true, the distribution of the various goodness-of-fit statistics described are all approx-

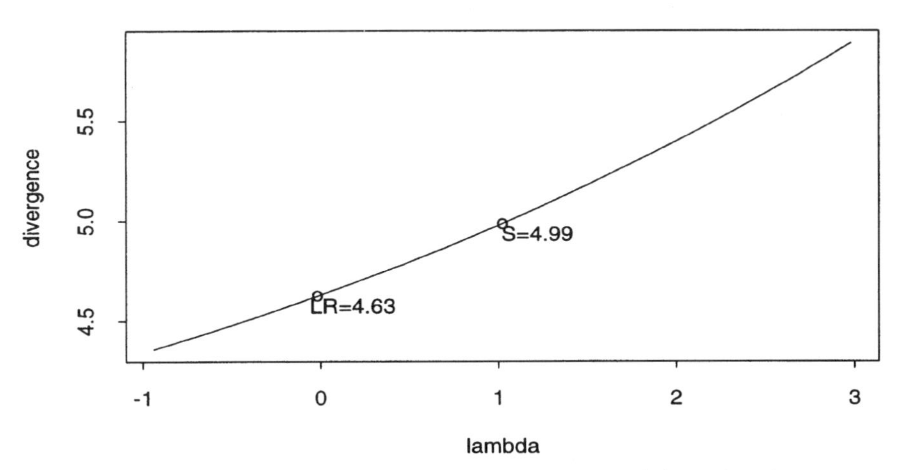

Figure 2.1. CR statistic for varying λ. Plot of power divergence statistics against the parameter λ for testing the goodness-of-fit of "double-birth rate in the January quarter" hypothesis.

*See p. 15 in Chapter 1 for an explanation.

imately χ^2_{k-p} provided the number of cells k is fixed while the expected values grow unbounded. This only holds in the limit however and the different statistics approach the limit at different rates under different conditions. Simulation studies have demonstrated that the Pearson statistic usually converges to chi-square more quickly than the LR statistic which tends to be too conservative.

It is by no means uncommon for data to be grouped into a large number k of cells and for the corresponding fitted values to be rather small. For instance, if data are classified according to several factors the resulting *contigency table* will have many cells each with a small expected and fitted value. Asymptotic results based on large expected values are thus of little relevance. Such data sets are called *sparse* and the null distributions will not be well described by the chi-square limit. There is again simulation evidence (see Further Reading in Chapter 1 p. 58) that the Pearson statistic is less affected by sparseness than the LR statistic, especially if the expected values are *uniformly* small across cells. The CR statistic is also less susceptible to sparseness effects than the LR statistic. However, for sparse enough tables the chi-square approximation breaks down for any of the asymptotically equivalent statistics we have seen.

One way to avoid these difficulties is to accumulate cells until their expected values become large enough for the chi-square limit to be accurate. We will need to subtract one degree of freedom for each accumulated cell. There are several shortcomings with recommending this in general. First, there may be several ways of accumulating the cells leading to different results. More importantly, information regarding the fit of the model to these cells separately will be lost. Second, the overall power of the test may be drastically reduced especially if there are few degrees of freedom to begin with. In principle it would seem preferable to accept the failure of the chi-square limit and to use a more accurate approximation to the null distribution without accumulating cells.

Example 2.4: Poor χ^2 Approximation The data in Table 2.9 are from Graubard and Korn (1987). They refer to 32,574 mothers, 93 of whom gave birth to babies with a congenital malformation. The data were further classified according to self-reported drinking frequency during pregnancy.

Analysis of this data set will illustrate several ideas. First, it is the first example of a two-sample test—a comparison of one multinomial distribution with

Table 2.9. Maternal Drinking and Congenital Malformation

	Drinks Per Day					
Malformation	0	<1	1–2	3–5	≥6	Total
Absent	17,066	14,464	788	126	37	32,481
Present	48	38	5	1	1	93
Total	17,114	14,502	793	127	38	32,574

another, and gives a concrete example where the LR and Pearson statistics take quite different values. Second, the chi-square approximation to both these statistics' null distributions is poor.

A natural question to ask is whether or not the drinking distribution is the same for mothers of malformed and normal babies. A preponderance of heavier drinkers for mothers of malformed babies might suggest an association of these two factors. Our data comprise two independent data sets, each giving the observed drinking distribution across five drinking categories. Under the full model there are 5 – 1 = 4 parameters for each set, giving 8 parameters in all. Call the vector of probabilities for the first row π_A and the vector of probabilities for the second row π_P. Then the null hypothesis of no association of the two factors is

$$\mathcal{H}_0 : \pi_A = \pi_P.$$

Under this model there is a single common probability vector π for both rows. Fitting the null model is quite easy. On writing down the likelihood we obtain a factor

$$\pi_1^{Y_1}\pi_2^{Y_2}\pi_3^{Y_3}\pi_4^{Y_4}\pi_5^{Y_5}$$

for each row and multiplying these together we get the likelihood

$$\pi_1^{17114}\pi_2^{14502}\pi_3^{793}\pi_4^{127}\pi_5^{38}.$$

This is exactly the same likelihood that we would obtain if we just fitted the full multinomial distribution to the bottom row. The estimates of the probability parameters would thus be the counts divided by the total 32,574. Expected values in the main body of the table are easily calculated. For instance, in the third cell of the second row the expected value under $\mathcal{H}_0$ is $93\pi_3$ and $\hat{\pi}_3 = 793/32{,}574$. Thus

$$\hat{e}_{23} = \frac{93 \times 793}{32{,}574} = 2.264$$

and this pattern (row total times column total divided by grand total) is repeated all over the table. Table 2.10 gives the fitted values for all 10 cells. To test the hypothesis that malformation and drinking are independent we could calculate any of the goodness-of-fit statistics described earlier and refer them to the χ^2 distribution. The degrees of freedom is 8 – 4 = 4. Contributions to the LR statistic and the Pearson statistic are also listed in Table 2.10, their totals being respectively, 6.20 and 12.08. The approximate P-values associated with these

Table 2.10. Fit of Independence Model to Malformation Data

	Drinks Per Day				
Absent ($\hat{e}_i$)	17065.1	14460.6	790.7	126.6	37.9
LR_{Pi}	−0.00	0.00	0.01	0.00	0.02
z_i^2	0.00	0.00	0.01	0.00	0.02
Present ($\hat{e}_i$)	48.9	41.4	2.3	0.36	0.11
LR_{Pi}	0.01	0.29	2.45	0.75	2.66
z_i	0.01	0.28	3.31	1.12	7.32

statistics are 0.184 and 0.017; the Pearson statistic indicates strong evidence against independence while the LR indicates almost none.

The P-values based on the χ_4^2 distribution are in fact extremely inaccurate. A simulation revealed that the true distribution of the Pearson statistic has a mean 4.716, standard deviation 5.316, and exceeds the 95% quantile of the χ_4^2 distribution with probability 0.102. Tests based on the Pearson statistic with chi-square approximation are therefore far too liberal. The true P-value associated with the observed value 12.08 is 0.067, not 0.017. The breakdown of the chi-square limit for the Pearson statistic is consistent with our earlier claim that the approximation is only accurate for sparse tables when the sparseness is uniform. In this data set there is a huge variation in the fitted values, from 17,065 down to 0.1. The same simulation revealed that the LR statistic has a mean 3.594 and standard deviation 2.427 and exceeds the 95% quantile of the χ_4^2 distribution with probability 0.025. Tests based on the LR statistic with chi-square approximation are therefore far too conservative. The true P-value associated with the observed value 6.20 is 0.133, not 0.185.

One way to make the two statistics agree better, both with each other and their notional chi-square distributions, is to accumulate the last three cells into a single cell. This leaves only two degrees of freedom. The values of the LR and Pearson statistics are then 4.979 and 7.007 with associated approximated P-values 0.083 and 0.030. There is still significant disagreement between the two statistics caused by the accumulated three cells still having a small fitted value of 2.73. If we accumulate the last four rather than three cells then the two statistics agree almost precisely, taking the common value 0.066. However, we have entirely lost the information regarding dependence of the drinking and malformation contained in the four accumulated cells. Accumulation does not seem a good idea for this data set bearing in mind the small degrees of freedom.

Notwithstanding their asymptotic equivalence, there is no avoiding the fact that the Pearson statistic is pointing more strongly against independence than the LR test. There is no way to resolve the disagreement and there simply is no correct answer. The example is revisited in Chapter 7 where it is reanalyzed using *conditional* methods.

Nonnull Distributions for LR and S The relative power of the two tests depends on the relative sizes of the noncentrality parameters in the nonnull distributions as described on p. 47. Suppose e_0 is the null mean of Y and e the nonnull mean. It can be shown that for $e - e_0 = O(n^{-1/2})$ the distributions of S and LR are noncentral chi-square with respective noncentrality

$$c_S^2 = \sum_{i=1}^{k} \frac{(e_i - e_{i0})^2}{e_{i0}}, \qquad c_{LR}^2 = 2 \sum_{i=1}^{l} e_i \log\left(\frac{e_i}{e_{i0}} \right). \tag{2.32}$$

The larger the noncentrality the greater will be the power. When the true model summarized by e is more remote from e_0 the statistics will still have the correct mean and variance for noncentral chi-square with the above parameters. It is apparent from these results that very general statements about the relative power of the two tests cannot be made. The results given here were proven by Mitra (1958) for the Pearson statistic and by Haberman (1974) for the LR statistic. More recent research has given results for more remote sequences of alternatives [see Drost et al. (1989)].

2.6 CONFIDENCE INTERVALS

Once a model is deemed adequate we will still need to describe and quantify our uncertainty about the fit of the model. Estimation of parameters has already been dealt with and is an automatic part of the fitting procedure. This section examines both exact and approximate methods of constructing confidence intervals for the parameters of Poisson and multinomial models. All the methods described are unified by the idea of *inverting a test*, and it is worthwhile to go back and read Section 1.5.8.

2.6.1 Exact Intervals

To construct an *exact* interval for ω requires a test statistic $T(Y, \omega)$ whose distribution $F(t; \omega)$ depends on ω only and on no other (nuisance) parameters of the model. For testing $\mathcal{H}_0 : \omega = \omega_0$ against $\mathcal{H}_1 : \omega < \omega_0$ the P-value is

$$\Pr(T \leq t) = F(t; \omega_0).$$

Recalling the relation of P-values and confidence intervals summarized in Eq. (1.41) in Section 1.5.8, the region

$$\{\omega : F(t; \omega) > \alpha_1\} \tag{2.33}$$

will be a $100(1 - \alpha_1)\%$ confidence set for ω. Typically $F(t; \omega)$ will be a decreasing function of ω and so the set is of the form $(-\infty, U(t)]$, in other words the solution of $F(t;\omega) = \alpha_1$ is an upper limit for ω. Similarly, for testing against $\mathcal{H}_1 : \omega > \omega_0$ the P-value is $\Pr(T \geq t)$. If t^- is the next lower possible value of T then this equals $1 - F(t^-;\omega_0)$. The region

$$\{\omega : 1 - F(t^-;\omega) > \alpha_2\} = \{\omega : F(t;\omega) < 1 - \alpha_2\} \tag{2.34}$$

will be a $100(1 - \alpha_2)\%$ confidence set for ω. If F is decreasing in ω then this set is of the form $[L(t), \infty)$ where $L(t)$ solves $F(t;\omega) = 1 - \alpha_2$. The intersection $[L(t), U(t)]$ of these two sets is a confidence interval for ω with coverage $1 - \alpha_1 - \alpha_2$.

As explained in Section 1.5.8, when the statistic is discrete the interval generated will be *conservative*, i.e., the coverage will be $\geq 1 - \alpha_1 - \alpha_2$. The name *exact* is unfortunate for this reason. However, the *worst* coverage of the interval (as a function of ω) will equal $1 - \alpha_1 - \alpha_2$ exactly. A further slight complication is that when the observed value of $\hat{\omega}$ is the most extreme possible then it will only be possible to solve one of the above equations leading to an exact one-sided interval.

The *shape* of the interval is determined by other considerations. A probabilistically symmetric interval has $\alpha_1 = \alpha_2$ and consequently the probability of the interval failing on the upper or lower end is the same. This is a very common but not necessarily best choice.

Example: Exact Interval for a Binomial Probability Suppose Y_1 has binomial distribution with parameters n and π_1. Since larger/smaller values of Y_1 indicate larger/smaller values of p_1, let us take $T = Y_1$ as our test statistic. To obtain the upper limit for π_1 we solve the equation $\Pr(T \leq t) = \alpha$. For instance, on observing $y_1 = 1$ success from $n = 10$ trials, a 2.5% upper limit for π_1 solves

$$\Pr(Y_1 \leq 1) = (1 - \pi_1)^{10} + 10\pi_1(1 - \pi_1)^9 = 0.025.$$

This nonlinear equation can be solved, for instance, using the Newton algorithm and gives the answer 0.4450. The next possible lower value of y_1 is $t^- = 0$. To obtain the lower limit for π_1 we solve

$$\Pr(Y_1 \leq 0) = (1 - \pi_1)^{10} = 0.975$$

which can be solved directly and gives 0.00253. Thus (0.00253, 0.4450) is the exact 95% confidence interval which accompanies the estimate 0.1 of π_1. When y_1 equals 0 or n then only a one-sided interval can be computed.

Example: Exact Interval for a Poisson Mean Suppose Y_1 is Poisson with mean parameter μ_1 and we observe $y_1 = 11$ events. Again take $T = Y_1$. The 5%

upper limit for μ_1 solves

$$\Pr(Y_1 \le 11) = e^{-\mu} \sum_{i=0}^{11} \mu^i/i! = 0.05$$

and has the solution 18.207. The lower 5% limit solves the equation

$$\Pr(Y_1 \le 10) = e^{-\mu} \sum_{i=0}^{10} \mu^i/i! = 0.95$$

with solution 6.169. Thus, an exact 90% interval for μ is (6.169, 18.207).

In practice, the distribution of a test statistic T is first very complicated and second depends on nuisance parameters. The exact construction can then not be applied. However, the nuisance parameters in the distribution can often be made to disappear by instead working with the conditional distribution given a sufficient statistic. The exact construction can then be applied using the conditional distribution but is often a very complicated calculation. Conditional methods are treated separately in Chapter 7.

2.6.2 Approximate Likelihood Based Intervals

The LR statistic has many attractive general properties as described in Chapter 1. For multinomial data it is given by

$$LR(Y, \omega_0) = 2 \sum_{i=1}^{k} \hat{e}_{1i} \log(\hat{e}_{1i}/\hat{e}_{i0})$$

where $\hat{e}_1$ are the fitted values under the assumed model and $\hat{e}_{0i}$ are the fitted values under the null hypothesis $\omega = \omega_0$. For Poisson data there is an extra $2\sum(\hat{e}_{1i} - \hat{e}_{i0})$ term [see Eq. (2.22)]. Since the null hypothesis specifies a single restriction on the parameters the approximate distribution is χ_1^2.

The approximate P-value for testing $\omega = \omega_0$ is $\Pr(\chi_1^2 \ge t(y, \omega_0))$ where $t(y, \omega_0)$ is the observed value of the LR statistic. This P-value is greater than α if, and only if, $t(y, \omega_0)$ is smaller than the upper $1 - \alpha$ quantile of the χ_1^2 distribution. Since χ_1^2 is the square of a standard normal the set

$$\{\omega_0 : LR(Y, \omega_0) \le z_{\alpha/2}^2\}$$

where $z_{\alpha/2}$ is a quantile of the normal distribution, and defines a $100(1 - \alpha)\%$ confidence region for ω_0. This set is usually, but not necessarily, an interval.

Likelihood based intervals have the logically attractive property of only including values of ω that are sufficiently supported by the data.

Example: Likelihood Interval for a Binomial Probability For a single observation $y = 1$ on the binomial distribution with parameter $n = 10$ and π, the LR statistics for testing $\pi = \pi_0$ is

$$LR(Y, \pi_0) = 2Y \log\left\{\frac{Y}{n\pi_0}\right\} + 2(n - Y)\log\left\{\frac{n - Y}{n(1 - \pi_0)}\right\}$$
$$= 2\log\left(\frac{1}{10\pi_0}\right) + 18\log\left(\frac{9}{10(1 - \pi_0)}\right)$$

since the null fitted values are $\hat{e}_1 = n\pi_0$, $\hat{e}_2 = n(1 - \pi_0)$. A 95% confidence interval for π is obtained by plotting this function and reading of the interval of values for which it is smaller than 3.841, being the 95% quantile of χ_1^2. This gives the interval (0.00597, 0.372).

It was pointed out in Section 1.5.3 that the LR statistic can be expressed

$$LR(\omega) = 2\{\ell(\hat{\omega}, \hat{\psi}) - \ell(\omega, \hat{\psi}(\omega))\} \tag{2.35}$$

where $\hat{\psi}(\omega)$ is the ML estimate of the nuisance parameters ψ when the interest parameter ω is assumed to be known. The partially maximized log-likelihood function $\hat{\ell}(\omega) := \ell(\omega, \hat{\psi}(\omega))$ was called the *profile log-likelihood*. It follows that a confidence interval for ω is obtained by reading off all those values for which the profile log-likelihood $\ell(\omega, \hat{\psi}(\omega))$ is within *half* a χ_1^2 quantile of its maximum value. We use this formulation in the next example.

Example: Likelihood Interval for a Poisson Mean Consider a single observation $y = 11$ with Poisson distribution and unknown mean parameter. Since there is no nuisance parameter the profile log-likelihood is just the ordinary log-likelihood, given by

$$\ell(\mu) = y \log \mu - \mu = 11 \log \mu - \mu. \tag{2.36}$$

The maximum value of this is 11 log 11 − 11 = 15.377. For a 90% interval we find all those values for which $\ell(\mu)$ is within 2.706/2 = 1.353, i.e., larger than 15.377 − 1.353 = 14.024. This gives the interval (6.407, 17.392).

Standard packages do *not* routinely compute likelihood intervals but they do output the deviance statistic of the fitted model. To compute a likelihood interval for ω within a model $\mathcal{M}$ first fit the model $\mathcal{M}$ and record the deviance $\mathcal{D}$. Then impose a known value for ω, for instance, using the `OFFSET` command in `GLIM`. This necessarily gives a larger deviance. By experimenting with the imposed value of ω one may find those values for which the deviance exceeds $\mathcal{D}$ by less than a χ_1^2 quantile. A **GLIM** macro is given by Aitkin et al. (1989, Appendix 4).

The profile log-likelihood summarizes our knowledge of the parameter ω and should be used in analysis reports where a parameter is of particular interest and the function is far from quadratic. It is rather rare that a confidence set for several parameters simultaneously is desired but in principle this presents no extra problems. The profile log-likelihood $\ell(\omega)$ is now a function of a vector ω. Values of the vector ω for which $\ell(\omega)$ is within some specified value of its maximum will define a confidence region with approximately the correct coverage.

2.6.3 Wald-Type Intervals

A confidence interval for ω may be based on the ML estimator $\hat{\omega}$ and its standard error $\hat{\sigma}$, often computed from the inverse information matrix. This is a special case of a more general family of methods. For any monotone increasing transform g, we showed in Section 1.6 that $g(\hat{\Omega})$ has approximate normal distribution with mean $g(\omega)$ and standard error $|g'(\hat{\omega})|\hat{\sigma}$. Thus

$$T(Y, \omega_0) = \frac{g(\hat{\Omega}) - g(\omega_0)}{|g'(\hat{\Omega})|\hat{\sigma}}$$

is approximately standard normal which gives the confidence interval

$$g(\hat{\omega}) \pm z_{\alpha/2}|g'(\hat{\omega})|\hat{\sigma} \tag{2.37}$$

for $g(\omega)$. Transforming this interval by the inverse transform of g will give an interval for ω. When g is the identity this gives the standard $\hat{\omega} \pm 1.96\hat{\sigma}$.

The question is how to choose the transform g. Asymptotically, it makes no difference, which transform is taken although for moderate samples it can make a considerable difference. The Wald-type interval Eq. (2.37) will be close to the likelihood based interval only if the profile log-likelihood $\hat{\ell}(\omega)$, when plotted against $g(\omega)$, is close to quadratic. This approach to finding a suitable transform g has been pursued by Sprott (1972) amongst others. Anscombe (1964) has considered transformations that produce a statistic with skewness as small as possible but the transformation involved can be complex. For binomial data, the inverse sine and inverse standard normal distribution functions have often been used, the latter popularized by Finney (1971).

Probability parameters are central to binomial models. Standard intervals can include values outside [0, 1]. Many transforms can be used to avoid this problem but the most common and convenient is the *logistic* transform. Nonnegative counts have nonnegative means. Applying the *logarithmic* transform produces an interval that cannot include negative values. These two transformations are illustrated in the following examples.

Example: Wald-Type Intervals for a Binomial Probability For a single binomial observation $y = 1$, from $n = 10$ trials, the estimate of the success probability π is $\hat{\pi} = 0.1$ with standard error $\hat{\sigma} = \sqrt{\hat{\pi}(1-\hat{\pi})/n} = 0.0948$. Thus, the standard Wald-type 95% interval is $(0.1 \pm 1.96 \times 0.0948) = (-0.086, 0.286)$ which includes a negative probability. The logistic transform, studied extensively in later chapters, is defined by

$$g(\pi) = \log\left(\frac{\pi}{1-\pi}\right).$$

The ML estimator of $g(\pi)$ is $\log(y/(n-y)) = -2.197$. The standard error of this estimator is $g'(\hat{\pi})\hat{\sigma} = 1/\sqrt{n\hat{\pi}(1-\hat{\pi})} = 1.054$. Thus, a 90% confidence interval for $g(\pi)$ is $(-2.197 \pm 1.645 \times 1.054) = (-4.263, -0.131)$. This can be turned into a confidence interval for π by applying the reverse transform $g^{-1}(x) = \exp(x)/(1+\exp(x))$. This gives the interval (0.0139, 0.467). Apart from the cases $y = 0$ or $y = n$ when the estimator of $g(\pi)$ is undefined, this method will *always* give an interval inside [0,1].

Example: Wald-Type Intervals for a Poisson Mean On observing $y = 11$ for a Poisson variable with unknown *positive* mean μ the ML estimate is 11 with standard error $\hat{\sigma} = \sqrt{\hat{\mu}} = 3.316$. Thus, the standard Wald-type 95% interval is $(11 \pm 1.96 \times 3.316) = (5.545, 16.455)$. Even though this does not include any negative values, it is clear that if we wanted an interval with much larger coverage, so that 1.96 would be replaced by a larger value, the above construction may violate the condition $\mu > 0$. The logarithmic transform $g(\mu) = \log(\mu)$ avoids this problem. We estimate $g(\mu)$ by $\log 11 = 2.398$. The standard error of this estimator is $g'(\hat{\mu})\hat{\sigma} = 1/\sqrt{\hat{\mu}} = 0.3015$. Thus, an approximate 95% interval for $\log \mu$ is $(2.398 \pm 1.96 \times 0.3015) = (1.902, 2.894)$. Exponentiating this gives the interval (6.699, 18.063) for μ.

Wald-type intervals can be constructed directly, whereas the exact and likelihood based intervals require interative computations. For the simplest cases of a binomial probability and Poisson mean illustrated above, the *Splus* functions `exact.ci.for.p`, `exact.ci.for.mu`, `profile.ci.for.p`, `profile.ci.for.mu`, found at the Wiley website, perform the required computations.

FURTHER READING

Section 2.5: Goodness-of-Fit

2.5.3 *The landmark paper of Karl Pearson (1900) introduced the statistic Eq. (2.25) to measure the overall discrepancy of a frequency distribution and an hypothesized distribution. More importantly, the distribution was given as χ^2 with degrees of freedom (a concept as yet unknown) one less than the number of cells. Unfortunately, Pearson thought that the degrees of freedom need not change when param-*

eters of the model were estimated. Fisher demonstrated that one degree of freedom was lost for each parameter estimated, pointed out that these estimates had to be efficient, and also gave necessary modifications when the expected cell frequencies were too small.

A modified version of Pearson's statistic reverses the roles of y_i *and* $\hat{e}_i$ *and is known as the Neyman statistic (Neyman, 1949). It is in fact the Wald statistic [see Cox and Hinkley (1970, p. 326)]. The null distribution of this statistic tends to the chi-square limit more slowly than the Pearson statistic and for this reason is not recommended.*

2.5.5 *In the context of testing the goodness-of-fit of multinomial models with* k *diverging, the normal limit has been demonstrated in Koehler and Larntz (1980). Cressie and Read (1989) invesigated the asymptotics of their power divergence statistics as both* k *and* n *grow unbounded but* k/n *converges, and found that they converge to* different *normal distributions whose means and variances depend on the particular statistic used. They also found that the LR statistic converged faster than the Pearson statistic to its normal limit. Further research has examined asymptotic normality of the null distributions when the number of cells* k *becomes unbounded [see Holst (1972), Morris (1975), Moore and Spruill (1975), and Horn (1977)]. McCullagh (1986) developed asymptotic approximations for large* k *but to the distributions of the LR and Pearson statistics* conditional *on the value of the ML estimator* $\hat{\theta}$ *assumed to be sufficient. This distribution depends on no unknown parameters at all and is an example of an* exact *procedure; see Chapter 7. The exact (conditional) distributions were shown to be asymptotically normal and second-order adjustments were given for enhanced accuracy in extreme cases.*

Section 2.6: Confidence Intervals

2.6.1 *The "exact" interval for the binomial parameter is called the* Clopper–Pearson *interval. A plot of the coverage of this interval against* π *reveals that the coverage is usually much larger than the nominal. This conservatism can be reduced in various ways at the cost of some computation while still retaining the property of "exactness" [see Blyth and Still (1983), Casella (1986), and Blyth (1986)]. "Exact" confidence for the difference or ratio of two binomial parameters has been studied by Santner and Snell (1980) and for the difference of two correlated proportions by Lloyd (1990). When nuisance parameters are present, one considers the worst coverage as the nuisance parameter varies which gives a generalized version of the exact construction [see Kabaila and Lloyd (1997)]. A simple approximation to these intervals has been shown to work well for simple logistic and log-linear models and is feasible to compute for quite complex models [see Kabaila and Lloyd (1998)].*

2.7 EXERCISES

2.1. For each of the following data sets, identify the most natural response variable and classify it as binary, categorical, ordinal, or numeric.

a. A large population of individuals is sampled and their political preference recorded as well as their sex and exact age.

b. The number of insects who survive at various concentrations of two alternative insecticides is measured with a view to comparing their efficacies.

c. Coal miners have the quality of their lung function measured on a five-point scale from normal to chronic, as well as their age and the number of years mining.

d. A random sample of individuals is classified as left–right handed, left–right footed, and the same statistics for their biological parents were recorded.

e. Samples of wild dogs from two closely related species are classified by the darkness of their fur color with a view to identifying systematic differences between them.

f. Samples of Iris plants from three closely related species have the average width and length of their petals and sepals measured with a view to predicting the species from these measurements alone.

2.2. In the general multinomial model for $Y_1, \ldots, Y_k$ the random variables have separate $Bi(n, \pi_i)$ distributions but are constrained to add up to n. We might anticipate a negative correlation. This exercise develops a formula for the correlation of Y_1 and Y_2.

a. Conditional on $Y_1 = y_1$, argue that the remaining variables $Y_2, \ldots, Y_k$ have the multinomial distribution with parameters $n - y_1$, $k - 1$, and probability parameters $\pi_i/(1 - \pi_1)$, $i = 2, \ldots, k$. Hence, Y_2 has the $Bi(n - y_1, \pi_2/(1 - \pi_1))$ distribution. By using the formula $\mathrm{E}(Y_1 Y_2) = \mathrm{E}(\mathrm{E}(Y_1 Y_2 | Y_1))$ show that $\mathrm{Cov}(Y_1, Y_2) = -n\pi_1\pi_2$.

b. Show that the correlation of Y_1, Y_2 is given in the equation below and that contrary to appearances the correlation cannot exceed one in absolute value.

$$\rho(Y_1, Y_2) = -\sqrt{\frac{\pi_1\pi_2}{(1 - \pi_1)(1 - \pi_2)}}.$$

c. What does the correlation equation give when $k = 2$? Explain.

2.3. Section 2.2.3 gave equations for the score function and expected information matrix in a general multinomial model.

a. Show that the score function has mean zero.

b. Show that the expected information matrix is also the covariance matrix of the score function.

2.4. The set of data in Table 2.11 relates to a sample of locusts who arrive randomly in the collection area and are sampled with uniform probability p. They are then classified according to sex (male or female) and age (adolescent, maturing, adult). Call the observed counts $Y_1, \ldots, Y_6$.

Table 2.11. Sex and Age Data for Locusts

Males			Females		
Adolescent	Maturing	Adult	Adolescent	Maturing	Adult
22	19	16	29	23	14

a. What is the distribution of the counts conditional on the total? Interpret the probability parameters and relate them to the means of the Y_i. What is the distribution of the counts for the three age groups among the 57 male locusts? Interpret the probability parameters and relate them again to the means.

b. Estimate the proportion of males in the population who are adolescents and test the hypothesis that this proportion is greater than $1/3$ at the time of sampling.

c. Assuming that the locusts arrived in the collection area according to a Poisson process what are the distributions of the Y_i?

d. Test the hypothesis that the proportion of females in the population sample is greater than 50%.

2.5. Fictitious data below refer to motor accidents serious enough to involve hospitalization in Victoria over the second week of 1985.

Type	Number
Fatal driver	18
Nonfatal driver	66
Fatal pedestrian	16
Nonfatal pedestrian	3

a. Test the hypothesis that the long run proportion of nonfatal pedestrian accidents among all motor accidents is 1 in 10 against the hypothesis that it is less than this.

b. Give an exact 95% confidence interval for μ_3, the mean number of nonfatal pedestrian accidents during the period.

c. Give an exact 95% confidence interval for the long run proportion of fatal pedestrian accidents among all serious pedestrian accidents.

2.6 The data in Table 2.12 list fictitious counts of fatal air crashes in Australia by quarter over a 10-year period. The second row of data is over a 20-year period including the first ten-year period. Consider the hypothesis that the accident rates are uniform across these four quarters.

Table 2.12. Fatal Air Crashes by Quarter

	January	April	July	October
10 years	6	4	4	3
20 years	12	8	7	8

a. Calculate maximized log-likelihoods under the hypothesis and the full model for the first data set. How much less likely are the data assuming the hypothesis than not assuming it?

b. Repeat the above for the second larger data set.

c. Why are the data so much less likely for the second data set than the first under either hypothesis? Why is this difference largely unimportant and how does it relate to the arbitrary constant in the usual definition of the likelihood function?

2.7. Three types of Iris species are identified in a sample of 48. There are 13 of type 1, 14 of type 2, and 21 of type 3. It is desired to test whether or not the species occur in equal proportions and to investigate the statistical properties of the test(s). Assume an independent Poisson sampling model for the three counts.

a. Calculate LR, Pearson, and CR statistics for testing the goodness of fit of the equal proportions model and give approximate P-values.

b. Assume that the mean number of plants in each species is 16. Simulate the LR, Pearson, and CR statistics, compare the closeness of the distributions to χ_2^2, and give true P-values for the observed data.

c. From the simulations find 95% quantiles for the three statistics. Why can't an exact quantile be found? What is the size of the test which rejects the equal proportions hypothesis when the statistics exceed these quantiles?

d. Using the simulated quantiles as critical values, use simulation to investigate the power of the three statistics when the true means are (i) 13, 13, 22 and (ii) 10, 16, 22.

2.8. A set of Poisson random variables Y_i have means μ_i related to a covariate x_i by $\mu_i = ab^{x_i}$.

a. Interpret the parameters a and b in words.

b. Using Eq. (2.12) show that the fitted values e_i satisfy

$$\sum_{i=1}^{k}(y_i - e_i) = 0, \sum_{i=1}^{k} x_i(y_i - e_i) = 0.$$

c. Find expressions for the entries of the observed information matrix and conclude that this is the same as the expected information.

d. If Y_i is observed for the four values $x_i = 0, 1, 2, 3$, then show that the distribution of these counts conditional on the total is multinomial with probability parameters

$$\pi_i = \frac{\beta^i}{(1+\beta)(1+\beta^2)}.$$

2.9. According to Mendelian genetics, the four possible combinations of a dihybrid cross occur in the proportions 9 : 3 : 3 : 1. The data in Table 2.13 relate to tomato plants, the two characteristics being tall/dwarf and cut/potato leaf. How well do these data fit the theory?

Table 2.13. Tomato Dihybrid Cross Data [from Devore 1982a)]

Tall, Cut	Tall, Potato	Dwarf, Cut	Dwarf, Potato
926	288	293	104

2.10. Scientists self crossed red-seeded sorghum to produce red, yellow, and white seeded plants. According to genetic theory, these should occur in the ratio 9 : 3 : 4. How well do the data in Table 2.14, fit the theory?

Table 2.14. Self-Cross Experiment on Sorghum [from Devore (1982b)]

Red Seeds	Yellow Seeds	White Seeds
195	73	100

2.11. The family of power divergence statistics $\chi^2(\lambda)$ was defined in Eq. (2.30). This family includes many well-known goodness-of-fit statistics. Show that provided the totals of the observed and fitted values agree

a. $\chi^2(1) = \sum_{i=1}^{k}(y_i - \hat{e}_i)^2/\hat{e}_i$;

b. $\chi^2(0) = 2\sum_{i=1}^{k} y_i \log(y_i/\hat{e}_i)$;

c. $\chi^2(-1) = 2\sum_{i=1}^{k} \hat{e}_i \log(\hat{e}_i/y_i)$;
d. $\chi^2(-2) = \sum_{i=1}^{k} (y_i - \hat{e}_i)^2/y_i$.

2.12. Two hundred American citizens were surveyed and asked to identify which of five items were most fearful to them. The result are broken down by sex in Table 2.15.

Table 2.15. Greatest Fear by Sex

	Public Speaking	Heights	Insects	Financial Problems	Sickness/ Death
Male	14	7	4	16	10
Female	11	15	10	6	12

a. Fit the model that assumes no difference between the sexes in the relative proportions within each category.
b. Calculate contributions to Pearson, LR, CR, and $\sqrt{}$-transform goodness-of-fit statistics.
c. Perform a simulation (using estimated values under H_0 for the unknown parameters) to assign correct P-values to these statistics.

2.13. Let X be binomial with $n = 25$ and $p = 0.1$.
a. Given approximate normal distributions of the following possible testing variables leaving the mean expressed as a function of p but $p = 0.1$ substituted for the variance: $T_1 = X/25$, $T_2 = \log((X+0.5)/(25.5-X))$, and $T_3 = \sin^{-1}(\sqrt{X/25})$
b. Simulate 1000 values of X and hence T_1, T_2, T_3 and look at their distributions. Which, if any, appear normally distributed?
c. Suppose that $x = 5$ is observed and we wish to test $H_0: p = 0.1$ against $H_1: p > 0.1$. Calculate exact and continuity corrected approximate P-values using T_1, T_2, and T_3.

2.14. For a particular plant the leaf shape (cut or potato) and size (tall or dwarf) are controlled by independent genes a and b. Mendelian genetics tells us that if two characteristics are distributed independently in a dihybrid cross then we expect the four phenotypes ab, Ab, aB, AB to occur in proportions 9/16, 3/16, 3/16, 1/16, respectively. The data in Table 2.16 comprise counts of the four possible phenotypes for samples of plants collected in three different districts. We are interested first, in whether the Mendelian theory is born out, and second, whether there are differences between plants in different districts.

Table 2.16. Mendelian Data from Four Districts

	Tall/Cut	Tall/Potato	Dwarf/Cut	Dwarf/Potato
District 1	926	288	293	104
District 2	467	151	150	47
District 3	693	234	219	70

a. Consider each row of the data to be multinomial. Fit the model which assumes the Mendelian theory is correct and calculate the deviance of this model.

b. If the population is not in equilibrium then different proportions to the Mendelian proportions are expected, namely $p_a p_b$, $(1 - p_a)p_b$, $p_a(1 - p_b)$, and $(1-p_a)(1-p_b)$. Estimate p_a, p_b and separately for each district give the deviance of the model.

c. Fit the model if p_a, p_b are the same for each district. Give the deviance of the model.

d. Test the fit of the three models as well as against each other.

2.15. The following are a set of counts of fatal car accidents for five consecutive Mondays.

$$10, 14, 13, 20, 18$$

Treat these as Poisson variables with respective means $\mu_1, \ldots, \mu_5$ and consider the following hypotheses:

$$\mathcal{H}_1 : \mu_i \text{ unrestricted}, \ \mathcal{H}_{01} : \mu_i \text{ equal}, \ \mathcal{H}_{02} : \mu_i = 3(2 + i)$$

a. Test $\mathcal{H}_{01}$ against $\mathcal{H}_1$ using the LR and Pearson statistics, using a chi-square approximation to the distribution of each. Specifically state the approximate P-values in each case.

b. Test $\mathcal{H}_{02}$ against $\mathcal{H}_1$ using the LR and Pearson statistics, using a chi-square approximation to the distribution of each. Specifically state the approximate P-values in each case.

c. Assuming $\mathcal{H}_{02}$ is true, generate five new data values and calculate both the LR and Pearson statistics. Repeat this 10,000 times. By counting up how many of the simulations are greater than the observed values of (b) state the exact P-values corresponding to these observed values.

d. Which statistic appears to have a distribution closer to χ_5^2?

2.16. In a study of obesity, samples of white noninstitutionalized women between the ages of 20 and 24 inclusive were obtained and their body weight classified according to the Quetelet index, which divides weight by height squared. The data in Table 2.17 summarizes the results for Britain, Canada, and the United States.

Table 2.17. Obesity Classification for Women from Three Countries

	Underweight	Normal	Overweight	Obese
Britain	63	153	44	13
Canada	148	249	30	8
United States	78	174	37	22

a. What is the relevance of the high total counts for Canada to assessing the relative obesity levels in the three countries?

b. Fit the model which assumes that the obesity distribution is identical for the three countries assuming a multinomial distribution. Give the deviance of the model and test the goodness-of-fit.

c. Fit the Poisson model that takes the mean value of cell (i,j) to be $e_{ij} = \tau_j \tau_i$. Show that the fitted values are exactly the same as part (b) and that the distribution of the data within each row is multinomial with probability vector π.

d. What important feature of the data is ignored in this analysis?

2.17. In order to find an "exact" confidence interval for a binomial probability it is necessary to solve equations of the form

$$f(\pi) = \sum_{y=0}^{y_1} \binom{n}{y} \pi^y (1-\pi)^{n-y} - q = 0$$

where n, y_1 are given integers and q a given quantile.

a. Show that the derivative of $f(\pi)$ is given by

$$f'(\pi) = -np_{n-1}(y_1; \pi)$$

where $p_n(\ ; \pi)$ is the binomial (n, π) probability function.

b. Find an "exact" 90% confidence interval when $n = 10$ and $y = 3$. Compare this interval with the standard symmetric interval generated by the estimate and its standard error.

c. Perform a simulation to find the actual 90% quantile of the deviance [see Eq. (2.35)] when $n = 10$ and $\pi = 0.3$ and compare this with the χ_1^2 quantile. What is the coverage of the likelihood interval when the χ_1^2 approximation is used and $\pi = 0.3$?

2.18. In order to find an "exact" confidence interval for a Poisson mean it is necessary to solve equations of the form

$$f(\mu) = \sum_{y=0}^{y_1} e^{-\mu}\mu^y/y! - q = 0$$

where n, y_1 are given integers and q a given quantile.

a. Show that the derivative of $f(\mu)$ is given by

$$f'(\mu) = -p(y_1; \mu)$$

where $p(\,; \mu)$ is the Poisson probability function.

b. Find an "exact" 90% confidence interval when $y = 10$. Compare this interval with the standard symmetric interval generated by the estimate and its standard error.

c. Perform a simulation to find the actual 90% quantile of the deviance when $\mu = 10$ and compare this with the χ_1^2 quantile. What is the coverage of the likelihood interval when the χ_1^2 approximation is used and $\mu = 10$?

2.19. a. From observing $y = 5$ on a $Bi(15, \pi)$ distribution, plot the profile log-likelihood and the Wald statistics using observed, expected, and estimated information. Use these to give 95% confidence intervals for the mean value 15π.

b. From observing $y = 5$ on a $Pn(\mu)$ distribution, plot the profile log-likelihood and the Wald statistics using observed, expected, and estimated information. Use these to give 95% intervals.

c. Give exact intervals for parts (a) and (b).

2.20. Let $Y_1, \ldots, Y_k$ be multinomial in the usual notation. Consider the model $\mathcal{H}_0 : \pi_1 = \pi$. Suppose that $Y_1 = y_1$.

a. Let p_j be the probability of an observation in category j given that it is not one of the y_1 in category 1. What is the joint distribution of $Y_2, \ldots, Y_k$ conditional on $Y_1 = y_1$?

b. Show that the fitted values under $\mathcal{H}_0$ are

$$\hat{e}_1 = n\pi, \qquad \hat{e}_j = \frac{n(1-\pi)y_j}{(n-y_1)}, \qquad j = 2, \ldots, k.$$

c. Show that the partially maximized multinomial log-likelihood Eq. (2.4) is identical to the profile log-likelihood based on the binomial distribution of Y_1.

d. Argue that when S is a sufficient statistic for an interest parameter ω whose distribution depends only on ω that

$$\ell(\omega, \lambda) = \ell_{Y|S}(\lambda) + \ell_S(\omega)$$

where the first factor is based on the conditional distribution of the data and the second on the marginal distribution of S. Conclude that the profile log-likelihood depends only on the second factor. Apply this theory to conclude part (c).

2.21. The aim of this question is to give explicit forms for the Score and Wald statistics for testing the goodness-of-fit of a multinomial model.

a. Recall the forms [see Eqs. (2.7) and (2.9)] for the score function and expected information, respectively, under the full model. The inverse of the expected information has diagonal elements $\pi_i(1 - \pi_i)/n$ and off diagonal elements $-\pi_i \pi_j / n$. By matrix multiplication show that $I_0^{-1} U_0$ is a vector with elements $(y_i - \hat{e}_i)/n$ where $\hat{e}_i$ are fitted values under $\mathcal{H}_0$. Then show that the Score statistic $U_0^T I_0^{-1} U_0$ is just Pearson's statistic.

b. Recall that the expected information matrix for a full Poisson model is diagonal with entries $1/\mu_i$. Show that the Wald statistic

$$W = \sum_{i=1}^{k} \frac{(y_i - \hat{e}_i)^2}{y_i}$$

where $\hat{e}_i$ are fitted values under $\mathcal{H}_0$. By the equivalence of multinomial and Poisson models this is also the Wald statistic for the multinomial model provided that the fitted values and observed values agree in their total.

2.22. A genetic theory of Mendel predicted two possible species to occur in the ratio 3 to 1. From a sample of 8023, 6022 were of one species and 2001 of the other.

a. Calculate the P-value for testing the goodness-of-fit of the theory. It makes little difference which statistical procedure you use.

b. Using the reproductive property of the χ^2 distribution, Fisher calculated chi-square goodness-of-fit statistics for many of Mendel's published data sets and added them together to obtain a value of around 42 with 84 degrees of freedom. What is the probability of obtaining as small a value as this? What does this suggest?

2.23. The data in Table 2.18 were obtained by classifying 156 dairy calves by whether or not they caught pneumonia within the first two months of life. Those who got a first infection were also classified by whether or not they got a second infection within two weeks after the first. Note that one cell of the table *necessarily* contains no counts as a calf cannot get a secondary infection with no primary infection. These data were published in Agresti (1990) and are credited to Dr. T. Tran and Dr. G. Donavan of the College of Veterinary Medicine, University of Florida.

Table 2.18. Data on Bovine Pneumonia Infections

	Secondary Infection	
	Yes	No
Primary infection	30	63
No primary infection	0	63

a. Let π_{ij} denote the probability of a calf being classified in cell (i,j). Consider the hypothesis that the chance of secondary infection given primary infection is the same as the chance of primary infection. Express this hypothesis in terms of the π_{ij}.

b. If Y_{ij} is the count in cell (i,j) then write down the likelihood for the probability of a primary infection π and show that the ML estimator is $(2Y_{11} + Y_{12})/(2Y_{11} + 2Y_{12} + Y_{22})$.

c. Use the LR and Pearson goodness-of-fit statistics to test the goodness of fit of this model.

d. From the variances and covariances of the three counts implied by the multinomial distribution and the formula for variance of a function of variables, compute a standard error for $\hat{\pi}$.

e. Compute an approximate confidence interval using the likelihood function in part b.

2.24. Recall the expressions for S and LR given on p. 88. By using the expansion for $\log(1+x)$ about $x = 0$ show that the two statistics are close when both D_1 and D_2 are small.

2.25. The theory for exact upper and lower limits was explained on p. 102.

a. Show that on observing $y = 0$ successes from n binomial trials with success probability p, an exact upper $(1 - \alpha)$ limit for p is $1 - \alpha^{1/n}$.

b. Show that on observing $y = n$ successes from n binomial trials with success probability p, an exact lower $(1 - \alpha)$ limit for p is $\alpha^{1/n}$.

c. Show that on observing $y = 1$ successes from n binomial trials, an exact lower $(1 - \alpha)$ limit for p is $1 - (1 - \alpha)^{1/n}$.

d. Show that on observing $y = n - 1$ successes from n binomial trials, an exact upper $(1 - \alpha)$ limit for p is $(1 - \alpha)^{1/n}$.

e. Show that on observing $x = 0$ events for a Poisson variable of unknown mean μ an upper $1 - \alpha$ limit for μ is $-\log(\alpha)$.

2.26. Suppose that over a certain time period, seven shooting stars are seen. Of these, $y_1 = 2$ break into fragments and the other $y_2 = 5$ "burn out." Suppose Y_1, Y_2 are Poisson with means μ_1, μ_2.

a. Let π be the probability that a shooting star fragments. Express π in terms of μ_1 and μ_2. What is the distribution of Y_1 conditional on $Y_1 + Y_2 = 7$? Give an exact 95% interval for π.

b. What is the distribution of $Y_1 + Y_2$? Give an exact 95% interval for $\mu_1 + \mu_2$.

CHAPTER 3

Binary Contingency Tables

The simplest categorical response is a binary response. Indeed the simplest possible random variable is a binary random variable. A binary random variable X is a categorical random variable on $k = 2$ categories. For convenience we label these categories success S and failure F. Then X has its distribution defined by the single probability π where

$$\Pr(X = S) = \pi, \Pr(X = F) = 1 - \pi.$$

A set of independent observations $X_1, \ldots, X_n$ on X will give frequencies Y and $n - Y$ of successes and failures, respectively. The distribution of Y will be binomial with parameters n and p and probability function

$$P_B(y; n, \pi) = \frac{n!}{y!(n-y)!} \pi^y (1 - \pi)^{n-y} \qquad y = 0, \ldots, n \tag{3.1}$$

which is a special case of the multinomial distribution. The random variable $n - Y$ will have the $Bi(n, 1 - \pi)$ distribution.

The probabilistic behavior of X is entirely specified by π and once this value is known, no more can be known about the random variable. In this chapter we are interested in infering values of π from samples on X where π varies systematically with other observable factors. The entire chapter is analogous to analysis of variance (anova) for normal data. We begin with the simplest case of the two sample problem, that of comparing two unknown probabilities π_1 and π_2. This is the analogue of the two sample t-test. We extend this to the comparison of k probabilities broken down by two factors rather than one.

In this and the following chapter, methods based on the full-likelihood and its sampling distribution are studied. Alternative methods based on conditional likelihood are treated separately in Chapter 7.

3.1 BINOMIAL DATA

The data structures and models in this chapter all have a common form. They assume that the data $Y_1, \ldots, Y_k$ are independent binomial variables with respective parameters n_i and π_i. In this section we will write down the log-likelihood and give general forms for the LR and Score statistics. We introduce the log-odds transformation which is central to many binomial models. We also discuss the different sampling schemes that may give rise to contingency table data and point out that the likelihood analysis is identical for all these sampling schemes.

3.1.1 Binomial Likelihood and Log-Odds

The likelihood for k independent binomial random variables is the product of the individual binomial probability functions namely

$$P_B(y_1, \ldots, y_k; \pi_1, \ldots, \pi_i) = \prod_{i=1}^{k} \binom{n_i}{y_i} \pi_i^{y_i}(1-\pi_i)^{n_i - y_i} \tag{3.2}$$

considered as a function of the π_i rather than y_i (in which case the combinatoric terms may be dropped). The ML estimate of π_i from this likelihood is the observed proportion Y_i/n_i. Models which restrict the π_i are usually of interest. For example, a common null hypothesis of interest is that the π_i have equal value π. The above probability function then becomes

$$\pi^t(1-\pi)^{n-t} \prod_{i=1}^{k} \binom{n_i}{y_i} \tag{3.3}$$

where $n = \sum n_i$, $t = \sum y_i$. The ML estimator is $\hat{\pi} = T/n$ and T is sufficient for π with $Bi(n, \pi)$ distribution. The fitted values for this model are $\hat{e}_i = n_i\hat{\pi}_i = n_i t/n$. For a general model with fitted values $\hat{e}_i$, the LR statistic

$$LR = 2 \sum_{i=1}^{k} \left\{ y_i \log\left(\frac{y_i}{\hat{e}_i}\right) + (n_i - y_i)\log\left(\frac{n_i - y_i}{n_i - \hat{e}_i}\right) \right\}, \tag{3.4}$$

is used for testing fit. When $y_i = 0$ or $y_i = n_i$, we define $0 \log 0 = 0$.

There are great advantages to transforming the probabilities π_i by

$$\text{logit}(\pi) := \log\left(\frac{\pi}{1-\pi}\right), \tag{3.5}$$

variously called the *logistic*, *logit*, or *log-odds* transformation. We denote logits by ψ_i and this notation will be respected from this point on. Unlike the π_i, the logits are unrestricted and can take any value between $\pm\infty$. It is helpful to think of the logits, rather than the probability parameters, as being analogous to the mean parameter in normal theory linear models because it is the logits that are usually modeled linearly, not the probability parameters themselves. Logits have an attractive symmetry property, namely

$$\text{logit}(1 - \pi) = -\text{logit}(\pi)$$

which means that if we decide to reverse the roles of the binary outcomes and let π be the probability of failure rather than success then the logit parameter simply changes sign. A logit of zero corresponds to $\pi = 0.5$.

The probability parameters may be recovered by "undoing" the log-odds transformation via the *expit* transformation

$$\pi_i = \exp\{\psi_i\}/(1 + \exp\{\psi_i\}).$$

In terms of the parameters ψ_1, ψ_2 the log-likelihood function (3.2) becomes

$$l_B(\psi_1, \ldots, \psi_k) = \sum_{i=1}^{k} \{y_i \log \pi_i + (n_i - y_i)\log(1 - \pi_i)\} \tag{3.6}$$

$$= \sum_{i=1}^{k} \{y_i\psi_i - n_i \log(1 + e^{\psi_i})\} \tag{3.7}$$

The general forms given above will be used continually throughout this and the next chapter.

3.1.2 Sampling Models for Association Data

Commonly we wish to test the association of two factors such as smoking and cancer, or survival and medical treatment. As an example, imagine taking a sample of n cancer patients and randomly assigning n_1 to be given treatment 1 and the remaining $n_2 = n - n_1$ to be given treatment 2. After one year, the number of survivals under each treatment Y_1, Y_2 are recorded and on the basis of these two random variables we will compare the relative merits of the two treatments. A general data set is shown in Table 3.1. Count data which has been classified in this way, by a row factor and a column factor, is called a 2×2 contingency table.

The sampling scheme we have been describing has the number of patients n_1, n_2 fixed in advance and the random number of survivals in each group observed and recorded subsequently. This sampling scheme is called a *prospective* study

Table 3.1. Paradigm Two-Sample Data Set

	Survived	Died	Total
Treatment 1	Y_1	$n_1 - Y_1$	n_1
Treatment 2	Y_2	$n_2 - Y_2$	n_2
Total	T	$n - T$	n

and, if the n individuals are randomly assigned to the two treatment groups, it is called a *clinical trial*. This is really the "gold standard" for valid inference; probability statements about the test statistics and estimators derived later are valid regardless of whether or not other extraneous factors affect an individual's response to treatment.

There are many other ways in which the data in a 2×2 table can arise. For instance, imagine we went through hospital records and randomly chose n of them to examine in detail, and that Table 3.1 is the result of classifying each by whether or not the patient survived and which treatment was applied. This is called a *cross-sectional* study and is more susceptible to systematic bias than the clinical trial. For instance, perhaps doctors decided to give treatment 2 mainly to the more serious cases as it is more invasive and potentially risky. A higher mortality rate for patients given treatment 2 may then simply reflect the fact that it was given to less promising patients. We study this issue in more detail in Section 3.6.1.

In a prospective study the treatment conditions are set for chosen individuals and then the response observed. *Retrospective* sampling refers to experiments where fixed size samples of individuals are chosen who have already given each response and the conditions preceding that response are recorded. The response or effect is observed before the cause. In the paradigm above we imagine randomly choosing $n - t$ records from the hospital register of deceased patients and t records from the hospital register of patients who are discharged and presumed to have survived. We then go through these records and note whether they were given treatment 1 or treatment 2. Thus, the row totals N_1 and N_2 are now random variables and the column totals t and $n - t$ are fixed. This sampling scheme, also known as a *case-control study*, is again vulnerable to the systematic bias described for the cross-sectional study.

Many sociological studies are retrospective or cross-sectional studies as it may not be possible, either physically, legally, or ethically, to choose which patients are to be given which treatment. For instance, the association of lung cancer and smoking has been revealed by examining records of individuals known to have lung cancer and comparing their smoking rates with other individuals known to be healthy. It would hardly be ethical to choose patients in advance to smoke 50 cigarettes per day in the interests of science.

Considering the accident data described and analyzed in Chapter 1 suggests a fourth natural way in which 2×2 contingency table data might arise. The total number of accidents there was random, and we supposed each count to be an independent Poisson random variable. Similarly, we might exhaustively examine all hospital records for a certain period of time and then classify them into the four cells. Row, column, and grand totals are all random under such *Poisson sampling*.

It is not unrealistic to imagine even more complex schemes. For instance, if treatment 2 is a new treatment then we might have to wait for volunteers, all other patients being given standard treatment 1. A statistician might have advised the medical researcher that at least n_2 patients should be given the new treatment if statistically valid results are to be achieved. In this case n_2 would be fixed in advance while the number of remaining patients n_1 would be random. Luckily, it makes virtually no mathematical difference which of these sampling schemes is employed although it does make a difference to the potential unseen biases in the results.

3.1.3 Variable Response Probability Does Not Matter

In concluding the binomial distribution, we usually suppose that each binary trial has the same probability π for success. It is a mathematical fact that, so long as the trials are *independent*, the distribution of the total successes Y is still binomial even if the π's vary randomly from trial to trial. The probability parameter of this binomial distribution will be the mean probability of success. In the context of a clinical trial, probabilities of responses *will* likely vary from individual to individual but, contrary to popular belief, this does not matter. The total number of successes is still binomial. The probability parameter π of this distribution is the probability of a randomly chosen individual responding. This may or may not be the probability of a given individual responding.

Consider the following situations. We take a random sample of n individuals from a population. (1) Everybody in the population has the same probability of survival, 0.5. (2) Half of the population carries an undetectable gene which means they will definitely die if given the treatment, and the other half will definitely live. It is a good exercise to think carefully about these two situations and convince yourself that they are probabilistically the same. Indeed, there is no way to distinguish these two models statistically, provided that each individual is tested only once; each individual either lives or dies. Count them up and it is binomial. It makes no difference whether the uncertain response comes from (1) some basic variability of the treatment and the body's response to it or (2) from our not knowing which of two distinct genetic groups we belong to. Either way, it is a toss of the coin. Of course, if we could detect the genetic marker that makes the treatment fatal the situation would be quite different.

Similarly, two treatments need not have a uniform effect to be compared. If $\pi_1 > \pi_2$ then an individual would be well advised to take treatment 1 to maximize the chance of survival, regardless of whether treatment 1 might actually be fatal to some individuals and a certain cure for others.

If the individual response probabilities vary *together* then the binomial distribution will *not* be appropriate because the binary trials will then not be independent. This leads to *over-dispersed* binomial data (see Section 4.4).

3.2 COMPARING TWO BINOMIAL SAMPLES

This section considers the comparison of $k = 2$ binomial variables. The data below comes from a clinical trial and is used throughout to illustrate the methods described. As already noted, other sampling schemes may give rise to contingency tables which can be analyzed using exactly the same methods.

Example 3.1: Mice Tumor Data Essenberg (1952) conducted a clinical trial to investigate the effect of tobacco smoke on tumor risk. To this end, 72 albino mice were randomly divided into two groups and one group kept in a chamber filled with a certain dose of tobacco smoke—one cigarette per hour. The other (control) group was kept in a chamber without smoke. After a year, the 55 surviving mice were sacrificed and autopsied for tumors.

Let the probability that a mouse develops a tumor be π_1 conditional on treatment and π_2 conditional on no treatment. We are primarily interested in comparing π_1 and π_2, in particular, in the null hypothesis that treatment with smoke has no effect on tumor rate. Parametrically this is expressed as $\mathcal{H}_0 : \pi_1 = \pi_2$ and a test of this hypothesis is called a test of *association* between the row factor (treatment) and the column factor (tumor).

Table 3.2. Tumor Prevalence Among Mice by Exposure to Tobacco Smoke

	Tumor	No Tumor	Total
Smoke treatment	21	2	23
Control	19	13	32
Total	40	15	55

3.2.1 Measuring Association by the Log-Odds Ratio

One could certainly measure the association of treatment and response by $\pi_1 - \pi_2$. If this is positive then response is positively associated with treatment 1. There are good reasons to look at the difference in logits instead. This gives the association measure

$$\Delta = \psi_1 - \psi_2 = \log\left\{\frac{\pi_1(1-\pi_2)}{\pi_2(1-\pi_1)}\right\} \tag{3.8}$$

called the *log-odds ratio*. This has exactly the same *sign* as $\pi_1 - \pi_2$, and the null hypothesis $\pi_1 = \pi_2$ corresponds to $\Delta = 0$. Some advantages of Δ over other measures of association are listed in Section 3.6.2. In final reports, it is better to express results in terms of the *odds ratio* $\phi = \exp(\Delta)$. For instance, if $\phi = 2$ then the odds of survival under treatment 1 is twice that under treatment 2. The log-odds ratio $\Delta = \log 2 = 0.693$ simply measures this on the log scale.

We can reparametrize the binomial model in terms of the association parameter Δ and a nuisance parameter ψ_2 which measures the chance of survival under treatment 2 and is often of no direct interest. We can test specific hypotheses about the relative effectiveness of the two treatments by testing nonzero values of Δ. In terms of these new parameters

$$\pi_1 = \frac{\exp\{\Delta + \psi_2\}}{1 + \exp\{\Delta + \psi_2\}}, \qquad \pi_2 = \frac{\exp\{\psi_2\}}{1 + \exp\{\psi_2\}} \tag{3.9}$$

and substituting these into the probability function Eq. (3.2) gives

$$P_B(y_1, y_2; \Delta, \psi_2) = \binom{n_1}{y_1}\binom{n_2}{y_2} \frac{\exp\{\Delta y_1 + \psi_2(y_1 + y_2)\}}{(1 + \exp\{\Delta + \psi_2\})^{n_1}(1 + \exp\{\psi_2\})^{n_2}}. \tag{3.10}$$

Considered as a function of (Δ, ψ_2) this defines the likelihood function.

3.2.2 Likelihood Estimates and Tests of No Association

The log-odds ratio Δ is related to the probability parameters π_1, π_2 according to Eq. (3.8). Substituting the ML estimators $\hat{\pi}_i = Y_i/n_i$ gives

$$\hat{\Delta} = \log\left\{\frac{\hat{\pi}_1(1 - \hat{\pi}_2)}{\hat{\pi}_2(1 - \hat{\pi}_1)}\right\} = \log\left\{\frac{Y_1(n_2 - Y_2)}{Y_2(n_1 - Y_1)}\right\} \tag{3.11}$$

the ML estimator of Δ. This is the logarithm of the ratio of observed odds.

Denote the expected values $E(Y_1) = n_1\pi_1$ and $E(Y_2) = n_2\pi_2$ by e_1 and e_2, respectively. From Eq. (3.6) the log-likelihood is

$$l(\pi_1, \pi_2) = c + \sum_{i=1}^{2} \{y_i \log e_i + (n_i - y_i)\log(n_i - e_i)\} \tag{3.12}$$

after absorbing the terms $n_i \log n_i$ into the arbitrary constant c. For a fitted model $\mathcal{M}$ we substitute fitted values $\hat{e}_i$ to give the maximized log-likelihood.

For the full model, the fitted values are the observed values so we substitute y_i for e_i to give the maximized log-likelihood. The LR goodness-of-fit statistic of model $\mathcal{M}$ is then given by

$$LR = 2\sum_{i=1}^{2}\left\{y_i \log\left(\frac{y_i}{\hat{e}_i}\right) + (n_i - y_i)\log\left(\frac{n_i - y_i}{n_i - \hat{e}_i}\right)\right\}$$

$$= 2\sum^{*} y_{ij}\log(y_{ij}/\hat{e}_{ij}) \tag{3.13}$$

where $\sum^{*}$ denotes a sum over all *four* cells of the table and y_{ij} and $\hat{e}_{ij}$ are observed and fitted values in cell (i,j). The degrees of freedom of this approximately chi-square statistic is 1 since we are testing a single parameter Δ. An alternative test statistic is the Pearson statistic

$$S = \sum_{i=1}^{2}\frac{(y_i - n_i\hat{\pi})^2}{n_i\hat{\pi}(1-\hat{\pi})}. \tag{3.14}$$

This does not look the same as the general Pearson statistic Eq. (2.25). However, a simple exercise in algebra shows that S can be expressed as a sum of terms $(Y_{ij} - \hat{e}_{ij})^2/\hat{e}_{ij}$ across all four cells of the 2×2 table. Both LR and S have approximate χ_1^2 distributions. In the special but common case of testing the null model $\Delta = 0$, the ML estimate of π is $\hat{\pi} = t/n$. Hence, the fitted values are $\hat{e}_1 = n_1 t/n$, $\hat{e}_2 = n_2 t/n$.

Example 3.1: (Continued) **Mice Tumor Data.** Estimates of the tumor rates are $\hat{\pi}_1 = 21/23 = 0.913$ and $\hat{\pi}_2 = 19/32 = 0.594$ and of the odds ratio $\hat{\phi} = (21 \times 13)/(19 \times 2) = 7.184$. We estimate that a mouse has around 7 times the odds of developing a tumor when exposed to smoke. The log-odds parameter Δ is estimated by $\log(7.184) = 1.972$.

To compute a test statistic we need the fitted values under the null hypothesis that tumor rate is unaffected by smoke treatment. Multiplying row and column totals and dividing by 55 gives the four fitted values 16.727, 6.273, 23.273, 8.727 to 3 decimal places. The LR statistic Eq. (3.13) is therefore

$$LR = 42\log\left(\frac{21}{16.727}\right) + 4\log\left(\frac{2}{6.273}\right)$$
$$+ 38\log\left(\frac{19}{23.273}\right) + 26\log\left(\frac{13}{8.727}\right)$$

giving 7.635, while the Pearson statistic is 6.878 using either Eq. (3.14) or

$$\frac{(21-16.727)^2}{16.727}+\frac{(2-6.273)^2}{6.273}+\frac{(19-23.273)^2}{23.273}+\frac{(13-8.727)^2}{8.727}.$$

Compared to the approximate χ_1^2 distribution this is strong evidence against the null hypothesis that smoke treatment has no effect on tumor rate. Since it is natural to consider a one-sided alternative hypothesis in this problem, the P-values are half the χ_1^2 tail probabilities giving 0.0028 for the LR statistic and 0.0044 for the Pearson statistic.

3.2.3 Confidence Intervals for Δ

To summarize what the data say about the difference in tumor rates one would, at the very least, report a confidence interval for Δ or ϕ. There are several alternative methods for doing this.

Wald-Type Intervals Wald-type intervals were described in Section 2.6.3. They are based on the transformed CLT [see Eq. (1.49)] which says that for any smooth transform g

$$g(\hat{\pi}_i) \xrightarrow{d} \mathcal{N}\left(g(\pi_i), \{g'(\pi_i)\}^2\, \frac{\pi_i(1-\pi_i)}{n_i}\right), \qquad i=1,2. \tag{3.15}$$

Since $\hat{\pi}_1$, $\hat{\pi}_2$ are independent we obtain the interval

$$\left(g(\hat{\pi}_1) - g(\hat{\pi}_2) \pm z_{\alpha/2} \sqrt{\frac{g'(\hat{\pi}_1)^2\hat{\pi}_1(1-\hat{\pi}_1)}{n_1} + \frac{g'(\hat{\pi}_2)^2\hat{\pi}_2(1-\hat{\pi}_2)}{n_2}}\right)$$

for the association measure $g(\pi_1) - g(\pi_2)$. Elementary statistics courses often give the interval for $\pi_1 - \pi_2$ based on the identity function g.

We have been stressing logits as preferable to probabilities. The logit parameters $\psi_i = \log(\pi_i/(1-\pi_i))$ have ML estimates

$$\hat{\psi}_i = \text{logit}(\hat{\pi}_i) = \log\left(\frac{Y_i}{n_i - Y_i}\right), \qquad i=1,2$$

called *empirical logits*. The derivative of this function is $g'(\pi) = 1/(\pi(1-\pi))$ and so from Eq. (3.15) the approximate distribution of the empirical logits is

$$\hat{\psi}_i \xrightarrow{d} \mathcal{N}\left\{\log\left(\frac{\pi_i}{1-\pi_i}\right), \frac{1}{n_i\pi_i(1-\pi_i)}\right\} \tag{3.16}$$

and so the approximate distribution of the ML estimator $\hat{\Delta} = \hat{\psi}_1 - \hat{\psi}_2$ is normal with mean $\psi_1 - \psi_2 = \Delta$ and variance

$$\text{Var}(\hat{\Delta}) = V_1 + V_2 = \frac{1}{n_1\pi_1(1-\pi_1)} + \frac{1}{n_2\pi_2(1-\pi_2)} \tag{3.17}$$

which is easily estimated. This leads to an approximate $100(1-\alpha)\%$ interval

$$(\hat{\Delta} \pm z_{\alpha/2}\sqrt{\hat{V}_1 + \hat{V}_2})$$

for the association measure Δ, where $\hat{V}_i = n_i/(y_i(n - y_i))$ estimates V_i. This interval is the direct analogue of the standard two-sample t-interval for a difference of means $\mu_1 - \mu_2$. Exponentiating the interval for Δ gives an interval for the odds ratio $\phi = \exp(\Delta)$.

Example 3.1: (Continued) **Some Alternative Intervals for Mice Tumor Data.** The estimates $\hat{\pi}_1 = 21/23$, $\hat{\pi}_2 = 19/32$ have standard errors $\hat{\sigma}_1 = 0.0587$, $\hat{\sigma}_2 = 0.0868$. An estimate of $\pi_1 - \pi_2$ is $\hat{\pi}_1 - \hat{\pi}_2 = 0.3193$ with standard error $\sqrt{0.0587^2 + 0.0868^2} = 0.1048$. An approximate 95% confidence interval for $\pi_1 - \pi_2$ is

$$(0.3193 \pm 1.96 \times 0.1048) = (0.1138, 0.5247).$$

The empirical logits are $\hat{\psi}_1 = \log(21/2) = 2.351$ with standard error $\sqrt{23/(21 \times 2)} = 0.740$ and $\hat{\psi}_2 = \log(19/13) = 0.379$ with standard error $\sqrt{32/(19 \times 13)} = 0.360$. An estimate of $\Delta = \psi_1 - \psi_2$ is $\hat{\Delta} = 2.351 - 0.379 = 1.972$ with standard error $\sqrt{0.740^2 + 0.360^2} = 0.823$. An approximate 95% confidence interval for Δ is

$$(1.972 \pm 1.96 \times 0.823) = (0.359, 3.585)$$

and exponentiating gives the interval (1.432, 36.04) for ϕ which accompanies the estimate $\hat{\phi} = 7.184$.

As a further illustration we compute an interval for the ratio π_1/π_2 by first computing an interval for $\log \pi_1 - \log \pi_2$. The derivative of $g(\pi) = \log \pi$ is $g'(\pi) = 1/\pi$ and so the asymptotic variance of $\log(Y_i/n_i)$ is $(1 - \pi_i)/\pi_i n_i$. The estimate of $\log \pi_1$ is $\log(21/23) = -0.0910$ with standard error $\sqrt{2/(21 \times 23)} = 0.0643$. The estimate of $\log \pi_2$ is $\log(19/32) = -0.5213$ with standard error $\sqrt{13/(19 \times 32)} = 0.1462$. Hence, the estimate of $\log(\pi_1/\pi_2)$ is 0.4303 with standard error $\sqrt{0.0643^2 + 0.1462^2} = 0.1598$. An approximate 95% confidence interval for $\log(\pi_1/\pi_2)$ is

$$(0.4303 \pm 1.96 \times 0.1598) = (0.1172, 0.7434).$$

Exponentiating this interval gives the interval (1.124, 2.103) for π_1/π_2 with ML estimate $\hat{\pi}_1/\hat{\pi}_2 = 1.538$. More refined approximate intervals for the difference and ratio of two probabilities are in Santner and Snell (1980).

Modified Empirical Logits One obvious problem with empirical logits is that they can equal $\pm\infty$ when Y_i/n_i is either 0 or 1. An alternative is to use *modified* empirical logits

$$\hat{\psi}_i^* = \log\left(\frac{Y_i + \frac{1}{2}}{n_i - Y_i + \frac{1}{2}}\right) \tag{3.18}$$

which are always finite and whose expectations exist. The addition of 1/2 rather than some other positive constant makes the expectation of the modified logits $\psi_i + o(n_i^{-1})$ and is optimal in this sense [see Haldane (1955) and Gart and Zweifel (1967)]. This is a very desirable property particularly when approximating more complex linear models for the logit parameters. It can also be shown that $|\hat{\psi}_i^*| \le |\hat{\psi}_i|$ from which it follows that $|\hat{\Delta}^*| \le |\hat{\Delta}|$. An estimate of the variance of the modified logits is

$$\hat{V}_i^* = \frac{(n_i + 1)(n_i + 2)}{n_i(Y_i + 1)(n_i - Y_i + 1)} \tag{3.19}$$

and again the constants added minimize the bias of this variance estimate.

Example 3.1: (Continued) **Mice Tumor Data.** The modified empirical logits are $\hat{\psi}_1^* = \log(21.5/2.5) = 2.152$, and $\hat{\psi}_2^* = \log(19.5/13.5) = 0.368$. Estimated variances for these estimators are

$$\text{Var}(\hat{\psi}_1^*) = \frac{24 \times 25}{23 \times 22 \times 3} = 0.395, \text{Var}(\hat{\psi}_2^*) = \frac{33 \times 34}{32 \times 20 \times 14} = 0.125.$$

The modified estimator of Δ is $\hat{\Delta}^* = \hat{\psi}_1^* - \hat{\psi}_2^* = 2.152$ with standard error $\sqrt{0.395 + 0.125} = 0.721$. An approximate 95% confidence interval for Δ is

$$(2.152 - 0.368 \pm 1.96 \times 0.721) = (1.784 \pm 1.414) = (0.370, 3.198)$$

and exponentiating gives the interval (1.448, 24.48) for ϕ which accompanies the estimate $\hat{\phi}^* = \exp(1.784) = 5.953$. A comparison of results from using modified and unmodified methods is shown in Table 3.3. The ratio of the estimate to its standard error, labeled z-stat, is also given. The associated P-values are already one-sided, and are less extreme than those obtained using the LR and Pearson statistics. Nevertheless, using any of the four approaches, evidence against $\Delta = 0$ is strong.

Table 3.3. Comparison of Empirical and Modified Logits

	$\hat{\psi}_1$ (se)	$\hat{\psi}_2$ (se)	$\hat{\Delta}$ (se)	z-Stat	P-Value
Unmodified	2.351 (0.740)	0.379 (0.360)	1.972 (0.823)	2.396	0.0083
Modified	2.151 (0.628)	0.368 (0.353)	1.784 (0.721)	2.474	0.0067

Likelihood Intervals Likelihood intervals were discussed in Section 2.6.2. These intervals are obtained by plotting the profile likelihood function $\hat{l}(\theta)$ and reading off values within half a χ_1^2 quantile of the maximum. To obtain a likelihood interval for Δ, we first write down an expression for the profile likelihood. This is just the log-likelihood [see Eq. (3.10)] with $\hat{\psi}_2(\Delta)$ substituted for ψ_2, giving

$$\hat{l}(\Delta) = y_1\Delta + t\hat{\psi}_2(\Delta) - n_1 \log(1 + e^{\Delta+\hat{\psi}_2(\Delta)}) - n_2 \log(1 + e^{\hat{\psi}_2(\Delta)}). \tag{3.20}$$

An expression for $\hat{\psi}_2(\Delta)$, the ML estimator of ψ_2 for known Δ, follows from writing down the derivative of $l(\Delta, \psi_2)$ with respect to ψ_2, which is

$$\frac{\partial l}{\partial \psi_2} = t - n_1 \frac{\exp\{\Delta + \psi_2\}}{1 + \exp\{\Delta + \psi_2\}} + n_2 \frac{\exp\{\psi_2\}}{1 + \exp\{\psi_2\}}.$$

The profile ML estimator $\hat{\psi}_2(\Delta)$ is obtained by equating this to zero. Letting $\phi = \exp\{\Delta\}$ and $\lambda = \exp\{-\psi_2\}$ the ML estimator of λ for given ϕ satisfies

$$t = \frac{n_1\phi}{\phi + \lambda} + \frac{n_2}{1 + \lambda}$$

which rearranges into a quadratic for λ. Thus, $\hat{\psi}_2(\Delta)$ is easy to compute.

Example 3.1: (Continued) **Profile Limits for Mice Tumor Data.** Substituting the values for t, n_1, n_2, the partial estimate $\hat{\lambda}(\phi)$ solves the quadratic

$$40\lambda^2 + \lambda\{40(\phi + 1) - 23\phi - 32\} - 15\phi = 0.$$

For given Δ, we thus have a closed form for $\hat{\lambda}(\phi)$ and $\hat{\psi}_2(\Delta) = -\log(\hat{\lambda}(e^{\Delta}))$. This is then substituted into Eq. (3.20). Figure 3.1 shows a plot, not of $\hat{l}(\Delta)$ but of $2\{\hat{l}(\hat{\Delta}) - \hat{l}(\Delta)\}$ against Δ. This deviance statistic is approximately χ_1^2 and a 95% interval is obtained by reading off values for which it is less than 3.841. Also plotted is the Wald statistic which is the square of $(\hat{\Delta} - \Delta)/\text{s.e.}(\hat{\Delta})$ where $\hat{\Delta} = 1.972$ and $\text{s.e.}(\hat{\Delta}) = 0.823$ were computed earlier. Reading off values for which this is less than 3.841 generates the earlier given interval (0.359, 3.585). The Wald statistic is perfectly quadratic on the logit scale, whereas the profile

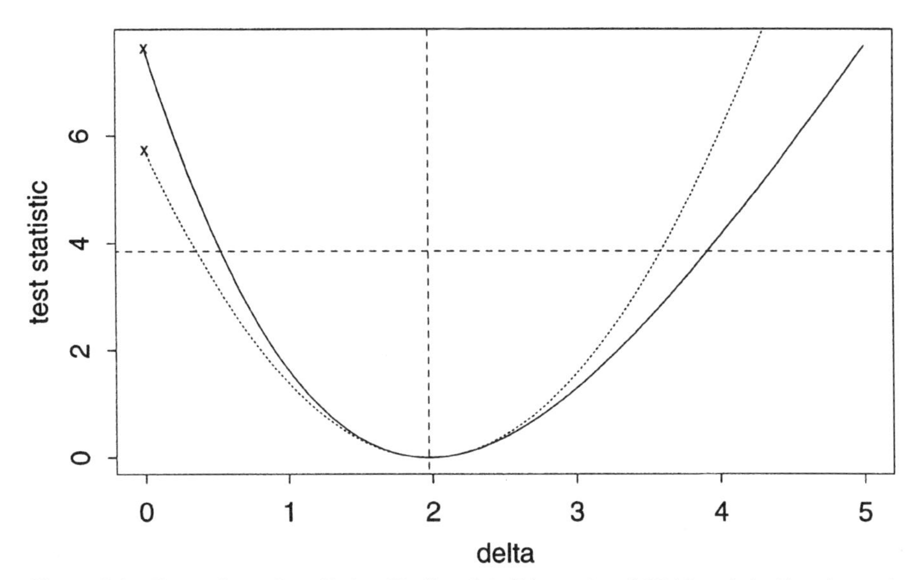

Figure 3.1 Comparison of profile log-likelihood (solid curve) and Wald statistic (dotted curve) against the log-odds parameter, for the mice-tumor data.

has a shape depending on the data, surely a desirable property. There is some disagreement between the curves. At the left-hand endpoints the values are 7.635 and 5.742, these being the LR and Wald test statistics for testing $\Delta = 0$.

Which Method? Even though the 2×2 table is an incredibly simple data structure, several methods have been listed for its analysis. All are in common use so it is important that you see all of them. Which if any is best? For moderately large n_i it will make very little difference which is used, unless some of the fitted values are very small. We saw this in the example. When the n_i are small, it seems that the conditional methods of Chapter 7 are appropriate. However, this is not a universally shared position. Among the unconditional methods given so far it is recommended to use the estimator $\hat{\Delta}^*$ and the profile likelihood-based confidence interval, though the interval based on modified logits is acceptable.

3.3 COMPARING SEVERAL BINOMIAL SAMPLES

Eyestrain is thought to be related to type of office work, specifically to prolonged exposure to visual display unit (VDU) screens. Reading and Weale (1986) classified 295 office workers into four groups, relating to their VDU exposure, and tested each for symptoms of eyestrain (Table 3.4). We use these data to illustrate most methods in this section. We consider a "success" to

Table 3.4. Presence of Eyestrain Classified by Type of Work for 295 Office Workers [from Reading and Weale (1986)]

Type of Work	VDU Data Entry	General VDU Use	Full-Time Typing	Standard Clerical Work
No eyestrain (S)	42	79	64	52
Eyestrain (F)	11	30	14	3
Total	43	109	78	55

be no symptoms of eyestrain, although this choice has no effect on our final results.

The notation used is shown in Table 3.5 below. A total of n individuals are classified into k groups, n_i being in group i. Within each group the number y_i responding is recorded. The total number of responses is t. The important question is whether or not the probability of response varies from group to group and, if so, how.

As in the two sample problem, there are several alternative sampling schemes that could give rise to the same counts. For instance, we might obtain records of n individuals and cross-classify them by response and group—a cross-sectional study. We might obtain a list of t responders and $n - t$ nonresponders and then determine their group membership—retrospective sampling. Individuals may appear randomly in time and be classified into the appropriate cell as they enter the study—Poisson sampling. As in the previous section, these different sampling models ultimately lead to identical statistical estimators and tests.

3.3.1 Independent Binomial Model

Within each group, each individual either responds or not and so the Y_i are binomial with means $e_i = n_i\pi_i$. The parameters π_i are the probabilities of response in group i. Note that the π_i do not have to sum to anything in particular since they each refer to different populations. Assuming independence of individuals, the joint distribution of $Y_1, \ldots, Y_k$ is a product of binomial probabilities as displayed in Eq. (3.2).

We are usually interested in whether or not these probability parameters are equal. In our example, the null hypothesis of interest is that the class of work

Table 3.5. Generic *k*-Sample Data Set

Group	1	2	·	·	k	Total
Response	y_1	y_2	·	·	y_k	t
No response	$n_1 - y_1$	$n_2 - y_2$	·	·	$n_k - y_k$	$n - t$
Total	n_1	n_2	·	·	n_k	n

is unrelated to the probability of eyestrain which is in general

$$\mathcal{H}_0 : \pi_1 = \pi_2 = \cdots = \pi_k = \pi,$$

k equaling 4 in our example. Under this hypothesis, the log-likelihood Eq. (3.6) simplifies, the ML estimate of the common value π is $\hat{\pi} = t/n$ and the fitted values are

$$\hat{e}_i = n_i \hat{\pi}_i = n_i t/n.$$

Fitted values in the cells corresponding to no eyestrain are $n_i - \hat{e}_i$. Considering all $2k$ cells in the table a general formula for the fitted value in cell (i, j) under the null model is

$$\hat{e}_{ij} = \text{row total } i \times \text{column total } j/\text{grand total}.$$

The LR statistic Eq. (3.4) for testing $\mathcal{H}_0$ is

$$LR = 2 \sum_{i=1}^{k} \left\{ y_i \log\left(\frac{y_i}{\hat{e}_i} \right) + (n_i - y_i) \log\left(\frac{n_i - y_i}{n_i - \hat{e}_i} \right) \right\} \tag{3.21}$$

which is a sum of $2k$ terms, each term being of the form $2y_i \log(y_i/\hat{e}_i)$ as in Eq. (2.21). The Pearson statistic is just Eq. (2.25) applied to all $2k$ cells. Labeling the count in row i and column j by Y_{ij}, alternative forms of this statistic are

$$S = \sum_{j=1}^{2} \sum_{i=1}^{k} \frac{(Y_{ij} - \hat{e}_{ij})^2}{\hat{e}_{ij}} = \sum_{i=1}^{k} \frac{n_i(\hat{\pi}_i - \hat{\pi})^2}{\hat{\pi}(1 - \hat{\pi})}. \tag{3.22}$$

The first form is naturally interpreted as a measure of the lack of fit of the fitted values $\hat{e}_{ij}$ to the counts Y_{ij}. The second form is naturally interpreted as a measure of the dispersion of the estimates $\hat{\pi}_i$. The degrees of freedom for these statistics is $k - 1$ since there are k probability parameters under the full model and a single common parameter π under the null model.

As for 2×2 tables, there are advantages to parametrizing in terms of log-odds. We define

$$\psi_i = \log\left(\frac{\pi_i}{1 - \pi_i} \right), \Delta_i = \psi_i - \psi_k, \qquad i = 1, \ldots, k - 1 \tag{3.23}$$

the parameters Δ_i now measuring the log-odds ratio for the 2×2 subtable made up of classes/columns i and k. An alternative parametrization (used by GLIM) compares each group not the group k but to group 1. Regardless of which group is chosen as the baseline, the hypothesis of no group/class differences now becomes the hypothesis that all the log-odds ratio parameters are zero. Maximum likelihood estimates of our Δ_i parameters are

$$\hat{\Delta}_i = \log\left\{\frac{Y_i(n_k - Y_k)}{Y_k(n_i - Y_i)}\right\} \tag{3.24}$$

and approximate variance expressions analogous to Eq. (3.17) are easily written down. In terms of the alternative parameter set $(\Delta_1, \ldots, \Delta_{k-1}, \psi_k)$ the null hypothesis specifies that each of the Δ_i are zero with ψ_k a nuisance parameter. The likelihood for these new parameters is

$$l(\Delta, \psi_k; y) = c + \sum_{i=1}^{k-1} y_i\Delta_i + t\psi_k - \sum_{i=1}^{k} n_i \log(1 + \exp\{\psi_i\}). \tag{3.25}$$

3.3.2 Dispersion Tests of No Association

Let g be any smooth function. For each group compute an estimate $g(\hat{\pi}_i)$ of $g(\pi_i)$. The π_i are equal if, and only if, the $g(\pi_i)$ are equal. This suggests a test based on the dispersion of the estimates $g(\hat{\pi}_i)$ about $g(\hat{\pi})$, the latter being the ML estimator of an assumed common $g(\pi)$. Such a statistic is

$$S_g = \sum_{i=1}^{k} \frac{n_i\{g(\hat{\pi}_i) - g(\hat{\pi})\}^2}{\hat{\pi}(1 - \hat{\pi})g'^2(\hat{\pi})} \tag{3.26}$$

which has asymptotic χ^2_{k-1} distribution. When g is the identity then it is easily shown that S equals $(y_{ij} - e_{ij})^2/e_{ij}$ totaled over all $2k$ cells of the table, as displayed in Eq. (3.22). This is of course the Pearson statistic which is identical to the Score statistic as stated in Chapter 2.

Example 3.2: (Continued) **Dispersion Tests for Eyestrain Data.** Partial calculations for computing the test statistic S_g for the eyestrain data are shown in Table 3.6, for g the identity and log-odds functions.

The Pearson statistic measures dispersion of the five estimates $\hat{\pi}_1, \ldots, \hat{\pi}_5$ and how consistent these values are with the underlying probability parameters being equal. The second statistic measures the variability of the five estimates $\hat{\psi}_1, \ldots, \hat{\psi}_5$ and how consistent these are with the underlying logit parameters being equal. Under the null hypothesis of parameter equality an estimate of the common value π is $237/295 = 0.803$ and of the common value ψ, logit(0.803) $= 1.407$. The contributions to the Pearson statistics require expressions for the

Table 3.6. Computation of Various Test Statistics for Testing Association of Job Class and Eyestrain

Job Class	Data	General	Type	Standard	All
No eyestrain (y_i)	42	79	64	52	237
Number (n_i)	53	109	78	55	295
$\hat{e}_i$	42.58	87.57	62.66	44.19	53
$\hat{\pi}_i$	0.792	0.725	0.820	0.945	0.803
Logit ($\hat{\pi}_i$)	1.340	0.968	1.520	2.853	1.407
S	0.040	4.265	0.145	7.027	11.478
S_{logit}	0.038	3.324	0.155	18.140	21.657

variances which are

$$\text{Var}(\hat{\pi}_i) = \frac{\pi_i(1 - \pi_i)}{n_i}, \qquad \text{Var}(\hat{\psi}_i) = \frac{1}{\pi_i(1 - \pi_i)n_i}$$

and substituting null estimates. The statistics $S = 11.478$ and $S_{\text{logit}} = 21.657$ differ substantially, and by examining the contributions to these statistics we see that the main disagreement comes from the fourth job class. The proportion with eyestrain here is 94.5% and the log-odds transformation is quite extreme near the boundaries of the unit interval.

Notwithstanding the difference in the two test statistics, the P-values are both extremely small, 0.0007 for S and 0.0000 for S_{logit}, to 4 decimal places. This is typical behavior of central-limit-type approximations as discussed in Chapter 2—when the probability being approximated is small the *relative* error can be quite large even though the *absolute* error is small. In practice, it makes little difference which P-value above we take since the hypothesis of equal eyestrain risk will be rejected in either case. The LR statistic Eq. (3.21) is found by summing

$$2y_i \log\{y_i/\hat{e}_i\} + 2(n_i - y_i)\log\{(n_i - y_i)/(n_i - \hat{e}_i)\}$$

across the whole $2 \times k$ table and the result is 13.340.

3.3.3 Poisson Modeling Approach

Binomial data can be analyzed from a different point of view, by assuming all counts are Poisson random variables. It may well be that the counts are Poisson, for instance, if they are obtained from watching some continuous time process for a fixed amount of time. However, it makes no difference whether or not the counts really are Poisson. We may always pretend that binomial data are Poisson, because we ultimately obtain the same results.

Table 3.7 displays an alternative notation for the data. The count in row i

Table 3.7. Poisson Approach to $2 \times k$ Table

Group	1	2	· ·	k	Total
Response	$Y_{11}\,(y_1)$	$Y_{12}\,(y_2)$	· ·	$Y_{1k}\,(y_k)$	T
No response	$Y_{21}\,(n_1 - y_1)$	$Y_{22}\,(n_2 - y_2)$	· ·	$Y_{2k}\,(n_k - y_k)$	$N - T$
Total	N_1	N_2	· ·	N_k	N

and column j is denoted Y_{ij} and is assumed Poisson distributed with mean μ_{ij}. The previous binomial notation is displayed in parentheses, and in a smaller font. The general relationship between multinomial and Poisson was discussed at length in Section 2.4 and it is worthwhile rereading it now. That discussion applies to each column of the table. For instance, in column j we were earlier assuming $Y_j \xrightarrow{d} Bi(n_j, \pi_j)$. We now have instead two independent Poisson variables Y_{1j}, Y_{2j}. Conditional on their total, $N_j = Y_{1j} + Y_{2j}$, the counts are again binomial. The connection between the parameters is

$$\pi_j = \frac{\mu_{1j}}{\mu_{1j} + \mu_{2j}}, \qquad \psi_j = \log\left(\frac{\mu_{1j}}{\mu_{2j}}\right), \qquad \Delta_i = \psi_i - \psi_k = \log\left(\frac{\mu_{1j}\mu_{2k}}{\mu_{1k}\mu_{2j}}\right).$$

A test of equality of the π_j then becomes a test of equal *ratios* of μ_{1j} to μ_{2j} or of equal *differences* of $\log \mu_{1j}$ and $\log \mu_{2j}$.

There is a more natural parametrization that exploits the two-way row and column structure of the data. Consider the model

$$\log(\mu_{ij}) = m + s_i + t_j + \gamma_{ij} \qquad i = 1, 2 \qquad j = 1, \ldots, k. \tag{3.27}$$

This is analogous to a two-way analysis of variance, with s_i, t_j being interpreted as row/column effects and the γ_{ij} as interactions between the row and column effects. The main difference is that the logged means rather than the means are being described. Simple algebra shows that

$$\psi_j = \log(\mu_{1j}/\mu_{2j}) = s_1 - s_2 + \gamma_{1j} - \gamma_{2j}.$$

Now we are usually interested in testing if the π_j, or equivalently, if the ψ_j are constant. Apparently the ψ_j are constant if, and only if, the interaction parameters γ_{ij} are all zero. In this case the common value is $\psi = s_1 - s_2$. This anticipates a general pattern that will be made explicit in Section 6.2.3, namely that main effects in the binomial model become interactions in the Poisson model. Here, the group effect becomes an interaction of group/column with response/row.

The model as given is over-parametrized and it is usual to impose restrictions so that the parameters number $2k$. A common set of restrictions is called *set to zero* constraints. Choose one row and one group to be the baseline or reference group. The package **GLIM** illustrated below always chooses row 1 and group

1. Then set those parameters to zero which have subscripts $i = 1$ or $j = 1$. In other words, $t_1 = s_1 = 0$, $\gamma_{1j} = \gamma_{i1} = 0$ leaving a mean parameter m, a single row parameter s_2, $k - 1$ column parameters $t_2, \ldots, t_k$, and $k - 1$ interaction parameters $\gamma_{22}, \ldots, \gamma_{2k}$. Estimates of these parameters are labeled on the later GLIM output. A little algebra shows that

$$\pi_j = \frac{\mu_{1j}}{\mu_{1j} + \mu_2} = \{1 + \exp(s_2 + \gamma_{2j})\}^{-1} \qquad \gamma_{2j} = -\log\left\{\frac{\pi_j(1 - \pi_1)}{\pi_1(1 - \pi_j)}\right\}. \quad (3.28)$$

Thus γ_{2j} is the logged relative odds of *non*response for group j compared to group 1. These parameters are different to the Δ_i parameters defined earlier that were log-odds of response for group j compared to group k. If group k is chosen as the baseline group, then the γ_{2j} would be identical to the Δ_j as we earlier defined them, but with the wrong sign.

The models described so far, both binomial and Poisson, are examples of a wide class of models called *generalized linear models*. There is no need at this stage to go into the full theory, but it is necessary to understand what a generalized linear model is. The next section places the previous models in this general framework and examines some computer output for the eyestrain data that uses exactly the parametrization described above.

3.3.4 Analysis Based on Generalized LM's

Generalized linear models were introduced by Nelder and Wedderburn (1972) to unify various types of "regression-like" statistical methods, not the least of which were the majority of statistical methods described in this book. A brief account is given here and a more detailed account in Section 4.6.

A generalized linear model for an independent set of data $Y_1, \ldots, Y_k$ comprises three elements. First, the error distribution or random component of the model must be a member of the so-called *exponential family*. This family includes normal, gamma, Poisson, and binomial distributions, hence its relevance to categorical data.

The second element of a generalized linear model is a link function denoted by η. Rather than model the mean $\mu = E(Y)$ of the distribution directly we model $\eta(\mu)$. For binomial, Poisson, and normal models the logit, log, and identity functions, respectively, are called the canonical link.

The third and germane element of a generalized linear model is an expression for $\eta(\mu)$ as a linear combination of parameters β, i.e.,

$$\eta(\mu_i) = \sum_{j=1}^{p} X_{ij}\beta_j$$

where $\mu_i = \mathrm{E}(Y_i)$, X_{ij} is a design matrix and $\beta_1, \ldots, \beta_p$ are unknown parameters numbering $p \leq k$. In matrix notation $\eta = X\beta$.

Solution of the ML equations for estimating β reduces to iteratively reweighted least squares. Tests of hypotheses are usually based on the LR statistic and its approximate chi-square distribution. Tests of a parameter being zero are often based either on comparing the estimate to its standard error (the Wald test) or on removing the parameter and seeing how the LR goodness-of-fit or deviance statistic changes (LR test).

The independent binomial model for a $2 \times k$ table is a generalized linear model. First, the underlying distribution is binomial. Second, the Eq. (3.9) defining Δ can be rewritten

$$\mathrm{logit}(\pi_i) = \psi_k + \Delta_i, i = 1, \ldots, k-1$$

being linear in the parameters Δ_i, ψ_k. The Poisson model Eq. (3.27) is also a generalized linear model; the underlying distribution is Poisson which is an exponential family and $\log(\mu_{ij})$ is modeled linearly in terms of s_i, t_j, and γ_{ij}.

Below is the GLIM output for the eyestrain data, using the Poisson modeling approach. The first model fitted is the full model with the "set to zero" parametrization described in the last section. The second model fitted is the null model, i.e., the model assuming that the π_j are constant. In the Poisson approach, this means setting the *interaction* parameters to zero so that the null model is a simple *additive* model **STRAIN+CLASS**. I have supplied a short syntax of **GLIM** and several other packages, accessible at the Wiley website.

```
[i] ? $SLENGTH 8$
[i] ? $VARIATE Y$FACTOR STRAIN 2 CLASS 4$
[i] ? $LOOK Y STRAIN CLASS$
[o]      Y      STRAIN     CLASS
[o] 1    42.000     1.000     1.000
[o] 2    11.000     2.000     1.000
[o] 3    79.000     1.000     2.000
[o] 4    30.000     2.000     2.000
[o] 5    64.000     1.000     3.000
[o] 6    14.000     2.000     3.000
[o] 7    52.000     1.000     4.000
[o] 8     3.000     2.000     4.000
[o]
[i] ? $YVARIATE Y$ERROR POISSON$                 Poisson error and log link
[i] ? $FIT STRAIN*CLASS$                                        Full model
[o] scaled deviance = 0. at cycle 2           Zero deviance for full model
[o]      residual df = 0
[o]
[i] ? $DISPLAY E$
[o]       estimate       s.e.     parameter
[o]  1       3.738     0.1543     1                                      m
```

```
[o] 2    -1.340   0.3387   STRAIN(2)                    s2
[o] 3    0.6318   0.1910   CLASS(2)                     t2
[o] 4    0.4212   0.1986   CLASS(3)                     t3
[o] 5    0.2136   0.2075   CLASS(4)                     t4
[o] 6    0.3715   0.4009   STRAIN(2).CLASS(2)      gamma22
[o] 7   -0.1801   0.4492   STRAIN(2).CLASS(3)      gamma23
[o] 8    -1.513   0.6836   STRAIN(2).CLASS(4)      gamma24
[o]
[i] ? $FIT STRAIN + CLASS$                          Null model
[o] scaled deviance = 13.340 at cycle 4        LR test of null
[o]       residual df = 3
[o]
[i] ? $DISPLAY E$
[o]        estimate      s.e.     parameter
[o] 1         3.751    0.1403     1
[o] 2        -1.408    0.1465     STRAIN(2)       -logit(237/295)
[o] 3        0.7211    0.1675     CLASS(2)
[o] 4        0.3864    0.1780     CLASS(3)
[o] 5       0.03704    0.1925     CLASS(4)
```

For the full model the `STRAIN` and interaction parameters transform directly to estimates of the π_i through Eq. (3.28). For instance

$$\hat{\pi}_2 = [1 + \exp\{-1.340 + 0.3715\}]^{-1} = 0.7248$$

which agrees with the data of 79 successes from 109 in class 2.

For the null model only the parameter `STRAIN(2)` has any meaningful interpretation and is simply the negative of the logit transform of the estimate $\hat{\pi} = 0.803$. The mean parameter (labeled "1") and the `CLASS(i)` parameters simply ensure that the fitted values in each column/class sum to the correct total n_j. The differences of these parameters are the logged ratios of the n_j. For instance

$$\texttt{CLASS(3)} - \texttt{CLASS(4)} = 0.3864 - 0.0370 = 0.5436 = \log(78/55).$$

By way of contrast the GLIM analysis using the independent binomial model is given below. Under the full model the estimates labeled `CLASS(i)` are now the logits for comparing class i with class 1. For instance, the estimate labeled `CLASS(2)` is

$$\hat{\psi}_2 - \hat{\psi}_1 = \text{logit}(79/109) - \text{logit}(42/53) = -0.3715.$$

These are not estimates of the earlier defined parameters Δ_i that compare column i with column k. If this parametrization was desired then the directive `FACTOR CLASS 5 (5)` sets the baseline to be class 5, rather than class 1. Finally, note that in both approaches the LR goodness-of-fit statistic for the null model is labeled deviance and takes the value 13.34, that we computed earlier on p. 135.

```
[i] ? $UNITS 4$VARIATE Y N$FACTOR CLASS 4$
[i] ? $DATA Y N CLASS$READ
[i] $REA? 42 53 1
[i] $REA? 79 109 2
[i] $REA? 64 78 3
[i] $REA? 52 55 4
[i] ? $YVARIATE Y$ERROR BINOMIAL N$!                        Binomial model
[i] ? $FIT 1$!                            Fit null model first this time
[o] scaled deviance = 13.340 at cycle 3                     LR test of null
[o]      residual df = 3
[o]
[i] ? $DISPLAY E$
[o]      estimate       s.e.     parameter
[o]  1     1.408     0.1464     1
[o]
[i] ? $FIT CLASS$!            Full model has zero deviance
[o] scaled deviance = 0. at cycle 2
[o]      residual df = 0
[o]
[i] ? $DISPLAY E$
[o]       estimate       s.e.     parameter
[o]  1       1.340     0.3387     1
[o]  2     -0.3715     0.4009     CLASS(2)     psi2-psi1
[o]  3      0.1801     0.4492     CLASS(3)     psi3-psi1
[o]  4       1.513     0.6836     CLASS(4)     psi4-psi1
```

3.3.5 Directed Tests of Contrasts and Trends

In the context of normal theory analysis of variance, tests of the existence of group differences is only a first step in the analysis. Similarly, a test of association or no difference in the probability parameters is far too blunt a tool on which to base the analysis and summary of data. If we decide there *is* an association then we need to describe it. For instance, are all the π_i different from each other or are only some of them different? When a test of association does not reveal significant group differences, this might be because we have not directed the test to be sensitive to the kind of differences anticipated. The main ideas presented in this section are from Berkson (1938), Cochran (1955), and Armitage (1955).

Tests of Trend When the group labels are ordinal, it makes sense to look for a systematic trend of the response proportions. For instance, we might test whether or not the logits ψ_i differ by the same amount Δ for adjacent groups. Such a *test of trend* will typically involve one degree of freedom and will have more power for detecting group differences than the general independence test with $k - 1$ degrees of freedom. The comparison is entirely analogous to that of linear regression with one-way analysis of variance.

Example 3.3: Male Voice Pitch and Testosterone Table 3.8 summarizes a survey of 195 professional or student male singers. These data were collected to test the hypothesis that men with deep voices have higher levels of testosterone. Consequently, one might expect their fathers to have higher levels. Since high levels of testosterone are associated with a higher chance of having male children one might expect singers with lower voices to have more brothers than sisters. The data for those with no siblings is uninformative and has been omitted.

Fitting the independence model to this 2×6 tables gives a deviance of 9.851 on five degrees of freedom which is quite moderate evidence (P-value = 0.089) against the null hypothesis of no difference between the singing groups. However, this test is sensitive to arbitrary differences between groups. Background knowledge suggests a systematic trend as we move from bass to counter-tenor and this is apparent in the raw proportions of males. Let us suppose that each step from the lowest pitch (bass = 1) to the highest pitch (counter-tenor = 6) decreases by Δ, the logit probability ψ_i of a sibling being male. We then have $\psi_2 = \psi_1 - \Delta$, $\psi_3 = \psi_1 - 2\Delta$, etc. In other words, $\psi_i = \psi_1 - (i-1)\Delta$. Substituting these into the general likelihood Eq. (3.7)

$$
\begin{aligned}
l_B(\psi_1, \Delta) &= \sum_{i=1}^{6} \{y_i\psi_i - n_i \log(1 + e^{\psi_i})\} \\
&= \sum_{i=1}^{6} \{y_i(\psi_1 + (i-1)\Delta) - n_i \log(1 + e^{\psi_1 - (i-1)\Delta})\} \\
&= \psi_1 \sum_{i=1}^{6} y_i - \Delta \sum_{i=1}^{6} (i-1)y_i - \sum_{i=1}^{6} n_i \log(1 + e^{\psi_1 - (i-1)\Delta}) \\
&= 231\psi_1 - 356\Delta - \sum_{i=1}^{6} n_i \log(1 + e^{\psi_1 - (i-1)\Delta})
\end{aligned}
$$

Table 3.8. Siblings Sex Ratios

Voice Group	Brothers (y_i)	Sisters ($n_i - y_i$)	Proportions	Fitted π_i
Bass	77	47	0.621	0.628
Bass-baritone	38	30	0.539	0.583
Baritone	75	58	0.564	0.536
Tenor-baritone	5	4	0.556	0.489
Tenor	27	37	0.422	0.442
Counter-tenor	9	15	0.375	0.396

The number of brothers and sisters of male singers by pitch of voice [from Lister (1984)].

that must be maximized with respect to the parameters (ψ_1, Δ). The estimate of Δ is 0.1896 with standard error 0.064. The Wald statistic for testing $\Delta = 0$ is $0.1896/0.064 = 2.962$ giving a one-sided P-value 0.0015. Alternatively, we can test $\Delta = 0$ by comparing the deviance of the null model (9.851 on 5 df) with the trend model (0.906 on 4 df) giving a difference 8.945 on 1 df. This is an approximate χ_1^2 statistic, and its value is approximately the same as the squared Wald statistic. Since $\exp(0.1896) = 1.209$ the trend model estimates that the *odds* of a sibling being male is 20.9% lower for each higher voice group. Assuming this model, the estimated probability of a sibling being male is given in the right-hand column of the table.

In fact, the above trend model is an example of a simple *logistic regression*, studied in the next chapter. The important point to take from his example is not the details of the trend model itself, but that a more refined model of association will often have greater power than a cruder and completely general goodness-of-fit test.

Partitioning of Chi-Square Consider the LR goodness-of-fit test of $\mathcal{H}_0 : \pi_i = \pi$. This test has $k - 1$ degrees of freedom, because it is equivalent to testing whether all of $\Delta_1, \ldots, \Delta_{k-1}$ are equal to zero. If we reject $\mathcal{H}_0$ then which of the Δ_i are responsible? Or is the main pattern of association described not by the Δ_i but say a few simple combinations of them?

As a first step we could test each $\Delta_i = \psi_i - \psi_k$ separately. How? By just looking at the 2×2 table comprising columns i and k and computing the LR test statistic that will have 1 degree of freedom. There are major problems with this. First the LR statistics for $\Delta_1, \Delta_2, \ldots, \Delta_{k-1}$ will *not* add up to the LR statistic for the full $2 \times k$ table. It is entirely possible that each of the individual LR statistics will be insignificant even if the overall LR statistic is highly significant. Moreover, the individual LR statistics will be highly correlated, because each involves a comparison with the same group k. What we ideally seek is a way of breaking up the question "are the π_i different?" into $k - 1$ separate and simpler questions whose answers do not influence each other. An overall significant χ^2 result may then be attributed to specific and simply explained sources of interest.

Let us use the notation $(124|56)$ to denote the 2×2 table obtained by adding columns 1, 2, 4 into a single column and columns 5, 6 into another column. We call a set of $k-1$ such subtables a *partition* if the corresponding LR statistics are asymptotically independent and sum to the LR statistic for the full table. The aim of a partition is to identify main sources of a large overall LR statistic by attributing it to differences between certain categories or groups of categories. One example of a partition for a 2×4 table is $(1|2)$, $(12|3)$, $(123|4)$. We may of course permute the column labels to obtain other partitions. The earlier parameters Δ_i correspond to the table $(i|k)$. These do *not* form a partition.

Example 3.2: (Continued) **Partition of Eyestrain Data.** For the eyestrain data, the LR test of no difference in groups was 13.340 on 3 df. It is left as an exercise for you to compute the LR statistics 5.89 for (1|4), 13.13 for (2|4), and 4.97 for (3|4). These add up to 23.99. A better approach is to use the partition (1|2), (12|3), (123|4). For (1|2) the table has entries (11, 42, 30, 79) and LR = 0.884. For (12|3) the entries are (41, 121, 14, 64) with LR = 1.655. For (123|4) the entries are (55, 185, 3, 52) with LR 10.790. These *do* add up to 13.340 and show that the overall significant difference between the four classes is mostly due to differences between class 4 and classes 1, 2, and 3. There is no evidence of differences between classes 1, 2, and 3.

How to Construct a Partition For readers familiar with the normal theory analysis of variance, partitions correspond to *orthogonal contrasts*. A contrast is a linear combination of logits whose coefficients sum to zero. For instance, with $k = 4$, the parameter $\Delta_1 = \psi_1 - \psi_4$ is a contrast having coefficients (1, 0, 0, −1). Two contrasts are *orthogonal* if their coefficient vectors are orthogonal. The contrasts Δ_1, Δ_2, Δ_3 are *not* orthogonal because the vectors (1, 0, 0, −1), (0, 1, 0, −1), (0, 0, 1, −1) are not orthogonal. The table (12|3) corresponds to the contrast $(\psi_1 + \psi_2)/2 - \psi_3$ with coefficients (1/2, 1/2, −1, 0, 0). It is very easy to check that (1|2), (12|3), (123|4) are orthogonal.

There are many other sets of orthogonal contrasts. The device of constructing coefficient vectors that are orthogonal allows you to dream up some of your own.

Multiple Comparisons. As in normal theory, when performing several independent tests we must deal with the problem of *multiple comparisons*. If a 95% critical value is used for each of the partition chi-square tests then the probability of at least one significant result will be $1 - .95^{k-1}$ which is approximately $0.05(k - 1)$. A simple rough adjustment for this is to take significance levels $\alpha/(k - 1)$ for each of the single degree of freedom chi-square tests. The test that rejects the hypothesis of no group differences when one of these tests is significant will then have the correct type 1 error probability. This test is not the same as the overall chi-square test, however. Rather, it is more powerful at detecting departures from zero of the specific contrasts chosen. This power advantage is purchased at the expense of lower power for detecting other types of associations.

In the eyestrain example, the LR statistic 10.79 for the contrast (123|4) has associated P-value 0.0010. This is 50 times smaller than 0.05 so there is no question of its being significant. If the most significant contrast instead had P-value 0.02, then this would *not* be counted significant at level 5%.

3.4 ANALYSIS OF SEVERAL 2 × 2 TABLES

In this section we take one more step towards a general theory for categorical responses broken down by factors. It is commonly the case that a binary response to treatment is tested under k different sets of conditions. This gives a set of k 2×2 tables often collectively called a $2 \times 2 \times k$ table. The data are said to be *stratified* by the conditions. The following example will be used to illustrate the theory of this section.

Example 3.4: Smoking and Lung Impairment Suppose random samples of individuals were chosen from the cities of Los Angeles, New York, and Washington, DC. Each individual is classified as smoker/nonsmoker. Lung capacity is tested by blowing into a standard measuring device. By comparing performance with norms for healthy individuals, each lung capacity measurement is converted into a binary response—healthy or impaired. These fictitious data form the $2 \times 2 \times 3$ Table 3.9. We imagine lung health to be the binary response and the smoking and city factors as explanatory variables. Using the methods of Section 3.2 we could statistically assess the effect of smoking on risk of lung impairment separately for each city. We might additionally want to

1. See if the effect of smoking is the same in each city. For instance, in cities with poor air quality (such as LA) one might find that smoking a given number of cigarettes has a more adverse effect because of an interaction with atmospheric pollutants.
2. Give an overall combined assessment of the effect of smoking using data from all three cities, either assuming or not assuming the effect is the same for each city as appropriate.

There are two further generalizations possible. First, the tables could be general $r \times c$ tables. Second, the different conditions under which these tables were collected could themselves be related to other factors. Treatment of such data requires the completely general theory of log-linear models studied in Chapter 6.

3.4.1 Independent Binomial Model

For definiteness, let us imagine collecting data on the survival or non-survival of a random sample of individuals in a particular demographic group, each randomly allocated to one of two different treatments. This process is repeated for several other demographic groups (say different sexes and agegroups) giving k

Table 3.9. Lung Health for Smokers/Nonsmokers in Three Cities

	Los Angeles		New York		Washington, DC	
	Impaired	Healthy	Impaired	Healthy	Impaired	Healthy
Smoker	12	3	9	5	6	2
Nonsmoker	6	10	12	14	12	6

contingency tables each of which carry information about the relative effects of the two treatments on that group. Let subscript $i(j)$ refer to treatment i applied within the jth group/table. Thus, for instance $Y_{2(3)}$ would be the number of survival-response of the $n_{2(3)}$ individuals from the third demographic group who were given treatment 2.

It is natural to suppose that each of the $2k$ responses have independent binomial distributions i.e.

$$Y_{i(j)} \overset{d}{=} Bi(n_{i(j)}, \pi_{i(j)}) \qquad i = 1, 2 \qquad j = 1, \ldots, k,$$

where $\pi_{i(j)}$ denotes the probability of a randomly chosen individual from group j responding when given treatment i. The logits of these parameters are denoted $\psi_{i(j)}$ and the log-odds ratio for the jth groups is

$$\Delta_{(j)} = \psi_{1(j)} - \psi_{2(j)} = \log\left\{\frac{\pi_{1(j)}(1 - \pi_{2(j)})}{\pi_{2(j)}(1 - \pi_{1(j)})}\right\}.$$

This measures the relative effectiveness of treatment 1 compared to treatment 2 on the log-odds scale for group/table j. There are thus $2k$ parameters in total which we choose to be $\Delta = (\Delta_{(1)}, \ldots, \Delta_{(k)})$ and $\psi_2 = (\psi_{2(1)}, \ldots, \psi_{2(k)})$ the first set being of primary interest in assessing the treatment. The relationship between parameters and the probability parameters is given by (3.9) with a subscript (j) on each symbol. The reverse relationship is

$$\operatorname{logit}(\pi_{1(j)}) = \psi_{2(j)} + \Delta_{(j)}, \qquad \operatorname{logit}(\pi_{2(j)}) = \psi_{2(j)}. \tag{3.29}$$

The $\psi_{2(j)}$ measure underlying response rates under treatment 2 for each of the k groups and are of no interest in assessing the *relative* effectiveness of the two treatments. In the smoking example they represent underlying impairment rates (on the log-odds scale) for nonsmokers in the three different cities. For testing the effect of smoking these are treated as nuisance parameters although they may well be parameters of interest in their own right if underlying rates for cities are to be compared.

The log-likelihood function for (Δ, ψ_2) is obtained from the general binomial log-likelihood Eq. (3.7) reproduced in the first expression below, and then substituting expressions for $\psi_{i(j)}$ giving

$$\begin{aligned}\ell(\Delta, \psi_2) &= \sum_{i=1}^{2}\sum_{j=1}^{k}\{y_{i(j)}\psi_{i(j)} - n_{i(j)}\log(1 + e^{\psi_{i(j)}})\}\\ &= \sum_{j=1}^{k}\left\{\Delta_{(j)}y_{1(j)} + \psi_{2(j)}T_{(j)} - \sum_{i=1}^{2} n_{i(j)}\log(1 + e^{\psi_{i(j)}})\right\}.\end{aligned} \tag{3.30}$$

We are primarily interested in the effects of the treatment, summarized in the interest parameter Δ. The log-likelihood function is simply the sum of the separate log-likelihoods from the k tables. Maximizing each of these separately, it is clear that the ML estimates in this full model are just the observed proportions, odds, or odds ratios in each table. For instance, the ML estimator of $\Delta_{(1)}$ is the observed log-odds ratio in table 1, the ML estimator of $\psi_{2(k)}$ is the observed log-odds $\log(y_{2(k)}) - \log(n_{2(k)} - y_{2(k)})$ for treatment 2 in table k.

The $2 \times 2 \times k$ table is actually the analogue of two-way analysis of variance for normal data. To see this consider a general two-way interaction model for the logits

$$\psi_{i(j)} = \mu + \alpha_i + \beta_k + \gamma_{ij} \qquad i = 1, 2 \qquad j = 1, \ldots, k.$$

As in two-way anova, this model is over-parametrized and we need to set some parameters to zero. Let us set μ, α_1, α_2 and $\gamma_{21}, \gamma_{22}, \ldots, \gamma_{2k}$ all equal to zero. Then it follows that

$$\psi_{1(j)} = \beta_j + \gamma_{1j}, \psi_{2(j)} = \beta_j.$$

Comparing this with Eq. (3.29) we see that $\Delta_{(j)}$ has the role of γ_{1j}. These are interaction parameters for treatment (row) and group (column) and measure the effects of treatment for the different groups. The parameters $\psi_{2(j)}$ have the role of the β_j. These measure column or group effects, in other words the underlying probabilities of response in the different groups, often of no direct interest. Since there is only one binomial observation per cell, the interaction model is the full model.

3.4.2 Inference Assuming Uniform Treatment Effects

Consider now the hypothesis that the treatment effects, measured by $\Delta_{(j)}$, are the same for each group. This is expressed parametrically by

$$\mathcal{H}_0 : \Delta_{(j)} = \Delta \text{ for all } j.$$

Called the *uniform treatment effects* model, it imposes essentially $k - 1$ restrictions on the full model and the likelihood Eq. (3.30) becomes

$$\ell(\Delta, \psi_2) = \Delta Y_{1+} + \sum_{j=1}^{k} T_{(j)}\psi_{2(j)} + \sum_{j=1}^{k} \log\{(1 + e^{\Delta + \psi_{2(j)}})(1 + e^{\psi_{2(j)}})\} \quad (3.31)$$

where $Y_{1+} = Y_{1(1)} + Y_{1(2)} + \cdots + Y_{1(k)}$. We are interested in the treatment effect Δ. Computing ML estimates of Δ and the $\psi_{2(j)}$ requires iterative methods. Once these are computed, Δ can be tested either by comparing the estimate to a standard error or, by refitting the model assuming $\Delta = 0$ and recording the change in deviance. The deviance of the fitted uniform treatments model will have $k-1$ degrees of freedom, and can be used to test the uniform treatments hypothesis itself.

The uniform treatments model is actually the analogue of the *additive* two-way analysis of variance model for normal data. To see this consider the two-way additive model

$$\psi_{i(j)} = \mu + \alpha_i + \beta_j \qquad i = 1,2\; j = 1,\ldots,k \tag{3.32}$$

where α_i describe row effects and β_j column effects. As in two-way anova, some further restrictions are required and we choose μ and α_2 to be set to zero. Then it follows that $\psi_{1(j)} = \alpha_1 + \beta_j$, $\psi_{2(j)} = \beta_j$ which is identical to the uniform treatments model after identifying α_1 with Δ and $\psi_{2(j)}$ with β_j. Thus, Δ can be thought of as a row effect, measuring the amount by which the proportions in row 1 exceed those in row 2. To allow different treatment effects in each group, we would include an interaction between row/treatment and column/table effects. This model is full and has deviance zero.

Example 3.4: (Continued) **GLIM Output for Smoking Data.** Both the uniform treatments model and the full model Eq. (3.29) are generalized linear models; the data are binomial and the logit probabilities are modeled linearly in terms of other parameters. We can use standard software to fit these models and `GLIM` is used below as an illustration. To emphasize the connection with two-way anova, it is helpful to set out the proportions of responses in a $2 \times k$ table, with the binomial proportion $Y_{i(j)}/n_{i(j)}$ in cell (i,j). For the smoking example we display the data as

	Los Angeles	New York	Washington, D.C.
Smoker	12(15)	9(14)	6(8)
Nonsmoker	6(16)	12(26)	12(18)

There are six binomial counts classified by a factor `SMOKE` on two levels and a factor `CITY` on three levels. The first model fitted is the full model, as it contains the interaction of `SMOKE` and `CITY`. The addition of `-SMOK-1` to the `FIT` command is included to give the parametrization in Eq. (3.29)]—the `-1` sets $\mu = 0$ and the `-SMOK` sets α_1, α_2 to zero. The parameters γ_{2j} are automatically set to zero. Setting redundant parameters to zero is sometimes called *aliasing*.

```
[e] ? $SLENGTH 6$VARIATE Y N$FACTOR SMOKE 2 (2) CITY 3$
[e] ? $DATA Y N$READ
[e] $REA? 12 15 6 16 9 14 12 26 6 8 12 18
[e] ? $DATA SMOKE CITY$READ
[e] $REA? 1 1 2 1 1 2 2 2 1 3 2 3
[e] ? $ERROR BINO N$YVARIATE Y$      independent binomial model
[e] ? $FIT CITY*SMOK-1-SMOK$                         full model
[o] scaled deviance = 0.00000 at cycle 4
[o]             d.f. = 0
[o]
[e] ? $DISP E$
[o]        estimate       s.e.     parameter
[o]   1     -0.5108     0.5164     CITY(1)                log(6/10)
[o]   2     -0.1542     0.3934     CITY(2)               log(12/14)
[o]   3      0.6931     0.5000     CITY(3)                log(12/6)
[o]   4       1.897     0.8266     CITY(1).SMOKE(1)          delta1
[o]   5      0.7419     0.6825     CITY(2).SMOKE(1)          delta2
[o]   6      0.4055     0.9574     CITY(3).SMOKE(1)          delta3
```

Because the model is full the deviance is zero. The parameters labeled `CITY(j)` are the column effects β_j, earlier identified with $\psi_{2(j)}$. The estimates are the observed log-odds of impairment for nonsmokers. The parameters labeled `CITY(j).SMOKE(1)` are interactions γ_{1j} that we earlier identified with $\Delta_{(j)}$. The estimates are the observed log-odds ratios for the three cities. The next model fitted is the uniform treatments model that we decided was additive in the row and column effect. The extra `-1` in the `FIT` directive sets $\mu = 0$, while α_2 is automatically alised by `GLIM`.

```
[e] ? $FIT CITY+SMOK-1$                 uniform treatments model
[o] scaled deviance = 1.749 at cycle 4
[o]         d.f. = 2
[o]
[e] ? $DISP E$
[o]       estimate       s.e.     parameter
[o]   1    -0.1665     0.4275     CITY(1)
[o]   2    -0.2599     0.3584     CITY(2)
[o]   3     0.5332     0.4447     CITY(3)
[o]   4      1.067     0.4635     SMOKE(1)                est of delta
[o]
[e] ? $PRINT %FV$                                    fitted values
[o]  10.66 7.336 9.680 11.32 6.656 11.34
[e] ?$PRINT %X2$                                 Pearson statistic
[o]     1.742
[e] ? $FIT CITY$                                 no smoking effect
[o] scaled deviance = 7.381 at cycle 4
[o]            d.f. = 3
```

The deviance of this model is 1.749 on two degrees of freedom. This can be used to test the uniform treatment hypothesis itself. There are two degrees of freedom because we are testing essentially two things, for instance, $\Delta_{(1)} = \Delta_{(2)}$, $\Delta_{(1)} = \Delta_{(3)}$. The estimate of Δ is 1.067 with standard error 0.4635 and so the Wald test of $\Delta = 0$ is $W = (1.067/0.463)^2 = 5.30$ and the P-value from χ_1^2 is 0.0213. For a one-sided test against $\Delta > 0$ the P-value is half this amount. An alternative test of $\Delta = 0$ follows from fitting the model with `CITY` only, which increases the deviance to 7.381 from 1.75, an increase of 5.63. This is an alternative to $w = 5.30$.

The parameters `CITY(j)` estimate underlying impairment rates for nonsmokers in the three cities. Converting these from logits to probabilities using the transformation $p = \exp(\psi)/(1+\exp(\psi))$ gives estimates 0.458, 0.435, 0.630 for the cities Los Angeles, New York, Washington, D.C., respectively. These differ quite appreciably from the observed rates of 0.375, 0.461, 0.666. In fact, according to our model, LA has a *higher* underlying rate than NY, even though the reverse is true in the raw data. How has this happened? The highest effect of smoking was in LA, since $\hat{\Delta}_{(1)} = 1.897$ was the highest in the full model. Under the uniform treatments hypothesis this is replaced by the lower value 1.067. The only way to make the impairment probabilities for LA consistent with this figure is to (i) increase the estimate of impairment for LA nonsmokers from 0.375 to 0.458, and (ii) decrease the estimate of impairment for LA smokers from 0.80 to 0.710.

The moral of this tale is an important one and was already stressed on p. 18 in Chapter 1. Uncritical application of a sequence of tests followed by inference based on the "best fitting" model is not the basis of good statistical analysis. In the present example if we were particularly interested in the underlying impairment rates for smokers in the three cities, then we would need to think very carefully about fitting the uniform treatments model since this reverses the pattern of our inference. The fact that the uniform treatments model fit well *overall* says nothing about its effect on *specific* parameters of interest. And the fact that it fits well does not mean it is true, especially for a small scale data set like this one.

```
[e] ? $FIT SMOK$     collapses tables over three cities
[o] scaled deviance = 4.1259 at cycle 4
[o]             d.f. = 4
[o]
[e] ? $DISP E$
[o]       estimate       s.e.     parameter
[o]  1      0.9933     0.3702     1
[o]  2      0.9933     0.4513     SMOKE(1)
```

The final model ignores the city entirely and includes `SMOKE` as the only explanatory variable. This corresponds to accumulating the data for the three cities into a single table. This is usually a statistically *invalid* thing to do. In this case the estimated effect of smoking is 0.993, rather close to the estimates obtained earlier. It should be noted however that the estimate of a treatment effect can change quite drastically when tables are accumulated. In fact, *there is no limit* to how much the estimate can change. We discuss this important issue at length in Section 3.6.1.

3.5 INFERENCE USING WEIGHTED LEAST SQUARES

Parameter estimates under the uniform treatments hypothesis require iterative methods and cannot be given directly. This is the case with most models for more complicated contingency tables. A different method of estimation, which can be computed directly, is based on weighted least squares. Such methods were much more popular several decades ago before packages for fitting generalized linear models were available. Nevertheless, these methods are (i) of some historical interest, (ii) easily calculated, and (iii) used as initial guesses in the iterative schemes used to compute exact ML estimates. Weighted least squares are considered in a more general context in Section 4.5.

3.5.1 Estimating Δ

We have estimates $\hat{\Delta}_{(j)}$ of $\Delta_{(j)}$ from the jth table. Their variances V_j are estimated from Eq. (3.16), applied separately to each table. Alternatively and preferably, we could use modified estimates $\hat{\Delta}^*_{(j)}$ and their estimated variances.

Under the uniform treatment assumption, each estimator independently estimates the same treatment effect Δ and an optimal linear combination of these is obtained by weighting inversely by the variance giving the weighted least-squares estimator

$$\hat{\Delta}_{WLS} = \frac{\sum \hat{\Delta}_{(j)}/V_j}{\sum 1/V_j}. \tag{3.33}$$

The minimized variance is $1/\sum V_j^{-1}$. Of course, the V_j are replaced by estimates both in $\hat{\Delta}_{WLS}$ and in its variance. Even though $\hat{\Delta}_{WLS}$ is generally different from the ML estimator of Δ, it can be shown to be consistent for Δ as the $n_{i(j)}$ increase for fixed k. Under these same conditions, each of the $\hat{\Delta}_{(j)}$ are approximately normal and so $\hat{\Delta}_{WLS}$ and its standard error can be used to form confidence intervals for Δ and to test Δ in the usual way. It is interesting that the normal limit for $\hat{\Delta}_{WLS}$ holds, even when the $\hat{V}_j$ and $\hat{\Delta}_{(j)}$ are highly correlated (see Exercise 3.23).

Example 3.4: (Continued) **WLS Inference for the Smoking Data.** We have calculated below (unmodified) empirical logits and their estimated variances for each of the three cities.

$$\hat{\Delta}_{(1)} = \log(120/18) = 1.897, \qquad \hat{V}_1 = 0.683 = \frac{10+6}{10 \times 6} + \frac{12+3}{12 \times 3}$$

$$\hat{\Delta}_{(2)} = \log(126/60) = 0.742, \qquad \hat{V}_2 = 0.466 = \frac{14+12}{14 \times 12} + \frac{5+9}{5 \times 9}$$

$$\hat{\Delta}_{(3)} = \log(36/24) = 0.405, \qquad \hat{V}_3 = 0.916 = \frac{6+12}{6 \times 12} + \frac{2+6}{2 \times 6}$$

Under the uniform treatment hypothesis, the estimated common value Δ is

$$\hat{\Delta}_{WLS} = \frac{\dfrac{1.897}{0.683} + \dfrac{0.742}{0.466} + \dfrac{0.405}{0.916}}{\dfrac{1}{0.683} + \dfrac{1}{0.466} + \dfrac{1}{0.916}} = 1.102$$

which implies that the odds of lung impairment is $\exp\{1.102\} = 3.01$ times higher for smokers than for nonsmokers. The variance of this estimator is 0.2126 being the inverse of the denominator of $\hat{\Delta}_{WLS}$, so the standard error is 0.4611. This should be compared with the ML estimator of 1.067 and its standard error 0.4635, computed in `GLIM` in the previous section.

The estimate and standard error generate the approximate 95% interval

$$(\hat{\Delta}_{WLS} \pm 1.96 \times \text{s.e.}(\hat{\Delta}_{WLS})) = (1.102 \pm 0.904) = (0.198, 2.006)$$

for Δ, the assumed common log-odds ratio. Exponentiating gives an interval (1.22, 7.43) for the odds-ratio ϕ. This accompanies the estimate $\hat{\phi} = 3.01$. Thus, the data indicate that the odds of impairment are between 20% and seven times higher for smokers than for nonsmokers. Our best point estimate is that it increases risk by a factor of 3. Using modified logits gives smaller estimates of Δ for each table and also when combined. The separate modified estimates are 1.752, 0.695, and 0.302 with variances 0.597, 0.435, and 0.768. The WLS estimate of Δ is 0.927 and of ϕ is about 2.5, instead of 3.

3.5.2 Testing for Nonuniform Treatment Effects

The uniform treatments hypothesis can be tested using the LR or Pearson goodness-of-fit statistics. We already did this for the smoking data and obtained the values LR = 1.749, S = 1.742 both on 2 df. But these statistics require fitting the model by maximum likelihood.

An alternative direct computation can be based on measuring the dispersion of the estimates $\hat{\Delta}_{(j)}$ about the assumed common estimate $\hat{\Delta}_{WLS}$. This is very similar to the idea behind the dispersion tests of no group effect in $2 \times k$ tables, given in Section 3.3.2. The statistic

$$BD = \sum_{j=1}^{k} \frac{\{\hat{\Delta}_{(j)} - \hat{\Delta}_{WLS}\}^2}{\hat{V}_{(j)}}, \tag{3.34}$$

is often called the *Breslow–Day* statistic [see Breslow and Day (1980)] and, like the LR and Pearson statistics it has approximate χ^2_{k-1} distribution.

Example 3.4: (Continued) **Breslow–Day Test for Smoking Data.** We have earlier computed estimates 1.897, 0.742, 0.405 of the log-odds ratio for each city with variances 0.683, 0.466, 0.916. The combined WLS estimate was 1.102. Hence,

$$BD = \frac{(1.897 - 1.102)^2}{0.683} + \frac{(0.742 - 1.102)^2}{0.466} + \frac{(0.405 - 1.102)^2}{0.916} = 1.734$$

which is hardly different from LR = 1.749, S = 1.742. Using modified logits instead, BD = 1.773. The approximate P-value from χ^2_2 is 0.420. In Section 7.5 we compute a P-value using an entirely different approach, with the aid of the package *StatXacT*, and obtain the answer 0.490. Any way we look at it, there is little evidence against uniform effects of smoking.

What if Treatment Effects are not Uniform? If the $\Delta_{(j)}$ really are unequal then we cannot talk about *the* effect of smoking—there are k separate effects. As in our discussion of partitioning chi-square, rejecting the hypothesis of uniform treatment effects still leaves the problem of describing what these effects are. It may be of interest to perform inference on certain linear combinations of the $\Delta_{(j)}$. For instance, is the effect of smoking the same in Los Angeles and New York, i.e., does $\Delta^{(1)} - \Delta^{(2)} = 0$? To estimate an arbitrary linear combination we use the approximate result

$$\sum_{j=1}^{k} c_j \hat{\Delta}_{(j)} \xrightarrow{d} \mathcal{N}\left(\sum_{j=1}^{k} c_j \Delta_{(j)}, \sum_{j=1}^{k} c_j^2 \hat{V}_{(j)} \right)$$

valid for large counts $n_{(j)}$ and fixed number of tables k.

Example 3.4: (Continued) **The Average Effect of Smoking.** Take the c_i to all equal $1/3$. Then the average effect of smoking is the average of 1.897, 0.742, 0.405 giving 1.015. The variance of this estimate is the sum of the individual variances 0.683, 0.466, 0.916 divided by 9, giving 0.2294. Thus, the standard error is 0.4790. Both the estimate and standard error happen to be close to the estimate and standard error for Δ under the uniform treatments hypothesis. This will not be the case if either the model fits poorly or if there are very different numbers of subjects in each group.

Measuring treatment effect by the "average" across the k different groups is a completely arbitrary measure; there is no particular reason for taking the simple average especially if some conditions are either more common or of more interest than others. Quoting any average measure, however, it is chosen, suppresses possibly strong associations in some groups. Residents of New York would presumably want to known about the risk of smoking in their city, not any kind of average effect across three cities. Planners in public health, on the other hand, might well be interested in the average measure.

3.6 SOME CAUTIONS AND QUALIFICATIONS

3.6.1 Hidden Factors and Simpson's Paradox

What would happen if we joined all the k tables into a single 2×2 table? When is it valid to do this? This second question is easy to answer. From the log-likelihood Eq. (3.30) we concluded that *all* the $Y_{1(j)}$ and $T_{(j)}$ are sufficient statistics for the parameters. A necessary and sufficient condition for us to replace the $Y_{1(j)}$ with their sum Y_{1+} without loss of information is that the $\Delta_{(j)}$ are all equal; a necessary and sufficient condition for us to replace the $T_{(j)}$ with their sum T, without loss of information, is that the $\psi_{2(j)}$ are all equal. In this case the log-likelihood for the sufficient statistics Y_{1+}, T is identical to the log-likelihood obtained from collapsing the k tables into a single 2×2 table. In this particular circumstance and no other, there is nothing statistically lost by accumulating the tables into a single table. In particular, *it is not enough* that the $\Delta_{(j)}$ be constant.

So much for the mathematical answer. What are the consequences of combining tables when, for instance, the $\Delta_{(j)}$ are equal but the $\psi_{2(j)}$ are not? In fact, the consequences can be drastic and it is entirely possible that the effect of treatment will appear to be the reverse of the effects seen separately in the k tables. This is not a problem specific to categorical data.

One very plausible reason this reversal can happen is that treatment 1 might be applied more often in sets of conditions unfavorable to response than treatment 2. The accumulated data would then show up an association of treatment 1 with no response and treatment 2 with response, however, this association depends entirely on how we assigned the treatments and not on their effectiveness. A simple example should serve to illustrate.

Example 3.5: A new cancer treatment gives the fictitious results tabulated. We will suppose that the data comprise hospital records over a certain fixed period of time and that therefore all counts are random.

	Survived	Died	Total
New treatment	117	104	221
Old treatment	177	44	221
Total	294	148	442

We see that about 53% survive under the new treatment and about 80% under the old treatment and so it seems that the new treatment is worse than the old. The estimated odds ratio is $\hat{\Delta} = -1.27$. Suppose now that records are examined more closely and it is noted that of the 442 patients 156 were terminal patients and 286 nonterminal. The data in the first table pointed to the superiority of the old treatment. Quite a different pattern is seen when the data are broken down by the prognosis of the patient. Within each of these tables separately, the new treatment gives a higher survival rate than the old treatment. The estimated log-odds ratios are 1.11 and 0.42 respectively, *both* being positive, whereas the estimate from the combined table was negative. The fact that the partial associations in each table may be of opposite sign to the association in the accumulated table, is called *Simpson's Paradox*.

	Terminal			Nonterminal		
	Survived	Died	Total	Survived	Died	Total
New treatment	17	101	117	100	3	103
Old treatment	2	36	38	175	8	183
Total	19	137	156	275	11	286

Since all the counts are random it makes sense to statistically compare not only survival rates for different treatments but survival rates for terminal and nonterminal patients as well as the rates at which the new and old treatments were given to terminal/nonterminal patients. Table 3.10 gives estimated log-odds ratios for the associations of treatment (T), survival (S), and prognosis (P). Each of these are given both separately for both levels of the other factor and for the table accumulated over the other factor.

Table 3.10. Comparison of Combined and Separate Log-Odds Ratios

Log-Odds		$S \times P$	$T \times P$	$T \times S$
Combined		−5.19	2.17	−1.27
Separate	Level 1	−5.29	2.70	1.11
	Level 2	−5.98	2.01	0.42

As expected, survival rates are much smaller for terminal patients than for nonterminal patients (odds ratios around -5) both separately for patients given either treatment and combined. The associations labeled $T \times P$ measure the extent to which the new treatment was given more to terminal patients. Patients have twice the odds of being given the new treatment if they are terminal. This figure varies to some extent when patients who survive or die are considered separately but not drastically. The association of most interest, labeled $T \times S$, changes drastically when broken down by the third factor P as we have seen.

Conditions under which a set of tables may be collapsed have been given by Bishop (1971) in the general context of log-linear models. These conditions will be given in Chapter 6. In the present example we would require either that (i) there is no difference in the survival rates for terminal and nonterminal patients, not only overall but when broken down by the treatment they were given *or*, (ii) there is no difference in the proportion given the new treatment for terminal and nonterminal patients, not only overall but when broken down by subsequent survival. Intuitively speaking the first condition means that prognosis is not associated with survival and the second that prognosis is not associated with treatment. The fact that there has to be no association not only overall but separately for all levels of the other factor *is* a special feature of categorical data analysis at least when treatment effects are measured with *odds*. For instance, when treatment effects are measured by mean response this stronger condition is not needed.

The problem of falsely combining tables is easily remedied by not doing it. Of more concern is the fact that tables may already be combined without the data analyst's knowledge. For instance, if we were unable to identify terminal and nonterminal patients we would not realize how misleading the original table was. Of course, a good analyst would try to ensure that data contained no systematic sampling biases, for instance, by asking "were treatments given equally to patients with different levels of health?" However, there may always be other factors that we might not think to ask about.

The best way of ensuring that all background factors, both visible and hidden, do not bias the results is to perform a randomized prospective study or clinical trial. The process of randomly deciding which individuals are to be given which treatment means that it is very unlikely that treatments will be applied very unequally to different sub-groups. In fact, when treatment allocation is randomized, then the binomial distributions of the response counts are formally correct with probability parameters being the mean of the distribution of the true probability parameters in the sampled population. This is not to say that complete randomization is the best experimental design. As a general principle the power of statistical tests is enhanced if as many treatments as possible are applied to as many homogeneous groups as possible. In practice this means that if we can identify two sub-groups likely to differ in their response then we should decide beforehand to allocate treatments to these groups equally rather than leave it to chance. For instance, we could decide to ensure that of the

156 terminal patients half are given each treatment but decide randomly which patients are given which.

3.6.2 Reasons for Using Log-Odds

In analyzing binomial data we are interested in comparing probabilities, for instance, π_1 and π_2. There are many ways to measure possible differences, say, by $\pi_1 - \pi_2$ or π_1/π_2. There are good reasons why the odds ratio

$$\phi = \frac{\pi_1}{1 - \pi_1} \bigg/ \frac{\pi_2}{1 - \pi_2}.$$

or simple functions of it such as log, are the preferred measure. We list these reasons beginning with those that are first, most important, and second, unique to odds ratio.

Estimability When Sampling is Retrospective Probably the most important reason for using log-odds is that the log-odds *ratio* can be meaningfully estimated from either restrospective or prospective studies. Imagine testing two treatments on a sample of patients. A prospective data set is one where the patients are chosen *prior* to treatment and treatments (1 or 2) randomly assigned to them. In a retrospective study, a number t of surviving patients and $n - t$ patients who died are chosen *after* treatment and it is found that Y_1 of the first group were given treatment 1 and Y_2 treatment 2. The variables t and $n - t$ are then fixed and it is Y_1 and Y_2 that are random. From retrospective data we can estimate quantities like $\Pr(T1|\text{survived})$ and $\Pr(T1|\text{died})$, not the reverse. The more interesting quantities $\Pr(\text{survived}|T1)$ and $\Pr(\text{survived}|T2)$ are completely inestimable. For instance, the proportion y_1/N_1 and y_2/N_2 will depend largely on how many survival records $t = y_1 + y_2$ were chosen; if many were chosen both these proportions will be large; if few were chosen both will be small. We can only estimate these two probabilities directly from a *prospective* study. Let us measure the difference between two probabilities by odds ratio. A simple exercise in conditional probability shows

$$\frac{\text{odds}(T1|\text{survived})}{\text{odds}(T1|\text{died})} = \frac{\text{odds}(\text{survived}|T1)}{\text{odds}(\text{survived}|T2)}.$$

Thus, even though the probabilities $\Pr(\text{survive}|T1)$ and $\Pr(\text{survive}|T2)$ are inestimable from a retrospective study the odds ratio corresponding to these two inestimable probabilities *can* be estimated. This property is unique to odds ratio, or monotonic functions of it. After fitting the logistic binomial model to retrospective data, the estimates of actual survival probabilities and their standard errors have no meaning at all. The estimator and standard error for Δ are meaningful and are numerically the same whether sampling is prospective, retrospective, or indeed Poisson. Of course, the fact that Δ is estimable does not mean that there is nothing to choose between prospective and retrospective sampling. Retrospective studies are subject to hidden sampling biases whereas prospective studies are not.

Admission of Exact Conditional Methods A second important and unique property of using log-odds is that it is only possible to construct exact conditional tests for functions of the odds ratio. More generally exact conditional procedures are available whenever the logistic parameters ψ_i are modeled linearly (see Chapter 7 for details). This is because $\Delta = \log(\phi)$ is the natural parameter of the binomial distribution. It is not possible to construct an exact conditional test of, or intervals for, $\pi_1 - \pi_2$ or π_1/π_2, for instance. Only monotonic functions of the odds ratio work.

Interpretation for Extreme Probabilities Odds ratio is a more meaningful and flexible measure of the relative values of two probabilities regardless of the absolute values. To see this, imagine two treatments which give respective probabilities of survival, $\pi_1 = 0.99$ and $\pi_2 = 0.995$. Both $\pi_1 - \pi_2$ and π_1/π_2 measure these treatments to be quite similar. If two other treatments gave respective survival probabilities $\pi_1 = 0.500$ and $\pi_2 = 0.505$, then the measures $\pi_1 - \pi_2$ and π_1/π_2 hardly change from the first example. Yet in the first example there is surely much greater reason to prefer treatment 2, the probability of dying being *half* that under treatment 1. The odds ratio on the other hand gives quite different measures in these two cases, as we would like, and strongly prefers treatment 2 in the first example. Odds is not the only measure which passes this test. Any function $g(\pi_i)$ whose derivative increases with $|\pi_i - 0.5|$ will do the job.

Invariance Since it is arbitrary which we call response and which we call nonresponse it is desirable that a model transform simply when the roles of success and failure are reversed. Log-odds simply changes sign. Thus, parameters estimated in say a linear model for the logits will simply change sign. The log-odds is not the only transform that has this property. If F is the cdf of any continuous variable distributed symmetrically about zero, then $F^{-1}(\pi_i)$ will have this same property. The complementary log–log transform is occasionally used in data analysis and does not have this property. Reversing the roles of response and nonresponse induces quite a complicated change in the parameter estimates for such models.

Unboundedness A fundamental reason why log-odds ratio is a useful transformation is that the range of this transform is the whole real line, even if one of the π_i is fixed. In contrast, $\pi_1 - \pi_2$ or π_1/π_2 range over sets which depend on the value of π_2. Thus, there is the prospect that ϕ or $\log \phi$ may be used as a measure of the difference of π_1 and π_2 regardless of their absolute values. For any monotonic transform g of π to the whole real line, $g(\pi_1) - g(\pi_2)$ has this property. It is possible to fit linear models to such functions without any restrictions on the range of parameters. Modeling probabilities π_i directly is difficult because they are confined to [0, 1]. For instance, the *dose-response* model that is the binomial data analogue of simple linear regression, has

$$\psi(x) = \alpha + \beta x$$

defining a logistic relation between $\pi(x)$ and the covariate x. These logistic curves are automatically bounded between 0 and 1. Other families can be used for dose-response models but they are not as simply fitted or interpreted. The parameter β gives the change in odds (on the log scale) of response as the covariate is changed by one unit.

Automatic Modeling of Interactions Suppose that experimental conditions are either very favorable or unfavorable to response. If we vary some covariate x then the effect on response probability will be quite limited—it might change from small to very small. There is not much room to move. On the other hand, if conditions are such that response probability is moderate, say around 0.5, then varying the same covariate x may have a more pronounced effect. There is much more room for $p(x)$ to vary in either direction. On the logistic scale the above arguments do not apply—there is plenty of room for logit(p) to move no matter what the underlying value because it is unbounded.

If we model such data on the probability scale then we will require some interaction terms—the effect of the covariate x will vary with the experimental conditions. Logistic models can automatically produce this behavior without any interaction terms. This offers the possibility of a simpler model and a simpler description of the data.

Improved Accuracy of Normal Approximation The ML estimate of Δ has a distribution closer to normal than say the difference $\hat{\pi}_1 - \hat{\pi}_2$. Closely related to this is the fact that the profile log-likelihood for Δ is closer to quadratic than the profile for $\pi_1 - \pi_2$. Log-odds is not unique in this. For instance, $\log(\pi_1/\pi_2)$ enjoys much the same properties.

3.6.3 Some Minor Difficulties with Log-Odds*

Combining Odds One unattractive feature of using odds or log-odds is that when a population is divided into two groups the (log)odds for the population is not a weighted average of the (log)odds for the two groups separately. For instance, the odds ratio for the male population might be 2.0 and for the female population 2.0 but it does not follow from this that the odds ratio for the combined population is 2.0.

*See p. 15 in Chapter 1 for an explanation.

As a simple example, imagine that we divide the population of a certain country into the oldest half and the youngest half. Suppose that the proportion of older individuals who have some sign of cardiovascular disease is 89% for nonsmokers and 94.2% for smokers. The association of smoking and cardiovascular disease as measured by odds ratio is then 2.00. Suppose that for younger individuals 13.7% of nonsmokers have some sign of disease compared to 24.1% of smokers. The odds ratio is again 2.00. Now, since we have divided the population into two equal-sized groups, the proportion of nonsmokers with some sign of disease is $(89.0 + 13.7)/2 = 45.35\%$ and the proportion for smokers is $(94.2 + 24.1)/2 = 59.15\%$. The odds ratio from these two figures is not 2.00 but is 1.745.

There is nothing particularly wrong with this when we realize that odds ratios tend to be larger for comparing extreme probabilities than nonextreme probabilities. The differences between the proportions for smokers and nonsmokers *is* the average of the differences for the two age groups. The odds ratio simply measures these differences in a different (and nonlinear) way. One consequence of this is that tables of count data can only be combined into a single table under more stringent conditions than is the case for continuous data analyzed through means. Combining the two age groups in the example above gives a different measure of the effect of smoking. This is not so much a problem with, as a property of, odds.

Exaggeration of Association Consider a $2 \times k$ table where the k categories are obtained by classifying some underlying continuous variable into categories. The resulting categorical variable is then ordinal. A single number summarizing the overall association of the response factor and the ordinal column factor may be defined as follows. Two individuals are said to be concordant if the diagonal joining their cells slopes down to the right and discordant otherwise. A generalization of the odds ratio to the $2 \times k$ table is the ratio of concordant to discordant pairs

$$\phi = \pi_c/(1 - \pi_c)$$

where π_c is the probability that two individuals are concordant. For a 2×2 table this reduces to the ordinary odds ratio [see Agresti (1980)]. The generalized measure extends to $r \times c$ tables where row and column factors are ordinal and a simple formula for π_c in terms of the cell probabilities π_{ij} is available.

The difficulty arises when we consider the effect on the odds ratio of different methods of categorizing the continuous variable underlying the ordinal column factor. For instance, what happens if we combine two columns? As an example consider the data of Rosenberg et al. (1988) relating nonfatal heart attacks in men to their coffee consumption. Using the first five columns the generalized odds ratio is 2.02, whereas categorizing individuals as either nondrinkers or drinkers and using the 2×2 table comprising the first and last four columns, the odds ratio is 2.56.

	Coffee Consumption (Cups/Day)					
	0	1–2	2–4	5–9	10	> 0
Controls	205	275	219	199	70	763
Cases	147	325	351	437	288	1401

One might say it is inappropriate to categorize the column factors into just two categories if this affects the odds ratio so much and that one should really be measuring the association as a function of the number of cups of coffee per day (see Section 4.3). On the other hand, the odds ratio 2.56 is of some interest and the estimator does correctly estimate the true underlying association of heart attack with drinking coffee at all. But it is worrying that the way the column factor is defined affects the odds ratio so much.

The general tendency is for the odds ratio to become larger as categories are combined. It would be useful to have a simple formula to correct for this and it is demonstrated in O'Gorman and Woolson (1993) that

$$\Delta\sqrt{\frac{k}{k+2}}$$

is a better measure than Δ of the underlying association with the continuous variable. Using this modified log-odds ratio, results will depend less on how many categories are chosen for the column factor. The adjustment works best when equal numbers of observations occur in each column and when the underlying continuous factor is normal.

The tendency of the odds ratio to exaggerate the association with an underlying continuous variable is not unique to odds ratios—most other measures of association display the same feature.

3.6.4 Sparse Tables with Many Parameters*

Asymptotic results tell us what happens in the limit. They provide a good working approximation only if we can argue we are close to this limit. For instance, if all fitted values are small, the limiting chi-square distribution for large fitted values will not be useful. Models that involve many parameters also present difficulties because the asymptotics of likelihood theory require the parameter to have fixed dimension p while n increases. If p is large and of similar order to n then it makes little sense to approximate distributions by a theoretical limit that assumes a fixed number of parameters. The consequences of having too large a parameter vector are potentially disastrous and both estimators and tests can be *inconsistent* in realistic cases. This is of particular relevance to contingency table data where the data are broken down by other factors.

Inconsistency of the ML Estimator for Twins Data The first example of inconsistent ML estimation was given by Neyman and Scott (1948). The following, quite realistic, example illustrates the extreme forms of theory breakdown that are possible, especially with binary data.

Imagine testing a set of identical twins, one twin being randomly chosen to be given treatment 1, the other given no treatment (treatment 2). Each twin either responds or does not and the data could be set out in an extremely sparse contingency table with $n_1 = n_2 = 1$ and with four possible count patterns in the table (1, 0, 1, 0), (0, 1, 1, 0), (1, 0, 0, 1) and (0, 1, 0, 1). As usual the treatment effect Δ will represent the log-odds ratio of response under treatment 1 to treatment 2 with ψ_2 being the log-odds of response under no treatment. This nuisance parameter measures the underlying response rate for this particular pair of twins. Now imagine testing k sets of twins in the same way and assuming a common treatment effect Δ. Every pair of twins tested adds two binary observations to the data set and one new parameter $\psi_{2(j)}$. This data set displays two extreme features hostile to the asymptotic theory. First, the sample sizes are small for each table and second there are many (nuisance) parameters.

By finding an expression for the profile ML estimator $\hat{\psi}_{2(j)}(\Delta)$, and substituting this in the score function $\partial l/\partial\Delta$, it can be shown that the ML estimator of Δ satisfies

$$\frac{\partial l}{\partial \Delta}(\Delta, \hat{\psi}_2(\Delta)) = Y_{1+} - M_2 - M_1 \frac{\exp\{\Delta/2\}}{1 + \exp\{\Delta/2\}} = 0 \tag{3.35}$$

where M_i is the number of tables for which $T_{(j)} = i$ [see Breslow (1981)]. What happens to the solution for large k? Well, Y_{1+}, M_2, and M_1 all approach their expectations by the Law of Large Numbers. Substituting these limits into the above equation, one can show that $\hat{\Delta} \rightarrow 2\Delta$. Details are in Exercise 3.25. The estimator $\hat{\phi}$ of the odds ratio thus converges not to ϕ but to ϕ^2. For instance, if the true odds ratio is 3 then as we acquire more and more twins and test them, the estimator will tend not to 3 but to 9!

Tables with $n_1 = n_2 = 1$ and k large are an extreme case though by no means uncommon. The analysis does suggest that in practice the ML estimator might be severely biased whenever there is a large number of tables each with a moderate amount of data, and this has been confirmed theoretically.

The primary reason the ML estimator is inconsistent when k grows large is that the unconditional ML estimating Eq. (3.35) for Δ does not have mean zero when ψ_2 is estimated. How then does one estimate Δ for twins data? The most common solution is to consider the likelihood conditional on the $T_{(j)}$. The resulting conditional ML estimator $\hat{\Delta}_c$ satisfies

$$Y_1 - M_2 + M_1 \frac{\exp\{\Delta\}}{1 + \exp\{\Delta\}} = 0.$$

Because this equation has mean zero, the conditional estimator is consistent; as more and more twins are tested the conditional estimator $\hat{\Delta}_c$ converges to Δ not to 2Δ. Conditional methods are covered in Chapter 7.

Deviance for Binary Data The sparse twins data set can be used to demonstrate another important theory breakdown. It is an example of a *binary* data set, i.e., data where each n_i equals 1. Standard goodness-of-fit statistics are completely worthless for testing goodness-of-fit of models to binary data. It can be shown that the deviance statistic for a nonempty logistic linear model reduces to

$$\mathcal{D} = -2 \sum_{i=1}^{k} \{\hat{\pi}_i \text{logit}(\hat{\pi}_i) + \log(1 - \hat{\pi}_i)\}. \tag{3.36}$$

This depends on the binary data y_i only through the fitted probabilities $\hat{\pi}_i$ and is uninformative about how close the $\hat{\pi}_i$ are to the y_i. Thus, no matter how wrong the model may in fact be, the LR test statistic will not detect it; see further in Section 4.1.2. Pearson's goodness-of-fit statistic fares no better, as it can be shown that $S = n$.

How should model fit be tested for binary data? The main problem is that all the fitted values are small (in fact smaller than 1). By grouping observations together one can compute a grouped deviance or Pearson statistic that *will* be informative and close to chi-square if all grouped fitted values are reasonably large. A systematic method of grouping the binary observations is to sort them from smallest to largest, suggested by Hosmer and Lemeshow (1980). One can also test binary models against specific departures by using *change* in deviance, i.e., we may fit null and alternative models to the binary data and even though the deviance $\mathcal{D}_0$, $\mathcal{D}_1$ do not measure fit of either model, their *difference* can still be used to test one model against the other.

FURTHER READING

Section 3.4: Analysis of Several 2 × 2 Tables

3.4.2 *When contingency table data for the association of factors X and Y are collected under several sets of conditions measured by a variable Z, the latter is often called a confounding variable. If it is not taken into account in the analysis, the association of Z with X or Y may bias or confound the estimated association of X with Y. This is what we have called Simpson's paradox. When the association of X and Y actually varies with Z the latter is often called effect modifying. This is called nonuniform treatment effects. A readable discussion on estimating odds ratios in the presence of confounding or effect modifying factors for retrospective data is given by Prentice (1976a) who illustrates the methods on endometrial cancer data.*

3.7 EXERCISES

These exercises form the main practice material for both Chapters 3 and 7. Exercises that refer to *exact* or conditional methods rely on material in Chapter 7. Exact methods are an alternative method for analyzing binary contingency tables, so it is pertinent to compare the methods on the same data sets.

3.1. Let X be $Bi(n, p)$ and consider the family of *modified* logistic transforms

$$Z_a = \log\left(\frac{X + a}{n - X + a}\right)$$

where a is to be chosen. Use the following facts:

a. $X/n = p + Z/\sqrt{n} + O_p(n^{-1})$ where $Z \stackrel{d}{=} N(0, p(1-p))$ by CLT,

b. $\log(1 + x) \approx x - \frac{1}{2}x^2 + O(x^3)$ for small x,

and show that only when $a = 1/2$ is

$$E(Z_a) = \log\left(\frac{p}{1-p}\right)$$

correct up to the $O(n^{-1})$ term [see Gart and Zweifel (1967)].

3.2. The data in Table 3.11 is from a study of 65 pregnant women at high risk of pregnancy-induced high blood pressure. They participated in a randomized clinical trial to determine the effects of a 100 mg daily dose of aspirin during the last three months of pregnancy.

Table 3.11. Cross Classification of 65 Pregnant Women by Appearance of High Blood Pressure and Treatment with Aspirin [from Schiff et al. (1989)]

	High Blood Pressure	Satisfactory Blood Pressure	Total
Aspirin treatment	4	30	34
Placebo treatment	11	20	31
Total	15	50	65

a. Estimate Δ, test $\Delta = 0$, and given an approximate 95% confidence interval for Δ using modified and unmodified empirical logits.

b. Repeat (a) using conditional methods.

3.3. The reliability of a production process is evaluated by taking a random sample of 10 items 7 of which are satisfactory. The production process is then reviewed in an effort to improve it. A subsequent sample of 7 items were all satisfactory; see Table 3.12.

Table 3.12. Production Reliability after Review

	Satisfactory	Faulty	Total
After	7	0	7
Before	7	3	10
Total	14	3	17

a. Under the hypothesis that there has been no change in the reliability of the process and that all tested units behave independently, list the conditional probability distribution of the number of satisfactory units in the "after" sample. Give the mean of this distribution.

b. Using the distribution in (a) give an exact P-value for testing whether or not the reliability has improved.

c. Estimate the ratio of the odds of satisfactory performance for the process after review compared to before, using unconditional empirical logits, modified logits, and the conditional likelihood.

d. Is it possible that the reliability has actually gotten worse? Specifically, test the hypothesis that the odds of a satisfactory unit after the change is half the odds of a satisfactory unit before the change. List the relevant conditional distribution.

e. Calculate an approximate chi-square statistic for testing the hypothesis in (a) and give the one-sided P-value.

3.4. Fisher (1935) described a tea-tasting experiment where a subject claimed to be able to detect whether milk or tea was added first to the cup. The subject was given eights cups and told that in four the milk was added first and in four the tea added first. Of the four cups with milk added the first three were correctly identified.

a. List the data in a table and interpret the odds ratio ϕ in words.

b. Why is the hypergeometric distribution the only sensible model for the number of correct identifications of the "milk-first" cups?

c. Estimate $\Delta = \log \phi$ and give the standard error using the recursive algorithm [see Eq. (7.8)]. Find an "exact" 95% interval for Δ and give the P-value for testing $\Delta = 0$ against $\Delta > 0$.

d. Let the exact interval in (c) be (Δ_l, Δ_u). Compute the likelihood ratio $L(\Delta_u)/L(\Delta_l)$. Is this answer generally true?

3.5. After a course on customer service, 30 customers are asked whether or not they were completely satisfied with their service. A year later the same experiment is conducted with a view to seeing whether or not the effects of the customer service course have waned. The data are given in Table 3.13.

Table 3.13. Customer Satisfaction after Course

	Satisfied	Unsatisfied	Total
1st sample	28	2	30
2nd sample	22	8	30
Total	50	10	60

a. Give the distribution of the number of satisfied customers in the first sample conditional on the total number of satisfied customers assuming there has been no change in service quality. Give a P-value for testing the hypothesis that the service quality has not changed.

b. Find the conditional ML estimate of the log-odds ratio.

c. Calculate empirical logits, their approximate standard errors, and 95% confidence intervals for the log-odds ratio. Repeat using modified empirical logits.

d. Report the LR statistic for testing that the quality of service has not changed using both the independent binomial and independent Poisson models. Identify relevant parameter estimates, standard errors, and test statistics in the two approaches.

3.6. It had been suggested that erosion of dental enamel is increased by exposure to chlorine in swimming pools. The data in Table 3.14 are from a retrospective case-control study. The explanatory factor is whether or not the subject swims for more than six hours per week. Is there evidence for a positive relationship between chance of dental erosion and exposure to chlorinated water?

Table 3.14. Case-Control Study of 49 Subjects with Dental Erosion and 245 Without, by Reported Exposure to Chlorinated Water [from Centerwell et al. (1986)]

	Erosion	No Erosion	Total
More than six hours swimming	32	118	150
Less than six hours swimming	17	127	144
Total	49	245	294

3.7. For two independent binomial trials with probabilities of success π_1, π_2, write down algorithms for computing the profile likelihood function for

π_1/π_2 and for $\pi_1 - \pi_2$. Using the mice-tumor data, compare the profile deviance function for $\theta = \pi_1 - \pi_2$ with the Wald-type statistic

$$(\hat{\pi}_1 - \hat{\pi}_2 - \theta)_2/(\hat{\pi}_1(1 - \hat{\pi}_1)/n_1 + \hat{\pi}_2(1 - \hat{\pi}_2)/n_2).$$

3.8. In a study of 65 patients who had received sodium aurothiomalate (SA) as a treatment for rheumatoid arthritis, Ayesh et al. (1987) examined the possibility that allergy to SA might be predicted by sulphoxidation (SI) (Table 3.15). Is there evidence for a relationship between SA and SI?

Table 3.15. Cross Classification of 65 Patients with Rheumatoid Arthritis by Allergic Reaction to SA Treatment and Normal or Abnormal SI [from Ayesh et al. (1985)]

	Toxic Reaction	No Reaction	Total
Abnormal SI	30	9	39
Normal SI	7	19	26
Total	37	28	65

3.9. Consider testing $\mathcal{H}_0 : \Delta = \delta_0$ against $\mathcal{H}_1 : \Delta > \delta_0$ using the conditional approach. The aim of this question is to justify rejection regions of the form $\{ y_1 \geq c \}$.

a. Show that the conditional mean $\mathrm{E}_c(Y_1 | t, \Delta)$ is a nondecreasing function of Δ. Thus, sufficiently large values of Y_1 point towards the alternative hypotheses.

b. If $\delta_1 > \delta_0$, show that the log-likelihood ratio

$$l_c(\delta_1) - l_c(\delta_0)$$

is a monotonic increasing function of y_1, using Eq. (7.4). Thus, for any δ_1, the most powerful test of δ_0 against δ_1 rejects if $y_1 \geq c$.

3.10. Suppose that $Y_1, \ldots, Y_k$ have the multivariate tilted hypergeometric distribution Eq. (7.12). Show that the marginal distribution of Y_i is tilted hypergeometric

a. By summing the distribution over the other arguments.

b. By noting that, before conditioning on their total, the Y_i are independent. (The order of conditioning has no effect.)

c. Show that when $k = 2$, the multivariate tilted Hg distribution reduces to the univariate tilted Hg distribution given in Eq. (7.2).

d. For the special case of a logistic regression, $\Delta_i = i\beta$. By substituting this into Eq. (7.13) find the sufficient statistic for β and an expression for the log-likelihood function.

3.11. Roth et al. (1975) carried out a study of 173 skin cancer patients to see if those who show no reaction to a contact allergen DCNB also show a negative response to croton oil, a skin irritant. They were also interested in whether the reaction to DCNB was related to the stage of the skin cancer. Data relating to these two issues are listed in Table 3.16. Is DCNB reactivity related to croton oil reactivity? Is it related to stage of skin cancer? What data would allow us to answer these questions better?

Table 3.16. DCNB Reactivity Experiment [from Roth et al. (1975)]

	Croton +	Croton −	Stage I	Stage II	Stage III
DCNB +	81	23	39	39	26
DCNB −	48	21	13	19	37
Total	129	44	52	58	63

3.12. Academic *inbreeding* is the practice of university departments hiring their own graduates as faculty. Too much inbreeding is considered bad for scholarly innovation and breadth. The data in Table 3.17 refers to a sample of academics, each of whom were classified by (i) whether they themselves were graduates of their own department, i.e., inbred, and (ii) the size of their department. Is there statistical evidence that the levels of inbreeding depend on the size of department?

Table 3.17. Academic Inbreeding

	Size of Faculty		
	<60	60–89	>89
Inbred	78	211	610
Not inbred	128	263	502

Academics classified by size of department and inbreeding [from Kornguth and Miller (1985)].

3.13. Recall the eyestrain data used for illustrative purposes in Section 3.3 and listed in Table 3.4. Break the total chi-square into three orthogonal contrasts, one of which compares VDU to non-VDU work.

3.14. The data in Table 3.18 are adapted from table 2/7 in Social Indicators III (1980), published by the Washington, DC Government Printing Office. It is a classification of admissions to a country mental hospital by diagnosis and gender.

Table 3.18. Mental Health Diagnosis and Gender

	Depressive Neurosis	Personality Disorders	Drug-Related Disorders	Childhood Disorders
Males	9058	16,999	15,373	8324
Females	14,036	7064	5759	4723

a. Is there sufficient evidence to conclude that the pattern of mental illness in males and females is different?

b. Bearing in mind the large sample, is a statistical test sensible?

3.15. Does growing up in a right-handed world exert an overwhelming bias toward righthandedness? I certainly would have thought so. Yet data relevant to this question were painstakingly gathered by Carter–Saltman (1980) and are displayed in Table 3.19. Some 808 children were sampled, 408 of these being adopted and 400 not. The handedness of the child and both parents was measured. If handedness is mostly biologically determined then one expects to find more right-handed children among those with right-handed parents but for adopted children, one expects to find the same proportions of right-handed children regardless of the handedness of the parent. Note that there were no left-handed parents couples.

Table 3.19. Handedness of Children and their Parents

Father	Mother	Biological Parents	Adoptive Parents
Right	Right	300/340	308/355
Right	Left	29/38	12/16
Left	Right	16/22	35/37

Each entry is the number of right-handed children from the total children in that cell [from Carter–Saltzman (1980)].

a. Test whether or not the handedness of the child is related to that of the biological parents.

b. Test whether or not the handedness of the child is related to that of the adoptive parents.

3.16. Table 3.20 lists fictitious data on 121 individuals who had their respiratory lung capacity measured and their smoking habits classified as none, occasional, regular, and heavy. Lung capacity is classified as either normal or impaired.

Table 3.20. Data on Smoking and Lung Capacity

	None	Occasional	Regular	Heavy	Total
Normal	36	25	28	5	94
Impaired	6	6	8	7	27
Total	42	31	36	12	121

a. Use the LR to test whether or not the level of smoking has any effect on the chance of lung impairment.

b. Perform the above test in GLIM using the Poisson distribution. In particular, identify the LR statistic and estimates of the change of lung impairment both under the hypothesis that smoking has no effect and otherwise.

c. Simulate four random variables from the binomial distribution with $p = 94/121$ and $n = 42, 31, 36, 12$, respectively. Calculate the LR statistic from these simulated data values for testing equality of the π_i. Repeat this 10,000 times and compare the distribution with χ_4^2. Give the true P-value for the observed value of the LR statistic.

d. Suggest two reasons why you might have been able to anticipate that the chi-square approximated P-value, calculated in part (a), was poor without doing the simulations.

3.17. A cancer study reported by Rosner (1986) produced the results shown in Table 3.21. The data were collected retrospectively from a sample of cancer patients and a random sample of controls who had not been diagnosed with cancer. Each individual was asked whether or not they smoked and moreover whether or not they were exposed to the passive smoke of others in their place of residence. Since it is already known that smoking increases the probability of smoking related cancers, the main interest is in the effect of passive smoking over and above the effect of active smoking.

Table 3.21. Data on Passive Smoking and Cancer [from Rosner (1986)]

	Nonsmokers		Smokers	
	Passive	Nonpassive	Passive	Nonpassive
Case	120	111	161	117
Control	80	155	130	124

a. Assuming the effect of passive smoking on probability of being a case is identical for smokers and nonsmokers, estimate the common log-odds using a weighted combination of modified estimates from each table and give a standard error.

b. Test the hypothesis that the odds ratio is identical for smokers and non-smokers. First use a weighted least squares approach and then compare the results with LR.

c. Assuming uniform effect of passive smoking, perform the Yates test of no effect and compute a 95% confidence interval for the common effect (you will need a program that computes mean and variance of tilted hypergeometric distributions).

d. What does one infer about the effect of passive smoking when active smoking is ignored? What other factors might have been taken into account to minimize such bias effects?

3.18. A new arthritis treatment was developed and given to some outpatients at Hospital A and some other patients in the aged section of Hospital B, during 1989. Later, records were examined but for ethical reasons only a portion of records could be released—215 from A and 99 from B. These records were classified into those who had or had not been given the new treatment. Subsequently, these 314 patients were evaluated for significant improvement as measured by increased bone calcium density and the results are given below. By this time authorities gave approval for a controlled study. In 1990, 400 patients from A and 240 patients from B were chosen and half of each group randomly assigned to be given the new treatment. The other patients were given the treatment appropriate to their conditions before the development of the new treatment. The results are listed in Table 3.22.

Table 3.22. Effectiveness of Arthritis Treatments

	Hospital A				Hospital B			
	1989		1990		1989		1990	
	Improved	Worse	Improved	Worse	Improved	Worse	Improved	Worse
New	50	26	68	132	26	57	38	82
Old	78	61	53	147	4	12	33	87
Total	128	87	121	279	30	69	71	169

a. It appears that there was a smaller proportion of improvements in 1990 than in 1989 for hospital A. The improvement rate was also much lower for hospital B than A in 1989. What do you think could be the explanation for this?

b. Assuming that the effect of treatment is the same in each of the four conditions, approximate the exact conditional P-value for testing whether or not the new treatment has a *beneficial* effect.

c. Estimate the log-odds ratio in each of the four tables and use these to first, test the uniform treatments hypothesis and second, to estimate an assumed common treatment effect. Give a confidence interval for this last parameter.

d. Fit the full model and uniform treatments model in GLIM using the independent binomial model. Test the uniform treatments hypothesis using Pearson chi-square and compare the estimate of the common treatment effect with the estimate in (c).

e. Combine the four tables into a single table and test if the new treatment is beneficial from this table. Try to explain intuitively why the data now indicates here that the new treatment is ineffective when we saw earlier that it was effective.

f. Fit the full model in GLIM and identify the estimated treatment effects. Also give the estimates of the $\psi_2^{(j)}$. Which, if any of these eight parameter estimates, have meaningful interpretations?

3.19. A sample of 146 five-year-old children had their teeth examined and those with decayed, missing, or filled teeth (dmft) were noted. From their address it was also determined whether their drinking water was fluoridated. Finally, on the basis of various sociological measurements their social class was designated as I, II, III, or unclassifiable. The data are given in Table 3.23.

Table 3.23. Fluoridation and Dental Health

	Class I	Class II	Class III	Unclassified	Total
Fluoridated	12/117	26/170	11/52	24/118	73/457
Unfluoridated	12/56	48/146	29/64	49/104	138/370

Note: Each entry gives the number of children with dmft from the total number of children in that cell [from Carmichael et al. (1989)].

a. Test if fluoridation effects depend on social class.

b. Assuming uniform fluoridation effects across social class, give a 95% interval for the relevant odds ratio and interpret in words.

c. Summarize the effect of social class, if any.

3.20. In Hong Kong, between August and October, 1996, there were 279 suicides or suicide attempts by people over the age of 60. Table 3.24 classifies each case by gender and the number of times suicidal intention was indicated to family and friends.

Table 3.24. Number of Times Suicidal Intention Expressed by Gender Among 279 Suicide Attempts

	Never	Once	Twice	3 Times	Several	Many
Males	119	26	4	1	1	0
Females	98	22	6	1	0	1

Source: From Elderly Suicide in Hong Kong (Chi, Yip, and Yu, 1997), Befrienders International.

a. Is there a difference between the genders? How reliable is the chi-square approximation?

b. Do an exact conditional test of independence.

3.21. A small clinical trial is performed on 60 patients, 40 of whom are male and 20 female. Within each sex group exactly half are chosen to be given a new treatment, the remainder being given the old treatment. Patients are then tested for improvement in their condition. For females, 4 of 10, improve under the new treatment and 1 of 10 under the old. For males, 18 of 20 improve under the new treatment and 12 of 20 under the old.

a. Show that the estimated odds ratio $\hat{\phi} = 6$ for each sex. Give the WLS estimate of the log-odds ratio $\Delta = \log \phi$ and its standard error. Compare with the ML estimator and its standard error.

b. Show that when tables are combined the estimate of the odds ratio is not 6. Give a standard error for $\hat{\Delta}$.

c. Repeat the calculations using modified logits and show that the estimate from the combined table does not lie between the estimates from the separate tables.

3.22. Let Y_{ij} denote the entries of a 2×2 table. Under the hypothesis of independence

$$\hat{e}_{ij} = \frac{Y_{i+}Y_{+j}}{Y_{++}}$$

where + denotes summation over a subscript.

a. Show that the row and column totals of the fitted values agree with the row and column totals of the data. Hence, show that the four differences $|Y_{ij} - \hat{e}_{ij}|$ for $i, j = 1, 2$ are all equal.

b. Show that the Pearson statistic [see Eq. (2.25)] which is a sum of four terms for the 2×2 table can be written as

$$S = \frac{n(Y_1 - \hat{e}_{11})^2}{\hat{e}_{11}\hat{e}_{22}}.$$

Hence, show that the Yates corrected Pearson statistic differs from the approximation to the conditional test [see Eq. (7.10)] by the factor $(n-1)/n$. Since this factor is so close to one, either are called the Pearson statistic with Yates correction.

3.23. Let $\hat{\Delta}_j$ be different estimators of a common value Δ, each having asymptotic normal distributions with mean Δ and variance $V_j \to 0$. Let $\hat{w}_j$ be consistent estimators of $w_j = 1/V_j$, i.e., $\hat{w}_j V_j \to 1$. Give an expansion of the estimator

$$\tilde{\Delta} = \sum_{i=1}^{k} w_i \hat{\Delta}_i \Big/ \sum_{i=1}^{k} w_i$$

about the estimator

$$\hat{\Delta} = \sum_{i=1}^{k} \hat{w}_i \hat{\Delta}_i \Big/ \sum_{i=1}^{k} \hat{w}_i$$

when the $\hat{w}_j$ are close to the w_j and notice that $\partial \hat{\Delta}/\partial \hat{w}_j = 0$ as the $\hat{\Delta}_j \to \Delta$. Hence, conclude that the asymptotics of $\tilde{\Delta}$ and $\hat{\Delta}$ are the same.

3.24. Table 3.25 summarizes the results of a survey of 218 Baltimore mothers of children displaying behavioral problems at school, classified by (i) whether or not they have previously lost a baby, and (ii) the birth order of the child. Children with no siblings were excluded from the study. A set of 147 mothers whose children did not display behavioral problems were used as controls.

Table 3.25. Relationship of Behavioral Problems to Birth Order and Previous Loss of Siblings [from Cochran (1954)]

Birth Order	2		3–4		5+		
Losses	Yes	No	Yes	No	Yes	No	Total
Problems	20	82	26	41	27	22	218
Controls	10	54	16	30	14	23	147
Total	30	136	42	71	41	45	365

a. Model the binary response of being a problem child in terms of birth order and loss. Let the effect of loss on this response be of primary interest. Test the uniform treatments hypothesis and summarize the average effect of loss on the odds of being a problem child.

b. Check to see if Simpson's paradox occurs when the data are collapsed over birth order.

3.25. Consider a $2 \times 2 \times k$ table where all the $n_{i(j)}$ equal 1. An example is twins data discussed on p. 161. Let Δ be the assumed common log-odds ratio and $\psi_{2(j)}$ the nuisance parameter in Table j. In this exercise the aim is to derive the ML estimating Eq. (3.35) and to show that the solution is inconsistent.

a. The score function with respect to $\psi_{2(j)}$ is $T_{(j)} - \pi_{1(j)} - \pi_{2(j)}$. Show that when $T_{(j)} = 0$ the ML estimator $\hat{\psi}_{2(j)} = -\infty$ regardless of the value of Δ. Similarly when $T_{(j)} = 1$ show that $\hat{\psi}_{2(j)} = +\infty$.

b. When $t_{(j)} = 1$ show that $\hat{\psi}_{2(j)}(\Delta) = -\Delta/2$.

c. Letting M_i be the number of tables with $T_{(j)} = i$, substitute $\hat{\psi}_2(\Delta)$ into the score function $\partial l/\partial\Delta$ and verify Eq. (3.35).

d. For large k, $Y_{1+}/k \rightarrow \sum \pi_{1(j)}$ and $M_i/k \rightarrow \Pr(T_{(j)} = i)$. Substitute these into Eq. (3.35). By removing the summation sign show that the solution $\hat{\Delta}$ satisfies $\text{expit}(\Delta) = \text{expit}(\hat{\Delta}/2)$. Since this solution is identical for each of the equations summed, it solves the summed equation and so $\hat{\Delta} \rightarrow 2\Delta$.

CHAPTER 4

Binomial Regression Models

Chapter 3 presented theory and methods for analyzing binary outcomes classified by one or two factors. In this chapter we extend the range to situations where the binary outcomes are explained by several classifying factors and several covariates. For normal data, the general linear model unifies analysis of variance, multiple regression, and analysis of covariance. For binary data, the unifying family of models is the family of generalized linear models. This family includes linear models for the log-odds, called logistic linear models. These have several advantages both conceptual and computational and will be emphasized throughout the chapter. The case where one or more of the explanatory variables is neither nominal nor interval but *ordinal* is discussed in chapter 6.

In this chapter we study both the theory and application of *linear binomial* models. Diagnostics for assessing both the local and overall fit are illustrated in Section 4.2. Models for so-called over-dispersed binomial data are studied in Section 4.4. Under certain conditions linear binomial models, including overdispersed versions, may be approximately fitted by weighted least squares and this connection is explored in Section 4.5.

4.1 LINEAR BINOMIAL MODELS

The notation for this chapter slightly extends that of previous chapters. Let the data comprise a set of binomial counts $Y_1, \ldots, Y_k$ these being counts of experimental units that respond *success* from possible counts $n_1, \ldots, n_k$. Associated with each response is a vector

$$x_i = (x_{i1}, x_{i2}, \ldots, x_{ip})^T$$

of explanatory variables of dimension $p \leq k$. The probability of response π_i for these n_i individuals is to be explained in terms of these variables through the *systematic component*

$$\pi_i = \pi(x_i) \qquad i = 1, \ldots, k.$$

The unknown regression function $\pi(x)$ is to be determined. What kinds of functions are plausible? The first thing we note is that $0 \leq \pi \leq 1$, so that linear models for $\pi(x)$ in terms of x are unlikely to be useful.

Unlike probabilities, logits are unconstrained so we could model them linearly. But there are other transforms besides log-odds that can also be used. Let g be a known function from [0, 1] onto $\mathcal{R}$, called the *link* function and consider regression functions of the form

$$g(\pi_i) = x_i^T\beta = x_{i1}\beta_1 + x_{i1}\beta_2 + \cdots + x_{ip}\beta_p \tag{4.1}$$

where the β_j are unknown parameters. This is the binomial data analogue of the multiple regression model. Together with the assumption of a binomial error distribution, the preceding model is a member of the class of *generalized linear models*, the theory for which is given in Section 4.6. When g is the logit transform the model is called logistic linear.

As with linear models for normal data, the mathematics of generalized linear models is both briefer and more illuminating when expressed in matrix notation. The linear predictor is

$$g(\pi_i) = (X\beta)_i$$

where X is the $(k \times p)$ design matrix set by the particular model and data structure we have. The model is specified by g and X. Whichever computer package is used to fit the model, both of these must be specified by the user, for instance by the `LINK` and `FIT` directives in `GLIM`.

Numeric explanatory variables such as age are called *covariates* and each covariate appears as a single column of X. Categorical explanatory variables such as sex are called *factors*. If there are l levels of the factor then there will be $l - 1$ columns of 1's and 0's in the design matrix X for that factor. There are several alternative ways of constructing these columns, which are also called *dummy variables*. An explanatory variable that is constantly equal to one corresponds to an overall mean term in the model and will appear as a column of ones in the design matrix X, by convention usually the first column. The number of linearly independent columns associated with an explanatory variable is its *degrees of freedom*. Thus, covariates all have one degree of freedom. In summary, the matrix algebra carries over directly from normal linear models and involves no new ideas at all. The underlying binomial distribution and the link function are the new additions.

4.1.1 The Logistic Linear Model

When the link function $g(x) = \log(x/(1-x))$ the model is called logistic linear. As in earlier chapters, we denote the logit probabilities or *logits* by $\psi_i = g(\pi_i)$. The regression function Eq. (4.1) becomes

$$\psi_i = \text{logit}(\pi_i) = x_{i1}\beta_1 + x_{i2}\beta_2 + \cdots + x_{ip}\beta_p \qquad i = 1, \ldots, k. \tag{4.2}$$

For binomial models, the logistic link function has special properties not shared by any other link function. For this reason it is called the canonical or natural link. This does not, of course, imply it is the "true" link.

Among the many advantages of the logistic link is the simple interpretation of the parameters. The parameter β_j measures the expected change in ψ when x_j changes by one unit, all other explanatory variables held constant. This is just like multiple regression except it is the logits rather than the means that change linearly. In communicating results it is preferable to use odds rather than log-odds. The odds $\phi = \exp(\psi)$ will increase by a factor $\exp(\beta_j)$ when x_j changes and the *percentage* increase will be $100(\exp(\beta_j) - 1)$. For example, suppose that in relating cancer rates to the number of cigarettes smoked per day we found an estimate $\hat{\beta} = 0.041$. Then we estimate that the odds of cancer is $100(\exp(0.041) - 1) = 4.2\%$ higher for every cigarette per day smoked.

The log-likelihood for the parameter β of a logistic linear model is obtained by substituting the relations [see Eq. (4.2)] into the general binomial log-likelihood Eq. (3.6). Reversing the order of summation and letting $S_j = \sum_{i=1}^{k} X_{ij}Y_i$

$$\begin{aligned} l(\beta) &= \sum_{i=1}^{k} y_i\psi_i - n_i \log(1 + \exp\{\psi_i\}) \\ &= \sum_{i=1}^{k} y_i \sum_{j=1}^{p} X_{ij}\beta_j - n_i \log(1 + \exp\{\psi_i(\beta)\}) \\ &= \sum_{j=1}^{p} \beta_j \sum_{i=1}^{k} X_{ij}y_i - n_i \log(1 + \exp\{\psi_i(\beta)\}) \\ &= \sum_{j=1}^{p} \beta_j S_j - \sum_{i=1}^{k} n_i \log(1 + \exp\{\psi_i(\beta)\}). \end{aligned} \tag{4.3}$$

Note that S_j is sufficient for β_j. Taken collectively the vector $S = (S_1, \ldots, S_p)^T$ can be written as $S = X^T Y$. Maximum likelihood estimates, their standard errors and fitted values will all be functions of the data only through p statistics $S_1, \ldots, S_p$. This is true only for the logistic link function and for other link functions the minimal sufficient statistic for β is the whole data vector Y. A few examples should clarify the notation.

Example 4.1: The 2 × 2 Table For a single 2×2 table there are $k = 2$ responses and the number of parameters is $p = 2$ when we allow treatments 1 and 2 to differ. In the notation of Section 3.2 with $\beta^T = (\psi_2, \Delta)$ the design matrix is

$$X = \begin{pmatrix} 1 & 1 \\ 1 & 0 \end{pmatrix} \Rightarrow X^T Y = \begin{pmatrix} Y_1 + Y_2 \\ Y_1 \end{pmatrix}$$

and so $T = Y_1 + Y_2$ is sufficient for the nuisance parameter ψ_2 and Y_1 for the interest parameter Δ. The general log-likelihood Eq. (4.3) reduces to

$$l(\psi_2, \Delta) = Y_1 \Delta + T\psi_2 - n_1 \log(1 + e^{\Delta + \psi_2}) - n_2 \log(1 + e^{\psi_2})$$

which agrees with the logarithm of Eq. (3.10) in Chapter 3.

Example 4.2: Uniform Treatments Model for 2 × 2 × k Table Following the notation of Section 3.3, let Δ be the common log-odds ratio and $\psi_{2(j)}$ the logit for treatment 2 in table j. Let $Y_{i(j)}$ be the number of successes for treatment i in table j. To write down the X-matrix we need to decide on an *order* for the data although this makes no difference to the final results. A natural order is $Y_{1(1)}$, $Y_{2(1)}$, $Y_{1(2)}$, $Y_{2(2)}$, etc. Then the linear predictor is

$$\begin{pmatrix} \psi_{1(1)} \\ \psi_{2(1)} \\ \psi_{1(2)} \\ \cdot \\ \psi_{1(k)} \\ \psi_{2(k)} \end{pmatrix} = \begin{pmatrix} \psi_{2(1)} + \Delta \\ \psi_{2(1)} \\ \psi_{2(1)} + \Delta \\ \cdot \\ \psi_{2(k)} + \Delta \\ \psi_{2(k)} \end{pmatrix} = \begin{pmatrix} 1 & 0 & 0 & \cdot & \cdot & 1 \\ 1 & 0 & 0 & \cdot & \cdot & 0 \\ 0 & 1 & 0 & \cdot & \cdot & 1 \\ 0 & 1 & 0 & \cdot & \cdot & 0 \\ \cdot & \cdot & \cdot & \cdot & \cdot & \cdot \\ 0 & 0 & 0 & \cdot & 1 & 1 \\ 0 & 0 & 0 & \cdot & 1 & 0 \end{pmatrix} \begin{pmatrix} \psi_{2(1)} \\ \psi_{2(2)} \\ \cdot \\ \cdot \\ \psi_{2(k)} \\ \Delta \end{pmatrix} = X\beta.$$

It can be checked that the general likelihood Eq. (4.3) reduces to Eq. (3.30) as given on p. 145. The sufficient statistics are

$$S = X^T Y = \begin{pmatrix} 1 & 1 & 0 & 0 & \cdot & 0 & 0 \\ 0 & 0 & 1 & 1 & \cdot & 0 & 0 \\ \cdot & \cdot & \cdot & \cdot & \cdot & \cdot & \cdot \\ 0 & 0 & 0 & 0 & \cdot & 1 & 1 \\ 1 & 0 & 1 & 0 & \cdot & 1 & 0 \end{pmatrix} \begin{pmatrix} Y_{1(1)} \\ Y_{2(1)} \\ \cdot \\ \cdot \\ Y_{1(k)} \\ Y_{2(k)} \end{pmatrix} = \begin{pmatrix} Y_{1(1)} + Y_{2(1)} \\ Y_{1(2)} + Y_{2(2)} \\ \cdot \\ \cdot \\ Y_{1(k)} + Y_{2(k)} \\ \sum_j Y_{1(j)} \end{pmatrix}$$

and so $Y_{1+} = \sum_j Y_{1(j)}$, being the total number of successes under treatment

1, is sufficient for Δ while the success totals $T_j = Y_{1(j)} + Y_{2(j)}$ for each table are sufficient for their nuisance parameters $\psi_{2(j)}$. In Chapter 7 we use these sufficient statistics to obtain exact conditional inference on Δ.

Example 4.3: Cancer Remission, Simple Regression What makes some cancer patients respond more quickly to treatment than others? Can remission be predicted at treatment or soon after? The data in Table 4.1 is extracted from a study by Lee (1974). The explanatory variable x relates to a measurement of cell activity after treatment. This variable is only roughly measured and has been rounded to the nearest even integer. There were 27 patients in the study and small numbers of patients for each distinct value of x. A simple plot of the estimated probability of remission Y_i/n_i as x_i varies does not produce a very illuminating picture (see p. 189). It is worth noting here that if the unrounded scores x_i are taken, then there are 27 distinct values and only one patient for each level of x_i. With each $n_i = 1$ it follows that y_i/n_i is either zero or one and so a plot against x_i would be even harder to interpret.

Table 4.1. Cancer Remission Data

x_i	8	10	12	14	16	18	20
Y_i/n_i	0/2	0/2	0/3	0/3	0/3	1/1	2/3
x_i	22	24	26	28	32	34	38
Y	1/2	0/1	1/1	1/1	0/1	1/1	2/3

Note: Numbers of remissions Y from number of patients n, by early cell activity predictor x [from Lee (1974)].

Nevertheless, the probability $\pi(x)$ of remission apparently increases with x. For instance, in the top half of the table there are 3 remissions from 17 while in the bottom half there are 6 from 10. Both from examining the data and from background knowledge, we expect that $\pi(x)$ should begin at some small value when x is small and then rise smoothly, increasing towards 1 as x increases. Such a family of curves is the *logistic* family defined by

$$\pi(x; a, b) = \frac{\exp\{a + bx\}}{1 + \exp\{a + bx\}}. \tag{4.4}$$

Plotted against x, this is a monotonic S-shaped curve whose rate of increase or decrease depends on b and whose location is determined by a. Another way to express these curves is

$$\psi(x) = \log\left\{\frac{\pi(x)}{1 - \pi(x)}\right\} = a + bx$$

which shows that it is a logistic linear model, in fact the simplest logistic regression model possible. The curves being fitted are not lines but logistic curves though these are indeed lines on the logit scale. Both the strengths and weaknesses of simple line fitting to normal data are inherited. In particular, cautions about extrapolating to remote x values and the influence of leverage points apply with equal force. In matrix notation the parameter vector $\beta^T = (a, b)$ and the linear predictor is

$$X\beta = \begin{pmatrix} a + bx_1 \\ a + bx_2 \\ \cdot \\ \cdot \\ a + bx_n \end{pmatrix} = \begin{pmatrix} 1 & x_1 \\ 1 & x_2 \\ \cdot & \\ \cdot & \\ 1 & x_n \end{pmatrix} \begin{pmatrix} a \\ b \end{pmatrix}.$$

Thus, the sufficient statistic for (a, b) is

$$X^T Y = \begin{pmatrix} Y_1 + Y_2 + \cdots + Y_k \\ x_1 Y_1 + x_2 Y_2 + \cdots + x_k Y_k \end{pmatrix}$$

and so $S_1 = \sum_{i=1}^{k} Y_i$ is sufficient for a and $S_2 = \sum_{i=1}^{k} x_i Y_i$ for b. This is again analogous to simple linear regression. Exact conditional inference on b could be based on the distribution of S_2 conditional on S_1 [see Cox (1970) and Chapter 7].

4.1.2 Likelihood Based Inference

The log-likelihood function Eq. (4.3) is of the form

$$\beta_1 S_1 + \beta_2 S_2 + \cdots + \beta_p S_p - \kappa(\beta)$$

where κ is a sum of log-exponentials. The derivative of this with respect to β_j is $S_j - \partial\kappa/\partial\beta_j$. Recall that each component of the score vector has mean zero. It follows then that

$$U_j(\beta) = \frac{\partial \ell(\beta)}{\partial \beta_j} = S_j - \mathrm{E}(S_j; \beta). \tag{4.5}$$

The ML estimators are obtained by equating the sufficient statistics S_j simultaneously to their mean values and solving for β. Since the mean values

$$\mathrm{E}(S_j; \beta) = \mathrm{E}\left(\sum_{i=1}^{k} X_{ij} Y_i\right) = \sum_{i=1}^{k} n_i X_{ij} \pi_i(\beta)$$

are nonlinear functions of β, these equations are solved iteratively. Necessary and sufficient conditions for ML estimates to be finite are outlined in Section 4.6; see also Exercise 4.20.

Maximizing the Likelihood A general algorithm for solving the likelihood equations is Newton–Raphson, outlined in Section 1.2.3. This requires the matrix of derivatives of the score vector U with respect to the parameter vector β, or equivalently the observed information matrix. Writing

$$\frac{\partial \pi_i}{\partial \beta_l} = \frac{\partial}{\partial \psi_i} \left\{ \frac{\exp(\psi_i)}{1 + \exp(\psi_i)} \right\} \frac{\partial \psi_i}{\partial \beta_l} = \pi_i(1 - \pi_i) X_{il}$$

the observed information matrix has entries

$$\mathcal{J}_{rs} = \sum_{i=1}^{k} n_i X_{ir} X_{is} \pi_i (1 - \pi_i) = (X^T W X)_{rs}$$

where W is the matrix with diagonal elements $n_i \pi_i (1 - \pi_i)$ often also called the *working weight* matrix. Note that $\mathcal{J}(\beta)$ is nonrandom and therefore is identical to the expected information matrix $I(\beta)$. Moreover, since the scores $U_j = S_j - \mathrm{E}(S_j)$, the expected information is just the variance matrix of the sufficient statistics $S_1, \ldots, S_p$. For link functions other than logistic, these two statements are false. However, the expected information matrix can always be written in the form $X^T W X$ for an approximate weight matrix W; see Section 4.6.

Example 4.3: (Continued) **Cancer Remission.** The design matrix X has first column $X_{i1} = 1$ and second column $X_{i2} = x_i$ where x_i are the values of the covariates. We saw that $S_1 = \sum Y_i$, $S_2 = \sum x_i Y_i$ and so the ML estimation equations are

$$\sum_{i=1}^{k} Y_i = \sum_{i=1}^{k} n_i \pi(x_i, a, b)$$

$$\sum_{i=1}^{k} x_i Y_i = \sum_{i=1}^{k} n_i x_i \pi(x_i, a, b)$$

where $\pi(x, a, b)$ are of the logistic form in Eq. (4.4). The information matrix is

$$\begin{pmatrix} \sum_{i=1}^{k} n_i \pi_i (1 - \pi_i) & \sum_{i=1}^{k} n_i x_i \pi_i (1 - \pi_i) \\ \sum_{i=1}^{k} n_i x_i \pi_i (1 - \pi_i) & \sum_{i=1}^{k} n_i x_i^2 \pi_i (1 - \pi_i) \end{pmatrix}$$

and it is easy to show that this is the variance matrix of the sufficient statistics $(\sum Y_i, \sum x_i Y_i)$. These nonlinear equations for (a, b) are solved using Newton–Raphson. Because of the special linear structure of generalized linear models, the algorithm simplifies to a kind of iterative weighted least squares. Details of this algorithm are given in Section 4.6. For the remission data, the parameter estimates are $\hat{a} = -3.77$, $\hat{b} = 0.145$ and so the odds of remission increases by a factor $\exp\{2\hat{b}\} = 1.336$ or 34% for each increase of 2 units in x_i. The variance matrix of the parameter estimates is

$$(X^T W X)^{-1} = \begin{pmatrix} 1.882 & -0.076 \\ -0.076 & 0.003487 \end{pmatrix}$$

and so the standard error of $\hat{b} = 0.145$ is $\sqrt{0.003487} = 0.0590$. The standard error of $\hat{a} = -3.77$ is $\sqrt{1.882} = 1.372$. The correlation of the two estimators is $-0.076/(1.372 \times 0.059) = -0.935$ which is very large. The parameters a and b are in fact highly *nonorthogonal*, a concept discussed in Chapter 1.

The parameter $\hat{a} = -3.77$ estimates $\text{logit}\{\pi(0)\}$ and so we estimate $\hat{\pi}(0) = \text{expit}(-3.77) = 0.0225$. Since no living patient has cell activity $x = 0$ this is rather meaningless. To obtain a more meaningful parameter, we could rescale the covariate x_i. For instance, if the cell activity of a normal healthy patient were 20 we could use $x_i^* = x_i - 20$ as the covariate. Then a would represent the logit probability of remission for a patient with cell activity 20, rather than 0. The slope parameter b remains unchanged, and would still be estimated by $\hat{b} = 0.145$. Figure 4.2 on p. 189 shows the fitted logistic curve (dotted) as well as two other curves that will be discussed later. The raw proportions y_i/n_i are plotted as points.

Standard Errors, Test, and Confidence Intervals The general asymptotic likelihood theory discussed in Section 1.3 applies to the logistic regression model. The asymptotic distribution of $\hat{\beta}$ is multivariate normal with mean vector the true value β and variance matrix $I^{-1} = (X^T W X)^{-1}$. Asymptotic here means that the expected/observed information is "large." Thus, a Wald-type confidence interval can be constructed for any of the parameters β_i. More generally, an interval can be constructed for any linear combination $c^T\beta$ of the β_i where $c = (c_1, \ldots, c_p)$ is a vector of coefficients. Using standard matrix algebra, the asymptotic variance of $c^T\hat{\beta}$ is $c^T(X^T W X)^{-1}c$ and a confidence interval for $c^T\beta$ is

$$\left(c^T\hat{\beta} \pm z_{1-\alpha/2}\sqrt{c^T(X^T \hat{W} X)^{-1}c} \right).$$

One important application is in giving an interval for predicted probabilities. Suppose we wanted to estimate $\pi(x)$ for a given vector of covariate values x. The logit probability is $\psi(x) = x^T\beta$ and so the above formula applies by using the covariate values $x_1, \ldots, x_p$ as the coefficients $c_1, \ldots, c_p$. This is converted to an interval for $\pi(x)$ by applying the reverse logit or *expit* transform to both endpoints. This interval accounts only for the ordinary statistical dangers of extrapolation and does not account for the model possibly breaking down for remote values of the covariates.

Model fit is assessed by computing the LR or deviance statistic that has the formula

$$2\sum_{i=1}^{k} y_i \log\left(\frac{y_i}{\hat{e}_i}\right) + 2\sum_{i=1}^{k}(n_i - y_i)\log\left(\frac{n_i - y_i}{n_i - \hat{e}_i}\right) = 2\sum^{*} y_i \log(y_i/\hat{e}_i) \quad (4.6)$$

where $\sum^{*}$ denotes summation over all successes and failures. An alternative is the Pearson statistic that instead sums terms $(y_i - \hat{e}_i)^2/\hat{e}_i$ or the Cressie–Read statistic. The asymptotic distribution is χ^2_{k-p}. For instance, the degrees of freedom is $k - 2$ for a simple regression model $a + bx$. The fitted values $\hat{e}_i$ are obtained from

$$\hat{e}_i = n_i\hat{\pi}(x_i) = n_i \operatorname{expit}\{\hat{\psi}(x_i)\} = n_i \operatorname{expit}(\hat{\beta}_1 x_{i1} + \cdots + \hat{\beta}_p x_{ip}).$$

There are, however, difficulties in assessing fit when the $\hat{e}_i$ are small and the above fit statistics are useless when the data are binary. For instance, we pointed out in Section 3.6.4 that the deviance for binary data is

$$\mathcal{D} = -2\sum_{i=1}^{k} \hat{\pi}_i \operatorname{logit}(\hat{\pi}_i) + \log(1 - \hat{\pi}_i)$$

a result due to Williams (1983) [see also Collett (1991, p. 64)]. This is a function of the fitted values $\hat{\pi}_i$ only and is apparently of no use for assessing fit. Note however that the individual contributions to the deviance *do* depend on the data point and the fitted value and so carry information about the fit of that point. Indeed these are squared residuals as discussed later in Section 4.2. Consequently, the *pattern* of these contributions can say something about the overall fit. Only their total says nothing.

Example 4.3: (Continued) **Prediction of Cancer Remission.** Assuming that the linear logistic model is correct, what is the probability of remission for a patient with cell activity $x = 20$?. We estimate $\hat{\psi}(20) = \hat{a} + 20\hat{b} = -0.880$ and so $\hat{\pi}(20) = 0.293$. The asymptotic variance of $\hat{\psi}(20)$ is

$$c^T(X^TWX)^{-1}c = (1 \quad 20)\begin{pmatrix} 1.882 & -0.076 \\ -0.076 & 0.003487 \end{pmatrix}\begin{pmatrix} 1 \\ 20 \end{pmatrix} = 0.247$$

and so the standard error of $\hat{\psi}(20)$ is $\sqrt{0.247} = 0.497$. A 95% confidence interval for $\psi(20)$ is $(-0.880 \pm 1.96 \times 0.497) = (-1.853, 0.0937)$. Applying the expit transform to both endpoints gives the 95% interval (0.135, 0.523) for $\pi(20)$. This is a rather wide interval, which is not surprising since there were only 27 patients in the data base.

To assess the fit of the model we could look at the deviance statistic LR = 15.662. Here there are $k = 14$ counts and $p = 2$ parameters and so the degrees of freedom is $14 - 2 = 12$. The deviance is a little high but not unusually so for the χ^2_{12} distribution. However, all 14 fitted values are smaller than 5. We could try to accumulate these into groups with $e_i > 5$ however the total of all the fitted values is 9, the same as the total of the y_i. Probably a more fruitful approach to assessing fit is to test whether further terms could be added to the linear predictor. For instance, we could fit the model $\psi(x) = a + bx + cx^2$. This leads to a deviance of 11.813 which is 3.839 smaller than the deviance of 15.662 for the simple linear model. The approximate distribution of this change in deviance is χ^2_1 and the 5% critical value for this distribution is 3.841. Thus, there is quite strong evidence that the simple linear model does *not* describe the data. The quadratic coefficient $\hat{c} = -0.016$ is negative indicating that $\pi(x)$ eventually decreases as x increases. This would need to be checked with medical experts.

Coding Binomial Data as Binary Binomial data can always be coded as binary; y successes from n trials are simply coded as y zeros and $n - y$ ones. Have we therefore concluded that the deviance statistic is useless for all binomial data?

The answer is of course no. When data are recorded as binary the deviance of the current model will increase from $2(\ell_f - \ell_c)$ to $2(\ell_b - \ell_c)$, where ℓ_c, ℓ_f, ℓ_b are the maximized log-likelihoods for the current, full binomial, and full binary models, respectively. The larger deviance will not be absolutely interpretable because of the discussion above, but the smaller binomial deviance will be absolutely interpretable if the n_i are not too small. Differences in deviances between two nested models are unaffected by whether data are coded as binomial or binary. Provided one is convinced that there is no systematic trend within each binomial group (i.e., that the full binomial model is correct in its systematic component) there would seem to be no reason for coding binomial data as binary. One practical reason against binary coding is that computations may be slowed.

4.1.3 Dose-Response Models

Suppose that experimental units are subject to larger and larger values of a covariate x. For instance, insects are subjected to a larger and larger dose of insecticide, and it is noted whether or not the exposure is fatal. Dose-response models are used to describe how the probability of response changes, usually increases, as dose increases. A wide class of models is

$$g\{\pi(x)\} = a + \beta_1 f_1(x) + \beta_2 f_2(x) + \cdots + \beta_p f_p(x) \tag{4.7}$$

where $\pi(x)$ is the probability of response, g is a link function, and $f_1(x), \ldots, f_p(x)$ is a set of known basis functions. This is a linear binomial model, where the covariates x_j are different functions $f_j(x)$ of a single covariate x.

This conceptual framework includes many common experimental situations. In toxicology experiments organisms are given various doses x of a poison and the response is death of the organism. The toxicity of the poison is measured by the rate at which $\pi(x)$ increases with x. In sociological investigations the age x of each individual is often related to the presence or absence of a particular trait. In quality control the load x on an experimental unit is related to its passing or failing some test of its function. A modern and definitive reference on dose-response models is Morgan (1992).

Tolerance Distributions What link functions g are likely to be useful? A simple and helpful concept here is that of the *tolerance distribution*. We imagine that each individual has an associated tolerance T_i that is unobservable to us. The experimenter applies a dose x_i and the individual responds if this dose x_i exceeds the unknown tolerance T_i. Hence,

$$\text{Pr(individual i responds)} = \pi(x_i) = \Pr(T_i < x_i).$$

The shape of the probability curve $\pi(x)$ depends on the distribution of the tolerance T across the population from which individuals are chosen to be tested. It is interesting to note that the individual's response is not considered random here; it is the tolerance T_i that is random and determines the binary response (cf. Section 3.2.1). In the toxicology framework tolerance could be naturally interpreted as genetic resistance to the toxin.

A family of tolerance distributions will determine a family of dose-response models. For instance, suppose the distribution function of T is $G\{(x - \mu)/\sigma\}$ where μ is a measure of location, σ of scale, and G is *continuous*. Then $\pi(x) = G\{(x - \mu)/\sigma\}$ which is equivalent to

$$G^{-1}\{\pi(x)\} = \frac{x - \mu}{\sigma} = a + bx$$

where $a = -\mu/\sigma$ and $b = 1/\sigma$. This is a linear binomial model and the link function is the inverse of the distribution function G.

When G is the standard normal distribution Φ, the induced link transform is $g = \Phi^{-1}$ called the *probit* transformation. The logistic distribution is another symmetric distribution often used to model continuous data. Unlike the normal distribution, its distribution function $G(x) = \exp(x)/(1 + \exp(x))$ can be given explicitly. The induced link transform $g = G^{-1}$ is then the logistic or log-odds transformation that we grew so fond of in Chapter 3. Both normal and logistic distributions are symmetric about zero, i.e., $G(x) = 1 - G(-x)$ which implies that

$$g(1 - \pi) = -g(\pi)$$

where $g = G^{-1}$. Since success and failure are often arbitrary labels, reversing their role should induce a simple transformation of the model. For models based on a symmetric tolerance distribution we simply change the sign.

An example of an asymmetric link is the so-called *complementary log–log* link based on the tolerance distribution

$$G(x) = 1 - \exp\{-\exp\{x\}\}, x \in \mathcal{R}$$

known as the *extreme value* or *Gumbel* distribution. This distribution is not symmetric about zero and the induced link is

$$g(\pi) = G^{-1}(\pi) = \log(-\log(1 - \pi)).$$

Another asymmetric link is the *log–log* transformation $-\log(-\log(\pi))$. Asymmetric link functions allow the probability $\pi(x) = G(a + bx)$ to approach the limits of zero and one at different (though not arbitrarily different) rates as x diverges to $\pm\infty$. Symmetric link functions implicitly impose identical rates of convergence at the extremes. Since tolerances are positive it is to be anticipated that the tolerance distribution would be skew in practice. One way to account for this is by applying a stabilizing transformation (such as log) to the dose. This is indeed very common in the analysis of dose response models and the logarithm of the dose is often called *dosage*.

While it might well be the case that an asymmetric link will fit a given set of data better than a symmetric one, from the viewpoint of systematic model building it seems more natural to begin with a symmetric link function and to build any asymmetry into the model as needed through appropriate functions of the explanatory variable x. An added advantage of this strategy is that any terms modeling asymmetry will appear explicitly and so must be purchased with degrees of freedom. There are situations where c-log–log links have a theoretical basis and two such applications are given in Exercises 4.11 and 4.12. Tests of the link function are discussed in Section 4.2.6.

Example 4.4: Germination of Horehound Seeds An experiment was carried out by researchers in the Department of Agriculture at La Trobe University with the aim of understanding the variability of germination rates of various seeds of a particular species of weed called *horehound*. One aim of the research was to predict areas where infestation of the weed was likely to be the greatest.

Batches of 50 seeds were germinated under conditions combining various temperatures and humidities. These conditions were mathematically combined into a single numerical score x_i, between 0 and 20, representing the favorability of the conditions to germination. The details of how this score is calculated are not important here but since it is only a rough summary of the experimental conditions it has been rounded to an integer between 1 and 20 inclusive. The data are given in Table 4.2 and the binomial denominators are of course all multiples of the seed batch size of 50.

Table 4.2. Seed Germination Data

x_i	1	2	3	4	5	6	7	8	9	10
y_i	9	25	32	46	51	59	60	70	71	75
n_i	500	350	450	200	200	200	250	200	150	100
x_i	11	12	13	14	15	16	17	18	19	20
y_i	108	97	69	149	84	169	119	127	277	185
n_i	200	150	100	250	100	200	150	150	300	200

Note: Number of seed germinations y out of number in batch n. The covariate x measures favorability to germination.

Whether or not a seed germinates is presumably controlled by its genetics. The tolerance T of a particular seed is the genetically determined point where conditions x just induce germination. It is hardly of interest to ask if germination rates change across the 20 conditions. Of course they do, and so a test of the 2×20 table for independence would be no more than a first step of the data exploration. A plot of the proportions Y_i/n_i against x_i gives a model free estimate of the response function $\pi(x)$. Unlike the cancer remission data, the n_i here are all large and so each proportion estimates $\pi(x_i)$ fairly precisely. The logits $\psi(x_i)$ are estimated by modified empirical logits $\log(y_i + 1/2) - \log(n_i - y_i + 1/2)$ and these are plotted on p. 193, suggesting a roughly linear increasing relationship.

Upon fitting the linear logistic model $\psi(x) = a + bx$ the parameter estimates are $\hat{a} = -2.901(0.083)$ and $\hat{b} = 0.278(0.0077)$. The odds of germination increases by $100(\exp(\hat{b}) - 1) = 31.9\%$ for each increase in the favorability index x. The goodness-of-fit statistic however is LR $= 130.5$ on 18 degrees of freedom. This bad fit could be due to any of several factors such as the wrong link, a missing quadratic term, or a few bad data points. But the plot of empirical logits does not suggest that the regression function is seriously wrong. The only other deficiency of the model could be the binomial distribution itself. We will consider these issues later.

4.1.4 Closeness of Probit and Logit Links

The logit and probit transformations are very close in shape. For x values not too extreme, the logistic distribution is close to normal with mean 0 and standard deviation $\pi/\sqrt{3}$. Only for extreme values of x can they be distinguished where

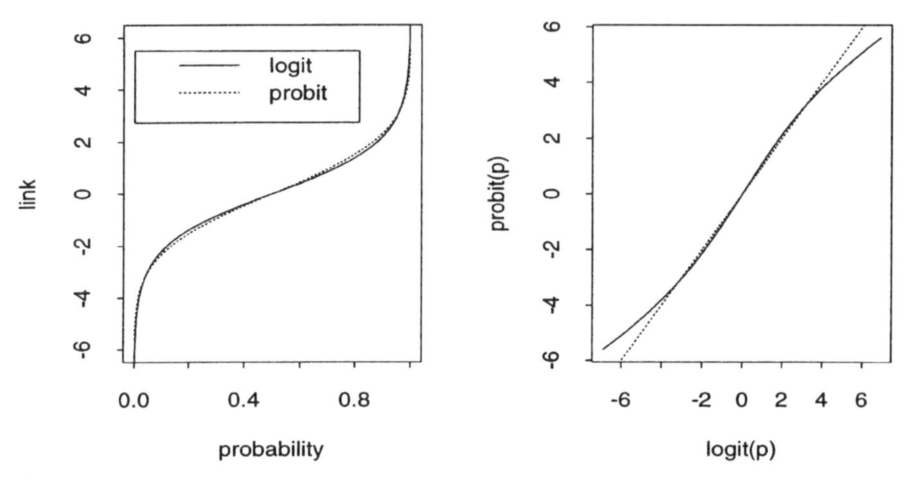

Figure 4.1. Probit and logit curves compared. The logit and probit (after rescaling) functions are almost identical except in the extremes. Except for data with extreme proportions of response, inferences from probit and logit models should be almost identical.

the logistic curve approaches 0 or 1 more slowly than the probit curve. Figure 4.1 compares logit and probit curves with the latter rescaled by the factor $\sqrt{3}/\pi$. The second part plots the probit against the logit.

Because of this closeness over a wide range of probabilities, probit and logistic models give similar fits to a given data set, provided that the fitted probabilities do not become extreme. Model predictions are also very similar except for the extremes. Not surprisingly, it is typically very difficult to discriminate between the probit and logit models through a formal test [see Chambers and Cox (1967)]. Since the models are so similar, the need for a test between them is not a pressing one.

Nevertheless, logistic models are preferable, if for no other reason than that the regression parameters $\exp(\beta_j)$ in a logistic linear model are easily interpreted as the increase in odds of response for each unit increase in x_j. Most people understand odds whereas probit transformed probabilities have no intrinsic meaning. Other reasons for using logits were listed in Section 3.6.2.

Example 4.3: (Continued) **Comparison of Links for Cancer Remission Data.** The logistic, probit and c-log–log linear models were fitted to these data in `GLIM` by using the `LINK` options `G`, `P`, and `C`. The deviances of the models were 15.662, 15.444, and 16.585, respectively. The fitted curves $\hat{\pi}(x) = G(\hat{a} + \hat{b}x)$ are compared in Figure 4.2.

Example 4.4: (Continued) **Comparison of Links for Seed Data.** The logistic, probit, and c-log–log linear models were fitted to these data giving deviances 130.5, 120.1, and 198.6, respectively each on 18 degrees of freedom. Clearly an

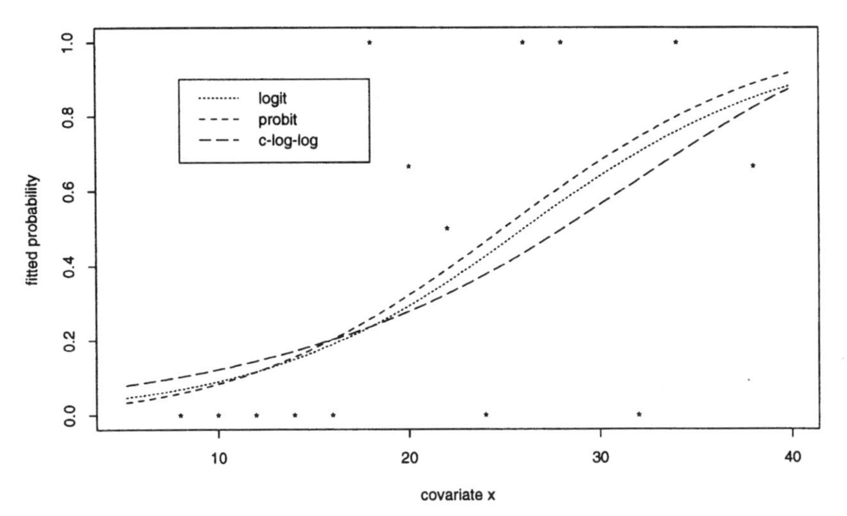

Figure 4.2. Comparison of three link functions. Three fitted models for the probability of cancer remission using (i) logit, (ii) probit, and (iii) c-log–log links. Empirical estimates are plotted as points.

alternative link function does not solve the problems of the logit model. However, the probit model fits considerably better, in essence, because the fitted probabilities converge to 0 and 1 faster than for the logit model and this faster convergence is supported by the data.

4.1.5 Estimating Effective Dose

In toxicology it is often of interest to determine the dose x at which the response probability equals a specified value, for instance 0.5. This is called the median *effective* or *lethal* dose, denoted LD50. Similarly, LD75 and LD90 measures are common. Under the model $g(\pi) = a + bx$, the p% lethal dose LD_p is $(g(p) - a)/b$. However, since doses are positive it is more convenient to work with

$$\theta = \log(\mathrm{LD}_p) = \log\{g(p) - a\} - \log\ b. \tag{4.8}$$

Note that for symmetric links like probit and logit, $g(0.5) = 0$. The ML estimator of θ simply replaces a and b by their ML estimators. Let V_{aa}, V_{ab}, V_{bb} be the variance and covariance of $\hat{a}$ and $\hat{b}$. Using the transformation Eq. (1.49) then gives

$$\mathrm{se}(\hat{\theta}) = \hat{b}^{-1}\sqrt{V_{aa}/\hat{\theta}^2 - 2V_{ab}/\hat{\theta} + V_{bb}} \tag{4.9}$$

and exponentiating the interval $\hat{\theta} \pm z_{1-\alpha/2} \times \mathrm{se}(\hat{\theta})$ gives an interval for LD_p.

One problem with estimating lethal doses in this way is that, for very small values of p, $g(p)$ will be large and negative and so LD_p can be negative. At least in the toxicology application this makes no sense and so it is customary to fit the basic logistic linear model using log(dose) often called *dosage*. The probability curves being fit are then of the form

$$\text{logit}\{\pi(x)\} = a + b \log\, x \Leftrightarrow \pi(x) = \frac{e^a x^b}{1 + e^a x^b}.$$

This implies that $\pi(0) = 0$ and this may not always be sensible, for instance, if natural mortality over the course of the experiment cannot be ignored. Fitting the model is no more difficult than fitting the ordinary model—one simply replaces the values of x_i by $\log\, x_i$. The lethal dose is $\exp\{(g(p) - a)/b\}$ and intervals are obtained using the previous method, and then exponentiating *twice* rather than once.

A plot of LD_p against p summarizes the effectiveness of the toxin. If dosage rather than dose is used in the linear model then this curve will pass through (0, 0) but otherwise it will not. There is of course no reason why one could not fit a quadratic or some other model to dose-response data. If one does, then the simple Eq. (4.8) will not apply. If the model is monotonic, then one can still plot the estimated LD_p against p by plotting the covariate x against the fitted probability $\hat{\pi}(x)$, with x on the vertical axis.

Example 4.4: (Continued) **Target Germination Rates.** For the seed data, response is germination rather than death so the word *lethal* is inappropriate. We can still ask what value of x is required for the germination rate to reach a target of p%. We will estimate LD_{50}. Recall the estimates $\hat{a} = -2.901$ and $\hat{b} = 0.278$ so that $\hat{\theta} = 2.3468$. The estimate of LD_{50} is exp(2.347) = 10.45. Using Eq. (4.9) the standard error is 0.1529 giving the interval (2.194, 2.500) for θ and (8.97, 12.18) for LD_{50}.

If we fit the model instead, using $\log(x)$ as the covariate then the deviance is 130.7, only slightly higher than the previous model. The parameter estimates are $\hat{a} = -4.742$, $\hat{b} = 2.175$, and so $\log(-a/b)$ is estimated by 0.7796. The variance of this estimate is 0.1202, again using Eq. (4.9), which gives the interval (0.654, 0.899). Exponentiating this interval and the estimate *twice* gives the interval (6.91, 11.69) to accompany the estimate 8.852 of LD_{50}.

Which of these models is to be preferred would depend on diagnostic methods to be described in following sections. It should be noted here though that *both* models are clearly wrong since the deviances of both are so high, and it would be highly misleading to report either of the above analyses, especially the confidence intervals. We will come back to these data later and make the appropriate adjustments.

4.1.6 Fitting Nonlinear Parameters

There is nothing in the asymptotic likelihood theory that requires parameters of a model for binary data to enter linearly through some known link. For instance, if neither linear nor quadratic functions appear to track a set of data we might consider fitting x^γ for unknown power parameter γ.

A general result from asymptotic likelihood theory is that the LR statistic for testing a single parameter θ has approximate χ_1^2 distribution. Expressed in terms of deviance the LR statistic is just

$$LR(\theta) = \mathcal{D}(\theta) - \mathcal{D}(\hat{\theta})$$

where $\mathcal{D}(\theta)$ is the deviance of the model with θ held fixed. This function is minimized at the ML estimate $\hat{\theta}$ and we obtain likelihood confidence intervals for θ, of the correct asymptotic size, by collecting together all values for which $\mathcal{D}(\theta)$ is within a χ_1^2 quantile of its minimum. The observed information is given in terms of the deviance by $\mathcal{J} = \mathcal{D}''/2$ and so

$$\mathrm{se}(\hat{\theta}) = \sqrt{2/\mathcal{D}''(\hat{\theta})}. \tag{4.10}$$

The second derivative of $\mathcal{D}(\theta)$ is necessarily positive at the minimum $\hat{\theta}$ and may be numerically approximated by computing

$$\mathcal{D}''(\theta) \approx \frac{1}{\delta^2}\left[\mathcal{D}(\theta+\delta) + \mathcal{D}(\theta-\delta) - 2\mathcal{D}(\theta)\right] \tag{4.11}$$

for a suitably small increment δ.

One common and flexible form for the predictor extends the general dose response model Eq. (4.7) by adding a single nonlinear term $f(x,\gamma)$ to the linear predictor giving

$$g\{\pi(x)\} = \alpha + b_1 f_1(x) + b_2 f_2(x) + \cdots + b_p f_p(x) + \beta f(x,\gamma). \tag{4.12}$$

When γ is fixed the last covariate $z_i = f(x_i,\gamma)$ can be computed and the model is linear in the unknown parameters. A generalized linear model package may then be used to maximize the likelihood with respect to the linear parameters and the deviance $\mathcal{D}(\gamma)$ recorded. It remains to minimize this deviance with respect to γ. One can of course vary γ over a grid of values and plot $\mathcal{D}(\gamma)$ against γ. This curve is minimized at the ML estimate $\hat{\gamma}$ and likelihood intervals (see Section 2.6.2) for γ can be read off as those values γ for which $\mathcal{D}(\gamma)$ is within a χ_1^2 quantile of the minimum.

Common choices for $f(x;\gamma)$ are the power curves x^γ, geometric curves γ^x, inverse linear curves $x/(x+\gamma)$, and shift curves $f(x-\gamma)$. As just one example, suppose that failure of a component only occurs when the load exceeds some threshold γ and for loads less than this the probability of response/failure is zero. Then the odds of response should be zero when $x=\gamma$ suggesting a linear model in $\log(x-\gamma)$ for the log-odds. Provided that the observed covariate values are larger than γ, this model, and any model of the form given in Eq. (4.12), can be fitted using the algorithm to be given below. Appropriate model forms should always reproduce any known limiting behavior such as zero probability of death when an individual is given zero dose of a toxin. This might indeed require a model involving nonlinear parameters.

An Iterative Algorithm The minimum of the deviance $\mathcal{D}(\gamma)$ can be found rather quickly without plotting the entire function. Let γ_t be a trial value assumed to be close to both $\hat{\gamma}$ and the true value γ. Then approximately

$$g\{\pi(x)\} = \alpha + b^T f(x) + \beta z_1 + \beta' z_2$$

where the *constructed variables*

$$z_1 = f(x;\hat{\gamma}_t), \qquad z_2 = f'(x;\hat{\gamma}_t)$$

and $\beta' = \beta(\gamma - \hat{\gamma}_t)$. A simple interative scheme for finding $\hat{\gamma}$ is to fit the covariates z_1 and z_2 and to compute the updated estimate $\hat{\gamma}_t + \hat{\beta}'/\hat{\beta}$ of γ. Convergence is not guaranteed for starting values $\hat{\gamma}_t$ far from the solution $\hat{\gamma}$. An appropriate starting value may be suggested by a constructed variable plot as given on p. 211.

It is unwise to try to fit more than one nonlinear parameter in this way particularly when, as is usually the case, the linear parameters are highly nonorthogonal to the nonlinear ones. For several nonlinear parameters no single algorithm will suit all cases but the right transformation of parameters will often result in (i) $\mathcal{D}(\gamma)$ being closer to quadratic in which case the algorithm will converge much faster, and (ii) greater orthogonality of parameters and correspondingly more circular deviance contours. A wide ranging discussion of nonlinear parameters, their interpretation, reparametrization and various fitting algorithms may be found in Gallant (1987), Seber and Wild (1989), and Ross (1990).

Example 4.4: (Continued) **Power Model for Seed Data.** The simple logistic linear model fitted the seed germination data very poorly. There is no reason a priori why logit-germination rates should vary linearly with the covariate x and we might consider fitting the model

$$\text{logit}\{\pi(x)\} = a + bx^\gamma$$

that includes a power term γ. If nothing else, such a model addresses the question of whether the log-odds increase with x more or less quickly than linearly. Figure 4.3 includes two plots summarizing the fits of the power model with γ taking values between 0.1 and 1.1 (with increment 0.2). The first plot superimposes the fitted probability curves $\hat{a} + \hat{b}x^\gamma$ for these six models on a plot of modified empirical logits. There are two conspicuous outliers in this data set neither of which could be accounted for by any reasonable model. The second figure plots the deviance $\mathcal{D}(\gamma)$ against γ. Each deviance value was obtained by computing the covariate $z_i = x_i^\gamma$ and fitting a linear dose-response model with the single covariate z_i.

The smallest deviance is 94.1(17) when $\gamma = 0.5$ indicating that a logistic linear model in $\sqrt{x}$ is a good model among the power models. The degrees of freedom is now 17 rather than 18 because an extra parameter, γ, has been estimated. Clearly, the deviance is still huge and there may be something fundamentally wrong with the binomial assumption itself. If so, then the deviance is not χ^2_{17} and changes in deviance are not χ^2_1. One very crude correction is to rescale $\mathcal{D}(\gamma)$ so that the minimum value is exactly 17. This and more refined techniques are discussed in Section 4.4 on over-dispersion.

Nevertheless, let us press on for illustrative purposes as if the deviance was distributed χ^2_{17}. Since the deviance for $\gamma = 1$ is 130.5 being 36.4 larger than the minimum value of 94.1, we would easily reject the hypothesis that $\gamma = 1$. The plausible values for γ have deviance within around $\chi^2_1(.99) \approx 7$ of the minimum value, i.e., values between about 0.3 and 0.7. The standard error of the ML estimate 0.5 is obtained by numerically approximating the second derivative of

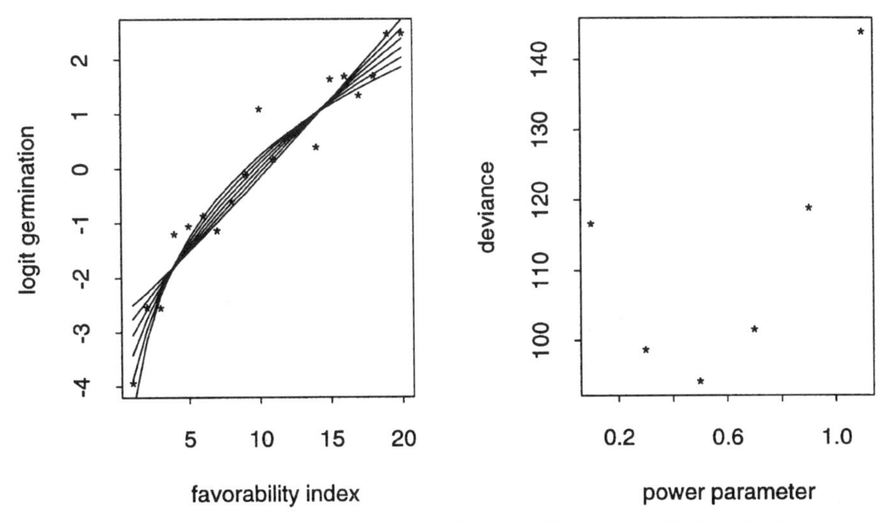

Figure 4.3. Fit of power model to seed germination data. The left plot displays six fitted power curves superimposed on the data, all on logistic scale. The right plot displays the deviances of the six fitted models. Note that the data are overdispersed and all models have inflated deviances.

$\mathcal{D}(\gamma)$. Some care is needed here that the deviance and its minimum are measured accurately and that the increment δ is not too small. In fact the minimum of $\mathcal{D}(\gamma)$ is at $\gamma = 0.47$ and the deviance was also computed at $\gamma = 0.45, 0.49$. This gave the estimate

$$\mathcal{D}''(0.47) \approx (\mathcal{D}(0.49) + \mathcal{D}(0.45) - 2\mathcal{D}(0.47))/(0.2)^2 = 305.5$$

and so the standard error is $\sqrt{2/305.5} = 0.081$ giving an approximate 99% confidence interval (0.288, 0.651) for γ. This differs only slightly from the likelihood interval because the deviance function is close to quadratic near the minimum. We again emphasize that since the deviances of all models are so large, these results are only for illustration and the above interval and standard error are not reasonable.

Intervals for Linear Parameters Confidence intervals for γ may be found from the deviance function $\mathcal{D}(\gamma)$ and requires a little work. Once we have found the ML estimator $\hat{\gamma}$, standard errors for the linear parameters β will come straight out of the package from which we can construct an interval. But wait a moment. Just fitting the linear model with $\hat{\gamma}$ substituted takes no account of the fact that γ has been estimated. The standard errors are *not* the same as we would obtain from the full information matrix for β and γ. Should we make some adjustment?

There are good reasons for *not* doing so. The interpretation of the slope parameter b is that it gives changes in the log-odds for each unit change in the effective covariate $x^{\hat{\gamma}}$. Whether or not $x^{\hat{\gamma}}$ is the "correct" covariate is really another issue and is addressed by testing values of γ. The standard error of $\hat{b}$ with γ left free mixes two sources of variation, the first the variability of $\hat{b}$ for fixed $\hat{\gamma}$ and second the additional (extreme) variability of $\hat{b}$ as γ varies across plausible values. Only the first component of variability has much meaning. As a general recommendation, inference on γ in Eq. (4.12) should be based on the deviance while inference on the linear parameters should be based on standard methods for linear models taking the nonlinear parameter as fixed at its estimated value.

Inference on *predicted values* should however take account of variability in all estimated parameters, linear, and nonlinear. Sometimes variability of the nonlinear parameters will have little effect on particular fitted values and lengthy calculations can be avoided, such as near the x-values of 4 and 14 in the seed germination example, see Figure 4.3.

4.1.7 Choosing Functions of x

The more general dose-response model Eq. (4.7) allows an arbitrary set of functions $f_1(x), \ldots, f_p(x)$ as the basis of the linear predictor. Some theoretical guidance as to the choice of the base functions $f_j(x)$ is provided by considering the probability function or density function of the covariate X conditional on the response Y taking value 1 or 0. Denote these conditional densities by $f(x|Y = y)$ and let $\pi(x) = \Pr(Y = 1|x)$ and $p = \Pr(Y = 1)$ be the conditional and unconditional probabilities of response. Then using the definition of conditional probability

$$\log\left\{\frac{f(x|Y=1)}{f(x|Y=0)}\right\} = \log\left\{\frac{\pi(x)}{1-\pi(x)}\right\} - \log\left(\frac{p}{1-p}\right).$$

Since p is not a function of x, the model for the logit probability differs from the model for the left hand side by a constant. If we have a good idea about the conditional distributions of the covariates, then we would know which functions of x to put in our logistic model for $\pi(x)$.

A wide class of distributions for the covariate X is the exponential family. Specifically, suppose that the density of X conditional on $Y = j$ is

$$f(x|Y=j) = \exp\{s^T(x)b(\theta_j) + a(x) - \kappa(\theta_j)\}$$

where the vector $s(X)$ is sufficient for the canonical parameter vector $b(\theta)$ and κ is the cumulant generating function. None of these functions vary with j but the parameter θ_j possibly does. In other words, the conditional distributions of X given $Y = j$ come from the *same* exponential family but with possibly differing parameters. Under this assumption

$$\log\left\{\frac{f(x|Y=1)}{f(x|Y=0)}\right\} = s^T(x)\{b(\theta_1) - b(\theta_0)\} - \kappa(\theta_1) + \kappa(\theta_0)$$

which is linear in the statistics $s_1(x), \ldots, s_p(x)$, provided $\theta_1 \neq \theta_0$. These function of x are then suggested in the linear logistic model for the distribution of Y. This result readily generalizes to more than one covariate. When the covariates are independent given the response the appropriate terms are just those that are appropriate for each of the individual explanatory variables. When the covariates x are not conditionally independent the assumption of a multivariate exponential family leads directly to sufficient statistics for the vector x. These will include appropriate product terms.

It is by no means common to have a good idea about the distribution of the covariates conditional on response and nonresponse. Indeed, it is often unreasonable to treat the covariate as random at all, for instance, when the values are determined by an imposed experimental design. Even when the covariates are random and we do know something about their distribution, then we may still not be satisfied with the exponential family assumption. Nevertheless the theory gives potential guidance where there is no other reason for choosing particular basis functions for the covariates. Table 4.3 lists the sufficient statistics for sev-

Table 4.3. Theoretically Suggested Covariate Transformations

Distribution	Sufficient Statistics
Poisson (λ)	x
Normal (μ, σ^2)	x, x^2
Gamma (r, α)	x, $\log x$
Beta (a, b)	$\log x$, $\log(1 - x)$

eral common exponential families. These functions of the covariates are likely to be important components of a good predictor in the general dose-response model. The theory of this section is based on Kay and Little (1987).

Example 4.5: Menstruation of Girls in Warsaw The data in Table 4.4 relate to a sample of girls in Warsaw, the response variable indicating whether or not the girl has begun menstruation and the explanatory variable age in years (measured to the month). These data were analyzed by Aranda–Ordaz (1981), Guerrero and Johnson (1982), and Kay and Little (1987).

A plot of the proportions Y_i/n_i against age x_i reveals a smooth monotonically increasing relationship, displayed on p. 198. Transforming these proportions by logit or probit link gives a roughly linear relationship but the c-log–log link does not. A simple linear regression on x_i with probit link gives an adequate fit as measured by overall deviance 22.887(23).

Why not stop here? Although the overall fit is adequate various diagnostic plots to be described in Section 4.2 reveal that the relationship of the linked probabilities with age may be nonlinear. Table 4.5 lists the deviances of various models relating the probability of menstruation to age. The three links logit, probit, and c-log–log are used. The first line compares the simple linear model in age for which we saw the probit link is preferred. The second line gives fits for

Table 4.4. Menstruation of Girls in Warsaw

x	9.21	10.21	10.58	10.83	11.08	11.33	11.58	11.83	12.08
y	0	0	0	2	2	5	10	17	16
n	376	200	93	120	90	88	105	111	100
x	12.33	12.85	12.83	13.08	13.33	13.58	13.83	14.08	14.33
y	29	39	51	47	67	81	88	79	90
n	93	100	108	99	106	105	117	98	97
x	14.58	14.83	15.08	15.33	15.58	15.83	17.58		
y	113	95	117	107	92	112	1049		
n	100	102	122	111	94	114	1049		

Numbers y who have reached menarche from n in age-group with center x [from Finney (1971)].

Table 4.5. Various Fits to the Menstruation Data

Model	df	Logit	Probit	c-Log–Log
x	23	26.703	22.887	118.821
x, x^2	22	23.202	15.149	34.966
$\log z, \log(1-z)$	22	14.658	14.449	23.765
z^λ	22	20.309 (0.58)	14.117 (0.51)	13.909 (−0.68)

models including a quadratic term in age. This is appropriate if the distributions of ages among those menstruating and those not menstruating are both normally distributed. The deviances are now all much lower, even lower than the degrees of freedom in the case of the probit link. The third line summarizes fits of linear models in $\log(z)$ and $\log(1-z)$, where $z = (x-9)/9$ is the age transformed to the interval [0, 1]. These are the appropriate predictors if the distributions of these standardized ages are from the beta distribution (see Table 4.3.) The deviances are much smaller than the degrees of freedom for the logit and probit links and about as expected for the log–log link.

Finally we consider power transformations of the standardized ages z_i using the Box–Cox family of transformations

$$f(z,\lambda) = \frac{z^\lambda - 1}{\lambda}$$

which is equivalent to fitting ordinary power terms z^λ but includes the model $\log(z)$ at $\lambda = 0$ in a smooth way. As in all models fitted the probit link gives a better fit than the logit. This is because the data includes several zero responses and several very high responses (1049 out of 1049 for instance) and the probit link happens to fit this tail behavior better. The estimates of λ are in parentheses. The c-log–log link leads to the fitted probability

$$\pi(x) = 1 - \exp\{-\exp\{5.614 + 2.313x^{-.68}\}\}$$

and the smallest deviance of any of the 12 models fitted. This model formula is difficult to interpret, however.

A good model that combines excellent fit with ease of interpretation is the logit model in $\log(z)$ and $\log(1-z)$. The fitted formula for *odds* is

$$\text{odds}(x) = 11.416z^{4.435}(1-z)^{-2.147} = 0.07485(x-9)^{4.425}(18-x)^{-2.147}.$$

In particular, the exponent 4.425 describes how quickly the odds of menstruation decrease to zero as the age decreases towards 9 and the exponent 2.147 how quickly the odds of menstruation increases as age increases to 18. Figure 4.4 plots

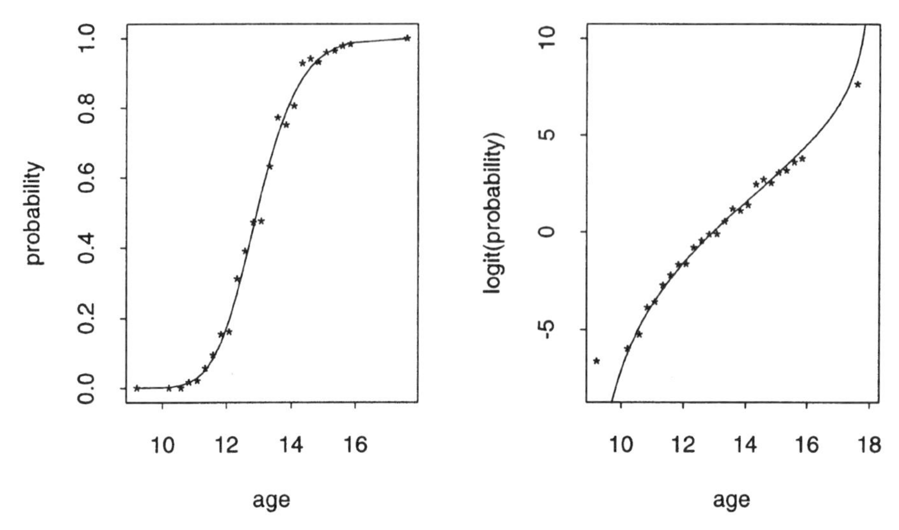

Figure 4.4. Logit model for menstruation data. Fit of the best logistic model plotted (i) on the probability scale with observed proportions and (ii) on the logit scale with modified empirical logits.

the raw proportions and empirical logits with the above model formula superimposed. The empirical logits are quite clearly nonlinear but are well tracked by the model except for the extreme point at $x = 8$. This corresponds to an estimated probability near 0 for which the variability of the empirical logit is diverging. This point automatically has low weight in the fitting procedure and it is therefore no surprise that the fitted curve misses it. Put plainly, the estimated logit of about -6 in the plot might just as well be -12 since both values correspond to incredibly small probabilities whose minute differences are visually exaggerated by the logit scale. The left plot, on the probability scale, tracks the extremes very well.

4.2 DIAGNOSIS OF MODEL INADEQUACY

A main aim of data analysis is the construction of models that are not so simple that they violate the spirit of the data nor so complex that they tell us nothing that the raw data do not tell us. *Diagnostic* statistics are navigational aids in our quest for a good model; they reveal exactly *how* as well as *if* a model is inadequate and point the way to a better model. Relying on goodness-of-fit tests alone is like mountaineering with an altimeter and a blindfold. Using diagnostics is like removing the blindfold, although admittedly the visibility may not always be good.

Binomial linear models assume a binomial error distribution, a link function, and a linear predictor any of which may be in error (and almost certainly will).

Diagnostics are used to detect the failure of any of these three components, typically under the assumption that the other two are correct. They are also used to identify possibly aberrant data points, called *outliers*. The link function and the possibility of small binomial denominators n_i make construction and use of these diagnostics a little harder than in normal theory although intuition gained from linear normal models is an invaluable theoretical guide. This section covers most diagnostics in common use and illustrates their application.

4.2.1 Residuals and the Fit of a Data Point

Unscaled or Raw Residuals Unscaled residuals, often called *raw* residuals, measure the raw difference between observed and fitted values. In classical regression, the discrepancy is almost universally measured by $y_i - \hat{e}_i$. In the context of binomial models, it is not clear that this is the most natural definition. For a linear model with ling g, inadequacy of the linear predictor $X\beta$ should be better revealed by looking at the differences between the raw data proportions and the fitted values on the link scale, i.e., by looking at $g(y_i/n_i) - g(\hat{\pi}_i)$. However, for all the link functions in common use $g(y_i/n_i)$ is undefined when y_i equals 0 or n_i. We may avoid this problem by defining the (raw) *link residuals* as

$$(y_i/n_i - \hat{\pi}_i)g'(\hat{\pi}_i), \tag{4.13}$$

obtained by Taylor expansion of $g(y_i/n_i)$ about $g(\hat{\pi}_i)$. These residuals are also suggested by the algorithm used for fitting generalized linear models [see Eq. (4.47) and the following comment]. For the logistic link we obtain

$$\frac{y_i - n_i\hat{\pi}_i}{n_i\hat{\pi}_i(1 - \hat{\pi}_i)} \approx \operatorname{logit}(y_i/n_i) - \operatorname{logit}(\hat{\pi}_i) \tag{4.14}$$

called *logit* residuals. These are very simple to compute and are commonly used as diagnostics for logistic linear models. Note that there is no $\surd$ on the denominator. Whereas raw residuals for normal data have exact mean zero, raw residuals for binomial models only have asymptotic mean zero.

Raw residuals are typically plotted against predictors, to identify inadequacies of the linear predictor. For an adequate model there should be no trend. However, the variability is not uniform and will be greater when n_i is small or when $\hat{\pi}_i$ is extreme. This can make recognizing trends in the plot more difficult than for classical normal models. Since they are not standardized, raw residuals are of little use for detecting outliers.

Scaled Residuals Large raw residuals flag poorly fitting data points, but how large is large? The LR and Pearson statistics measure global agreement of the fitted and observed values and are already scaled having approximate chi-square distribution. So-called *Pearson* and *deviance* residuals are defined as

$$r_{Pi} = \frac{y_i - \hat{e}_i}{\sqrt{n_i\hat{\pi}_i(1-\hat{\pi})}}$$

$$r_{Di} = \text{sign}(y_i - \hat{e}_i)\left[2y_i \log\left(\frac{y_i}{\hat{e}_i}\right) + 2(n_i - y_i)\log\left(\frac{n_i - y_i}{n_i - \hat{e}_i}\right)\right]^{1/2}$$

and the sum of squares of these are the Pearson and LR statistics, respectively. Both have approximate variance 1.0 provided that π_i is not too extreme and that p is much smaller than k. Pearson residuals generalize to

$$r_{gi} = \frac{g(y_i/n_i) - g(\hat{\pi}_i)}{g'(\hat{\pi}_i)\sqrt{\pi_i(1-\pi_i)/n_i}}. \tag{4.15}$$

A special case is the residual of Anscombe (1964) based on a rather complicated transformation to reduce skewness. The function $g(y) = \sin^{-1}(\sqrt{y})$ has also been used since $g(Y_i)$ has asymptotic variance $1/(4n_i)$ which is free of π. If the numerator of Eq. (4.15) is replaced by its Taylor approximation then this reduces to the Pearson residual and so the choice of g typically has little effect on the overall pattern of residuals, except for extreme fitted values where the scaling is unreliable anyway.

Standardized Residuals While scaled residuals have asymptotic mean zero, their asymptotic variance may be significantly less than one if the number of parameters p in the model is too large compared to k. This is clear from the fact that S and LR have asymptotic χ^2_{k-p} distributions, not χ^2_k. The direct reason is most clearly seen in the Pearson residuals. The denominator $n_i\hat{\pi}_i(1-\hat{\pi}_i)$ estimates the variance of $y_i - e_i$, not the variance of $y_i - \hat{e}_i$ which tends to be smaller because of the natural tendency for fitted values to follow data values. In fact, the asymptotic variance of the raw residual vector $Y - \hat{e}$ is

$$\text{Var}(Y - \hat{e}) = \text{Var}(Y)[I - W^{1/2}X(X^TWX)^{-1}X^TW^{1/2}]$$

where W is called the working weight matrix of the linear model. This matrix is diagonal with the approximate inverse variances of the link residuals on the diagonal. For the logistic link these weights are $n_i\pi_i(1-\pi_i)$. The matrix

$$H = W^{1/2}X(X^TWX)^{-1}X^TW^{1/2} \tag{4.16}$$

is called the *hat matrix* because for normal models it is the matrix that maps the data vector Y to the vector of fitted values often denoted $\hat{Y}$. Thus H puts the hat on Y although this is not exactly true for binomial models. The hat matrix is readily available in the major statistical packages and allows the Pearson and deviance residuals to be properly *standardized* by dividing them by $\sqrt{1 - H_{ii}}$. We will use

$$r_{Pi}^{*} = \frac{r_{Pi}}{\sqrt{1 - H_{ii}}}, \qquad r_{Di}^{*} = \frac{r_{Di}}{\sqrt{1 - H_{ii}}}$$

to denote these standardized Pearson and deviance residuals. These do have asymptotic variance 1. The diagonal elements H_{ii} of H are called *leverages*. It can be shown that their average value is p/k and so only for models with many parameters compared to the number of data points will the standardization adjustment be important. Standardized residuals are still correlated in general, since the off-diagonal elements of H are typically nonzero. This correlation cannot be readily removed but does not seem to be a great impediment to their use.

Standardized residuals are correctly scaled for their statistical variability, and are used to identify individually aberrant data points called *outliers*. Especially when n_i is large, r_i^{*} can be treated like a standard normal variable. A value of 3 or larger flags a point that is statistically unlikely to be governed by the model. However, when the fitted probabilities $\hat{\pi}_i$ are close to 0 or 1, comparison with normal deviates will be highly misleading. Plotting scaled residuals against fitted values can also be useful for assessing the correctness of the mean-variance relation, implied by the error distribution, but their instability for extreme fitted values makes it much harder to assess whether the spread is uniform than for normal data. Smoothing, as described in Section 5.4.3, can aid interpretation.

Example 4.6: The 1980 Presidential Election Data displayed in Table 4.6 were reported by Clogg and Shockey (1988) and relate to voting in the 1980 U.S. Presidential election. A random sample of individuals classified by race as white or nonwhite had their political views determined through a questionnaire and reduced to a 7-point scale with 1 representing extremely liberal and 7 representing extremely conservative. Their voting intention was reduced to a binary response, namely Reagan or Carter/other. The first figure is the number intending to vote for Reagan and the second figure is the number surveyed in that group.

Table 4.6. 1980 Presidential Voting Data

View	1	2	3	4	5	6	7
White	1/13	13/70	44/115	155/301	92/153	100/141	18/26
Nonwhite	0/6	0/16	2/25	1/32	0/8	2/9	0/4

In as much as the Republican and Democratic parties are associated with conservative and liberal policies, respectively, we expect a trend in voting intentions, i.e., we expect the proportion voting for Reagan to increase with their political view rating, although not necessarily in a linear manner. The shape of this trend is better revealed by looking at residuals than the raw data. Indeed as a general rule, the relation of the residuals to a particular factor will give a better idea of the involvement of the factor than a plot of the raw data because the residuals have the effects of other factors removed.

As a first step we fit the model with only race as a predictor. In this case there are only two parameters π_W and π_N being the probability of voting for Reagan for whites and nonwhites. The ML estimates are $\hat{\pi}_W = 423/819$ and $\hat{\pi}_N = 5/100$ and the fitted values are just these probabilities multiplied by the sample size n_i. The residuals are displayed in Table 4.7, the upper figures being Pearson and the lower deviance residuals. Neither of these residuals have been standardized by dividing by $\sqrt{1 - H_{ii}}$ because the H_{ii} are actually constant for this model and so standardizing does not alter the pattern. There is a clear increasing trend in the residuals for whites but no apparent trend for nonwhites. The overall proportion of Reagan supporters among nonwhites is much lower than for whites and it would be difficult to pick up any trend with such small counts in any case. There is good agreement between the two sets of residuals and the choice seldom makes much difference to the overall pattern. The analysis of the closeness of Pearson and LR statistics on p. 93 implies that Pearson and deviance residuals will be close when fitted values are large and raw residuals small.

Table 4.7. Pearson (Upper) and Deviance (Lower) Residuals

View	1	2	3	4	5	6	7
White	−3.17	−5.54	−2.87	−0.05	2.09	4.58	1.79
	−3.42	−5.73	−2.88	−0.05	2.11	4.66	1.82
Nonwhite	−0.56	−0.92	0.68	−0.49	−0.65	2.37	−0.46
	−0.78	−1.28	0.64	−0.52	−0.91	1.78	−0.64

Likelihood Residuals An elegant residual which turns out to be a compromise between the Pearson and deviance residuals is based on the idea of deleting the ith data point and recording the reduction in deviance and called a *deletion* or *likelihood* residual. This is rather intensive in computation but luckily a close approximation to the deviance change is

$$H_{ii}[r_{Pi}^*]^2 + (1 - H_{ii})[r_{Di}^*]^2$$

and the corresponding residuals are

$$r_{Li} = \text{sign}(y_i - \hat{e}_i)\sqrt{H_{ii}[r_{Pi}^*]^2 + (1 - H_{ii}[r_{Di}^*]^2}.$$

For models with few parameters the H_{ii} will be small and deletion residuals will be close to standardized deviance residuals but for models with many parameters they will be closer to Pearson. Likelihood residuals have a lovely and unique property. Let $\hat{\pi}_{(i)}$ be the estimate of π_i when the ith data point is omitted and let $\Pr(y_i|y_{(i)})$ be the probability $\Pr(Y_i = y_i)$ computed using the binomial distribution with parameters n_i and $\hat{\pi}_{(i)}$. The tail probability using this distribution is a natural probabilistic measure of how extreme, i.e., unlikely a data point is. For binary data we simply compute $\Pr(y_i|y_{(i)})$ itself [see Davison (1988a)]. This *cross-validation* probability turns out to be well approximated by

$$\Pr(y_i|y_{(i)}) = \exp\{-r_{Li}^2\}\sqrt{1 - H_{ii}}.$$

Example 4.3: (Continued) **Looking for Outliers in Cancer Remission Data.** The remission data involved very small binomial denominators n_i. There is no loss in taking these data to be 27 binary observations instead of 14 binomials. Table 4.8 lists standardized Pearson, deviance, and deletion residuals as well as cross-validation probabilities for each of the binary observations, the fitted model being the simple linear logistic model in the covariate x. For brevity repeated pairs (x_i, Y_i) have been omitted. There is close agreement between de-

Table 4.8. Detection of Extreme Points for Example 4.3

x_i	y_i	r_{Pi}	r_{Di}	r_{Li}	$\Pr(y_i\|y_{(i)})$
8	0	−0.271	−0.376	−0.370	0.843
10	0	−0.310	−0.428	−0.421	0.809
22	0	−0.691	−0.883	−0.875	0.454
14	0	−0.405	−0.551	−0.543	0.722
16	0	−0.462	−0.623	−0.615	0.666
18	1	1.892	1.744	1.752	0.045
20	1	1.655	1.624	1.626	0.069
20	0	−0.604	−0.789	−0.781	0.530
22	1	1.448	1.504	1.501	0.103
22	0	−0.691	−0.883	−0.875	0.454
24	0	−0.789	−0.984	−0.975	0.376
26	1	1.108	1.266	1.257	0.200
28	1	0.970	1.151	1.138	0.263
32	0	−1.348	−1.439	−1.428	0.122
34	1	0.649	0.839	0.815	0.478
38	1	0.497	0.664	0.640	0.607
38	0	−2.012	−1.800	−1.836	0.031

viance and likelihood/deletion residuals since the leverages H_{ii} are all rather small. No single point stands out as aberrant.

Multiple Comparisons of Residuals How large must a residual be before the corresponding data point is called an outlier? Since we have k residuals the probability of at least one residual being large will increase with k. Computing the probability of the most extreme residual will then exaggerate the lack of fit of the point. A very rough adjustment is based on assuming the residuals are asymptotically independent standard normal which will not be a good assumption if the model involves many parameters, say if $p/k > 0.1$, or if the fitted values are too small. With this assumption if R_{max} is the largest absolute residual then

$$\Pr(R_{max} \geq r) = 1 - [2\Phi(r) - 1]^k \approx 2k\Phi(-r) + O(\Phi(-r)^2)$$

where Φ is the standard normal distribution function and $\Phi(-r)$ is small. Ignoring the error term, an approximate $(1-\alpha)\%$ cut-off value for the largest absolute residual is

$$\Phi^{-1}(0.5[(1-\alpha)^{1/k}+1]) \approx \Phi^{-1}\left(1 - \frac{\alpha}{2k}\right) \tag{4.17}$$

valid for small α. When p/k is not very small this adjustment is inaccurate.

Examples 4.4 and 4.5: (Continued) We illustrate the use of scaled and unscaled residuals on the menstruation and seed data. The first two plots in Figure 4.5 are for the menstruation data. We tried several models for these data, one of which is summarized in Figure 4.4. Is there really anything wrong with the simple logistic model with age included linearly? The left-hand plot shows a plot of logit residuals for the simple linear–logistic model whose deviance was 26.703(25). While the pattern looks a little peculiar, especially the outlier at the top right, it is unclear whether or not there is a trend. The smooth curves will be explained in Chapter 5. The solid curve describes the trend with confidence bands (dotted). The latter suggests that any trend is of borderline significance. Deletion residuals plotted at right are on an absolute scale and for these data are, if anything, smaller than to be expected for standard normal variables. These have been plotted against age but could just as well have been plotted against their index. No point stands out as an outlier. In summary, fitting the simple linear model is probably statistically defensible from these plots.

The lower two plots repeat this residual analysis for a logistic model fitted to the seed data with the single predictor x. Recall that we identified two aberrant points in the data plots given in Figure 4.3. The bottom left figure plots logit residuals against the covariate again with a smooth trend estimate. The trend, if any, is cubic in shape and is of borderline statistical significance.

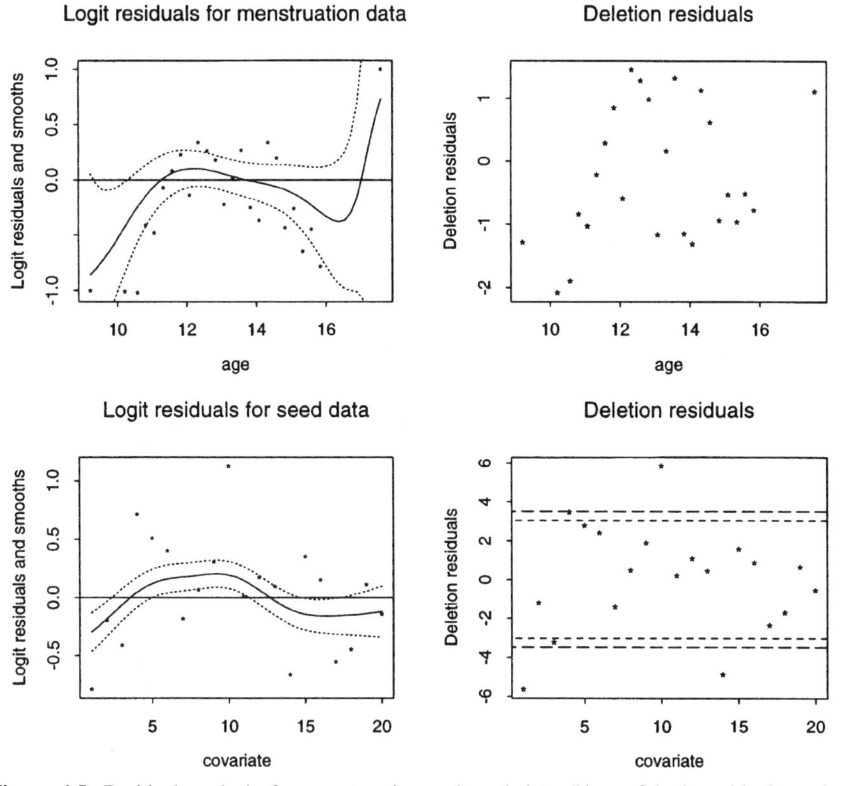

Figure 4.5. Residual analysis for menstruation and seed data. Plots of logit residuals against covariate (left) with smooth trend estimate and confidence bands. Plots of deletion residuals with outlier bands based on standard normality (right).

The plot at right displays deletion residuals together with 95 and 99% critical values for the largest of 20 absolute residual, as given in Eq. (4.17). Two points lie way outside these bands, the same two points that stand out in the data plot. In addition, several other points are identified as extreme. Note however that these residuals are collectively too large—the deviance is 130.54(18)—and the model is surely wrong. If this were not the case, then the plots would suggest two or three aberrant (possibly incorrect) points. However, we will see below that these points are not outliers when compared to the distribution of the residuals as a whole.

In both of these examples there is a single covariate and points of special interest can be just as easily identified by plotting an appropriate response such as empirical logits against that covariate. This allowed us to confirm the residual analysis by looking at the data plot. When several covariates are present, it is difficult to get an overall feel for the relation of the response to these covariates. Residuals are an invaluable although not unique graphical technique for getting such a feel.

4.2.2 Probability Plots

Data points not governed by the model will tend to stand out from the pattern of normality. How do we decide if, say, the three largest residuals are improbably large? To complicate matters, even when the model is correct the normal distribution will be unreliable for points with extreme fitted values or small n_i. If the candidate outliers have fitted values that are moderately extreme, how confident can we be that they are aberrant?

Probability plots give a graphical comparison of the distribution of a set of residuals with their theoretical null distribution. When none of the fitted values are extreme, the theoretical distribution is standard normal. There are, of course, dozens of global tests of standard normality available. The advantage of probability plots is that they visually display how the distribution departs from normality. For instance, are the residuals uniformly too large, or are only a small number too large? When the normal approximation is unreliable, the theoretical distribution may be replaced by the *simulated* null distribution and its *envelope*.

Normal and Half-Normal Plots Let $Z_1, \ldots, Z_k$ be k independent standard normal variables, and $Z_{(i)}$ the ith largest of these. It can be shown that

$$\mathrm{E}(Z_{(i)}) \approx \Phi^{-1}\left(\frac{i - 3/8}{k + 1/4}\right), \qquad \mathrm{E}(|Z_{(i)}|) \approx \Phi^{-1}\left(\frac{i + k - 1/8}{2k + 1/2}\right).$$

Let $r_{(i)}$ be ordered residuals and $|r|_{(i)}$ ordered absolute residuals, from a fitted model. Under the null hypothesis that the model is correct and that the r_i are standard normal, a plot of $\Phi(r_{(i)})$ against $(i - 3/8)/(k + 1/4)$ should be linear with slope one and intercept zero. This is called the *normal* plot. A plot of $\Phi(|r|_{(i)})$ against $(i + k - 1/8)/(2k + 1/2)$ should also be linear with slope one and intercept zero. This is called a *half-normal* plot. Outliers stand out as points far from these theoretical lines.

Simulation Envelopes Individual data points are assessed by comparing their standardized residuals with the normal distribution. This is done because if the model is true then the residuals are approximately standard normal. However, if n_i or $\hat{e}_i$ are too small then the distribution will not be standard normal, even if the model is true. An extreme residual may then be caused not by the data point being aberrant but by the extreme conditions of the model. This issue does not occur for normal models—standardized residuals have the student t distribution exactly if the model is correct. Even when conditions are ideal for the normal limit, the question of what constitutes a significant deviation from linearity in the half-normal plot is not easy to answer.

These issues are addressed by simulating the distribution of the residuals taking estimated model parameters as true values. This gives an estimate of the true joint distribution of the residuals if the fitted model is true, in particular, the expected pattern $E(R_i)$ of residuals as well as confidence limits about this pattern, often called a *simulation envelope*. If the purpose of the plot is to detect outliers then the choice of residual is not critical and it is quite acceptable to use unscaled residuals—the simulation automatically adjusts for scale. The algorithm is as follows:

1. Decide on which residuals are to be simulated. These may be raw or standardized, ordered or unordered, signed or unsigned, but scaled, ordered, unsigned residuals are preferred.
2. Using estimated model parameters $\hat{\pi}_i$ for $i = 1, \ldots, k$ simulate new data from the $Bi(n_i, \hat{\pi}_i)$ distribution.
3. Fit the model to the simulated data and compute the vector of residuals of the type chosen in 1.
4. Repeat 2 and 3 B times.
5. For each i there are B simulated residuals. Use a measure of location μ_i of these to estimate the expected value of the ith residual and quantiles to estimate upper and lower limits.

If the residuals do not follow the simulated mean trend then the model is probably structurally deficient. For instance, if the absolute residuals are systematically larger than the mean trend then the data might be overdispersed (see Section 4.4). This will have already caused an inflated deviance. But the plot gives further information about the source of this lack of fit, in particular, whether the fit is uniformly poor or whether a few aberrant points are responsible. Probably the best displays plot sorted residuals against their simulated mean trend μ_i. Departure from the model appears as departure from linearity, significant departures being calibrated by the envelope. While the more simulations the better, Atkinson (1981) suggests $B = 19$ such simulations is sufficient for the exploratory purposes of residual analysis. In this case the maxima and minima of the simulations of each residual give upper and lower limits with associated 95% confidence.

Examples 4.4 and 4.5: (Continued) **Simulation Envelopes.** Figure 4.6 shows half-normal probability plots of absolute Pearson residuals for the menstruation and seed germination data. The deviation from a standard straight line appears visually significant. The dotted line is the mean μ_i of 19 simulations ordered from smallest to largest. The solid line is the identity function. In both cases the dotted line is close to the solid line which indicates that if the model were true the residuals are distributed close to standard normal—in other words, the conditions for approximate normality of the residuals do hold for this model and the half-normal plots are *not* misleading.

Interpretation of the half-normal plots is aided by the simulation envelopes at right. The solid lines are the estimated 5%, 50%, and 95% quantiles of the

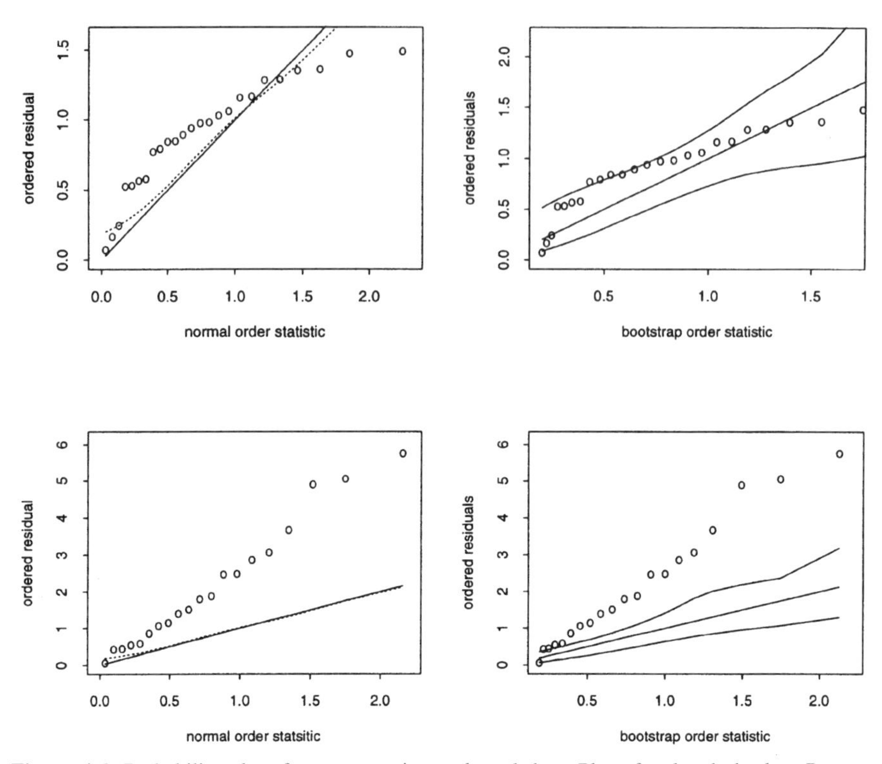

Figure 4.6. Probability plots for menstruation and seed data. Plot of ordered absolute Pearson residuals against normal order statistics for menstruation data (above) and seed data (below). The menstruation data deviate from the theoretical line in shape (left) and from the simulation envelope (right). The seed data plots are approximately linear but with too large a slope indicating over-dispersion.

null distribution of the residuals. The μ_i are plotted on the x-axis but as we just saw these hardly differ from the normal order statistics. Both plots indicate significant deviation from normality. For the menstruation data, there are too many residuals in the moderate range around 0.75. The largest residuals are in fact a little smaller than expected. For the seed data, all residuals are above the upper envelope. As the plot is still linear, it seems that while the residuals are approximately normal and independent they have variance uniformly larger than 1.0, a phenomenon called over-dispersion. The residual plot shown in Figure 4.5 suggested three outliers. These points are not in fact outliers when an account is taken of the pattern of the residuals as a whole. True outliers would have appeared in the upper right of the plot, the rest of the points following the expected trend.

While simulation envelopes are perhaps too much trouble in routine data analysis, the simple half-normal plot is often useful even without the simulation envelope.

4.2.3 Residuals and the Shape of the Predictor

Suppose we have fitted a certain model including computing residuals, and are now considering whether it might be worth adding the covariate z to the linear predictor. For instance, suppose we have fitted the linear model $X\beta$ with link g but that the true model has the form

$$g(\pi) = X\beta + f(z)\gamma$$

for some covariate z. We might expect that plotting raw linked residuals against z would estimate the function $f(z)\gamma$ and so reveal the deficiency of the model. In fact, it can be shown that the link residuals

$$r_{Li} = (y_i - n_i\hat{\pi}_i)g'(\hat{\pi})$$

consistently estimate not $f(z)\gamma$ but

$$\mathrm{E}(R_L) = W^{-1/2}(I - H)W^{1/2}f(z)\gamma \tag{4.18}$$

where the hat matrix H and W are from the model with linear predictor $X\beta$. Only when the vector $f(z)$ is orthogonal to the design matrix X does this necessarily reduce to $f(z)$. Otherwise, the residuals may give a quite biased estimate of $f(z)$, depending on the size of the hat matrix H. This implies first that relationships even imperfectly suggested by raw residual plots might be worth pursuing. However, we would prefer to modify the plot so that the mean does reveal the true relationship, and we would like to achieve this without going to the trouble of actually fitting another model. The plots in this section are due to Wang (1985, 1987).

Added Variable Plots The plot will be easier to interpret if the variance of each point is approximately constant. The linked residuals have variances $1/W_{ii}$, and since they are proportional to $y_i - n_i\hat{\pi}_i$ the scaled linked residuals $r_{Li}\sqrt{W_{ii}}$ are identical to the Pearson residuals. Premultiplying Eq. (4.18) by $W^{1/2}$ we obtain

$$\mathrm{E}(R_P) = (I - H)W^{1/2}f(z)\gamma.$$

An *added variable plot* is a plot of r_P against $(I - H)W^{1/2}f(z)$. The elements of $(I - H)W^{1/2}f(z)$ are called *residual carriers*. For models with normal errors, added variable plots have the following attractive properties. (1) The least squares estimate of the slope of the plot is the least squares estimate of γ, and (2) the correlation of the plot measures the significance of $\hat{\gamma}$. For other generalized linear models, these are only approximately true.

Added variable plots reveal both how z should be involved in the linear predictor and how individual data points deviate from the main pattern. If $f(z)$ is correct, the plot should be linear with slope γ and the strength of the relationship estimates the importance of $f(z)$ in the model. Systematic deviations from linearity suggest that $f(z)$ is not the correct transform. A plot displaying no relationship suggests that z is not needed in any form in the predictor. An *Splus* function `added.variable` is on the Wiley website.

Constructed Variable Plots The last plot considered in this section is more generally directed towards revealing whether a covariate z should be replaced by some member of a specific *family* of functions of z. Let $f(z, \lambda)$ be a nonlinear family of transforms of z and suppose that $f(z, \lambda_0) = z$. A simple example is the power family $f(z, \lambda) = z^\lambda$ with $\lambda_0 = 1$. We would like to know whether z should enter the model linearly or nonlinearly. Suppose we have fitted the linear model with linear predictor $X\beta + \gamma z$ and are considering extending this to $X\beta + \gamma f(z, \lambda)$. Taking a Taylor expansion of f about $\lambda = \lambda_0$ gives an approximate linear predictor

$$X\beta + \gamma z + \gamma(\lambda - \lambda_0)f'(z, \lambda_0)$$

where f' is the derivative with respect to λ. The so-called *constructed* variable is

$$u_i = \hat{\gamma} f'(z, \lambda_0) \tag{4.19}$$

where $\hat{\gamma}$ is the slope estimate for the covariate z in the linear model. For the specific case of the power transformations the constructed variable is $u_i = \hat{\gamma} z_i \log z_i$. A *constructed variable* plot is an added variable plot for the constructed variable u. Specifically, we plot the Pearson residuals r_P of the linear model against the vector $(I - H)W^{1/2}u$. The strength of association in the plot reflects the importance of adding a function from the chosen family. The slope of the plot estimates $\lambda - \lambda_0$ and can be used as a starting value for proper maximum likelihood fitting of the nonlinear term.

Example 4.5: (Continued) **Menstruation Data.** We begin with the simple linear probit model. Adding the term x^2 decreases the deviance from 22.887 to 15.149 (see Table 4.5), and the estimated quadratic coefficient is -0.048. However, the raw residual plot in the top left panel of Figure 4.5 hardly cries out for a quadratic term. We argued however that raw residuals are *not* unbiased for the true missing term $f(z)$. Could $f(x) = x^2$ have been discovered from an added variable plot?

The left plot of Figure 4.7 is the added variable plot for adding x^2 and shows a clear decreasing trend. The slope of the least squares line is -0.055. The reduction in residual sum of squares due to fitting the line is 7.12, quite close to the drop in deviance $22.887 - 15.149 = 7.738$ from adding the x^2 term to the probit model. So the association displayed in the added variable plot gives a

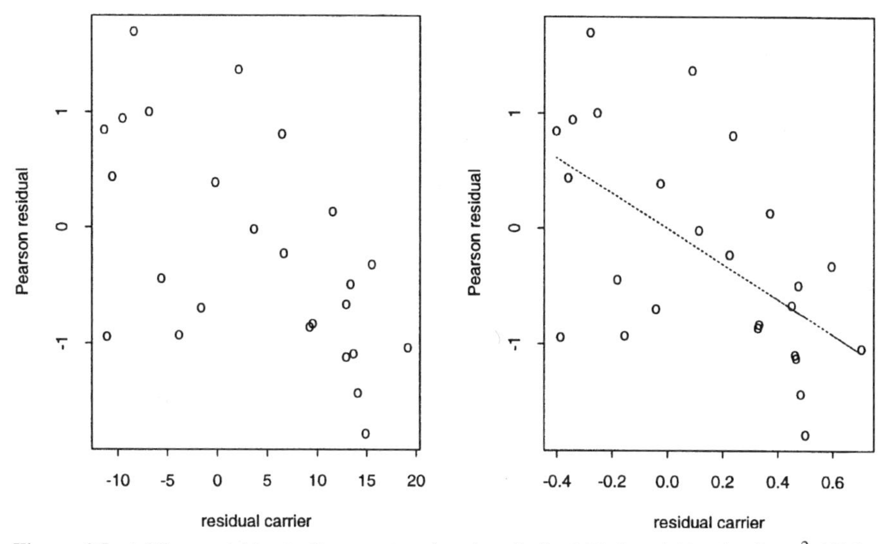

Figure 4.7. Adding variables to the menstruation data. Left: Added variable plot for x^2. Right: Constructed variable plot for power term.

good idea both of the estimated coefficient and its statistical significance without actually fitting the quadratic model. One could also look at added variable plots for many other functions such as $\log x$, $\log(x-9)$, $\sqrt{x}$, etc., and this is a natural step after fitting the linear term.

The added variable plot looks like it may have some curvature, suggesting that x^2 is not the best function of x to add. Consider *replacing* x by a power term x^λ. The right-hand plot displays the constructed variable plot for this. The Pearson residuals have been used for the added variable plot. There is a clear linear relationship with negative slope -1.531 indicating a power transformation with exponent around -0.53. The ML estimate of this power parameter is -0.51 and gives a deviance 14.516 being 8.37 smaller than for the linear model. In fact, this deviance is even lower than the quadratic model which had two covariates, x and x^2.

The added and constructed variable plots look very similar in this example, except for the differing x-scales. The reason is that x^2 and the constructed variable $x \log x$ are extremely highly correlated ($\rho = 0.998$) which would not be the case if, for instance, the x-values were smaller in magnitude. Note also that we earlier found a good probit model with covariate $(x-9)^{0.51}$ and we are now suggesting the covariate $1/x^{0.51}$. The equal exponents here are a coincidence. Both models fit the data equally well.

4.2.4 Leverage and Influence

Standard errors quantify the unreliability of estimators and tests due to random variations in the data. Wrong conclusions may also be generated by observations that are misrecorded or otherwise corrupted and the effects of these may sometimes be much larger than the effects of random variation. A data point is said to be *influential* if its deletion has a relatively large influence on the inference drawn. While influential data points are not necessarily wrong or even outliers, we must admit that inferences based largely on just one observation are not to be trusted. Standard error expressions will not reflect this unreliability and will in fact indicate more accuracy when influential observations are present.

A direct measure of influence is obtained by fitting a model with a particular data point excluded and seeing how much the inference changes. It is too much computation to do this for every data point and model fitted. Luckily the change to the ML estimator $\hat{\beta}$ when the ith data point is deleted can be approximated by

$$\hat{\beta}_{(i)} - \hat{\beta} = -\frac{Y_i - \hat{e}_i}{\sqrt{1 - H_{ii}}} (X^T WX)^{-1} x_i \tag{4.20}$$

where $\hat{\beta}_{(i)}$ is the parameter estimate with the ith data point excluded, H_{ii} are the *leverages*, and x_i is the covariate vector for the ith data point. For normal linear models W is the identity and the expression is exact. Nevertheless, the expression gives a good indication of the change for nonnormal models and can be used to assess the influence of each point without further model fitting. The leverages H_{ii} themselves measure the distance of the covariate vector x_i from the centroid of the covariate vectors adjusted by the effective weights W. It is simple to show that the H_{ii} all lie between 0 and 1 and add up to the number of fitted parameters p. Their average value is thus p/k and so we define standardized leverage as $L_i = kH_{ii}/p$ which is 1.0 on average and considered large if greater than 2.0.

A data point Y_i may have high leverage but have no influence on the fit if it happens to agree exactly with the fitted value $\hat{e}_{(i)}$ when Y_i is omitted. Thus, influential points will need to have high leverage *and* be far from $\hat{e}_{(i)}$ which implies that they are outliers both in their x and Y values. Cook's distance is defined as

$$D_i(\hat{\beta}) = \frac{1}{p} (\hat{\beta} - \hat{\beta}_{(i)}) X^T WX(\hat{\beta} - \hat{\beta}_{(i)}) \approx \frac{L_i r_i^{*2}}{k}$$

where the approximation is based on Eq. (4.20) and r_i^* is one of the standardized residuals discussed earlier. The l.h.s is a quadratic form measuring the overall difference between the vector $\hat{\beta}$ and $\hat{\beta}_{(i)}$ and values around 1 or higher are considered high. This corresponds to the parameter estimates changing by about one standard error. The quantity D_i will rank points correctly in terms of their effect on the fit but it may underestimate the effect of the most influential points, as noted by Davison and Tsai (1992, Section 4).

Cook's distance explicitly depends on the measure of leverage L_i and the measure of extremeness r_i^*. A useful display is to plot the data indices i at the points (L_i, r_i^*), in other words to plot r_i^* against L_i using the index of the data point as the plotting symbol. Points with extreme leverage and/or influence are then easily identified [see Williams (1982b)]. An even better plot is $(r_i^*)^2$ against L_i with the hyperbolae $(r_i^*)^2 L_i = k$, $2k$ superimposed. These curves correspond to Cook's distance 1 and 2. Points of high influence appear in the upper-right corner of the plot, well separated from the others. An *Splus* function `influence.plot` for doing this is at the Wiley website.

Frequently we will be interested in specific parameters and hypotheses rather than the overall fit of the model. In this case it is more relevant to measure or approximate the influence of data points on these specific inferences rather than on the estimate of β as a whole. Several useful influence diagnostics for this purpose are listed in Collett (1991, Section 5.5).

Example 4.4: (Continued) **Influence Diagnostics for Seed Data.** The logistic linear model fitted to the seed data gave deviance 130.5 on 18 degrees of freedom which when combined with the earlier probability plot of the residuals strongly suggests these data are over-dispersed. The intercept and slope estimates from this model are $\hat{a} = -2.907(0.084)$ and $\hat{b} = 0.277(0.0077)$. Figure 4.8 displays various plots relevant to the assessment of fit. The first superimposes modified empirical logits on the linear predictor and the second plot gives likelihood residuals. Several points (1, 10, 14) have extreme residuals but it was decided earlier that they were not outliers but consequences of systematic over-dispersion. The third plot shows Cook's distances and one point in particular stands out as influential, namely the first $y_1 = 9$. The fourth display plots squared residuals against leverage with contours corresponding to Cook's distance of 1.0 and 2.0. This plot gives direct visual information about the residual size, leverage and influence of each data point. The first data point has both high leverage (around 1.9) and high residual. The 10th data point has moderate influence even though it has a larger residual. This is caused by its small leverage (around 0.5) which in turn results from its being in the center of the x-space.

The squared residual associated with y_1 is 31.596 which predicts a drop in deviance to 98.94 on omitting the point. When the model is refitted with the first data point removed the deviance is actually 99.06 with new parameter estimates $\hat{a} = 2.713$ and $\hat{b} = 0.263$. While these changes do not seem large they are large compared to the standard errors of the original parameter estimates. In fact, the high mean deviance of $130.538/18 = 7.25$ means that the squared residuals and influence should all roughly be divided by this factor whereupon the first data point is not judged to be overly unusual or influential.

4.2.5 Nonorthogonality

Superficially the process of model building is one of finding as simple a model as possible that is true to the spirit of the data. However, data analysis is usually

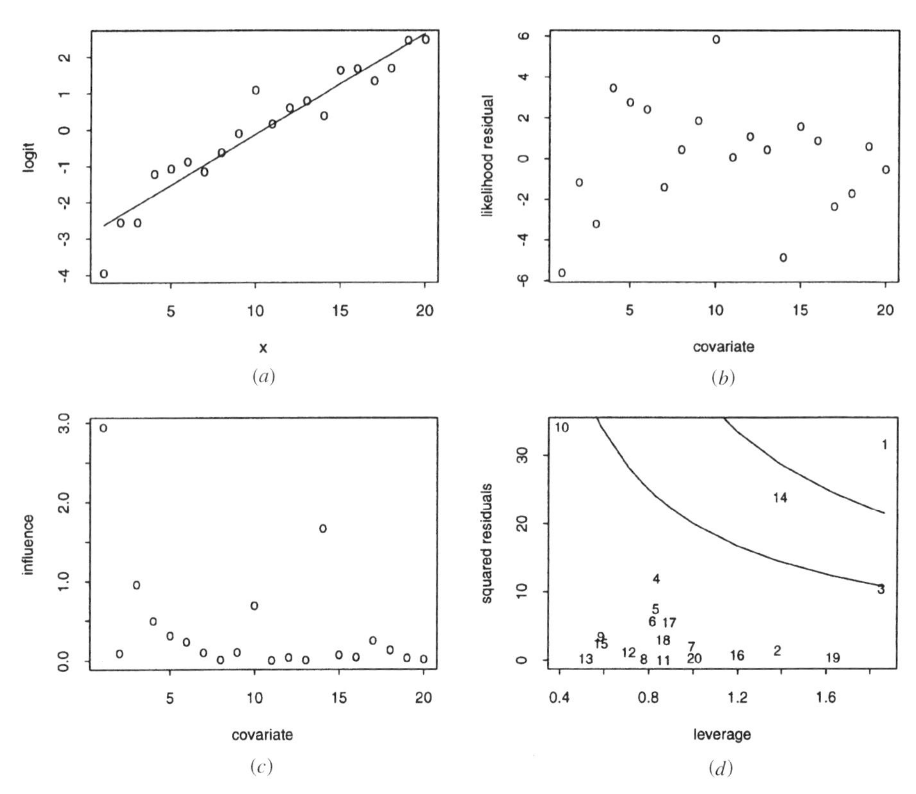

Figure 4.8. Influence diagnostics for seed data. (*a*) Comparison of fit with empirical logits, (*b*) likelihood residuals, (*c*) Cook's distance plot (two points have large influence), and (*d*) residuals against leverage with Cook's distance contours (data points 1 and 14 stand out).

performed with specific questions in mind such as the equality or otherwise of response probabilities for several specific groups of interest. Simplification of the model may have either a large or small effect on inferences to be made about specific parameters of interest regardless of whether global measures of fit such as LR support the simplification. One way of measuring the effect on specific parameters is *nonorthogonality* defined on p. 26.

Letting I denote expected information, the nonorthogonality of parameters θ and ϕ was defined in Eq. (1.24) to be

$$CORR^2(U_\theta, U_\phi) = \frac{I_{\theta\phi}^2}{I_{\theta\theta} I_{\phi\phi}}. \tag{4.21}$$

For several parameters these squared correlations are easily collected into a matrix. The information matrix (or its inverse which estimates the variance matrix of the estimates) is returned by most statistical packages.

Table 4.9. Fictitious Insecticide Data

Dosage (mg/l)		1	2	3	4	Total
1	y_i	2	2	7	19	30
	n_i	10	9	12	22	53
2	y_i	1	4	8	5	18
	n_i	13	11	10	6	40

High correlations have two important statistical interpretations. First, imposing a value for ϕ far from its ML estimate (for instance fitting the model with $\phi = 0$) will produce a large change in $\hat{\theta}$. If θ is of interest then at the very least we should be aware that our estimate of $\hat{\theta}$ is greatly affected by our modeling decisions, over and above what the data say about θ. Second, the standard error of $\hat{\theta}$ will be greatly *reduced*. While small standard errors are usually considered a good thing, this small standard error is generated by a modeling decision and may be misleadingly small.

Example 4.7: Insecticide Toxicity Data The data displayed in Table 4.9 are fictitious and refer to the number of mortalities for a set of 93 insects, 53 of which were given various dosage levels of insecticide 1, the others being given insecticide 2. The death rate apparently increases with dosage for both insecticides 1 and 2. The question of interest is which of the two insecticides is better in terms of killing a greater proportion of insects.

In order to directly compare the two insecticides we must suppose that the response function $\pi(x)$ has the same shape for both, otherwise one insecticide might be better at lower doses and the other at higher doses. This assumption can of course be checked by a goodness-of-fit test. Suppose that

$$\text{logit}\{\pi_i(x_j)\} = a_i + b_i x_j \qquad i = 1, 2 \qquad j = 1, 2, 3, 4.$$

The parameters a_i measure the underlying mortality rates when no insecticide is applied on the logistic scale and the parameter b_i the rate of mortality increase for each unit increase in dose. High values of b_i mean a more effective insecticide. Fitting this model gives the deviance 2.010(4) which provides no evidence against the assumption of a logistic linear rate of mortality increase.

```
[i] ? $LOOK Y N DOSE TYPE $
[o]          Y         N       DOSE      TYPE
[o]  1    2.000    10.000    1.000     1.000
[o]  2    2.000     9.000    2.000     1.000
[o]  3    7.000    12.000    3.000     1.000
[o]  4   19.000    22.000    4.000     1.000
[o]  5    1.000    13.000    1.000     2.000
```

```
[o]  6     4.000     11.000    2.000    2.000
[o]  7     8.000     10.000    3.000    2.000
[o]  8     5.000      6.000    4.000    2.000
[o]
[i] ? $FACT TYPE 2$VARI DOSE
[I] ? $VAR Y$ERROR BINO N$FIT TYPE*DOSE$DISP EV$
[o] scaled deviance = 2.0098 at cycle 4
[o]             d.f. = 4
[o]         estimate       s.e.      parameter
[o]  1       -3.120      1.001      1
[o]  2      -0.6113      1.525      TYPE(2)
[o]  3        1.193     0.3295      DOSE
[o]  4       0.3703     0.5782      TYPE(2).DOSE
[o]
[o] (Co)variances of parameter estimates
[o]  1         1.002
[o]  2        -1.002       2.325
[o]  3       -0.3108      0.3108     0.1085
[o]  4        0.3108     -0.8226    -0.1085     0.3343
[o]             1            2          3          4
[o]
[i] ? $FIT DOSE.TYPE+DOS$DISP E$
[o] scaled deviance = 2.1720 at cycle 4
[o]             d.f. = 5
[o]         estimate       s.e.      parameter
[o]  1       -3.400     0.7532      1
[o]  2        1.280     0.2588      DOSE
[o]  3       0.1554     0.2057      DOSE.TYPE(2)
```

In the natural GLIM parametrization, the logit response probability for treatment 1 is $\mu + bx$ and for treatment 2 is $\mu + a(2) + [b+b(2)]x$, i.e., the parameters $a(2) = a_2 - a_1$ and $b(2) = b_2 - b_1$ measure differences in intercept and slope. This is probably the most statistically meaningful parametrization. The parameter estimates are $\hat{\mu} = -3.12(1.00)$, $\hat{a}(2) = -0.61(1.5)$, $\hat{b} = 1.19(0.33)$, and $\hat{b}(2) = 0.37(0.58)$ and the covariance matrix is given above. Inverting this and converting to a matrix of squared correlations gives the nonorthogonality matrix

$$R = \begin{pmatrix} 1.000 & 0.407 & 0.874 & 0.307 \\ 0.407 & 1.000 & 0.266 & 0.877 \\ 0.874 & 0.266 & 1.000 & 0.303 \\ 0.307 & 0.877 & 0.303 & 1.000 \end{pmatrix}$$

There are two high correlations first between μ and b ($R_{13} = 0.874$) and second between $a(2)$ and $b(2)$ ($R_{24} = 0.877$). What is the explanation for these high nonorthogonalities? Clearly if we changed the assumed value μ (which in the GLIM parametrization is the intercept for insecticide 1) then the slope b of the line of best fit to the insecticide 1 data would change greatly. So would the variability decrease as one end of the line would be "pinned down" so to speak. Since we would not likely be removing μ from the model or assuming a value for it this nonorthogonality is only of academic interest. The nonorthogonality between $a(2)$ and $b(2)$ is of practical interest because these parameters measure differences between the two insecticides. Assuming a value for $a(2)$ might have a drastic effect on $\hat{b}(2)$ while drastically reducing its standard error. Again we can reason why this is so. If we supposed that $a(2) = 0$, i.e., that the intercepts of the two lines were equal, then the fitted lines for the two insecticides are forced to be equal at dose zero. This first makes it more difficult for the slopes to differ and so we will expect $b(2)$ to be closer to zero. Indeed we see that the estimate decreases from 0.3703 to 0.1554 in the `GLIM` output above. The variability of the differences in the two slopes must also be reduced as there is no freedom for the lines to disagree at the zero end. Again this is confirmed in the `GLIM` output where the standard error decreases from 0.5782 to 0.2057, a factor of 2.8 which is extremely close to the $(1 - p^2)^{1/2} = 2.85$.

Assuming equal intercepts for the two lines does not change the major conclusions in this example, namely that the insecticides are effective (test $b > 0$) nor that they appear to be equally effective (test $b(2) \neq 0$). The strength of the conclusion is also hardly affected, the z-values for testing $b(2) \neq 0$ being 0.640 when the intercepts are left free and 0.755 when they are set equal. Part of the reason is that the estimate of $a(2)$ is relatively small and assuming it is zero does not conflict greatly with the estimate. In other examples conclusions could be completed altered. The question remains, however: should we assume $a(2) = 0$ or not? In this example, the intercepts measure natural mortality of the insects and if the two groups of insects were chosen randomly then the mortality for both groups *must be equal* when no dose is applied. Knowledge of the experiment here indicates that equal intercepts be assumed, regardless of nonorthogonality. Deciding whether or not to set some parameters to zero in a model may often be helped by such extra-statistical knowledge, as may the entire modeling process.

4.2.6 Goodness of Link

The link function g is a critical ingredient in the systematic part of a model and misspecification may introduce significant distortion into our conclusions. In this section we consider the problem of identifying an appropriate link function, the linear predictor $X\beta$, and the error distribution being known.

Consider the problem of testing whether the fitted link function is the correct one. The "parameter" g under test here is a function and the parameter space of possible link functions is very large, so we might expect the power to be rather poor unless we choose a particular alternative link function of interest. Let g_0 denote the "null" link and g_1 the alternative. The difference in deviances for models assuming these two links may be computed however, since the two models are not nested the distribution will not be χ^2. Indeed the change in deviance statistic may take positive or negative values and in general simulation will be required to obtain the null distribution. A further problem with this test and other tests to be described below is that it is not clear which link function should be taken as the null and contradictory results might be obtained if the roles of g_0 and g_1 are reversed.

In order to derive a test statistic that is asymptotically χ^2 we embed the problem in a parametric family. Let $g(;\theta)$ be a family of monotonic functions containing the null link $g(;\theta_0)$ to be tested. Suppose also that this family contains the true link function g which is close enough to g_0 that the Taylor expansion

$$g(\pi;\theta) \approx g(\pi;\theta_0) + (\theta - \theta_0)^T \left. \frac{\partial g(\pi;\theta)}{\partial\theta} \right|_{\theta=\theta_0}$$

is reasonably accurate. Then the true model $g(\pi) = X\beta$ can be approximated by

$$g(\pi;\theta) = X\beta + \gamma Z$$

where Z is the matrix $\partial g/\partial\theta$ evaluated at $\theta = \theta_0$ and $\pi_i = \hat{\pi}_i$ with $\gamma = \theta_0 - \theta$. When θ is a single parameter Z will be a single vector and is another example of a constructed variable. Since the null link function corresponds to $\gamma = 0$ we test the link function by observing the drop in deviance when the constructed covariates Z are added to the model. An estimate of a more appropriate link function is also obtained from the estimate $\hat{\theta} = \theta_0 - \hat{\gamma}$. An added variable plot for the columns of Z will reveal which data points are providing the most evidence if any against the null link. Even though we have described this theory in terms of a probability parameter π it applies to any generalized linear model.

For binomial data the most common links are logit, probit, and complementary log–log. Since probit and logit are very difficult to distinguish, the most pressing need is for a data-based method of choosing between the last two. Two families are available for testing the logit link. The family

$$g(\pi;\theta) = \log\left\{\frac{(1-\pi)^{-\theta} - 1}{\theta}\right\}$$

includes the logit when $\theta = 1$ and the c-log–log when $\theta = 0$. The constructed variable for testing the null hypothesis of a logistic link $\theta = 1$ is then

$$z_i = \frac{\log(1-\hat{\pi}_i)}{(1-\hat{\pi}_i)^{\theta} - 1} = \frac{1}{\theta} = -\left(1 + \frac{\log(1-\hat{\pi}_i)}{\hat{\pi}_i}\right)$$

proposed by Aranda–Ordaz (1981). A useful two parameter family is

$$g(\pi_i; \tau, \kappa) = \frac{\pi^{\tau - \kappa} - 1}{\tau - \kappa} - \frac{(1 - \pi)^{\tau + \kappa} - 1}{\tau + \kappa}$$

proposed by Pregibon (1980). The parameter κ measures asymmetry of the link (or corresponding tolerance distribution) and τ the rate of divergence as π approaches one or zero. When both parameters are zero the link is logistic but the family does not contain the c-log–log. For testing the logistic link against this family of alternatives the constructed variables are

$$z_{1i} = (\log^2 \hat{\pi}_i - \log^2(1 - \hat{\pi}_i))/2, \qquad z_{2i} = -(\log^2 \hat{\pi}_i + \log^2(1 - \hat{\pi}_i))/2.$$

Since this procedure simultaneously tests for asymmetry and heavy tails it will be less powerful at detecting the specific c-log–log alternative. Adding the variable z_2 only provides a test of the heaviness of the tails of the link assuming symmetry, while adding the variable z_1 only provides a test of symmetry assuming standard logistic weight in the tails. Brown (1982) gave a test of the logistic link against the more general link of Prentice (1976b).

4.3 MODELING RELATIVE RISK

A statistical model for binary data relates the probability π of response to a set of covariates. Not uncommonly the response is something that individuals would wish to avoid such as disease, death, or injury and one of the covariates is of more interest than others perhaps because individuals or planners can influence the level of this covariate. For example, people have a choice about how many cigarettes they smoke, how much alcohol they consume before driving, and to some extent their levels of blood cholesterol. In this context it is of interest to express the effect of the particular covariate relative to some baseline level, often zero. For example, how much higher is lung cancer risk for someone smoking 20 cigarettes per day compared to a nonsmoker, taking into account other relevant covariates such as age and sex?

Let the probability $\pi(x, z)$ of response depend on a covariate x and possibly other explanatory variables denoted z. The *(relative) risk function* for a covariate x is defined here to be the relative *odds* of response at the level x compared to the baseline $x = 0$, i.e.,

$$r_z(x) = \frac{\pi(x, z)/(1 - \pi(x, z))}{\pi(0, z)/(1 - \pi(0, z))} = \frac{\text{odds}(x, z)}{\text{odds}(0, z)} \tag{4.22}$$

and we note that the risk function is 1.0 when $x = 0$. The log-risk is then a difference between the log-odds at x compared to zero and the log-risk function will equal 0 at $x = 0$. The relationship between log-risk and logits is discussed in

Section 4.3.2. In general, the effect of x on probability of response will depend on all other covariates and so the risk function $r_z(x)$ for x depends also on z. Some authors define the risk function as a ratio of probabilities but since probabilities are often nonestimable with sociological data, for instance with retrospective studies, it is preferable to work with odds. For small probabilities there will be little difference between probability and odds.

The risk function is a useful summary of the effect of a particular covariate on the probability of response. For models involving several covariates in a complicated fashion a statement of the estimated coefficients in a linear predictor is unlikely to give much feel for what the data and model are saying. Plots of risk functions for the important covariates (for different levels of other factors perhaps) convey much more clearly the important conclusions to be drawn from the data. In this section various methods of risk function estimation are developed.

4.3.1 Point Estimates and Confidence Intervals

The risk function is defined in Eq. (4.22) as the relative odds of response when the covariate takes value x compared to zero. Suppose for simplicity that there are no other relevant covariates z. If $Y(x)$ denotes the number of responses and $n(x)$ the number of experimental units at the covariate value x then the natural point estimator of $r(x)$ is the observed odds ratio for the data, namely

$$\hat{r}(x) = \frac{Y(x)/(n(x) - Y(x))}{Y(0)/(n(0) - Y(0))}. \tag{4.23}$$

This in turn is the estimate of the odds ratio in a 2×2 contingency table with rows categorizing response or nonresponse and columns categorizing the covariate value x or 0. The theory of point and interval estimation of odds ratios given in Chapter 3 applies directly. For instance, a more unbiased estimate of the *logged* risk is obtained by using a modified estimate with 1/2 added to the counts in Eq. (4.23). The estimated variance is

$$\text{Var}(\log\{r(x)\}) = \frac{n(x)}{Y(x)(n(x) - Y(x))} + \frac{n(0)}{Y(0)(n(0) - Y(0))}$$

with the adjustments given in Eq. (3.19) if a modified estimator is used. The estimate $\log\{\hat{r}(x)\}$ and its standard error can simply be computed, for all x for which data exists. The estimated risk function can be thought of another way: letting $\overline{Y}(x) = n(x) - Y(x)$ be the number of nonresponders (or controls) at covariate level x, we see that the function $\hat{r}(x)$ is simply the ratio of $Y(x)$ and $\overline{Y}(x)$ standardized to equal 1.0 at $x = 0$. Thus, the estimated risk function is the (standardized) ratio of the covariate distributions of responders and nonresponders. A plot of $\hat{r}(x)$ against x with confidence bands generated by the standard error gives a visual estimate of what the data say about the effect of x on the odds of response.

Example 4.8: Drinking and Fatal Motor Accidents Table 4.10 lists data relating to the blood alcohol concentration (BAC measured in mg/100 ml) of male automobile drivers broken down into four age groups. The case data comprise records of drivers who died in a motor accident in New South Sales, and the control data is extracted from a random survey of drivers in Adelaide, South Australia. The data are *retrospective* since groups of responders (case) and non-responders (control) have been chosen and their BAC levels measured subsequently. A prospective study would require us to administer specific doses of alcohol to drivers and then to send them out on the roads and observe who returned, all in the interests of science. Unfortunately for the statistician, we must make do with retrospective data with its inherent susceptibility to bias. There are indeed several biases in these data not discussed here and a more detailed analysis may be found in Lloyd (1991).

Table 4.10. Blood Alcohol Concentration Distribution for Automobile Drivers[a]

Blood Alcohol	Control Drivers					Case Drivers				
Concentration	<21	21–30	30–50	>50	All	<21	21–30	30–50	>50	All
0	549	1906	2444	955	5854	46	42	36	24	140
5–15	28	116	135	50	329	4	3	1	1	9
15–25	25	109	126	50	310	2	2	0	5	9
25–35	26	92	110	37	265	2	2	2	2	8
35–45	13	71	83	27	194	7	1	3	0	11
45–55	15	51	57	22	145	3	3	4	0	10
55–65	8	37	43	17	105	3	3	3	2	11
65–75	4	23	24	18	65	5	2	1	0	8
75–85	3	24	24	5	56	1	3	2	0	6
85–95	5	26	23	4	58	5	2	4	0	11
95–105	3	9	28	7	47	6	3	2	3	14
105–115	5	9	10	3	27	4	2	2	1	9
115–125	3	9	7	7	26	7	4	3	1	15
125–135	0	8	7	0	15	12	3	3	2	20
135–145	0	5	5	0	10	10	7	8	3	27
145–155	0	6	3	4	13	2	4	4	2	12
155–205	2	10	16	7	35	32	50	22	9	113
205–255	1	0	3	1	5	7	38	28	4	77
>255	3	6	0	0	9	5	25	37	4	71
All	693	2517	3148	1214	7572	163	199	165	64	591

[a]Cross-tabulation of age by blood alcohol content for a random sample of 7572 male drivers in Adelaide and 591 male drivers killed in New South Wales, 1984.

The relative risk function is simply the relative odds of response at covariate level x compared to zero. The special reversal property of odds means that risk can also be interpreted as relative odds of having covariate x for subjects that respond compared to those who do not. Indeed, as noted above $\hat{r}(x)$ is the ratio of the observed BAC distributions for cases and controls, standardized to equal one when $x = 0$. Thus, *relative* risk can be estimated from retrospective data although the absolute probability of response cannot.

Figure 4.9 plots point estimates of the logged relative risk function for all age-groups combined. The totals for case and control had $1/2$ added to each count and the distribution for cases was divided by the distribution for controls as in Eq. (4.23). The confidence bands were obtained via the modified standard error Eq. (3.19). The broad linear trend here suggests that the risk function is exponential. However, since the age factor has been ignored and age is a known predictor of driving behavior, the estimate obtained from collapsing over age may be misleading.

Simpson's Paradox Revisited The problem of hidden factors was discussed in Section 3.6.1 and it was noted there that the apparent effect of a factor when collapsed over another factor may be the reverse of its actual effect. This applies to estimating risk functions as well. To illustrate, in Table 4.11 we have taken data up to BAC of 80 collapsed over age and broken them into two *fictitious* groups, one group of reckless drivers who are largely involved in accidents and who also drink a lot and a group of sensible drivers who are in less accidents and drink much less. Computing point estimates of relative risk for these two groups separately produces two relatively flat functions indicating no involvement of

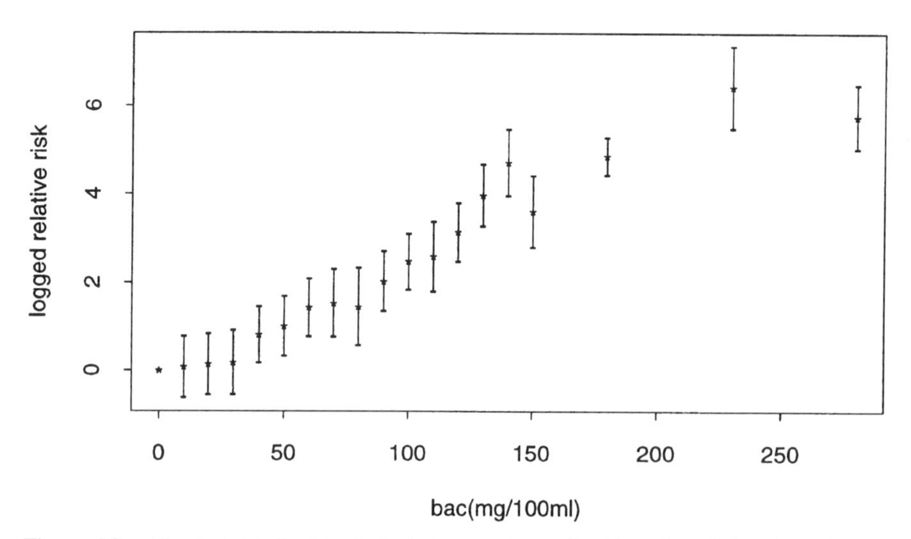

Figure 4.9. Estimated risk for blood alcohol concentration/accident data. Pointwise estimates based on modified empirical logits with pointwise 95% intervals.

Table 4.11. Fictitious Breakdown of Blood Alcohol Count/Accident Data

	Reckless Drivers			Sensible Drivers		
BAC	Case	Control	$\hat{r}(x)$	Case	Control	$\hat{r}(x)$
0	80	179	1.00	69	5675	1.00
10	5	11	1.02	4	318	1.03
20	5	12	0.94	4	298	1.10
30	5	13	0.86	3	252	0.98
40	9	27	0.75	2	167	0.98
50	9	37	0.54	1	108	0.76
60	10	24	0.93	1	81	1.01
70	7	11	1.42	1	58	1.42
80	6	13	1.03	0	43	0.00

alcohol. When the two data sets are combined we have just seen that the opposite is the case. The reason is that reckless drivers drink heavily and drive badly *regardless of BAC* and this produces a bogus association in the collapsed data. Since we cannot deny the possibility that there are two such groups of reckless and sensible drivers in the population this might temper our enthusiasm for making dogmatic statements about what these data say about alcohol and the risk of accident involvement. Just to drive the point home it is possible to further divide "reckless' and "sensible" drivers into fictitious sub-groups within which the risk again appears to increase systematically with BAC.

4.3.2 Binary Regression Approach

Equation (4.23) only gives estimates at observed values of x. Further, if $n(0)$ is small the estimated function will be very imprecise. If we are to extrapolate or estimate risk at all when no data is available at $x = 0$ then some smoothness assumption about $r(x)$ must be made.

There is a close relation between relative risk modeling and logistic models for $\pi(x)$ with a free intercept. Taking the logarithm of Eq. (4.22) gives

$$\log\{r_z(x)\} = \text{logit}\{\pi_z(x)\} - \text{logit}\{\pi_z(0)\} \tag{4.24}$$

and so the logged-risk function is just the logit probability standardized to pass through the origin as a function of x. The adjustment $\text{logit}\{\pi_z(0)\}$ is the intercept term of the logistic model. For instance, consider the logistic model with systematic component

$$\text{logit}\{\pi_z(x)\} = \alpha + f(x) + g(z)$$

where f and g are to be estimated and without loss of generality we may assume that $f(0) = 0$. Provided that the other covariates z do not involve x, substitution into Eq. (4.24) implies the logged risk function $f(x)$, which does not depend on z. A single function f, describing the regression of $\pi(x)$ on x with intercept removed, thus describes the effect of the covariate x on response probability. An estimate of the risk function is $\exp\{\hat{f}(x)\}$ where $\hat{f}(x)$ might be based on a linear model $x^T\beta$ or perhaps some smoothing technique.

Confidence bands are normally set on the log scale first because the standard error of $\hat{f}$ is usually directly available and second since risk is always positive. The intercept parameter α measures the absolute rather than relative odds of response and will not be estimable for retrospective data sets but this does not prevent estimation of relative risk. For a linear model $f(x) = x^T\beta$

$$\mathrm{Var}(\log\{r(x)\}) = x^T(X_1^T W X_1)^{-1}x$$

where X_1 is the design matrix *excluding the first column*. For instance, a quadratic regression with $\mathrm{logit}\{\pi(x)\} = a + b_1x + b_2x^2$ gives the logged risk estimate $\hat{b}_1x + \hat{b}_2x^2$ with variance

$$\mathrm{Var}(\hat{b}_1)x^2 + 2\,\mathrm{Cov}(\hat{b}_1, \hat{b}_2)x^3 + \mathrm{Var}(\hat{b}_2)x^4$$

and this is used to plot confidence bands about the estimated curve. When x is involved in the vector z_x then the log-risk function will involve an additional term in z giving

$$\log\{r(x)\} = f(x) + g(z_x) - g(z_0).$$

Thus, product terms such as x_1x_2 in a model mean that the risk function for x_1 will depend on x_2 and vice versa. As a simple example, a logistic linear model for the probability of lung cancer which included age, sex, and number of cigarettes in the formula `AGE+SEX+CIGS` would imply a single risk function for both sexes and all ages. If an interaction term such as `AGE*CIGS` were included in the logistic model then the risk function would vary with age.

Example 4.8: (Continued) **Logistic Models for Drunk-Driving Data.** To investigate the involvement of age in the association of alcohol and accidents, take the number of accidents in each age/BAC category as response and the total records in each age/BAC category (i.e., case+control) as the binomial denominator. A logistic linear model `AGE*BAC` was fitted, which has linear predictor $a_i + b_ix$ for age-group i. The parameters b_i represent the rate at which the odds of an accident increase with increasing BAC. The covariate x was coded as the middle of the range and as 280 in the last category. The deviance was 131.9(76) with parameter estimates below.

	Under 21	21–30	30–50	Over 50
$\hat{b}_i$	0.0261(0.0021)	0.0269(0.0014)	0.0280(0.0015)	0.0225(0.0022)
$\hat{a}_i$	−2.423(0.134)	−3.982(0.139)	−4.398(0.151)	−3.689(0.180)
$\hat{a}_i(0)$	−2.560(0.152)	−3.837(0.156)	−4.232(0.168)	−3.669(0.203)
$\hat{a}_i(b)$	−2.440(0.115)	−3.957(0.116)	−4.301(0.122)	−3.889(0.159)

The $\hat{a}_i$ measure nothing in themselves but their differences measure differences in the odds of an accident at $x = 0$. For instance, the odds of an accident for a young driver is $\exp(-2.423 + 4.398) = 7.206$ higher than for middle-aged drivers. The estimates $\hat{a}_i(0)$ have been estimated from the $x = 0$ data alone and are quite similar although with higher standard errors. Imposition of the risk model enhances precision since the entire weight of the data is brought to bear on all parameter estimates. A test of the equality of the a_i is obtained by fitting `AGE*BAC-AGE` and observing the much larger deviance 213.4(71). Different underlying accident rates for different age-groups is of course to be anticipated. Note however that the statistical difference detected here is *not* due to there being more younger drivers on the roads since this has been adjusted for by the control data.

The estimates of the separate logged risk functions are all lines of the form $\hat{b}_i x$ with standard error $\text{se}(\hat{b}_i)x$. The estimated curve and bands on the log-scale are thus four lines all passing through the origin and are not displayed here. On fitting a common slope parameter to the four age-groups (model formula `AGE+BAC`) the deviance is 136.2(79) which is an insignificant increase from 131.9(76). As is apparent from the estimates $\hat{b}_i$ and their standard errors, there is little statistical evidence for a difference in the BAC risk functions across age-groups. Thus, while the underlying accident rates for different age-groups appear to differ, the impairing effect of alcohol seems uniform across age-groups. Taking this last model as the best description of the data, the estimate of the common slope is $\hat{b} = 0.0266(0.0008)$ and so the common risk function for all age-groups is

$$\hat{r}(x) = e^{0.02659x} = 1.0269^x$$

indicating that the odds of an accident rises 2.7% for every mg/100 ml of alcohol in the driver's blood. Thus, at the legal limit of 50 mg/100 ml, the risk is apparently 3.78 times higher than for a completely sober driver. The estimates of the underlying rates in this simplified model are listed as $\hat{a}_i(b)$ and have the same pattern as earlier estimates. This will not always be the case and as was seen earlier the differences in intercepts and differences in slopes are highly nonorthogonal in general. As an epilogue to this analysis, there is actually quite strong evidence (P-value = 0.031) that the risk functions do differ across age-groups (Lloyd, 1991) but this is only revealed when an additional important factor, namely day of week, is taken into account.

4.4 MODELING OVER-DISPERSION

All the models so far in this chapter are based on the binomial distributions for the observations Y. Violation of this assumption has potentially serious consequences on the validity of the inferential procedures described. What could possibly go wrong in assuming that the number of successes from n trials is binomial?

This question is answered in this section and uses the following notation. Let X_j be the indicator variable that takes value 1 if trial j is a success and value 0 otherwise. Then

$$Y = \sum_{j=1}^{n} X_j.$$

It is beyond dispute that each X_j has the $Bi(1, \pi_j)$ where $\pi_j = \Pr(X_j = 1)$ being the probability of success on the jth trial. For Y to have a binomial distribution it is sufficient to assume: (1) the π_j are equal, (2) the π_j are nonrandom, and (3) the X_j are independent. If the first assumption is violated then the distribution of Y will exhibit less variability than a binomial variable with the same expectation. If (2) is violated the distribution of Y exhibits more variability and if (3) is violated then Y will exhibit more(less) variability than a binomial variable if the correlations of the X_j are positive(negative). Combinations of these assumption violations can conceivably produce a variable with the same variability as a binomial. In fact we will identify a situation where (1) and (2) are both false but the distribution is exactly binomial.

In this chapter we examine models for over-dispersion arising from failure of (1), (2), or (3). We consider how over-dispersion is identified and how over-dispersion can be accounted for in computing standard errors, confidence intervals and tests.

4.4.1 Some Models for Over-Dispersion

In this section we justify the claims made in the previous section concerning the failure of the binomial assumptions on the variability of Y. This requires further assumptions about the nature of these violations.

Unequal π's Lead to Under Dispersion If assumption (1) is violated then the distribution of Y depends on the precise values of all the π_j. Let $\bar{\pi}$ and s_π^2 denote the sample mean and variance of the probability parameters $\pi_1, \ldots, \pi_n$. Then

$$\mathrm{E}(Y) = \sum_{j=1}^{n} \mathrm{E}(X_j) = \sum_{j=1}^{n} \pi_j = n\overline{\pi} \tag{4.25}$$

and so the mean of Y is the same as if each of the X_j had the same probability of success $\overline{\pi}$. The variance however is

$$\mathrm{Var}(Y) = \sum_{j=1}^{n} \mathrm{Var}(X_j) = \sum_{j=1}^{n} \pi_j(1 - \pi_j) = n\{\overline{\pi}(1 - \overline{\pi}) - s_\pi^2\} \tag{4.26}$$

which is at most the usual binomial variance. In the most extreme case, each of the π_i are either zero or 1 and so each X_j is deterministic and the variance of Y is zero. This phenomenon is called under-dispersion and is rather rare in practice since if the probabilities π_j differ then they will likely differ from experiment to experiment and so the π_j cannot be treated as constants.

For structured data sets we will have many such variables Y_i to be related to a set of covariate measurements. Each Y_i may violate the assumption (1) and have its own particular sequence of probability parameters $\pi_{i1}, \ldots, \pi_{i,n_i}$. The distribution of an estimator of some regression parameters would be a function of the Y_i and so all these π_{ij} are involved in the distribution. Clearly, some simplifying approach is needed to investigate the effect of violation of assumption (1) on such estimators.

Randomly Varying π's One attractive approach is to assume that for each observation Y the set of probability parameters are random in a manner to be specified. This randomness accounts for lack of control in experimental conditions. We assume that conditional on the probability parameters, the binary variables X_j are independent. The mean and variance expressions in Eqs. (4.25) and (4.26) are then treated as conditional on $\pi_1, \ldots, \pi_n$.

The unconditional mean of Y (ignoring the subscript i) is the mean of Eq. (4.25) which gives

$$\mathrm{E}(Y) = \mathrm{E}\{\mathrm{E}(Y|\pi_1, \ldots, \pi_n)\} = n\mathrm{E}(\overline{\pi}) \tag{4.27}$$

and the unconditional variance is

$$\begin{aligned}\mathrm{Var}(Y) &= \mathrm{Var}\{\mathrm{E}(Y|\pi_1, \ldots, \pi_n)\} + \mathrm{E}\{\mathrm{Var}(Y|\pi_1, \ldots, \pi_n)\} \\ &= n^2\mathrm{Var}(\overline{\pi}) + n\{\mathrm{E}(\overline{\pi}(1 - \overline{\pi})) - \mathrm{E}(s_\pi^2)\} \\ &= n\mathrm{E}(\overline{\pi})(1 - \mathrm{E}(\overline{\pi})) + n\{(n - 1)\mathrm{Var}(\overline{\pi}) - \mathrm{E}(s_\pi^2)\}. \end{aligned}\tag{4.28}$$

This variance may be larger or smaller than that of the $\mathrm{Bi}(n, \mathrm{E}(\overline{\pi}))$ distribution. Further investigation requires us to say more about the joint distribution of the π_j.

Identical Independent π's Lead to the Binomial Distribution Let us suppose that $\pi_1, \ldots, \pi_n$ are independent random variables with the same mean π for all j. In fact the independence assumption is stronger than necessary—it is sufficient that the π_j be uncorrelated (see Exercise 4.15). Now it is straightforward to show that

$$(n-1)\mathrm{Var}(\overline{\pi}) = \mathrm{E}(s_\pi^2).$$

This is easiest so see when each π_j has the same variance V when it is well known that

$$\mathrm{Var}(\overline{\pi}) = V/n, \mathrm{E}(s_\pi^2) = (n-1)V/n.$$

From Eq. (4.28) the last term vanishes and the variance of Y is the same as that of a $\mathrm{Bi}(n, \pi)$. In fact Y is binomially distributed. This can be shown as follows. First

$$\begin{aligned}\Pr(X_1 = x_1, \ldots, X_n = x_n) &= \mathrm{E}\{\Pr(X_1 = x_1, \ldots, X_n = x_n | \pi_1, \ldots, \pi_n)\} \\ &= \mathrm{E}\left(\prod_{j=1}^{n} \pi_j^{x_j}(1-\pi_j)^{1-x_j}\right) \\ &= \prod_{j=1}^{n} \mathrm{E}(\pi_j^{x_j}(1-\pi_j)^{1-x_j})\end{aligned}$$

where the last line follows by independence of the π_j. Similarly

$$\Pr(X_j = x_j) = \mathrm{E}(\pi_j^{x_j}(1-\pi_j)^{1-x_j}) \tag{4.29}$$

and so we conclude that $X_1, \ldots, X_n$ are independent random variables unconditionally. From Eq. (4.29)

$$\Pr(X_j = 1) = \mathrm{E}(\pi_j)$$

and so, provided only that each of the π_j have the same mean, the X_j are independent binary variables with the same probability of success and so Y has $\mathrm{Bi}(n, \mathrm{E}(\pi))$ distribution.

There is a simpler probabilistic argument when the π_j have the same distribution. Imagine populations $\mathcal{P}_\omega$ comprising a proportion π_ω "successes" where ω is an index, and let $\mathcal{P}$ be the union of these populations. The collection of π_ω's we suppose follows some distribution with mean π which is also the proportion of successes in the super-population $\mathcal{P}$. Thus, an independent random sample from $\mathcal{P}$ clearly produces $Bi(n, \pi)$ successes. But this sample can be thought of another way—as n single observations X_j. The probability parameters of these X_j are random and follow the distribution of the π_ω across the different populations. This is precisely the situation described above.

A Common Random π Leads to Over-Dispersion Suppose now that all the π_j are equal taking the common but random value P with mean π and variance V. This is almost the opposite of the independence case just considered. In this case the statistical variability of P induces an increase in the variability of Y. To see this, simply note that $\overline{\pi} = P$ and so $\text{Var}(\overline{\pi}) = V$ (c.f. previous subsection) whereas $s_\pi^2 = 0$ and so $\text{E}(s_\pi^2) = 0$. Then Eq. (4.28) gives

$$\text{Var}(Y) = n\pi(1 - \pi) + n(n - 1)V. \tag{4.30}$$

This is always greater than the binomial variance provided that $V = \text{Var}(P) > 0$, i.e., provided that there is some randomness in the common probability parameter P. If a set of variables $Y_1, \ldots, Y_k$ each had this *over-dispersed binomial distribution*, then they would range further from their means than do binomial variables. If a model were fitted, even with the correct systematic component, then the residual variation of the data about the fitted values would be larger than expected from binomial data and so the deviance of the model, computed under the assumption of a binomial distribution, would tend to be larger than the usual χ^2 variable. This is the phenomenon of over-dispersion and is quite common in practice.

The most common reason for data being over-dispersed is that experimental conditions are not perfectly under control and thus the unknown probability parameters vary not only with measured covariates but with unseen and uncontrolled factors. The assumption that P is random simply treats the effects of these uncontrolled factors on P as random variation.

Example 4.9: Thinking Further About Binary Trials Imagine that we have 10 insects that are subjected to some stimulus and Y is the number responding. Suppose that p_j = Pr(respond) varies from insect to insect. Then the number Y responding is *not* binomially distributed. Since we must admit that insects could be genetically different, is the binomial distribution ever justified?

In saying Y is not binomial we are taking the p_j as fixed constants. This is only appropriate if these probabilities stay the same in repetitions of the experiment. One obvious way to arrange this would be to repeatedly test *the same* 10 insects, obtaining observations $Y_1, Y_2, \ldots, Y_k$. If we did this, and if the p_j were not constant, then we could certainly find that this sample would be *less* dispersed than the binomial. For instance, if half the insects have $p_j = 0$ and the other half $p_j = 1$ then $Y_i = 5$ would be observed every time. In practice we would usually not test the same insects repeatedly but if we did then under-dispersion would likely result.

Suppose now that the stimulus is a dose of poison and response is death. We can certainly not test the same 10 insects again now. We might then take a further sample of 10 insects. Each set of 10 insects would have a different set of 10 response probabilities and give a different observation Y_i. If each set of insects is a simple random sample from some large population then the response probabilities $p_1, \ldots, p_{10}$ are an independent sample on a common distribution. We conclude that each of the Y_i *are* binomially distributed. The experimental procedure of randomly choosing subjects for each treatment is extremely common. One reason for such randomization is to remove bias. It is not always well appreciated that random choice of subjects, in this case insects, has the added effect of ensuring an exact binomial distribution regardless of population heterogeneity.

Finally, suppose that the dosage of poison cannot be controlled precisely or that other uncontrollable factors affect the efficacy of the poison. For simplicity, suppose also that the insects are identical (although this is not essential). On each experimental occasion, the set of 10 insects have identical response probability but this common probability varies randomly from one occasion to the next as the dose varies. Then the random variables $Y_1, \ldots, Y_k$ are *not* binomial and will be over-dispersed. To take an extreme case, suppose we decided to use no poison with probability one half and a definitely lethal dose with probability half. Then Y_i would take two possible values, 0 or 10, each with probability 1/2. The variance of this random variable is 10 times higher than that of the Bi(10, 0.5) distribution.

In practice it will commonly occur that uncontrollable factors will vary between different sets of experimental conditions. Thus, if we are trying to detect a relationship of the probability π of response with some covariate, the probability will be randomly perturbed from one set of conditions to the next. Ultimately, this will manifest as a large deviance even when the correct systematic part of the model is fitted. The precise form of the random variations of p has relatively little effect on the final results. Two common families are discussed below.

Parametrizing the Distribution of P It is worth pausing to consider what form the variability of P might take. Since P is confined to [0,1] the variance of P becomes small if E(P) is close to zero or 1. A plausible relation between the mean and variance of P is

$$V = \phi\pi(1 - \pi) \tag{4.31}$$

where ϕ is known as the over-dispersion parameter. While this is not the only possibility it produces a simplification

$$\mathrm{Var}(Y) = n\pi(1 - \pi)(1 + \phi(n - 1)) \tag{4.32}$$

of Eq. (4.30). The over-dispersion, $1 + \phi(n - 1)$, is linear in the sample size. In principle, a plot of squared standardized residuals against n_i would be approximately linear under this model although the variability of squared residuals can be rather large. Notice that when $n = 1$ over-dispersion will be undetectable. This is because a single binary trial is binomial whether or not the probability parameter is considered random.

An alternative assumption to Eq. (4.31) is that P has the beta distribution with parameters a, b. In this case, Y has the beta-binomial distribution whose first two cumulants are

$$\mathrm{E}(Y) = \frac{a}{a+b}, \qquad \mathrm{Var}(Y) = \frac{ab}{(a+b)^2}\left(1 + \frac{n-1}{a+b+1}\right).$$

This is a fully parametric model and can be fitted by maximum likelihood.

Positively Correlated Trials Lead to Over-Dispersion Suppose that the $\pi_j = \pi$ are constant and nonrandom but that the X_j have

$$\mathrm{Cov}(X_j, X_k) = \rho\pi(1-\pi) \qquad j \neq k$$

and hence correlation ρ when taken in pairs. Thus assumption (3) is violated but not (1) or (2) and the variance of Y is

$$\begin{aligned}
\mathrm{Var}(Y) &= \sum_{j=1}^{n} \mathrm{Var}(X_j) + \sum_{j\neq k} \mathrm{Cov}(X_j, X_k) \\
&= n\pi(1-\pi) + n(n-1)\rho\pi(1-\pi) \\
&= n\pi(1-\pi)(1+(n-1)\rho) \qquad (4.33)
\end{aligned}$$

the first sum containing n terms and the second $n(n-1)$.

Equation (4.33) is mathematically identical to Eq. (4.32) but with the correlation parameter ρ replacing the over-dispersion parameter ϕ. If $\rho > 0$ then there will be over-dispersion equivalent to that caused by the probabilities P varying with over-dispersion parameter ρ. Indeed, variations in a common probability P induce correlation in the binary variables (see Problem 4.15). One important distinction between Eqs. (4.33) and (4.32) is that the former is meaningful when $\rho < 0$. Note however that $\rho > -1/(n-1)$ and so there is a limit to the amount of pairwise negative correlation that a sequence of binary observations can exhibit.

Negative correlation is probably the most plausible source of underdispersion in real data. Such data tends to follow the mean value more closely than would binomial data and so deviances tend to be smaller than the usual χ^2. This means that changes in deviances also tend to be smaller and so effects appear to be less statistically significant.

4.4.2 Detecting Over-Dispersion

As the name suggests, over-dispersion (or indeed under-dispersion) is detected by observing that the data are too widely dispersed (not dispersed enough) about their mean values. Since the mean values are unknown we look at how closely the data follow the estimated mean values or fitted values and measure this by one of the goodness-of-fit statistics. If the statistical model with p parameters is correct then the goodness-of-fit statistic is approximately χ^2 with $k - p$ degrees of freedom. There are several reasons why the observed goodness-of-fit statistics may differ from the mean value $k - p$. These are listed and discussed below.

1. Small expected values: If too many of the $\hat{e}_i$ are small then the deviance and other asymptotic equivalents are not well approximated by chi-square. Ideally, a simulation could be performed using estimated parameter values and a better approximation to the P-value obtained. Alternatively, observations together with their fitted values can be pooled until the chi-square approximation is more adequate. Only then, if the P-value is still small, can it be said that "the model does not fit well." This leaves open the reason for the poor fit.
2. Wrong regression function: If the wrong shaped regression function is being fitted then the fitted values will systematically differ from the data values and give a poor fit. Wrong regression functions are detected by looking at residuals and plotting them against regressors.
3. A few bad data points: If the poor global fit measure is due to only a few data points then a more plausible explanation for the poor fit is that these data points do not follow the model but the others do. This situation is often identified from a nonlinear normal probability plot of standardized residuals.
4. Bad luck: If an accurate P-value has been used *and* the regression function is of the right shape *and* the normal probability plot is fairly linear—then a poor fit may still have occurred by bad luck. The P-value measures how unlucky one can consider oneself assuming the model is true. If this is too small then we would eliminate this possibility and conclude.
5. Over-dispersion: Other plausible explanations of poor fit having been considered and rejected, the larger than usual deviance is attributed to the data Y_i not being binomial.

In summary, a model is said to be over(under)-dispersed when some global measure of lack of fit is improbably large(small) after having made every effort via residual analysis and outlier identification to come up with a good fit. In practice, this means that the largest reasonable model has a large(small) deviance after removing outliers and pooling sparse cells.

Example 4.4: (Continued) **Over-Dispersion of the Seed Data.** Residuals from a logistic linear model for the seed data were analyzed in Section 4.2.1 and shown in Figures 4.5 and 4.6. While the first residual plot highlighted some apparent outliers there was no apparent trend and the half-normal plot suggested rather uniform over-dispersion. It is therefore appropriate to fit the over-dispersed model to these data, with perhaps quadratic and cubic terms in the covariate just to be sure that the systematic part of the model contributes nothing to the higher than expected dispersion. A method for doing this is given in the next section.

4.4.3 Methods of Accounting for Over-Dispersion

In the random probability model (4.31) only the mean and variance of P is involved, not the entire distribution. Similarly, in the correlation model only pairwise correlations of the X_j are considered since only these are relevant to $\text{Var}(Y)$. The actual distribution of Y is not determined in either case and so the over-dispersed binomial distribution is not really a distribution at all but a statement about the mean-variance relationship. Consequently, it makes no sense to talk about ML estimation.

Williams' Procedure The simplest approach (Williams, 1982a) is to note that if Y_i has the over-dispersed binomial distribution with parameters n_i, π_i, and ϕ then letting $w_i(\phi) = 1/(1 + \phi(n_i - 1))$

$$Z_i^2 = \frac{w_i(Y_i - n_i\hat{\pi}_i)^2}{n_i\hat{\pi}_i(1 - \hat{\pi}_i)}$$

is approximately χ_1^2 if n_i is reasonably large. The Pearson statistic obtained from fitting a generalized linear model with weights w_i is

$$S(w) = \sum_{i=1}^{k} \frac{w_i n_i (Y_i - \hat{e}_i)^2}{\hat{e}_i(n_i - \hat{e}_i)}$$

and only if the weights w_i equal the ideal $w_i(\phi)$ will the distribution of $S(w)$ be χ_{k-p}^2. Williams suggests finding that value of ϕ for which the expected value of $S(w)$ is exactly $k - p$. An algorithm for doing this is based on the approximate result

$$\text{E}\{S(w)\} = \sum_{i=1}^{k} w_i(1 - v_i d_i)(1 + \phi(n_i - 1)) \tag{4.34}$$

where w_i are the presently chosen weights, v_i are the diagonal elements of the working weight matrix which for a logistic model are $w_i n_i \pi_i(1 - \pi_i)$, and d_i is the variance of the linear predictor $X\hat{\beta}$. These are both readily accessible from a GLIM fit in the system vectors `%WV`, `%VL`. From a current estimate of ϕ the model is fitted with weights $\hat{w}_i = w_i(\hat{\phi})$ and the new estimate of ϕ obtained from

$$\hat{\phi} = \frac{S\{\hat{w}_i\} - \sum_{i=1}^{k} \hat{w}_i(1 - v_i d_i)}{\sum_{i=1}^{k} \hat{w}_i(n_i - 1)(1 - v_i d_i)}. \tag{4.35}$$

At the first step we take weights $\hat{w}_i = w_i(0) = 1$ and so the initial estimate

$$\hat{\phi} = \frac{S - (k - p)}{\sum_{i=1}^{k} (n_i - 1)(1 - v_i d_i)} \tag{4.36}$$

where S is the ordinary unweighted Pearson goodness-of-fit statistic. At the end of this procedure we have a presumably positive estimate of ϕ and estimated weights $\hat{w}_i = w_i(\hat{\phi})$ that are less than one when $n_i > 1$.

The Pearson statistic from the final weighted fit will equal $k - p$ exactly and the deviance will usually not differ much from this. The standard errors of parameter estimates will all be larger roughly by the square root of $S/(k - p)$ although some may be affected more than others. The parameter estimates and fitted values will also have been slightly changed by the reweighting. Pearson residuals will be the signed roots of the contributions to the weighted Pearson statistic [see Eq. (4.34)]. These are smaller than the original residuals by the square root of $w_i(\hat{\phi})$. Other residuals will be scaled down in a similar manner.

An Splus function `od.binomial` for fitting the over-dispersed model is at the Wiley website and a GLIM macro may be found in Collett (1991, p. 353).

Simplification for Constant n_i When the $n_i = n$ are constant, the weights $w_i(\phi)$ are constant and so

$$S(w) = \frac{1}{1 + \phi(n - 1)} S.$$

Estimation of ϕ is then simply achieved by equating $S(w)$ to $k - p$ in a single step and gives

$$\hat{\phi} = \frac{1}{n-1}\left(\frac{S}{k-p} - 1\right).$$

This is actually identical to Eq. (4.36). The estimated weights are then

$$w_i(\hat{\phi}) = \frac{k-p}{S}$$

and for this weighted fit the Pearson and deviance statistics will be $w_i(\hat{\phi})$ smaller (and so the Pearson equals $k - p$ exactly) and standard errors of parameter estimates will be $\sqrt{S/(k-p)}$ larger. The fitted values and parameters estimates will not have changed at all.

Testing After Fitting the Over-Dispersed Model It should be clear from Section 4.4.2 that the over-dispersed model should only be fitted when the regression function is very likely to be correct or close to correct. Accordingly, the Williams procedure is applied to the largest reasonable model we are prepared to consider, from now on called the *fullish* model. The fullish model contains more parameters than we would probably like and its main use is for detecting and fitting the over-dispersion. We would then proceed to test whether the parameters of this model can be reduced.

The weights $w_i(\hat{\phi})$ should not be changed as parameters are tested and perhaps dropped; rather the weights are obtained once and for all from the fullish model. Particular parameters in a given model may be tested by comparing estimates to their standard errors *with the weights from the fullish model still active*. Ratios of estimates to standard errors are referred to the t distribution on $k - p$ degrees of freedom where the fullish model has p parameters. Two hypotheses $\mathcal{H}_0$ and $\mathcal{H}_1$ both contained in the fullish model may be tested by recording the deviances of both models and computing

$$F = \frac{(\mathcal{D}_0 - \mathcal{D}_1)/(p_1 - p_0)}{\mathcal{D}_F/(k-p)}$$

where $\mathcal{D}_F$ is the deviance of the fullish model. The denominator will be very close to one and may be dropped if desired and the null distribution of the test statistic is approximately F with degrees of freedom $p_1 - p_0$ and $k - p$.

Example 4.4: (Continued) **Over-Dispersion of the Seed Data.** For the seed germination data the n_i are not constant and so the general Williams procedure will need to be employed. Fitting a cubic function to logit probability of response gives the deviance 85.119 on 16 degrees of freedom. The Pearson statistic is 84.136. Since the fitted values are all large there is evidence of rather extreme over-dispersion. The Pearson residuals range from −3.51 to

4.636. The parameter estimates for the linear, quadratic, and cubic components of the model are 0.816(.091), −0.0529(.0097), and 0.00148(0.00031), respectively, from which it appears that all terms are highly significant.

Using the Williams procedure the estimated value of ϕ is 0.0242. Notice that since the n_i are quite large, this implies quite a sizable over-dispersion effect. For instance, with $n_i = 200$ the variance of the data would be $(1 + 0.0242 \times 199) = 5.8$ times larger than for a binomial variable. This is not surprising as the original deviance was over 5 times its degrees of freedom. With $\hat{\phi} = 0.0242$ the weighted Pearson statistic exactly equals its degrees of freedom 16 and the adjusted Pearson residuals now lie between −1.609 and 2.317. The parameter estimates are 0.753, −0.0452, and 0.00121 with standard errors 0.252, 0.02584, and 0.0007952. The corresponding t statistics are 2.985, −1.748, and 1.520 with two sided P-values 0.0087, 0.0995, and 0.147. The fitting of an over-dispersed model thus leads us to remove quadratic and cubic terms and accept the linear logistic model for the regression function. This is typically the outcome of allowing for over-dispersion: data patterns will appear less significant when compared to the larger assumed random variation.

If we fit the simpler linear model using the over-dispersion weights then the linear parameter is estimated to be 0.2626 which means that the odds of germination increases around 30% for every unit increase in the covariate. The deviance of this model is 21.78 on 18 degrees of freedom. Recall that for the binomial model, the estimate was 32% but with deviance 130.5.

There is a final important point to make regarding the fitting of the over-dispersed model to these data. According to Eq. (4.30) the variance increases with n, linearly under the special model Eq. (4.32). This should be detectable by plotting the squared residuals against the binomial denominators. For this data set there is no apparent trend and if anything the larger residuals occur for smaller n's. Thus, the over-dispersion model Eq. (4.30) does not seem appropriate for these data and it might be better to assume that the variance is unrelated to n. In this case we just estimate the over-dispersion factor by $S/(k - p)$ and use the reciprocal as weights. This is identical to the procedure for constant n_i outlined earlier.

4.5 APPROXIMATION BY NORMAL GLM's

In this section a description of how to analyze a generalized linear model is given, using a normal approximation, with or without a free-scale parameter. This gives very similar results (sometimes identical) and is useful when a package such as `GLIM` is unavailable. In addition, it allows us to use some of the more structured and informative features of standard regression packages (such as ANOVA tables and model selection procedures). The normal approximation can be used to approximate *over-dispersed* binomial data by including a free-scale parameter, leading to the usual classical inference based on t and F distributions. The methods are illustrated on two data sets; one a set of sociological survey data and a second larger data set on seed germination rates.

4.5.1 Inverse Variance Weighting

Several times we have seen a normal approximation to the distribution of $\mathrm{logit}(Y_i/n_i)$ when Y_i is binomial. The approximate mean is $\psi_i = \mathrm{logit}(\pi_i)$ and approximate variance $V_i^* = 1/(n_i\pi_i(1-\pi_i))$. The logistic linear model supposes that ψ_i is a linear function of parameters. Thus, we have approximately

$$Y_i^* = \log\left(\frac{Y_i}{n_i - Y_i}\right) \stackrel{d}{=} \mathcal{N}((X\beta)_i, V_i^*(\beta)) \tag{4.37}$$

which is a normal linear model. Consider the estimator of β given by

$$\hat{\beta} = (X^T W X)^{-1} X^T W Y$$

called the *weighted* LS estimator. When W is the identity matrix this gives the usual LS estimator. A common choice for the weight matrix W is diagonal with the reciprocals of the variances V_i^* which are interpreted as weights on the data points Y_i^*. It seems logical to give less weight to observations with higher variance. This estimation procedure is called weighted least squares.

More generally suppose that the data $Y_1, \ldots, Y_k$ follow a generalized linear model with link function η and design matrix X. If μ_i is the mean of Y_i then the model assumes that the $\eta(\mu_i)$ are a linear function of parameters. An approximate analysis is possible by taking a function of the data that has approximate mean $\eta(\mu_i)$. The simplest such function is

$$Y_i^* = \eta(Y_i)$$

called the *empirical link transform*, however slight modifications are sometimes useful. For instance, for logistic–binomial models we use modified empirical logits to avoid infinite values and also because the mean of Y_i^* is then closer to $\log(\pi/(1-\pi))$. Similarly, for log–Poisson models it is common to add 1/2 to counts before taking the logarithm. The variance of the empirical link transform is

$$V_i^* = \{\eta'(\mu_i)\}^2 V(\mu_i)$$

which will be a function of β through $\mu_i(\beta)$ and $V(\mu_i) = \mathrm{Var}(Y_i)$. We have assumed here that there is no free-scale parameter in the variance function, which includes the main error distributions of interest in this book, binomial and Poisson. Under certain conditions it will be approximately true that

$$Y_i^* = \eta(Y_i) \stackrel{d}{=} \mathcal{N}(\eta(\mu_i), V_i^*). \tag{4.38}$$

If this approximation is taken to be exact then the minimum variance linear unbiased estimator on β is the weighted LS estimator with $W = \text{diag}(1/V_i^*)$. In the special case that the V_i^* are free of β, this is the ML estimator.

Fitting Algorithm Fitting of the model proceeds iteratively since the weights $1/V_i^*(\beta)$ generally depend on the unknown regression parameters. First, an unweighted model is fitted and β estimated by (the ordinary LS estimator) $\hat{\beta}_0$. This estimate is then used to define estimated weights $1/V_i^*(\hat{\beta}_0)$. A subsequent fit using these weights gives the next estimator $\hat{\beta}_1$ of β. The process continues until convergence. For some models, for instance, a fully crossed analysis of variance model, only one iteration is necessary as the estimates are not affected by the weights. How many iterations should be performed? If the model Eq. (4.38) is taken to be true then only one or two iterations are necessary for the estimator to have the same asymptotics as the infinitely iterated estimator. The asymptotic here is in k, the number of data points. If the n_i are not large then equation (4.38) may not be a good approximation to the generalized linear model.

The major difference between the output of an unweighted and weighted model will be the standard errors and residuals SS which will tend to be larger if the weights are larger than one and smaller if the weights are smaller than one. The estimate $\hat{\beta}$ will often not change drastically and may not change at all. It should be noted that the weighted LS estimator with weights unknown and hence estimated is *not* the ML estimator; the ML estimator is quite complicated and involves estimating equations in the residuals and the squared residuals. While these will lead to a more efficient estimator of β, the weighted LS estimator is still consistent and normally distributed under model Eqs. (4.38) or (4.40) given later.

Effect of no Scale Parameter Notice that there is no scale parameter σ^2 to be estimated in model (4.38). Consequently, inference is particularly straightforward. The weighted LS estimator has distribution

$$\hat{\beta} \stackrel{d}{=} N(\beta, (X^T W X)^{-1})$$

when the weights are known and very close to this when the weights are estimated. Note again the absence of a free-scale parameter σ^2 that is common in general linear normal models. Tests of a single parameter may be based on comparing $(\hat{\beta}_i - \beta_i)/\text{s.d.}(\hat{\beta}_i)$ to the standard normal distribution. The residual

SS from a weighted fit is defined as

$$Y^T W Y - \hat{\beta}^T X^T W Y$$

which has exact distribution χ^2_{k-p} under model Eq. (4.38). The residual mean square estimates 1.0 rather than a free-scale parameter σ^2 and may be used as a goodness-of-fit statistic. Nested models are compared by computing the change in RSS and comparing to chi-square, rather than the proportional change in RSS and comparing to F. The usual standardized residuals for normal linear models have the form

$$\frac{Y_i^* - \eta(\hat{e}_i)}{\sqrt{V_i^*(\hat{e}_i)}} \tag{4.39}$$

and are identical to the transformed residuals [see Eq. (4.15)] with $g = \eta$ defined on p. 200 except that the fitted values $\hat{e}_i$ here are obtained from the approximating rather than exact model. Standardized residuals returned from a package will have these residuals divided by the estimate of σ.

Of course, the simple inference methods described above will be completely invalid if the normality approximation in Eq. (4.38) is invalid. This certainly requires checking. For binomial data we are approximating a binomial distribution by a normal and require the n_i to be reasonably large and the probabilities not too extreme. For Poisson data we require the Poisson distribution to be close to normal which requires the mean to be large. The vast majority of data points will have to satisfy Eq. (4.38) if it is to be a useful approximation to the generalized linear model.

4.5.2 Weights and Over-Dispersion

It is certainly more common for normal data to have variance of unknown scale and for this scale parameter σ^2 to be estimated. Inclusion of a free scale parameter in Eq. (4.38) gives the model

$$Y_i^* = \eta(Y_i) \stackrel{d}{=} N(\eta(\mu_i), \sigma^2 V_i^*) \tag{4.40}$$

and is a normal approximation to the original generalized linear model with constant over-dispersion, i.e., where $\text{Var}(Y_i) = \sigma^2 V(\mu_i)$ rather than $V(\mu_i)$. In the binomial case, this is only really appropriate when the n_i are constant. Nevertheless, over-dispersion will be detectable fairly generally provided the model approximation Eq. (4.40) is reasonably correct.

A formal test of over-dispersion can be based on comparing the residual SS to the degrees of freedom and concluding that there is over-dispersion if it is large for a χ^2_{k-p} variable and other sources of model error have been eliminated (see Section 4.4.2). The residual mean square estimates σ^2.

Testing under the model Eq. (4.40) follows the usual normal theory prescription—parameter tests are based on the student-t distribution and testing of nested models may be based either on quadratic forms in the implied linear hypotheses or proportional change in residual SS compared to the Fisher or F-distribution (with identical results). The usual standardized residuals from the over-dispersed model have the form

$$\frac{Y_i^* - \eta(\hat{e}_i)}{\hat{\sigma}\sqrt{V_i^*}}$$

which differs from Eq. (4.39) only by the scale parameter $\hat{\sigma}$ on the denominator. These residuals *are* automatically returned by a weighted LS package.

4.5.3 Example 4.9: Survey on the Role of Women

The data in Table 4.12 are from a survey carried out by the University of Chicago's National Opinion Research Center. Respondents were asked to agree or disagree with the statement: "Women should take care of running their homes and leave the running of the country up to men." The gender and the number of years of completed formal education of each respondent were recorded. The data listed in Table 4.12 actually combine two surveys about a year apart and a small number of nonresponse records have not been included. These data have also been listed by Haberman (1974).

It is apparent from the data that as educational level increases the proportion of respondents agreeing with the proposition decreases. It is less clear (i) if gender is an important factor and (ii) if the effect of educational level is the same for both genders. The data have been processed and analyzed in `MINITAB`. All 42 data points are loaded into a single vector labeled `y`, while `educ` contains educational levels 1 through 20 and `sex` contains labels 1,2 for males/females, respectively. The first step computes empirical logits and then fits an unweighted regression with interaction between `sex` and `educ`, the latter treated as a covariate. This model assumes that the log-odds of agreeing with the proposition is a linear function of educational level, both intercept and slope depending on sex. The sole purpose of this fit is to obtain fitted values that are used to compute weights at Step 2 and the output in the ANOVA table and parameter estimates should be ignored, especially the standard errors.

```
MTB > let 'ly'=log(('y'+.5)/('n'-'y'+.5)) STEP 1: Empirical logits
MTB > glm 'ly'='sex' 'educ' 'sex'*'educ';            unweighted fit
SUBC> cova 'educ';
SUBC> fits 'fv'.
```

```
Analysis of Variance for ly

Source      DF    Seq SS    Adj SS    Adj MS       F        P
educ         1    55.129    55.129    55.129   84.31    0.000
sex          1     0.057     0.017     0.017    0.03    0.874
sex*educ     1     0.000     0.000     0.000    0.00    0.990
Error       38    24.847    24.847     0.654
Total       41    80.033

Term            Coeff      Stdev    t - value     P
Constant       1.6387     0.2459       6.66    0.000
educ(sex)
         1   -0.19251    0.02969      -.648    0.000
         2   -0.19302    0.02969      -6.50    0.000
educ         -0.19277    0.02099      -9.18    0.000
educ*sex 1    0.00026    0.02099       0.01    0.990
```

The weights have been calculated and vary from zero (for the data cell where nobody was surveyed) to 134 (for the cell where most people were surveyed). After including the weights in the fit the results change drastically, in particular the effect of sex is now quite significant. This factor measures the difference in intercepts of the two lines (on the log-odds scale) and is quite hard to interpret unless the lines have identical slope. For each level of educational attainment, the odds of agreeing decreases by 20.8% for males and 27.3% for females (from the parameter estimates −0.2334 for males and −0.3187 for females). The hypothesis that these effects are identical has associated P-value 0.019, however this, and all the P-values quoted by **MINITAB**, are based on F-tests which in turn are based on an over-dispersed model. The test based on the binomial model requires us to compare the relevant SS with the χ^2 distribution. For instance, the above test of equal educational effect for males or females is performed by computing

$$\Pr(\chi_1^2 > 9.41) = 0.0021$$

which is much stronger evidence than previously (mainly because the 9.41 has not been divided by the residual mean square of 1.57). The chi-square P-values in the column headed `CHI-P` have been manually added.

```
MTB > let 'fp'=exp('fv')/(1+exp('fv')) # STEP 2: Compute weights
MTB > let 'wgts'='n'*'fp'*(1-'fp')

'wgts'

         0.839    0.316    0.846    1.752    2.115    4.526    8.094
        10.335   30.973   14.436   18.748   22.224   79.525   20.726
        20.090    5.421   18.870    4.257    3.366    1.506    1.983
```

```
     0.796    0.151    0.000    1.312    2.050     4.636    5.144
    10.224   31.608   16.236   30.020   26.657   134.122   23.036
    19.245    7.226   20.185    4.323    2.563     0.318    0.546
```

```
MTB > glm 'ly'='sex' 'educ' 'sex'*'educ'; # STEP 3: Weighted fit
SUBC> cova 'educ';
SUBC> weig 'wgts';
SUBC> fits 'fv';
SUBC> sres 'sres'.
```

Analysis of Variance for ly

Source	DF	Adj SS	F	P	CHI – P
educ	1	393.52	250.99	0.000	0.000
sex	1	8.35	5.33	0.027	0.004
sex*educ	1	9.41	6.00	0.019	0.002
Error	37	58.01			

Term		Coeff	Stdev	t – value	P	
Constant		2.5615	0.2042	12.55	0.000	
educ(sex)						
	1	–0.23340	0.02397	–9.74	0.000	*Slope for males*
	2	–0.31877	0.02531	–12.60	0.000	*Slope for females*
educ		–0.27608	0.01743	–15.84	0.000	*Averaged slope*
educ*sex	1	0.04268	0.01743	2.45	0.019	*Slope difference*

```
'fv'
    5.340    1.730    3.043    7.206    7.607    14.315   22.643
   25.712   68.889   28.851   33.833   36.380   118.620   28.292
   25.204    6.276   20.239    4.246    3.133     1.313    1.624
    5.724    0.938        *    6.220    8.529    16.974   16.588
   28.991   78.530   35.155   56.282   42.977   184.794   26.992
   19.116    6.076   14.373    2.611    1.316     0.139    0.205
```

Is there evidence of over-dispersion? Notice that the residual mean square is 1.57 on 37 degrees of freedom with P-value

$$\Pr(\chi^2_{37} > 58.01) = 0.0152$$

for testing goodness-of-fit. The residual plots (not displayed) look to be adequate and so one would put this down either to over-dispersion or to a failure of the normal assumption. Indeed there are 11 data values out of the 42 for which the fitted values are smaller than 5 and the goodness-of-fit statistic is not to be trusted under these circumstances.

The analysis is repeated below in **GLIM**, the logistic binomial model being fitted rather than the approximating normal distribution. In addition, a table as similar to the above **MINITAB** table as possible has been added so that the results may be compared. Again, there is evidence of over-dispersion in the **sex*educ** model since the deviance is 57.848 on 37 degrees of freedom (P-value = 0.0157) and rather different results are obtained depending on whether we use χ^2 change in deviance tests or F-tests under the over-dispersed model. The parameter estimates are quite comparable, the odds of agreeing with the statement decreasing by 20.9% for males and 27.1% for females for each level of education. The standard errors for these are a little smaller in the **GLIM** analysis than in the approximate **MINITAB** analysis. Nevertheless, the broad agreement between the estimates and main conclusions is satisfying bearing in mind the many small expected values.

```
[i] ? $fact sex 2$yvar y$erro bino n$
[i] ? $fit sex*educ$
[o] scaled deviance = 57.85 at cycle 4
[o]          d.f. = 37
[o]
[i] ? $fit sex. educ$
[o] scaled deviance = 64.13 at cycle 4!             6.28 less than 57.85
[o]             d.f. = 38
[o]
[i] ? $fit sex+educ$
[o] scaled deviance = 64.68 at cycle 4!             6.83 less than 57.85
[o]             d.f. = 38
[o]
[i] ? $fit sex$
[o] scaled deviance = 451.71 at cycle 3!          393.86 less than 57.85
[o]             d.f. = 39
[i] ? $fit sex*educ-%gm-educ$!                    Nicer parameterization
[o] scaled deviance = 57.85 at cycle 4
[o]             d.f. = 37
[o]
[i] ? $disp e$
[o]        estimate       s.e.      parameter
[o]  1        2.103     0.2360      SEX(1)!                         a1
[o]  2        3.003     0.2724      SEX(2)!                         a2
[o]  3      -0.2344     0.02023     SEX(1).EDUC!                    b1
[o]  4      -0.3154     0.02365     SEX(2).EDUC!                    b2
[o]
```

Analysis of Deviance for y

Source	DF	Adj SS	F	P	CHI-P
educ	1	393.86	251.92	0.000	0.000
sex	1	6.28	4.02	0.052	0.012
sex*educ	1	6.83	4.37	0.043	0.009
Error	37	57.85			

4.5.4 Example 4.10: Seed Germinations

In 1992, researchers in the Department of Agriculture conducted an experiment on native grass seeds common in semiarid areas of New South Wales. Each trial comprised the placing of 25 seeds in a petri dish and observing the number of germinations after 25 days. The water potential (which relates to the moisture available to the seed for germination) was controlled at one of six possible levels. In addition, the temperature was controlled at two levels for twelve hours each day to simulate day/night temperatures. There were four different seed `types`.

The aim of the experiment was largely to model the effect of temperature on seed germination, in particular, the effect of differences in day and night temperatures. The conjecture was that large differences in temperature serve to crack the seed which allows moisture to enter and aids germination. There was no question that higher levels of water potential (corresponding to less available water) would produce lower germination rates and that the average temperature would have a strong effect on germination rates. Since seeds require a certain amount of warmth but not too much, the average temperature effect should be roughly concave in shape.

The design of the experiment is as follows. In total there are 1152 binomial observations available. These comprise 48 observations on each of the 24 possible combinations of `water` potential and seed `type`. These 48 combinations comprised four replicates at 12 different combinations of high temperature and low temperature. Unfortunately, the design was not well thought out in these variables. There were 6 levels of high temperature and 7 of low temperature but of the 42 possible combinations only 12 were used. It is more convenient to express each combination of low and high temperature in terms of a difference in temperature (`difftemp`, of which there are four distinct levels) and an average temperature (`avetemp`, of which there are 8 distinct levels. The table below shows which of the possible combinations of these are covered by the design, levels of `difftemp` labeling rows and levels of `avetemp` labeling columns, both in degrees celcius. The design is certainly unfortunate if the overall shape of the effects of the two factors `difftemp` and `avetemp` on response is the focus of interest.

The first step of an approximate analysis is to convert the counts into modified empicial log-odds and to fit as full a model as possible so as to identify any incorrect observations or over-dispersion. For the three-way model with

Table 4.13. Temperature Combinations Covered by Design

	15	17.5	20	22.5	23	25	27	27.5	30
0	●		●		●	●	●		●
5				●				●	
10			●			●			
15		●		●					

full interaction in `water` potential, seed `type`, and the twelve temperature combinations treated as a factor, the residual SS on the first unweighted fit is 221.9 on 864 degrees of freedom which means very little since we have not used any weights. The fitted values are saved to obtain such weights and the RSS changes to 539.64. This model assumes that each set of four replicates (taken under apparently identical experimental conditions) have the same mean and variance but with the specific relationship between mean and variance exhibited by empirical logits. Subsequent iterations produce no change (in this case of the full model) as the fitted values each experimental group are just the group mean regardless of the weights. While it appears that the data are under-dispersed, the deviance of 539.64 on 864 degrees of freedom cannot be trusted as 520 of the 1152 fitted values are too small (less than 2) or large (greater than 23). On removing these data points the weighted residual SS reduces to 494.48 on 864 − 520 = 344 degrees of freedom and the mean deviance of 1.437 is significant evidence of over-dispersion. We therefore divide the previous weights by this number, giving the weights

$$w_i = \frac{\bar{y}_i(25 - \bar{y}_i)}{25 \times 1.437}$$

in the ith of 288 distinct experimental groups, $\bar{y}_i$ being the mean of the four counts in group i. This reweighting does not affect estimates, standard errors, or tests but does affect residuals; the residual SS decreases to 375.53(864) and 344.0 for the 344 data points with nonextreme fitted values.

The main output listed below is a **MINITAB** analysis of variance table that has been slightly altered. The model fitted has a simple linear effect in `difftemp`, a quadratic effect in `avetemp` with `water` and `type` included as interacting factors. The line labeled `lack of fit` is the difference between the residual SS under this model and the initial full model, the weights from which have been retained for all subsequent fits. The F-statistics all appear significantly high which makes simple model building difficult. The main problem here is that there is too much data and relatively small effects are being precisely identified. The adjusted SS themselves give an idea of the *magnitude* of the effects. The model fails a goodness-of-fit test, which means that the linear effect in `difftemp` and quadratic effect in `avetemp` are not sufficient to model the data and the adjusted SS of 1013.56 is quite a large proportion of 6092.54 and means that the descriptive and explanatory power of the model decreases from 93.8% to 77.2%. Nevertheless such a description might be useful as a broad summary of the shape of the effects.

```
Analysis of Variance for y

Source  DF    Adj SS   Adj MS        F      P
type     3     55.49    18.49    14.86  0.000
water    5   3102.80   605.60   498.56  0.000
```

type*water	15	315.71	21.05	16.91	0.000
difftemp(type)	4	186.30	46.57	37.42	0.000
avetemp(type)	4	51.08	12.77	10.26	0.000
avetemp2(type)	4	43.70	11.75	8.78	0.000
Error	1116	1389.09	1.24		
Lack of fit	252	1013.56	4.02	9.25	
Error(full)	864	375.53	0.434		
Total	1151	6092.54			

Term		Coeff	Stdev	t − value	P
Constant		−5.2680	0.7799	−6.75	0.000
difftemp(type)					
	1	0.01460	0.01491	0.98	0.328
	2	0.073622	0.008924	8.25	0.000
	3	0.062595	0.009394	6.66	0.000
	4	0.06659	0.01106	6.02	0.000
avetemp(type)					
	1	0.9133	0.1856	4.92	0.000
	2	0.4242	0.1125	3.77	0.000
	3	0.1333	0.1270	1.05	0.294
	4	-0.1583	0.1287	−1.23	0.219
avetemp2(type)					
	1	−0.018070	0.004018	−4.50	0.000
	2	−0.009183	0.002524	−3.64	0.000
	3	−0.003431	0.002811	−1.22	0.223
	4	0.001176	0.002898	0.41	0.685

	Coeff	Stdev	t − value	P
Constant	−3.6124	0.7462	−4.84	0.000
difftemp	0.061146	0.005476	11.17	0.000
avetemp	0.19171	0.06838	2.80	0.005
avetemp2	−0.004674	0.001523	−3.07	0.002

Parameters estimated are listed below the table and indicate a strong positive effect in `difftemp`; the odds of germination increase by 6.3% for each increase in the day/night temperature differential. There is quite a large difference in this effect for the four different seed types (adj SS = 186.30); the effects are all positive and are 1.4%, 7.6%, 6.4%, and 6.9%, respectively, for the four different seed types. The linear and quadratic coefficients in `avetemp` are 0.1917 and −0.004674 and there is less evidence that a different function is needed for the four seed types than for `difftemp`. The separate estimates for the seed types seem unreasonably variable, especially for type 4 which has the opposite sign to the others. There is apparently interaction in the `type` and `water` effects which are best summarized by profile plots of the cell means.

Figure 4.10 displays the broad effects of the four factors `type`, `water`, `difftemp`, and `avetemp` on germination odds. The first plot gives 95% confidence intervals for the main effects of `water` and `type` on the log-odds scale, standardized to have

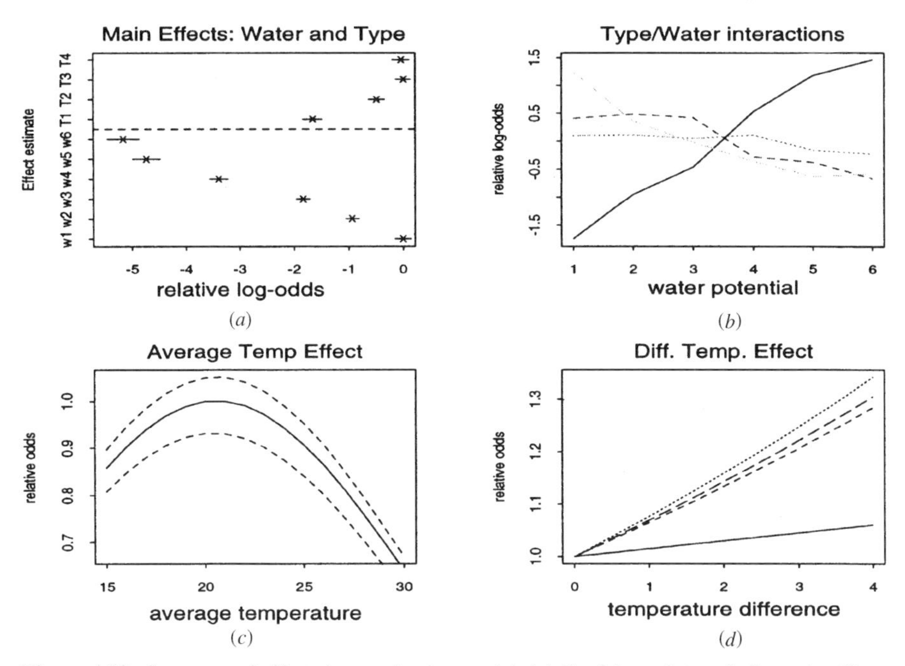

Figure 4.10. Summary of effects in germination model. (*a*) Confidence intervals for main effects of water and seed type, (*b*) water effects for different seed types, (*c*) (quadratic) effect of average temperature, and (*d*) (linear) effect of temperature difference for each seed type.

maximum value zero. The trend is systematically decreasing in water potential and seed type 1 has much lower germination rates than the other three seed types. The second plot gives profiles of the departure of the fitted values from the simpler model without interaction of `water` and `type`, on the log-odds scale. The solid line is seed type 1 which again stands out from the rest. The third plot gives the effect of average temperature which was assumed quadratic on the log-odds scale but is displayed on the odds scale standardized to have maximum value one. The last plot gives effects of `tempdiff` for the four seed types, again on the odds scale but standardized to have minimum value one. Seed type 1 is apparently different from the other three for which the effect of temperature differences on germination is greater.

4.5.5 Variable Selection Procedures

It is common with retrospective data that large numbers of covariates are measured along with each response. Step-by-step modeling of a response in terms of 30 predictors becomes an impossibly time consuming procedure. One advantage of approximating by a normal model is that variable selection procedures, supplied with most regression packages, are available.

Probably the safest method of choosing an appropriate subset of m possible predictors is to consider models with all possible combinations of these predictors (2^m in all) and to somehow measure the quality of each of these models. There are several such criteria, the best known being adjusted-r^2 that measures the correlation of the fitted and actual values penalized for the number of parameters in the model and Mallow's C_p that measures the mean prediction error and should equal the number of parameters in the model if the model is true. Selection procedures that look at all possible regressions are called *all subsets regression* and largely supercede the traditional backwards elimination procedure that can lead to extremely poor choice of predictors especially when they are highly correlated.

The command `BREGRESSION` in `MINITAB` can be used for this purpose but has two limitations not shared by the *Splus* function `leaps` which incorporates two important features. First, data points can be weighted explicitly which is essential to categorical data applications since our transformed data do not have equal variance. Second, the function fits the full model (with all predictors) robustly and then downweights outliers relative to this full fit when fitting all the other models. Thus `leaps` can find suitable regressions even when outliers are present.

A less formal graphical method of assessing which elements of the linear predictor are significant is the so-called *standardized normal plot*. This is constructed as follows. For each parameter estimate $\hat{\beta}_i$, divide it by its standard error to obtain a standardized estimate. If the corresponding parameter $\beta_i = 0$ then the standardized estimate should be close to standard normal and if all parameters β_i are zero then the collection of standardized estimates are a sample from the approximate standard normal distribution. Plotting the standardized estimates against the normal order statistics, the more significant parameter estimates will appear above the line $y = x$. Note however that the estimates may be correlated in which case removing one parameter may alter the significance of other parameter estimates.

4.6 INFERENCE ON GENERALIZED LINEAR MODELS*

A generalized linear model is a model for independent data $Y_1, \dots, Y_k$ with two essential components. First it is assumed that each Y_i has a distribution of the form

$$\log\{f(y;\theta,\phi)\} = \frac{y\theta_i - b(\theta_i)}{a_i(\phi)} + c(y,\phi) \tag{4.41}$$

where f denotes a density or probability function depending on whether the data are continuous or discrete. The class of distributions defined by Eq. (4.41) is called the *exponential family* and includes normal, gamma, Poisson, and binomial. For these last two distributions $a_i(\phi) = 1$. It can be shown that

*See p. 15 in Chapter 1 for an explanation.

$$\mathrm{E}(Y_i) = b'(\theta_i), \qquad \mathrm{Var}(Y_i) = b''(\theta_i)a_i(\phi). \tag{4.42}$$

It follows that b'' has the one sign and so b' is a monotonic function. Second, it is assumed that for some monotonic function η

$$\eta(\mu_i) = \sum_{j=1}^{p} \beta_j x_{ij}. \tag{4.43}$$

The function η is called the link function. In the special case that $\eta(\mu_i) = \theta_i$, i.e., when η is the functional inverse of b', the link is called canonical.

Likelihood Theory Let ℓ be the log-likelihood function. By writing

$$\frac{\partial \ell}{\partial \beta_j} = \sum_{i=1}^{k} \frac{\partial \ell}{\partial \eta_i} \frac{\partial \eta_i}{\partial \beta_j} = \sum_{i=1}^{k} \frac{\partial \ell}{\partial \theta_i} \frac{\partial \theta_i}{\partial \mu_i} \frac{\partial \mu_i}{\partial \eta_i} \frac{\partial \eta_i}{\partial \beta_j}$$

it follows simply that the components of the score function $\partial \ell / \partial \beta$ are

$$U_j(\beta) = \frac{\partial \ell}{\partial \beta_j} = \sum_{i=1}^{k} \frac{(y_i - \mu_i)}{\mathrm{Var}(Y_i)} x_{ij} \frac{\partial \mu_i}{\partial \eta_i}. \tag{4.44}$$

The Fisher information matrix is given by

$$I(\beta) = \sum_{i=1}^{k} \frac{x_i x_i^T}{\mathrm{Var}(Y_i)} \left(\frac{\partial \mu_i}{\partial \eta_i} \right)^2. \tag{4.45}$$

The ML estimator is defined by equating the score function to zero.

Canonical Links Recall that $\mu(\theta) = b'(\theta)$. When $\eta(\mu) = \theta$ the link is said to be canonical. For the $Bi(n_i, \pi_i)$ distribution the canonical link is $\log\{\pi_i/(1 - \pi_i)\}$, and for the $Pn(\lambda_i)$ distribution the canonical link is $\log \lambda_i$. In this case only

$$\frac{\partial \mu_i}{\partial \eta_i} \propto \mathrm{Var}(Y_i)$$

and provided that $a_i(\phi)$ does not depend on i the scores simplify to

$$U_j(\beta) = \sum_{i=1}^{k} (y_i - \mu_i)x_{ij} = S_j - \mathrm{E}(S_j; \beta)$$

where $S_j = \sum_{i=1}^{k} Y_i x_{ij}$ are respectively sufficient for the parameter β_j. In matrix notation the sufficient statistic $S = X^T Y$. It follows immediately that the observed information matrix is nonrandom and therefore identical to the expected information. For noncanonical links the matrices are different, and the minimal sufficient statistic for β is the whole data.

IRLS Algorithm The score and expected information have simple matrix representations. Let the diagonal matrix W and the vector r have elements

$$W_{ii} = (\partial\mu_i/\partial\eta_i)^2/\mathrm{Var}(Y_i), \qquad r_i = (y_i - \mu_i)\frac{\partial\eta_i}{\partial\mu_i}$$

called the working weight matrix and working residual vector, respectively. Notice that W_{ii} is the reciprocal of the asymptotic variance of r_i. As usual, let $X = (x_{ij})$ be the design matrix. Then

$$U(\beta) = \sum_{i=1}^{k} w_{ii} r_i x_{ij} = X^T W r, \qquad I(\beta) = \sum_{i=1}^{k} x_i x_i^T w_{ii} = X^T W X. \tag{4.46}$$

The Fisher scoring algorithm is just the Newton–Raphson algorithm for solving the vector equation $U = 0$ but with the derivative matrix $\partial U/\partial\beta$ replaced by its expected value. The algorithm can be written

$$\hat{\beta}_{k+1} = \hat{\beta}_k + I^{-1}(\hat{\beta}_k)U(\hat{\beta}_k)$$

where β_k is the kth approximation to the ML estimator $\hat{\beta}$. Using the matrix expressions in Eq. (4.46) for $U(\beta)$ and $I(\beta)$, this becomes

$$\begin{aligned}\hat{\beta}_{k+1} &= \hat{\beta}_k + (X^T W X)^{-1} X^T W r \\ &= (X^T W X)^{-1} X^T W X\hat{\beta}_k + (X^T W X)^{-1} X^T W r \\ &= (X^T W X)^{-1} X^T W (X\hat{\beta}_k + r) \end{aligned} \tag{4.47}$$

Thus, $\hat{\beta}_{k+1}$ is the weighted least squares estimate of β in the regression of the vector $z_k = X\hat{\beta}_k + r$ on X (see Section 4.5.1 or any linear models textbook). The *working* dependent variable z_k is just a linear approximation to the link transformed data $\eta(y)$ as is seen by writing

$$\eta(y_i) \approx \eta(\mu_i) + (y_i - \mu_i)\eta_i' = (X\beta)_i + r_i$$

and r_i are just the link residuals defined in Eq. 4.13 in Section 4.2.1. Recall from Section 4.5 that generalized linear model fits were approximated by regressing $\eta(y)$ itself on $X\beta$. The exact ML estimates are obtained by iteratively regressing a linear approximation to $\eta(y)$ on $X\beta$.

The IRLS algorithm is useful even more widely than in the context of generalized linear models. Generalized additive models are nonparametric models where the linear predictor is an additive combination of smooth functions. Such models can be fitted by replacing the above weighted least squares operation by a smoothing operation. Smoothing operators are a natural generalization of the hat matrix (see Section 5.1.3).

Existence of ML Estimates For binomial models, Silvapulle (1981) provides definite results, provided the link function η (i) is strictly increasing and (ii) is such that the functions $\log \eta^{-1}$ and $\log(1 - \eta^{-1})$ are both concave (i.e., have negative second derivative). This includes logistic and probit links. Since logistic models are equivalent to certain log-linear models, his results characterize existence of ML estimates for these special log-linear models.

The basic idea is that the set of covariate values corresponding to successes and failures (sets s and f below) must have some overlap. Collect together all the covariate vectors x_i corresponding to the binary successes $y_i = 1$. Let S be the set obtained by taking all positive linear combinations of these vectors, i.e.,

$$S = \left\{ \sum_{y_i = 1} k_i x_i : k_i > 0 \right\}$$

often called the convex cone generated by the x_i. Let s be the interior of this set. Similarly, let F be the convex cone generated by the covariate vectors x_i corresponding to the failures $y_i = 0$ and f the interior of this set. Then the ML estimator $\hat{\beta}$ is unique and bounded if, and only if, s and f have nonempty intersection. An example of a data set that fails Silvapulle's conditions is given in Exercise 4.20.

FURTHER READING

Section 4.1.3: Dose Response Models

4.1.6 *The constructed variable algorithm described here was suggested by Box and Tidwell (1962) and is closely related to work of Stevens (1951) and the more recent iteratively reweighted least squares algorithm of Green (1984).*

Section 4.2: Diagnosis of Model Inadequacy

4.2.3 *Atkinson's (1985) book on* Plots, Transformation and Regression *describes the basic added variable and probability plots described in this subsection but in the context of least squares only. So-called* partial *residual plots for use in logistic models were considered by Landswehr, Pregibon and Shoemaker (1984).*

Section 4.4: Modeling Over-Dispersion

Sums of independent nonidentical *binary variables were studied by Poisson (1837) after whom the Poisson–binomial distribution is named. More recent work of significance is found in Hoeffding (1956) and most recently by Wang (1993) who proved unimodality and that the variance increases with the heterogeneity of the probability parameters.*

Section 4.5: Approximation by Normal GLM's

4.5.1 *The belief that it is sufficient to iterate the generalized least squares algorithm only once or twice has been part of the statistical folklore. It has been proven by Jobson and Fuller (1980) that, provided one begins with a $\sqrt{k}$-consistent estimate $\hat{\beta}$, the one-cycle estimator has the same asymptotic variance as the infinite-cycle estimator. When the mean is modeled linearly [which is the case for the linear model Eq. (4.40) for the empirical logits] Carroll et al (1988) showed that the second-order correction to the variance of the estimator is the same for all iterations* after *the first, supporting the common practice of taking two iterations only. For moderate samples, the effect of the number of cycles, and indeed of weighting at all, depends mainly on the coefficient of variation of the data [see Chapter 2 of Carroll and Ruppert (1988)].*

4.7 EXERCISES

4.1. One thousand people were chosen at random from a population and surveyed on their smoking habits. All of these people were about the same age. They were followed for over 15 years and the number surviving recorded. These fictitious data are in Table 4.14.

Table 4.14. Cross-Sectional Data on Smoking and Mortality

Cigarettes/Day	0–10	11–20	21–30	31–50	51+
Number	426	312	137	87	38
Deceased	47	45	37	26	22

a. Why is it important that the subjects be of the same age? How could subjects differing widely in age bias the results? If the subjects had been from all age groups, could the differing ages be taken into account?

b. Calculate empirical logits at each smoking level. Plot the empirical logits against number of cigarettes smoked per day and also display the standard errors as vertical bars about these points.

c. Perform a normal approximation to the exact test (see Chapter 7) of the hypothesis that cigarette smoking has no affect on mortality.

d. Fit a least squares regression of the empirical logits on number of cigarettes smoked using the inverse of the variance estimates as weights. Report the important estimates and P-values for any hypotheses of interest. Investigate the effect of using the weights.

e. Fit a logistic binomial regression to the number of subjects dying and report the important estimates and P-values for any hypotheses of interest. Using this model, estimate the probability of dying in 15 years for a nonsmoker and give an approximate 95% confidence interval.

4.2. Caesarean births are often indicated for women with small pelvises. This in turn is related to other body measurements that are easier to obtain such as shoe size. The data in Table 4.15 tabulates 351 births by the shoe size of the mother and whether or not a Caesarean section was performed.

Table 4.15. Caesareans and Shoe Size[a]

Shoe Size	<4	4	4.5	5	5.5	≥6	Total
Caesarean	5	7	6	7	8	10	43
Natural	17	28	36	41	46	140	308

[a]Shoe sizes of mothers from 43 Caesarean sections and 308 natural births [from Frame et al. (1985).]

a. Does the chance of a Caesarean section depend on shoe size?

b. By giving each shoe size category a score, perform a test of trend of the chance of Caesarean section with shoe size.

c. Give a 95% confidence interval for the probability that a woman with a size 4 shoe will need a Caesarean section.

4.3. Norton and Dunn (1985) carried out a survey relating frequency of snoring to various diseases including heart disease. In Table 4.16, 2374 individuals without heart disease were classified, by their spouse, into one of four snoring groups based on how often they snore. The same classification was done by the spouses of 1210 hospital patients with heart disease.

Table 4.16. Snoring and Heart Disease[a]

Frequency	Never	Occasional	Almost Always	Always	Total
Heart disease	24	35	21	30	110
No heart disease	1355	603	192	224	2374

[a]Classification of 2484 subjects by heart disease status and frequency of snoring.

a. Test independence of the snoring/disease classifications.

b. Replace the snoring frequency labels by the numbers 1, 2, 3, 4 and test whether or not there is a trend in the probability of heart disease with snoring frequency.

c. Repeat part (b) with the scores 1, 2, 5, 6 and comment on how much of the differences between the groups can be attributed to trend.

d. Give a 95% confidence interval for the factor by which the odds of heart disease for a severe snorer is larger than that for a nonsnorer. Compare the results for the two different scorings.

e. Compute a confidence interval for the probability of heart disease for nonsnorers. Why doesn't this estimate mean anything?

4.4. A random sample of 14-year-old adolescents was taken, 1670 boys and 1799 girls. Over a one-week average, each was asked to report the number of hours spent in bed. This was divided by 7 and rounded up to the nearest half an hour to give a daily rate. The data, from MacGregor and Balding (1988), are given in Table 4.17.

Table 4.17. Sleep Patterns of Adolescents[a]

	≤7	7.5	8.0	8.5	9.0	9.5	10.0	>10.0
Boys	88	109	210	324	359	313	182	85
Girls	92	108	217	349	436	334	198	65

[a]Classification of 3469 adolescents by gender and daily hours spent in bed.

a. Test whether or not the sleep distributions are different for the two genders.

b. Is there a trend apparent in the data? Fit an appropriate model and compare the weight of evidence with that in (a).

4.5. Samples of a certain alloy fastener used in aircraft construction were subjected to different loads. The data in Table 4.18 give numbers of failures out of numbers tested for various accurately measured loads in units of 100 pounds per square inch. The data displayed are described by Montgomery and Peck (1982).

Table 4.18. Failure Testing of Alloy Fasteners

Load	Failures	Sampled	Load	Failures	Sampled
2.5	10	50	2.7	17	70
2.9	30	100	3.1	21	60
3.3	18	40	3.5	43	85
3.7	54	90	3.9	33	50
4.1	60	80	4.3	51	65

a. Plot modified empirical logits of failure against load and indicate standard errors.

b. Fit the logistic linear regression model $a + bx_i$ to the logit probability of failure. Compare results with fitting a weighted normal model.

c. What is the probability of failure when the load is zero? Hence, what is the likely true value of a? Suggest a transformation of the load covariate that avoids this problem.

d. Fit the model that supposes the functional form $\pi(x) = 1/(1 + a/x^b)$. In particular give approximate confidence intervals for a and b.

4.6. A survey was carried out to determine the opinions of people on whether or not Australia should become a republic; see Table 4.19.

Table 4.19. Attitude Towards a Republic

		Males		Females	
Code	Age	Opposed	Surveyed	Opposed	Surveyed
21.5	18–25	5	22	10	20
30	25–35	18	44	23	39
40	35–45	10	26	22	33
50	45–55	25	39	42	46
60	55–65	18	21	14	17
70	65+	11	13	15	16

a. Let $\pi_j(x)$ be the proportion of males (j = 1) or females (j = 2) aged x who oppose a republic. Consider the model

$$\text{logit}\{\pi_j(x)\} = \alpha_j + \beta_j x.$$

Interpret the parameters $\alpha_1, \alpha_2, \beta_1, \beta_2$ and give the values of sufficient statistics for these parameters.

b. Plot the logits of the proportions opposed against age-group for males and females on the same axis. Use the coded values for age as values of the age covariate. Consider separate least squares lines. If these lines were the true lines in the model of (a) then interpret what they mean in terms of the opinions of people of different sex and age-groups.

c. Approximate the exact P-value for testing whether or not older males tend to oppose republicanism more than younger males.

d. Analyze the data in GLIM. In particular, test whether age or sex have an effect on opinions and whether age has the same effect on opinion for males and females. Test these hypotheses using both change in deviance and the estimate compared to its standard error.

e. For the model that you believe describes the data best, calculate deviance residuals and plot them against age. Comment if appropriate. Calculate the sum of squares of these residuals and relate it to any earlier statistic.

4.7. Suppose that a model has been fitted to pure binary data. Let $\hat{\pi}_i$ denote the fitted value for unit i for which the true probability of response is π_i and suppose that the fitted value is hardly affected by whether $y_i = 0$ or 1.

a. Let $r_i = y_i - \hat{\pi}_i$ be the raw residual. Treating $\hat{\pi}_i$ as fixed, what are the mean and variance of r_i?

b. When do Pearson residuals have asymptotic mean 0 and variance 1?

c. Describe a situation where $\hat{\pi}_i$ is significantly affected by the value of y_i, is not consistent for π_i, and where the residuals consequently do not have the correct asymptotic mean and variance.

4.8. A survey was taken of 364 individuals for detectable lung impairment. The age of each patient was recorded and the self-reported number of cigarettes smoked per week (divided by 7). The fictitious results have been categorized in Table 4.20 and the covariate value replaced by an effective covariate value. Only a subset of the total data are given. Most records with very high smoking levels have been omitted.

a. Ignore the age variable. Plot estimates of the *logarithm* of the "risk function"

$$R(x) = \frac{\text{odds(Impaired}|x \text{ cigs/day)}}{\text{odds(Impaired}|5 \text{ cigs/day)}}$$

and indicate standard errors with vertical bars. Fit a logistic linear regression and draw the estimated log-risk function on top of the pointwise estimates. Surround this curve by 90% confidence bands to indicate the uncertainty.

b. Fit a logistic regression including age as an additive effect. Estimate

Table 4.20. Smoking and Lung Health by Age

Age	Cigarettes	Impaired	Sampled	Age	Cigarettes	Impaired	Sampled
15	25	0	5	55	5	1	14
65	5	4	15	45	35	1	7
35	25	3	17	45	25	15	26
45	5	4	17	15	5	1	12
55	35	6	8	65	35	3	6
25	5	1	14	65	15	2	8
80	15	1	4	55	25	9	32
65	25	7	21	25	15	2	15
35	15	3	18	45	15	0	9
80	5	5	8	15	15	0	11
35	5	1	15	25	25	2	22
25	35	1	7	35	35	0	9
55	15	2	12	80	35	1	2
35	60	8	12	80	25	7	16
15	35	1	2				

the logged-risk function and identify any important differences with (a) if there are any.

c. Suppose a person smokes 50 cigarettes per day and is 15 years old. Estimate the probability of having lung impairment and give a 95% confidence interval.

d. Investigate the effect of removing the single record with smoking level 60.

4.9. Refer again to the data in Table 4.20.

a. Fit the logistic regression model

$$\text{logit}\{\pi(x, a)\} = \mu + \beta x^{\alpha} + \gamma a$$

where x is number of cigarettes, a is age, and μ, α, β, and γ are unknown parameters to be estimated. Give a standard error for $\hat{\alpha}$ and a 90% confidence interval for α from the deviance graph.

b. Repeat the analysis of (a) but remove the record in the data corresponding to 60 cigarettes per day. Plot the estimated log-risk function. How do you regard the $x = 60$ point?

c. Again under the model with the $x = 60$ record removed, give a 90% confidence interval for the relative risk of lung impairment for 60 cigarettes per day relative to 50, i.e.,

$$r(60, 50) = \frac{\text{odds}(x = 60)}{\text{odds}(x = 50)}$$

first, assuming $\alpha = 1$ and second, assuming that α equals its estimated value. What is the moral of this calculation?

4.10. The data in Table 4.21, refer to 130 young pine trees that were kept in an artificial environment for 8 years. Listed are the numbers of plants showing signs of dieback for different concentrations of sulfur in the atmosphere. A binomial distribution was assumed for the number of trees with dieback and a linear logistic model was fitted to the probability, $p(x)$, of a tree showing dieback when sulfur concentration is at level x. This model was

$$\text{logit}\{p(x)\} = a + bx$$

for unknown parameters a and b. The estimates were $\hat{a} = 0.4719$ and $\hat{b} = 0.5465$.

Table 4.21. Association of Dieback and Sulfur Levels

Sulfur (ppm)	0.00	0.50	1.00	1.50
Dieback	25	16	20	29
Number	41	27	24	38

a. Calculate fitted values under this model.

b. Calculate Pearson residuals under this model.

c. Fit the model that assumes that sulfur concentration has no effect whatsoever on probability of dieback. In particular, list the fitted values and the Pearson residuals.

d. By comparing the sums of squares of residuals under the two models test the hypothesis $H_0 : b = 0$ against $H_1 : b > 0$. In particular, give an approximate P-value.

e. Perform an exact test of Cox (1970) of the hypotheses in (d), see Chapter 7.

4.11. One experimental design where the complementary log–log link is theoretically justified is *dilution assay* described by Fisher (1922). A solution has a very small number N of organisms per unit volume which cannot be counted directly. Samples of the solution are diluted by 1/2 several times and after the d_ith dilution the number of organisms has a Poisson distribution with mean $\mu_i = N/2^{d_i}$. Each of these diluted samples are then placed under favorable conditions and a culture of the organism will grow if at least one was present. The data finally comprises a sequence Y_i of binary responses (the sample grew or did not) and the number of dilutions.

a. Show that the probability that a sample contains no organisms is $p_i = 1 - \exp\{-\mu_i\}$.

b. Show that the complementary log–log model is appropriate for these data with unknown intercept $\log N$ and known slope $-\log 2$.

4.12. The complementary log–log link is theoretically justified in *seriological testing* [see Collett (1991) and Draper, Voller, and Carpenter (1972)]. An individual is infected with a disease with probability λ on any day. Infection can be detected by the presence of antibodies. In a population of N individuals it is desired to estimate the annual infection rate.

a. Show that probability of an individual aged A_i years having been infected is approximately $p_i = 1 - \exp\{-365A_i\lambda\}$ and hence that

$$\log\{-\log(1 - p_i)\} = \log(365\lambda) + \log A_i.$$

b. Show that the annual mean infection rate $h = N(1 - \exp\{365\lambda\})$.

c. The data in Table 4.22 relate to Amazonian villagers in 1971 and their infection by malaria. Replace the age ranges by their mid-points and by 30 for the last range. Fit a complementary log–log model to the probability of testing positive, using the OFFSET command to ensure that the slope coefficient for the covariate $\log A_i$ is 1. Hence, find a confidence interval for 365λ and for the yearly infection rate h per thousand villagers.

Table 4.22. Seriological Data for Brazilian Villagers

Age Group	A_i	y_i	n_i
0–11 months	0.5	3	10
1–2 years	1.5	1	10
2–4 years	3.0	5	29
5–9 years	7.0	39	69
1–14 years	12.0	31	51
15–19 years	17.0	8	15
≥20 years	30.0	91	108

4.13. Cox (1970) describes an experiment where ignots are subjected to different heating and soaking times in their preparation and then the binary response variable, ‘readiness for rolling’, measured. The data are given in Table 4.23. The numbers tabulated are the numbers out of the number in parentheses.

a. Fit the model which says that the logit probability of response is an additive function of a term depending on heating time and a term depending on soaking time. Be careful with the empty cells.

Table 4.23. Readiness of Ingots for Rolling by Heat and Soak Times

	Heating Time				
Soaking	7	14	27	51	Totals
1.0	10(10)	31(31)	55(56)	10(13)	106(110)
1.7	17(17)	43(43)	40(44)	1(1)	101(105)
2.2	7(7)	31(33)	21(21)	1(1)	60(62)
2.8	12(12)	31(31)	21(22)	0(0)	64(65)
4.0	9(9)	19(19)	15(16)	1(1)	44(45)
Total	55(55)	155(157)	152(159)	13(16)	365(377)

b. Can either heating or soaking time be removed from the model?

c. Fit the model which says that the logit probability of response is, for given soaking time, a linear function of heating time and, for given heating time, a linear function of soaking time.

d. Give 95% confidence intervals for the percentage increase in odds of an ingot being ready for rolling for each extra minute of (i) heating time and (ii) soaking time. On the basis of this, fit an even simpler model. Plot residuals against heating time and soaking time to determine if some other function of these variables is more appropriate.

4.14. The data in Table 4.24 are from Wilner et al. (1955) and are analyzed in Chapter 3 of Goodman (1978). The data are a cross-classification of survey results for 608 white women living in public housing projects by (i) their proximity to a black family, with $P = +$ indicating close, (ii) frequency of contacts with blacks, with $C = +$ indicating high frequency, and (iii) general local attitudes towards blacks, with $A = +$ indicating a more positive attitude.

Table 4.24. Racial Sentiments of White Women by Proximity, Contacts, and Local Attitudes [from Wilner et al. (1955)]

			Sentiment (S)	
Proximity (P)	Contact (C)	Attitudes (A)	+	–
+	+	+	77	32
+	+	–	30	36
+	–	+	14	19
+	–	–	15	27
–	+	+	43	20
–	+	–	36	37
–	–	+	27	36
–	–	–	41	118

a. Compute odds ratios of the $S \times P$ classification for the four different combinations of C and A. Compare these odds ratios with the $S \times P$ odds ratio collapsed over the other factors.

b. Find a logistic model for sentiment in terms of P, C, and A.

c. Find a model for attitude in terms of P, S, and C.

4.15. The data in Table 4.25 are from the famous Framingham Heart Study. Some background is given in Section 6.2.1. The binary responses here is presence of heart disease, to be related to cholesterol and blood pressure.

Table 4.25. Framingham Heart Study[a]

Heart Disease	Serum Cholesterol	Blood Pressure <126	126–145	146–165	>165
Present	<200	2	3	3	4
	200–219	3	2	0	3
	220–259	8	11	6	6
	≥260	7	12	11	11
Absent	<200	117	121	47	22
	200–219	85	98	43	20
	220–259	119	209	68	43
	≥260	67	99	46	33

[a]Cross-tabulation of individuals by cholesterol levels (in mg/100 cc), blood pressure (systolic in mm Hg) and absence of presence or heart disease [from Cornfield (1962)].

a. Fit the model that says that log-odds of heart disease depends additively on blood pressure and cholesterol, treated as categorical variables.

b. Fit the logistic model that says that the log-odds of heart disease depends additively on blood pressure and cholesterol, treated as covariates. Give confidence intervals for the effects of unit increases in blood pressure and cholesterol on the odds of heart disease.

c. Suppose that average serum cholesterol is 210 and average systolic blood pressure is 130. Give a confidence interval for the odds of heart disease for an individual with cholesterol of 250 and systolic blood pressure of 160, relative to an average person.

4.16. The data in Table 4.26 classifies applicants to the University of California at Berkeley by their gender, the department applied for, and finally their success which we may think of as the response [see Freedman et al. (1978, p. 14)]. Denote these factors by D = department, G = gender, and S = success on 6, 2, and 2 levels.

Table 4.26. Berkeley College Applications[a]

	Males		Females	
Department	Admitted	Refused	Admitted	Refused
A	512	313	89	19
B	353	207	17	8
C	120	205	202	391
D	138	279	131	244
E	53	138	94	299
F	22	351	24	317

[a]The number of college applications admitted and refused by gender and for which department they applied.

a. Ignoring which department is applied for, test whether or not males have a higher success rate than females.

b. Show that when the department applied for is taken into account, there is no statistical evidence that males and females have differing success rates.

c. Explain Simpson's paradox in terms of the association of gender with department. Hint: Which departments do males tend to apply for and how hard are these departments to enter?

4.17. Let $Y \overset{d}{=} Bi(n, p)$ conditional on p where p has the beta distribution with parameters a, b, i.e., density function

$$f(p) = \frac{p^{a-1}(1-p)^{b-1}}{\beta(a, b)} \quad 0 \le p \le 1.$$

a. What is the distribution of p given $Y = y$?

b. A heterogeneous population is sampled on t occasions. The probability that individual j is sampled on any occasion is p_j and the collection of these p_j follow a uniform distribution. One particular individual, call it k, is observed to have been sampled on every occasion. What is the distribution, mean, and variance of that individual's sampling probability p_k?

4.18. If $X_i \overset{d}{=} Bi(1, \pi)$ and if the correlation of X_i with X_j is ρ for $i \neq j$, then the variance of $Y = \sum_{i=1}^{n} X_i$ was shown to equal

$$\operatorname{Var}(Y) = n\pi(1-\pi)\{1+(n-1)\rho\}.$$

Suppose that P is a random variable with mean π and variance V taking values on [0, 1]. Suppose also that conditional on P, we have a collection of independent binary random variables X_i each with $Bi(1, P)$ distribution.

a. Show that unconditionally $E(X_i) = \pi$. Since X_i takes values 0 or 1, deduce the unconditional distribution of X_i.

b. Show that unconditionally, $\operatorname{Cov}(X_i, X_j) = V$. Thus the variability of the probability P common to the binary variables X_i induces a correlation between them. For instance, if P is uniformly distributed on [0,1], what is the induced correlation of the binary variables? The correlation is 1.0 only if $V = \pi(1-\pi)$. Show that this implies that $E(P^2) = E(P)$ which in turn implies that P is identically 0 or 1.

c. Using (b) and the equation in Exercise 4.18 show that

$$\operatorname{Var}(Y) = n\pi(1-\pi) + n(n-1)V$$

which was formulated on p. 229, by a more general method.

d. What is the distribution of Y given P? Hence, determine the equation above using yet another method.

e. Suppose that $i \neq j \neq k$. Give a formula for the correlation of X_i with $X_j + X_k$. Show that the condition for this correlation to be 1.0 is again $V = \pi(1-\pi)$.

4.19. What happens when there is variability and correlation of the probabilities in a binary sum? Suppose that $X_j \stackrel{d}{=} Bi(1, P_j)$ and are independent conditional on P_j. Suppose further that each P_j has the same mean π and variance $\phi\pi(1-\pi)$ and that the correlation of P_i and P_j is r, for $i \neq j$.

a. Show that the correlation of X_i and X_j is $r\phi$.

b. Then, give a formula for the variance of $Y = \sum_{i=1}^{n} X_i$.

c. When $r = 0$, show that X_i and X_j are independent. Since each is unconditionally $Bi(1, \pi)$, deduce the distribution of Y.

4.20. Table 4.27 gives the numbers of males and females who are deemed to be a "psychiatric case" according to a standard psychiatric interview. The individuals are also classified by their General Health Questionnaire (GHQ) score.

a. Fit a linear logistic or probit regression where the slope parameters are equal for males and females.

b. Try to fit separate logistic models to the males and females. Because this data set violates the necessary and sufficient conditions for existence of the ML estimate of Silvapulle (1981), you should receive some kind of error message.

Table 4.27. Health Score and Psychiatric Conditions, by Gender

	Males		Females	
GHQ Score	Case	Noncase	Case	Noncase
0	0	18	2	42
1	0	8	2	14
2	1	1	4	5
3	0	0	3	1
4	1	0	2	1
5	3	0	3	0
6	0	0	1	0
7	2	0	1	0
8	0	0	3	0
9	0	0	1	0
10	1	0	0	0
11	0	0	0	0
12	0	0	0	0
Total	8	27	22	63

c. Read Silvapulle's conditions described on p. 251. The covariate vectors here are of the form $(1, s_i)$ where s_i is the score. For males the set s is bounded by the vectors (1, 2) and (1, 10) and f by the vectors (1, 0) and (1, 2). The intersection of these sets is then zero. Show that the estimator for the females does exist because the sets s and f overlap.

CHAPTER 5

Smoothing Binomial Data

Models described in Chapter 4 were without exception parametric. For instance, the general dose response model

$$g\{\pi(x)\} = a + \beta_1 f_1(x) + \beta_2 f_2(x) + \cdots + \beta_p f_p(x)$$

imposes a specific functional form on $\pi(x)$ as a function of x, at least once the link g and basis functions $f_i(x)$ are chosen. There are several reasons why assuming a particular parametric family might not be a good idea, not the least of which is that the assumption might be wrong. On the other hand, summarizing the main features of a relationship with a small number of parameters is a way of simplifying the relationship suggested by the data. There are no completely general statements to be made about the relative advantages of parametric and nonparametric modeling. However, nonparametric methods are clearly underutilized for categorical data analysis at present. In this chapter the main ideas of nonparametric smoothing are presented and illustrated and their advantages are highlighted.

5.1 SMOOTHING A CONDITIONAL MEAN

In the simplest case consider a set of ordered pairs (Y_i, x_i) summarized in a scatter plot. The shape of this plot gives us information about how the observed y_i, and implicitly about how the mean values $E(Y_i|x_i)$, vary with x_i. Smoothing is an alternative to parametric curve fitting for (i) summarizing the relationship of the observed y_i and x_i and (ii) inferring the underlying relationship between $E(Y_i|x_i)$ and x_i.

While it may not always be the case that we want to model the *means* $E(Y_i|x_i)$ in terms of x_i this is certainly a very common and important case and smoothing is most easily introduced and understood in this context. For binary data Y_i the means are the π_i and smoothing reveals the relationship of π_i to x_i.

This leads us to a nonparametric alternative to the parametric binomial models outlined in the previous chapter.

5.1.1 Why Smooth?

Suppose we want to know the relationship of $E(Y|x) = \mu(x)$ to x. The most fundamental reason for fitting a smooth relationship is that we believe the relationship is, in fact, smooth. By smooth we do not mean the usual mathematical definition of derivatives existing up to a certain order. We have in mind a more common notion, that the curve $\mu(x)$ against x is regular rather than erratic and that its shape is largely summarized by a small number of features such as the overall steepness, the number and position of local maxima, and minima, etc.

Parametric modeling is based on the prior belief that a curve is not only smooth but has a specific shape, such as linear or quadratic. If this assumption is true then there are large measurable gains in statistical accuracy that are enjoyed because the entire weight of the data is felt in the estimate of the curve at each and every point. To a lesser extent, the same efficiency gains flow from an assumption of smoothness alone. On the other hand, the parametric and/or smoothness assumption may be incorrect and if so the curve estimate is biased towards curves satisfying the erroneous assumption, towards overly smooth curves in the case of nonparametric modeling. In summary, parametric and nonparametric models involve the same kinds of potential gains and losses.

Notwithstanding this general statement however, smooth curve estimators have important advantages over naive parametric estimators. First, they can fit data of almost arbitrary shape. The most evocative phrase used in support of nonparametric methods is that they "allow the data to speak for themselves." While this might not be strictly true, it captures the flavor and philosophy of nonparametric curve fitting. Second, they are automatically robustified against influential points and there is hardly any need for leverage or influence measures when smooth methods are used. This derives from the local weighting that forms the basis of most smoothers. Erroneous or outlying responses will also have a limited effect on the fit.

5.1.2 Local Polynomial Smoothing

Consider a random variable Y whose distribution depends on another variable x, random or otherwise. On the basis of a set of independent observations (Y_i, x_i) the mean $\mu(x)$ as a function of x is to be estimated. Clearly Y_i itself is an unbiased estimator of $\mu(x_i)$. However, without further assumptions we cannot estimate $\mu(x)$ at unobserved values of x. If $\mu(x)$ were constant then the mean of the data $\overline{Y}$ estimates this (flat) curve. If $\mu(x)$ were (almost) flat over a *region* of x values then the average of the data over this region would estimate this flat section of the curve. To implement such a *local average* we need to specify over what sized regions the curve can be considered reasonably flat.

Local Weighting An idea that unifies all useful smoothers is local weighting. A local weight function $w(x, x_i)$ is a decreasing function of $|x - x_i|$, such as

$$w(x, x_i) = \phi\left(\frac{x - x_i}{h}\right) \tag{5.1}$$

where ϕ is called the kernel function. Often ϕ is the standard normal density although any symmetric unimodal function may be used. Note that we are not assuming normality of any variable here but are simply using the normal density function ϕ as a local weight function. When $x_i = x$ we assign the maximum weight $\phi(0)$ and as x_i recedes from x the weight reduces to zero. The constant h determines how quickly it reduces, small values of h corresponding to a faster reduction and a *more local* weight function.

Locally Weighted Mean The weighted mean of a set of random variables Y_1, ..., Y_k is a statistic of the form

$$\hat{\mu} = \sum_{i=1}^{k} w_i Y_i \Big/ \sum_{i=1}^{k} w_i \tag{5.2}$$

where the weights $w_1, \ldots, w_n$ are nonrandom numbers, usually positive. We could use a weighted average to estimate the mean curve $\mu(x)$ at the point x by choosing a set of weights $w_1(x), \ldots, w_n(x)$. When these weights $w_i(x)$ equal local weights $w(x, x_i)$ then we call the estimator a locally weighted mean. This can be done at any desired point x and gives the curve estimator

$$\hat{\mu}(x) = \sum_{i=1}^{k} w(x, x_i) Y_i \Big/ \sum_{i=1}^{k} w(x, x_i). \tag{5.3}$$

When the local weights are of kernel form Eq. (5.1), then this is called *kernel* smoothing, introduced independently by Nadaraya (1964) and Watson (1964). When the smoothing parameter $h \rightarrow \infty$, the estimated curve approaches the constant function $\hat{\mu}(x) = \overline{Y}$, a very smooth function indeed. When $h \rightarrow 0$, the estimated curve approaches a step function passing exactly through the points (x_i, Y_i) and jumping to the next value half way between these points—a very unsmooth function. Intermediate values of h give estimated curves that follow the data smoothly without actually passing through every point.

While the theoretical properties of kernel smoothers are the simplest to derive, they are now known to have bias problems both at the endpoints and in areas of unevenly distributed x_i values. The curve estimator Eq. (5.3) is implemented in the *Splus* function `ksmooth` for various local weight functions.

Locally Weighted Line Locally weighted averages are motivated by the idea that the true curve $\mu(x)$ may be considered flat locally. A weaker assumption is that the $\mu(x)$ is locally a line. By analogy with the local mean, we could estimate the curve at the point x by fitting a linear regression weighted by the local weights, and then take the value of the estimated line at x.

Explicitly, let X be the $n \times 2$ design matrix with ith row $(1, x_i)$ and W_x be the diagonal matrix with elements $w(x, x_i)$. Then the "x-local" least squares estimate of the intercept and slope is given by the vector

$$(\hat{\alpha}(x), \hat{\beta}(x))^T = (X^T W_x X)^{-1} X^T W_x Y \tag{5.4}$$

and the estimate $\hat{\mu}(x)$ at the point x is $\hat{\alpha}(x) + x\hat{\beta}(x)$. Since to estimate $\mu(x)$ at other values of x the weight matrix W_x is changed, this sounds like a lot of computation but a single matrix formula can be given to compute $\hat{\mu}(x)$ at a grid of x-values [see Wand and Jones (1993, p. 119)].

When $h = \infty$ the local linear smoother fits the unweighted least squares line through the whole data. In contrast to the local mean, even the most smooth local linear fit will respect the gradient of the data. When $h = 0$ the local linear smoother is piecewise linear and passes exactly through the points (x_i, Y_i).

Is local linear smoothing then better than local mean smoothing? Except when $\mu(x)$ is constant, local lines can be shown to have lower asymptotic bias when the x-values are distributed asymmetrically in the local weighting range. This is especially important at the ends of the curve, i.e., in the extremes of the x-space see Hastie and Loader (1993) for a discussion as well as Wand and Jones (1993, Section 5.5).

Figure 5.1 illustrates and compares the local mean and local line smoothing algorithms. The data, given by Marsden (1987), are proportions of individuals in Bongono Zaire who are sero-positive to malarial infection, against age-group. The smoothing constant is $h = 5$ and the weight function is plotted along the x-axis about the point $x = 30$. The local mean and local line at this point are also plotted. The fitted curves are quite similar except at the endpoints. If the x-values were more asymmetric then there would be more disagreement between the curves.

One can easily generalize to local polynomials of higher order, but on the basis of fairly subtle theoretical work, local lines are probably as good as any in practice [see Further Reading]. The smoother `lowess` in *Splus* implements these estimators with downweighting of outliers, the degree of the locally fitted polynomial being chosen by the user.

5.1.3 Linear Smoothers

A smoothing operation is said to be linear if, for every point x, $\hat{\mu}(x)$ is a linear combination of the data values, i.e., if

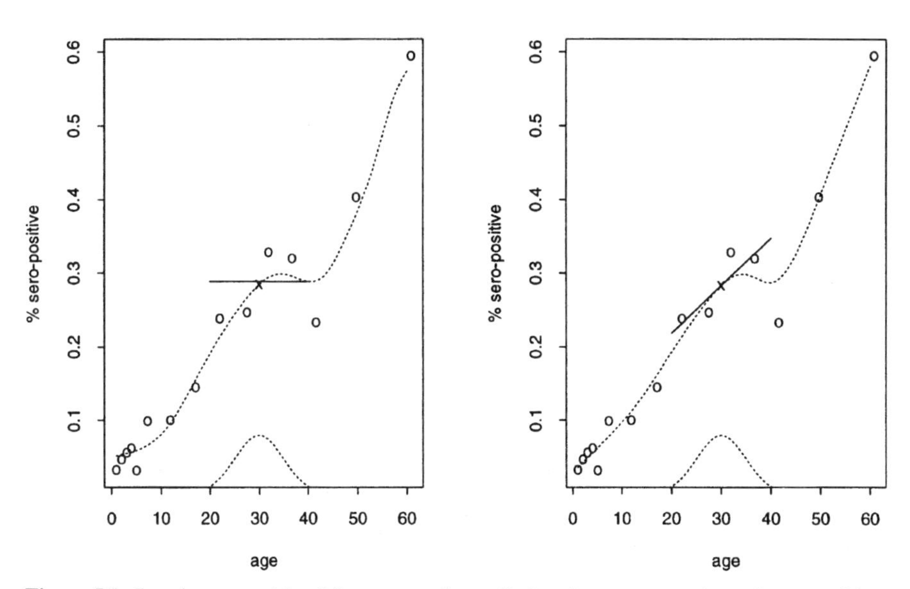

Figure 5.1. Local mean and local linear smooths applied to data on proportions of sero-positives against age. Also plotted are the Gaussian smoothing weights used, as well as the local fits around age 30.

$$\hat{\mu}(x) = H(x)^T Y \tag{5.5}$$

where $H(x)$ is a nonrandom k-vector and Y is the data vector. In addition $H(x)$ must sum to one or the smooth of a constant function will not equal the constant function. There are many examples. For a locally weighted mean smoother [see Eq. (5.3)],

$$H(x)^T = (w(x, x_1), w(x, x_2), \ldots, w(x, x_k)) \Big/ \sum_{j=1}^{k} w(x, x_j).$$

For a locally linear smoother

$$\hat{\mu}(x) = (1, x)(X^T W_x X)^{-1} X^T W_x Y$$

where W_x is a diagonal "x-local" weight matrix.

As usual, we define the fitted values to be the estimated mean trend at the observed values x_i and we denote them $\hat{\mu}(x_i) = \hat{Y}_i$. For a linear smoother, the fitted values are all linear functions of the data values and we write

$$\hat{Y} = HY$$

where the ith row of the $k \times k$ matrix H is $H(x_i)$. This provides an elegant unification with parametric inference where H is the hat matrix, a symmetric idempotent matrix of rank p. For nonparametric smoothers H will not be idemptoent—applying the smooth twice gives a different and smoother curve than applying it once. For the locally weighted mean curve

$$\hat{Y}_i = \sum_{j=1}^{k} w(x_i, x_j) Y_j \Big/ \sum_{j=1}^{k} w(x_i, x_j)$$

and so the smoothing matrix H has components

$$H_{ij} = \frac{w(x_i, x_j)}{\sum_{j=1}^{k} w(x_i, x_j)}.$$

For local linear fits it can be shown that

$$H_{ij} = \frac{K_{ij} T_{i2} + K_{ij}(x_i - x_j) T_{i1}}{T_{i2} T_{i0} - T_{i1}^2}$$

where

$$K_{ij} = \phi\{(x_i - x_j)/h\}, \quad T_{ij} = \sum_{l=1}^{k} K_{il}(x_l - x_i)^j.$$

The provided *Splus* functions `H.mean` and `H.line` compute the smoothing matrix for local mean and local line smoothing, respectively.

A theoretical advantage of linear smoothers is that their moments follow the usual linear algebra. Let the data vector Y have true mean vector μ and dispersion matrix V. Then the mean of the fitted values vector is $H\mu$. Only when $\mu(x)$ is invariant with respect to H, will the fitted values from the smooth be unbiased—for instance, if H is a locally weighted mean smoother and $\mu(x)$ is truly constant, or if H is a locally weighted linear regression smoother and if $\mu(x)$ is truly linear, or if H is the hat matrix from a linear model and the model is correct. However, usually the fitted values are biased for the true means. The bias of $\hat{Y}$ must be traded off against the dispersion of $\hat{Y}$ which, by the standard linear algebra, is given by

$$\text{Var}\{\hat{Y}\} = H \text{V} H^T. \tag{5.6}$$

This may be used to compute pointwise standard errors to accompany the smooth curve $(x_i, \hat{Y}_i)$ although one should keep in mind that confidence bands produced this way ignore bias in the curve estimate. The variance of the smooth at a particular point x is

$$\text{Var}\{\hat{\mu}(x)\} = H(x)^T \sum H(x). \tag{5.7}$$

5.1.4 Effective Degrees of Freedom

It would be useful to attach an effective number of degrees of freedom or number of parameters to a linear smooth summarized by its matrix H. This is useful for (i) calibrating how much smoothing we are doing on an absolute scale, e.g., with two parameters we are doing as much smoothing as fitting a line and (ii) for comparing different smoothers holding the amount of smoothing fixed. Three common measures are

$$tr(H) \leq tr(HH^T) \leq tr(2H - HH^T).$$

These come from an analogy with the corresponding least squares expression for (i) mean square error, (ii) Mallow's C_p, and (iii) residual error. For idempotent and symmetric H they agree but in general they are ordered as above. For local polynomial smoothers all three measures give the answer k when $h = 0$—as required since the fitted values then equal the data values. When $h = \infty$, the measures all give the answer $p + 1$, where p is the degree of the local polynomial – 1 for a local line and 0 for a local mean.

It is a little unfortunate that these measures depend on the x-values unless H is idempotent and symmetric. For nonlinear smoothers effective degrees of freedom can also be defined but will involve the values of the response as well. The *Splus* command `gam` inputs an effective degrees of freedom as a parameter of the smooth based on trace(H).

5.1.5 How Smooth?

When fitting polynomials we have the problem of model selection—what order polynomial should I fit? For nonparametric smoothers, we must similarly choose the smoothing parameter h. Ultimately, we would hope that the estimated curve $\hat{\mu}(x)$ is as close as possible to the true curve $\mu(x)$. One measure of closeness that combines bias and variance in a natural way is

$$\text{MPE}(h) = \sum_{i=1}^{k} \text{E}\{\hat{\mu}(x_i) - \mu(x_i)\}^2,$$

called mean prediction error (MPE). The value of h minimizing this gives the best curve estimator in this special sense.

Theoretical expressions for MPE unfortunately depend on (i) the underlying error variance, (ii) the exact spacing of the x-values, and (iii) the second derivative of the true curve $\mu(x)$. It is the last of these that is most problematic since the curve $\mu(x)$ is unknown. A solution to this problem is provided by *cross-validation* which is a very simple idea. We use the data to estimate MPE. Let us omit data point (x_i, Y_i) and then try to estimate Y_i from smoothing the remaining points. Denote this estimated data value by $\hat{\mu}_i^-(x_i)$. The cross-validation error criterion is

$$\mathrm{CV}(h) = \sum_{i=1}^{k} (\hat{\mu}_i^-(x_i) - y_i)^2.$$

The algorithm requires us to compute $\mathrm{CV}(h)$ for a range of values and find the value giving the smallest CV. While this may sound computationally prohibitive there are very simple formulas available for linear smoothers that do not require explicit refitting. Although it is uncommon in practice, CV can be used to choose the degree of a parametric polynomial p in exactly the same manner.

The basis of the CV method is that $\mathrm{CV}(h)$ is an approximately unbiased estimator of $\mathrm{MPE}(h)$ although a stronger justification for the procedure would be that the minimizer of CV is close to the minimizer of MPE. It is essential to this result that we use the fits $\hat{\mu}_i^-(x_i)$ with the data point omitted. If we instead use fitted values $\hat{\mu}(x_i)$ then CV is the residual sum of squares that is always minimized at $h = 0$.

5.2 SMOOTHING BINARY DATA

For continuous data Y_i, a plot of Y_i against x_i graphically summarizes the relationship of $\mu(x)$ to x. Such plots will suggest an appropriate model and ought to be tracked by the fitted values of a good model. Departures of the fitted curve from the plot will suggest additions and modifications to the model. In short, a plot of what the data themselves say about $\mu(x)$ is an indispensable aid to both model building and refinement.

For binary data Y_i the mean is $\pi(x)$ but a plot of binary data will be quite uninformative about the shape of this curve. Even for binomial data Y_i with small denominators n_i, plots of proportions Y_i/n_i against x_i are obscure, and the argument for smoothing in a way that "lets the data speak for themselves" is even more compelling than for continuous data.

Recall the cancer remission data described in Example 4.2. The n_i were all very small and the plot of the empirical probabilities on p. 189 gave little idea about how $\pi(x)$ might vary with x. The (modified) empirical logit plot is equally obscure and is not displayed. Figure 5.2 plots the raw 0–1 observations against

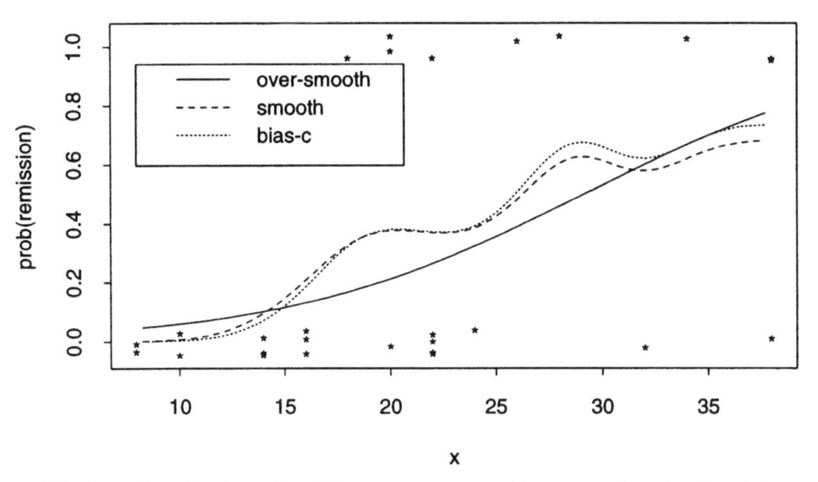

Figure 5.2. Smooth estimation of $\pi(x)$ for remission data. Three smooth estimates of the probability of remission as a function of x.

their covariate after adding a little random noise to separate identical observations (for instance there are two zeros at $x = 8$). This is called "jittering." Also plotted are three smooth estimates of $\pi(x)$ which may, if desired, be transformed to the logit scale. This may suggest a model or be used to judge an already fitted model. This section is concerned with how such smooths of binary data may be computed and statistically calibrated.

5.2.1 Smoothing a Simple Binary Regression

Consider a binary variable Y depending on a single covariate x with mean

$$\Pr(Y = 1|x) = \pi(x).$$

A random sample on the ordered pair (x, Y) is at hand from which the curve $\pi(x)$ is to be estimated. We may apply the local mean smoother [see Eq. (5.3)] directly giving the smooth estimator

$$\hat{\pi}_h(x) = \sum_{i=1}^{k} Y_i \phi\left(\frac{x - x_i}{h}\right) \bigg/ \sum_{i=1}^{k} \phi\left(\frac{x - x_i}{h}\right). \tag{5.8}$$

For $h > 0$, this estimates the probability relationship $\pi(x)$, for any real x, observed or not. The constant h controls the amount of smoothing. For small values of h the smooth just interpolates the data. This estimator is unbiased but has a large variance for sparse data. For h extremely large all data points are included in $\hat{\pi}_h(x)$ which converges to the function constant at the observed proportion of successes in the entire data. Obviously this is much too smooth. The same bias will to some extent persist even for moderate h.

If there are repeated x values then it will be computationally efficient to recode the binary data as binomial, i.e., Y_i successes out of n_i when $x = x_i$. In this case the smooth may be expressed as

$$\hat{\pi}_h(x) = \sum_{i=1}^{k} Y_i \phi\left(\frac{x - x_i}{h}\right) \Big/ \sum_{i=1}^{k} n_i \phi\left(\frac{x - x_i}{h}\right) \tag{5.9}$$

where the sum now extends over the distinct number of binomial observations rather than the (perhaps much larger) number of binary observations.

Standard Errors and Effective Sample Size The variance of the smooth estimate $\hat{\pi}_h(x)$ is a decreasing function of h since more data is averaged as h increases. An exact expression is Eq. (5.7) where $\sum$ is diagonal with elements $\pi(x_i)$. A useful approximation to this variance is obtained by expanding each $\pi(x_i)$ about $\pi(x)$ which leads to

$$\mathrm{Var}\{\hat{\pi}_h(x)\} \approx \left\{ \frac{\sum_{i=1}^{k} n_i \phi^2\left(\frac{(x - x_i)}{h}\right)}{\left[\sum_{i=1}^{k} n_i \phi\left(\frac{x - x_i}{h}\right)\right]^2} \right\} \pi(x)(1 - \pi(x)). \tag{5.10}$$

The inverse of the term in braces we denote $\tilde{n}_h(x)$. It measures the effective number of data points "in the region of x." With this notation, the variance is then of the classical binomial form $\pi(1 - \pi)/\tilde{n}$. As h increases from 0 to ∞, it is easily shown that $\tilde{n}_h(x)$ increases from "the number of data points at x" to the total number $n = \sum n_i$ of data points. When x is in a region containing little data, $\tilde{n}_h(x)$ will be small reflecting the dangers of extrapolation to areas with no data.

Bias Correction Local means are biased towards flatter curves. While local linear smoothers have better bias properties, there are reasons against their use with binary data. One reason is very simple—the fitted values may be outside [0,1] and when the data contain extreme proportions this does occur in practice, even after choosing h with cross-validation.

Local linear smoothing is similar to the following two-step procedure. (1) Compute the least squares line and the residuals from it. (2) Compute the local mean smooth of these residuals and add this smooth to the linear fit. It is the "line part" of this algorithm that is poorly suited to estimating probabilities. A better result could be obtained with the alternative: (1a) Compute the logistic linear regression line and compute residuals given in Eq. (4.14), and then do (2). The resulting estimator is bias corrected in the direction of the linear logistic which is desirable if the true curve is close to linear logistic. Copas (1983) described a similar simple bias-correction in favor of the probit model. A third algorithm incorporates smoothing directly into the GLIM fitting algorithm; see next section.

All of these bias corrections have the effect of magnifying any tendency of the fitted curve to monotonically increase or decrease, i.e., it biases the curve away from the characteristic flatness of local means. This is illustrated in Figure 5.2, where the local mean smoother (dashed curve) with $h = 3$ has been corrected for bias (dotted curve). As $h \to \infty$, the bias-corrected curve converges to an estimate of the probit linear model (solid curve). In practice, it is worth examining a range of bias corrected smooths. The supplied *Splus* function `binomial.smooth` incorporates the Copas bias correction.

Confidence Bands One use of the variance expression for $\hat{\pi}_h(x)$ is to compute a standard error and surround the estimated curve by the bands $\hat{\pi}_h(x) \pm \text{se}\{\hat{\pi}_h(x)\}$. There are problems associated with using these as formal confidence bands. First, if h is chosen to optimize mean prediction error, say by cross-validation, then the bias and standard error will be of comparable size. The bands are therefore not bands for $\pi(x)$ but rather for $\text{E}\{\hat{\pi}_h(x)\}$. This is a recurrent difficulty with all smoothing procedures.

A second problem is that the bands may violate the interval [0,1]. This is remedied by transforming to a link scale, for instance, estimating $\psi(x) = \text{logit}\{\pi(x)\}$ by

$$\hat{\psi}_h(x) = \text{logit}\{\hat{\pi}_h(x)\}.$$

Based on a first-order Taylor expansion the variance is approximately

$$\text{Var}\{\hat{\psi}_h(x)\} = \frac{1}{\tilde{n}_h(x)\pi(x)(1 - \pi(x))}$$

mirroring the formula for empirical logits seen in Chapter 3. Computing bands on the logit scale and back transforming produces bands entirely between zero and one. However, it in no way solves the bias problem just mentioned. Notice also that for $\pi_h(x)$ close to zero or one, the bands for $\pi(x)$ using this method become the entire interval. While this might seem a large price to pay for keeping the interval inside [0,1] the diverging bands do serve as a reminder that bands cannot be trusted for extreme probabilities. As a rough rule, bands should not be trusted on any scale unless the smaller of $\tilde{n}_h(x)\hat{\pi}_h(x)$ and $\tilde{n}_h(x)\{1 - \hat{\pi}_h(x)\}$ is at least 5.

Example 5.1: Smooths can Reveal Trend Consider the models A and B below. 150 independent binary observations were generated for these models with the covariate values x sampled from the uniform distribution on $[-3,3]$.

$$\pi_A(x) = \begin{cases} 0.9 & x < -1 \\ 0.1 & -1 < x < 1, \\ 0.5 & x > 1 \end{cases} \quad \pi_B(x) = \frac{\exp\{\text{sgn}(x)x^2\}}{1 + \exp\{\text{sgn}(x)x^2\}}$$

Figure 5.3 shows the results of various smooths of these simulated data which are displayed in the left plots after "jittering." The upper two plots summarize data generated from model A and the lower two from model B. The purpose of these plots is to convince the reader that (i) almost nothing can be gleaned from a plot of binary data and (ii) the smoother is capable of revealing the true underlying trend with reasonable precision, although their usefulness is limited unless the sample size is at least $k = 100$.

For $h = 0.3, 0.4, 0.5, 0.6, 0.7$ the upper-left plot gives bias-corrected smooths, as well as the true curve (dotted) for model A. These curves give some idea of

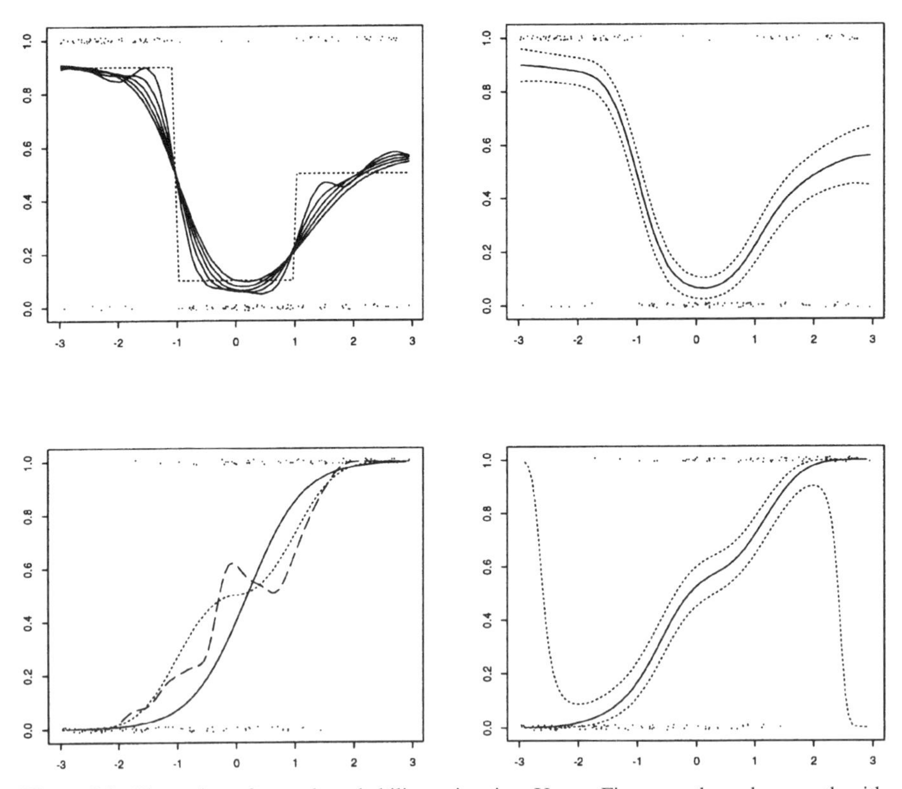

Figure 5.3. Illustration of smooth probability estimation. Upper: Five smooths and a smooth with bands on π-scale for model A. Lower: Five smooths and a smooth with bands on ψ-scale for model B.

the true curve but do not get very close to the shape of a step function; there is large bias in these smooth estimates near the steps because the true curve is not smooth! Smaller values of h would reduce this bias but increase the variance. To the right of this plot is the smooth with $h = 0.3$ surrounded by approximate confidence bands obtained by adding and subtracting one standard error. Note that these bands do not include the true step function. This is because of the bias in the smooth.

The bottom left plot displays two smooths (i) an oversmooth (solid curve) with $h = 2$ and (ii) a slight undersmooth (dashed curve) with $h = 0.3$ superimposed on the true relationship (dotted curve) of model B. The bias we saw in case A is much less apparent here, primarily because the true curve is smooth. The bottom right plot gives bands about the $h = 1$ smooth obtained by adding and subtracting one standard error to the logit estimate and then transforming back to the probability scale. Note the divergence of the bands at the extremes. Quite different bands are obtained if we compute them directly on the probability scale, at least at the extremes.

Finally, the supplied *Splus* function `binomial.smooth` produces an object with five components containing x-values, the smooth, the bias-corrected smooth, the standard error, and the effective sample size. A similar alogrithm could be programed in GLIM or many other packages. *Splus* has been chosen since it has superior graphics that show off this essentially exploratory technique to the greatest advantage. All the plots in this chapter were produced in *Splus*.

5.2.2 Smoothing a Multiple Binary Regression

There are two limitations of the kernel smoother [see Eq. (5.8)] whether corrected for bias or not. First, the distribution of the response (binomial here) is not used; it would be nice if a smooth estimator could be derived from an appropriate (binomial) likelihood. Second, the kernel smoother has only been designed for a single covariate. It is certainly possible to smooth a response Y_i in several covariates. One method uses *multivariate* weight functions combined with means or multiple regressions. Another method based on a simplifying assumption is to be presented in this section.

Generalized additive models (GAM's) are a family of models that do (i) make use of the distribution of the response and (ii) allow smoothing in several covariates. Formally a GAM is defined by two assumptions: (1) That the response variables Y_i are independent and have distributions from the same exponential family but with possibly differing means μ_i, (2) that for a suitable transform, η, called the link function

$$\eta(\mu) = \alpha + \sum_{j=1}^{p} f_j(x_j)$$

where $x_1, \ldots, x_j$ is a set of predictors and the f_j are unknown functions. The book by Hastie and Tibshirani (1990) has been influential in popularizing these models; see especially Section 4.5 for application to binomial data. When each $f_j(x_j) = \beta_j x_j$ the generalized additive model reduces to a generalized linear model. There is no reason why one cannot smooth nonparametrically in some variables and parametrically in others.

The curves f_j cannot be estimated without supposing some smoothness conditions. The smoothness is introduced into the reweighted least squares step of the fitting algorithm for GLIM's [see Eq. (4.47)] specifically by replacing the regression step on the residuals by a smooth. This algorithm can be formally motivated as the maximizer of a log-likelihood appropriately penalized for curves f_j that are too erratic. The result of the fitting procedure is a set of estimated functions $\hat{f}_j$ that are smooth and summarize the dependence of the mean response on the jth predictor. Inference can be based on the usual likelihood ratio statistic where the degrees of freedom in each component f_j is determined by the effective degrees of freedom used for smoothing the IRLS residuals in x_j.

The major simplifying assumption in a GAM is that the true response surface is an *additive* combination of the functions $f_j(x_j)$. If it is believed that there are interactions between predictors then extra terms, perhaps a product of two predictors, need to be included. This is also true of GLM's.

5.2.3 The *Splus* Command `Gam`

Implementations of mixed parametric/nonparametric models will no doubt appear in standard packages in the future but the only current implementation known to the author is in *Splus*. This command is quite slow (as are most significant computations in *Splus*) and can also be quite fragile. The *Splus* command `gam` fits a generalized additive model specified by (i) an error distribution for the response and (ii) a list of terms in the additive model either parametric or nonparametric. It inherits many features from the command `glm`. The syntax of the command is

```
gam(formula, family, weights)
```

although other more advanced arguments may be supplied. The `family` argument specifies a distribution such as `binomial` or `gaussian`. The `weights` argument is a vector of nonnegative numbers and allows more or less weight to be attached to an observation if desired. An unweighted model is fitted by default. Most of the model's substance is in the `formula` which is a list of model terms best illustrated by example. The model formula

```
germination~factor(type)+s(htemp,3)+lo(ltemp)+water
```

specifies an additive model for `germination` with a categorical factor `type` in the predictor as well as the covariate `water`. The term `s(htemp,3)` specifies that predictor `htemp` is modeled nonparametrically and uses 3 effective degrees of freedom, i.e., as many as a cubic term in `htemp`. The $s(.)$ operator indicates that a *smoothing spline* is used for the smooth. Other smoothers such as kernel smoothers or lowess can also be used. For instance, the term `lo(ltemp)` specifies that `ltemp` enters the model nonparametrically and uses the robust smoother lowess mentioned in the previous section. A peculiarity of the formula specification in *Splus* is that for binomial data a two-column matrix comprising the responses in the first column and the nonresponses in the second column is supplied as the response, e.g., `cbind(y,n-y)~x` would fit a simple linear logistic regression to y successes out of n.

The result of the `gam` is an object containing many of the quantities necessary for inference and data summary. The `coefficients` component lists parameter estimates for the terms of the model, parametric and nonparametric. The nonparametric coefficients estimate a simple linear relation in the smoothed predictors. The truly nonparametric part of the fit appears in the component `smooth` which gives the estimated smooth over and above the linear trend, stored in columns of a matrix. The usefulness of the smooth terms over and above the linear trend are summarized in the component `nl.chisq` which contains a type of score test statistic for testing the smooth model against the simple linear model. The degrees of freedom (in `nl.df`) is the trace of the implicit smoother matrix in the particular covariate minus 1 (parameter for the slope of the line). Fitted values on the data scale are in `fitted.values` and on the link scale in `additive.predictors`. The log-likelihood ratio or deviance is in the component `deviance` and has the same use and interpretation as for generalized linear models. Further details of generalized additive modeling in *Splus* are in the Splus manual at the Wiley website.

5.2.4 Analysis of Seed Germinations

The seed germination data described in Section 4.5 is ideal for illustrating the simplicity and usefulness of smoothing models. Recall that the response is the number of germinations out of 25, replicated four times. The covariates are water potential at six levels, simulated night temperature, and simulated day temperature. The existence of effects for water potential, night temperature, and day temperature were not in question but the exact shape of these effects was of primary interest. In particular, a major aim of the experiment was to determine how the *variation* in temperatures, i.e., the difference between day and night temperature, affects germination rates. It was hypothesized that such differences may serve to crack the seed and aid germination. The seeds in question are a native grass common in semiarid areas of New South Wales, where the daily variation of temperatures can become extreme. For simplicity, the germination pattern for seed type 2 (of the four types) will only be analyzed.

The first step is to recode the high and low temperature into two new variables, the average temperature labeled `ave` and the variation or difference in temperatures, labeled `vary`. We are anticipating that germination rates will be a smoothly increasing function of `vary`. The relationship with `ave` we anticipate will be smooth but concave, germination being retarded when average temperature is either too low or too high. A smooth decreasing relationship in `water` was already well known. Notwithstanding our preconceptions, smooth regression in these three covariates can be used to model both concave and monotonic curves provided we supply enough degrees of freedom to the smooths (3 or 4).

There are 288 observations, each binomial with $n = 25$. There are four replicates each of 72 experimental conditions comprising all combinations of the six water potential levels with 12 day/night temperature combinations. These 12 temperature combinations (see table) are not balanced in the four levels of temperature difference (0,5,10,15) and eight levels of average temperature (15,17.5,22.5,23,25,27,27.5,30). This is clearly a very poor design for discovering the response surface as a function of `vary` and `ave` and we must make some assumptions about the shape if we are to have any success at all. Let us assume that the effect of temperature variation is the same (on the logistic scale) for different levels of average temperature. This is the additive model assumption. Even if untrue, the marginal effects give an idea of the effects averaged over the observed distribution of the other covariate.

ROWS:AVE COLUMNS:VARY

	0	5	10	15	ALL
15.0	24	0	0	0	24
17.5	24	0	0	0	24
20.0	48	0	0	0	48
22.5	48	0	0	0	48
23.0	0	0	0	24	24
25.0	0	24	24	0	48
27.0	0	0	0	24	24
27.5	0	0	24	0	24
30.0	0	24	0	0	24
ALL	144	48	48	48	288

Since `vary` is on only four levels two degrees of freedom were used for the smooth, three degrees of freedom for `water` being on six levels, and four degrees of freedom for `ave` being on eight levels. In the `gam` notation two degrees of freedom is equivalent to fitting a quadratic, etc. Below is a summary of the output from the *Splus* command

```
gam(germination~s(water,3)+s(ave,4)+s(vary,2),"binomial")
```

```
> fit$coefficients
 (Intercept) s(water, 3) s(atemp, 4) s(dtemp, 2)
  2.654     -1.382     0.00538     0.0899
```

```
> summary(fit)
Deviance Residuals:
    Min       IQ     Median      3Q     Max
  -3.713  -0.775    -0.208    0.702  4.506

  Null Deviance: 4485.7 on 287 degrees of freedom

Residual Deviance: 396.6 on 278.0 degrees of freedom

Number of Local Scoring Iterations: 5

DF for Terms and Chi-squares for Nonparametric Effects

               Df Npar    Df Npar    Chisq     P(Chi)
(Intercept)     1
s(water, 3)     1          2         47.44     0.000
s(ave, 4)       1          3         51.66     0.000
s(vary, 2)      1          1          7.03     0.009
```

Notice that the mean deviance of 396.6/278 = 1.42 is significantly larger than expected. As mentioned in Section 4.5, this is due to variability between the seeds and so we should really be allowing for this over-dispersion. Since the n_i are constant we may use the simple mean deviance correction described in Section 4.4. The table gives chi-square statistics for testing the nonparametric component of each covariate as well as the degrees of freedom used in the smooth, which differs slightly from the target degrees of freedom supplied in the command. Dividing each of the chi-square statistics by 1.42 gives the values 33.40, 36.38, 4.95, all still highly significant for χ_1^2.

There is very strong evidence that the water effect is nonlinear despite the plot below indicating that there is little difference in the linear and nonparametric curves. This is because the magnitude of the water effects, both linear and nonlinear, are so large. The linear effect is summarized by the regression parameter -1.382 indicating that the odds of germination decreases by a factor of $\exp(1.382) \approx 4$ for each increase in water potential.

There is overwhelming evidence that the effect of average temperature is nonlinear which is unsurprising as a concave effect was anticipated. The linear coefficient, 0.00538, is very small and does not summarize the shape of the average temperature effect. There is strong evidence that the effect of `vary` is nonlinear. The linear part of this effect, returned in the component `$coefficients` of the `gam` object, is 0.0899 and indicates that the odds of germination increases by 9.4% for each increase in the day/night temperature differential.

While the object allows fitted values to be computed, since the observed values of `water`, `ave`, and `vary` are so discrete it is necessary to extrapolate to produce a pleasing display of the smooths. The command `predict.gam` was used to compute the smooths at an even grid of coordinates covering the observed values and the results are displayed in Figure 6.4. This command is described in the supplied *Splus* manual. Superimposed are parametric fits obtained from a model linear in `water` and `vary` but quadratic in `ave`. The curves are on the logistic scale shifted to have maximum value 0.0. The scale of the y-axis indicates the size of the effects on the logistic scale, `water` clearly being the most significant effect.

The Pearson residuals for all 288 data points have also been computed. The sum of squares of these residuals is 416.96 on 278 degrees of freedom however many of the fitted values are very small so this is not a reliable goodness-of-fit statistic. Accumulating cells with small fitted values into cells with fitted value at least 2, the residuals then have average squared size $379.4/239 = 1.58$: in other words, the effective standard deviation of the residuals is $\sqrt{1.58} = 1.256$. The residuals have been plotted against the ranks of the fitted values. There seems to be no trend in mean or dispersion although it is difficult to detect trend in dispersion visually. For identifying outliers, use the 95% limits

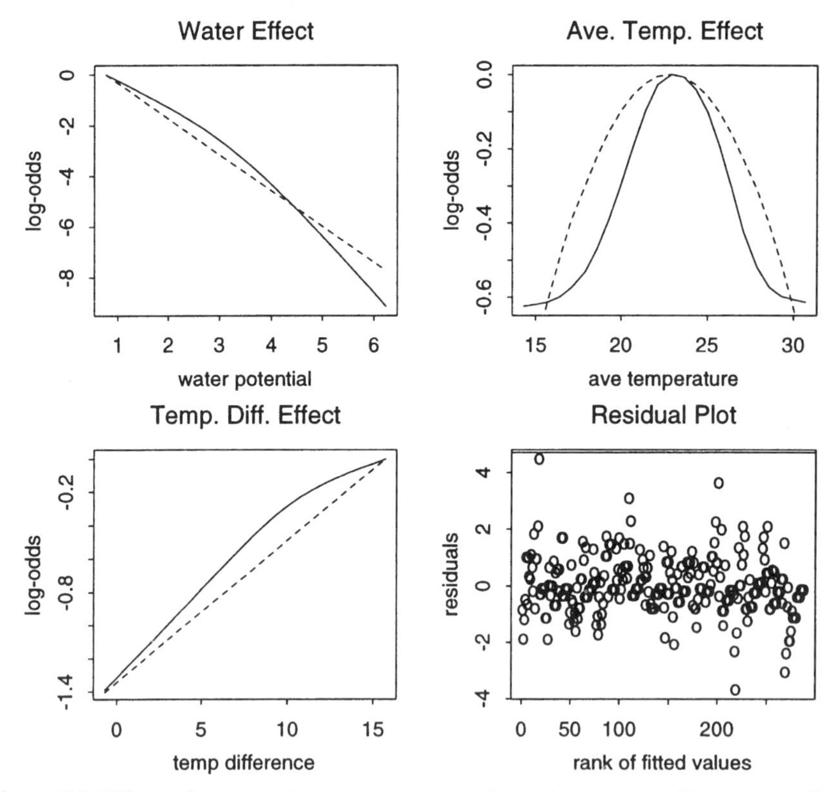

Figure 5.4. Effects of water and temperature on seed germination rates. Both a generalized additive model and a parametric model have been fitted. The first three plots compare the smooth fits with linear, quadratic, and linear fits, respectively. The final plot is of Pearson residuals from the generalized additive model against ranks of fitted values.

of 4.71 beyond which the most extreme of 288 normal variables with standard deviation 1.256 should not lie. There are no points violating this limit but one point in the top-left corner comes close. The fitted value corresponding to this point however is quite small and so it is unreliable.

A plot of the same residuals against the fitted values themselves rather than the ranks shows some peculiar "arm-like" structures especially for small fitted values. This is a consequence of the mild discreteness of the residuals (here all $n_i = 25$) and is much more serious for binary data (see Section 5.4). In addition, the fitted values are distributed very unevenly across the x-axis which introduces further structure into the plot and makes informal evaluation even more difficult. On balance, it seems better to plot residuals against ranks of fitted values in this instance.

5.3 ESTIMATING ATTRIBUTABLE RESPONSE

The relative risk function can be estimated nonparametrically. Recall Eq. (4.24) displaying the logged risk function as the logit probability function $\psi(x)$ with $\psi(0)$ subtracted. A smooth estimator of the risk function, is

$$\hat{r}(x) = \exp\{\hat{\psi}_h(x) - \hat{\psi}_h(0)\}$$

where $\hat{\psi}_h(x)$ is a nonparametric estimator of $\psi(x)$. Two natural possibilities are (i) the logit transform of the kernel estimator [see Eq. (5.8)] of $\pi(x)$ and (ii) the linear predictor $\hat{\alpha} + \tilde{f}(x)$ from a generalized additive model.

It is obvious that the probability of response at $x = 0$ plays an important role in such estimation. Indeed, if there is no data at $x = 0$ then the extrapolated estimate $\hat{\psi}_h(0)$ may be quite sensitive to the type and amount of smoothing. One strange consequence is that the plotted point estimates $(x_i, \hat{r}(x_i))$ may not follow the linear or smooth fit. This occurs when the empirical logit at $x = 0$ deviates from the general pattern.

Example 5.2: Smooth Risk May Not Fit Pointwise "Data" An artificial example is given in Figure 5.5. Empirical logits have been plotted against a covariate and two models superimposed, one linear logistic with deviance 16.4(17) and one smoother. The corresponding log-risk estimates were obtained from the logit estimates by subtracting respective estimates of $\psi(0)$ or, in other words, by moving the curves up until they pass through the origin. None of these curves appear to follow the "data" very well although the smooth does a better job than the line.

The reason is that the empirical logit at $x = 0$ is larger than its neighbors and appears to have been higher than the general trend. This sampling error results in the *entire* pointwise risk estimate being much lower than it should. This illustrates a strength and a weakness of modeling. If the linear model is wrong

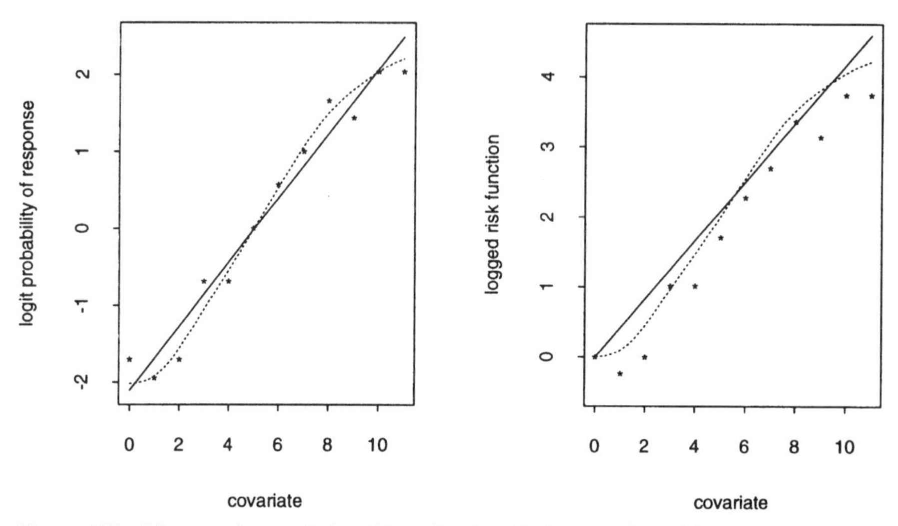

Figure 5.5. Linear and smooth log-risk estimation. Estimates of empirical logits (left) and implied log-risk (right) obtained from (i) empirical log-odds, (ii) logistic linear (solid line), and (iii) kernel smooth of empirical probabilities (dotted curve). Notice that the fitted curves do not pass through the data.

then the low extrapolated probability at $x = 0$ in the left hand plot is wrong and so the linear risk function is badly wrong. On the other hand if the linear model is correct then it gives a *much* more precise estimate of $\hat{\psi}(0)$ than either the raw data or the smooth (which is somewhere in between) since all of the data is used in its estimation. The linear risk function will then be a much more accurate estimate despite its disagreement with the pointwise "data." These issues arise in any modeling but are magnified in relative risk estimation by the sensitivity to $\psi(0)$.

5.3.1 Attributable Response

The estimated risk function calibrates the statistical association between response and the covariate of interest. One cannot usually infer a causal connection from a statistical association. Nevertheless, for purposes of interpretation it is often enlightening to assume a causal connection and to see what this assumption says about the number of responses that are due to the covariate. For instance, we might estimate that 40% of lung cancers are "due to smoking cigarettes" by which we mean if nobody smoked then the model predicts a 40% lower rate of lung cancer. To do this we (i) assume that the relative risk $r(x)$ applies to each and every individual at covariate level x and (ii) combine the risk function with an estimate of the distribution of the covariate x across the population.

Suppose that in a finite population, $M(x)$ individuals have covariate level x and respond independently with probability $\pi(x)$ an *increasing* function of x.

On average there will be $\mu(x) = M(x)\pi(x)$ responders from this group and so

$$\mathrm{E}(Y) = \int M(x)\pi(x)dx$$

in total. Now imagine that all individuals whose covariate level is $x \geq \gamma$ instead had covariate level γ. Then $N(x, \gamma) = M(x)[\pi(x) - \pi(\gamma)]$ fewer responses at x will be expected and we call this quantity the *attributable response*. Since direct estimates of $\pi(x)$ are not available for retrospective sampling we reexpress this in the form

$$N(x, \gamma) = \mu(x)\delta(x, \gamma) \tag{5.11}$$

where

$$0 < \delta(x, \gamma) = \frac{\pi(x) - \pi(\gamma)}{\pi(x)} = \frac{r(x, \gamma) - 1}{r(x, \gamma)} [1 - \pi(\gamma)] \qquad x \geq \gamma$$

with $r(x, \gamma)$ the odds ratio of response for x compared to γ. The correction factor in brackets cannot be estimated from retrospective data but may be effectively ignored provided that $\pi(\gamma)$ is small. When $\gamma = 0$ the quantity $\delta(x, 0)$ measures the proportional reduction in probability of response when the covariate is reduced from x to 0. It is tempting to interpret this as the probability that a given response at level x was caused by the covariate. For instance, if $r(x) = 3$ then of every three responses at x one is due to the baseline risk and the other two to the covariate and so $\delta(x, 0) = 2/3$. The *total response attributable* to covariate levels x exceeding γ is

$$N(x \geq \gamma) = \int_{x \geq \gamma} N(x, \gamma)dx. \tag{5.12}$$

5.3.2 Parametric and Nonparametric Estimation

To estimate $N(x, \gamma)$ we must estimate two quantities, the expected number $\mu(x)$ of responders at covariate level x and the relative risk $r(x, \gamma)$. Different estimates of these lead to different estimates of the attributable response.

The relative risk function will be estimable from retrospective or prospective samples either through some linear model, by point estimates (in which cse the integral becomes a sum) or by a smoothing technique. The expected number of responders in the population will also often be directly estimable. For example, with the BAC/accident data in Section 4.3.1, the responding (i.e., accident involved) individuals are all sampled and so we may use $Y(x)$ itself, or a suitable smooth, as an estimate. Where the response group is not exhaustive we could inflate this by dividing by the sampling fraction. If this is unknown then only relative values of $\mu(x)$ will be estimable but may be divided by the total number of responses to estimate the *proportional* rather than the absolute attributable response.

Estimating $r(x)$ pointwise by Eq. (4.23) and $\mu(x)$ by $Y(x)$ the estimate of the attributable response [see Eq. (5.11)] becomes

$$\hat{N}(x,\gamma) = Y(x) - \frac{Y(\gamma)}{n(\gamma) - Y(\gamma)}\,[n(x) - Y(x)]$$

and summing over $x \geq \gamma$ estimates the total attributable response. At the other extreme we may estimate $r(x)$ by a linear model and $\mu(x)$ by some smooth such as *loess* or the weighted kernel smooth (5.3) giving

$$\hat{N}_h(x,\gamma) = \hat{\mu}_h(x)\left(\frac{\hat{r}_L(x) - 1}{\hat{r}_L(x)}\right)$$

which when integrated over $x \geq \gamma$ estimates total attributable response. The estimated function $\hat{N}(x,\gamma)$ may take negative values in some circumstances indicating that an increase in response is predicted from reducing the covariate. If we are absolutely convinced that the risk function is increasing then $\hat{N}(x,\gamma)$ should be truncated at zero.

For fixed γ, a plot of $\hat{N}(x,\gamma)$ against x can be used to identify sub-groups of the population whose high covariate values are largely contributing to the total response. If these individuals could be persuaded to reduce their covariate levels then a large reduction in total response is anticipated. Since $\hat{N}(x,\gamma)$ combines the risk with the distribution of the covariate, conclusions based on it will not mimic conclusions about the risk. For instance, a certain age-group might have the most sharply increasing risk but the lowest attributable response if very few individuals actually have a high level of the covariate. Standard errors for $\hat{N}(x,\gamma)$ are fairly complex [see Lloyd (1996)].

Example 5.3: Drinking and Fatal Accidents, Revisited For the BAC/accident data listed on p. 221, calculational details of estimating attributable risk are shown in Table 5.1. The first three columns give pointwise estimates and total response attributable to $x > 0$. The total number of deaths attributable to alcohol impairment is 389 out of 591. The estimate $\hat{\delta}(x,0)$ increases more or less steadily from 0 to 1 as x increases and measures the proportion of deaths attributable to alcohol impairment at BAC $= x$. The second three columns give details of the smoothed estimate of total attributable response. The first column is a smooth estimate of $\mu(x)$. There is a slight complication because of the unequal bin sizes and the data in the last three BAC bins must be scaled down by a factor of 5 since they cover a range of BAC values 5 times greater than the other bins (the scaling factor of 5 for the last bin is arguable). These rescaled

Table 5.1. Attributable Response Estimates

	Pointwise			Smoothed		
BAC	$Y(x)$	$\hat{\delta}(x,0)$	$\hat{N}(x,0)$	$\mu_h(x)$	$\hat{\delta}_L(x,0)$	$\hat{N}_h(x,0)$
10	9	0.07	0.63	9.02	0.23	2.08
20	9	0.12	1.11	9.11	0.41	3.72
30	8	0.16	1.25	9.09	0.55	4.96
40	11	0.55	6.06	8.99	0.65	5.85
50	10	0.63	6.31	8.89	0.73	6.51
60	11	0.76	8.33	8.70	0.79	6.91
70	8	0.78	6.24	8.57	0.84	7.21
80	6	0.78	4.57	8.36	0.88	7.34
90	11	0.87	9.52	8.50	0.91	7.79
100	14	0.91	12.80	10.42	0.93	9.67
110	9	0.92	8.31	12.47	0.94	11.78
120	15	0.96	14.34	13.98	0.96	13.38
130	20	0.98	19.62	15.43	0.97	14.92
140	28	0.99	27.74	16.82	0.97	16.39
150	12	0.97	11.67	17.82	0.98	17.48
180	113	0.99	112.11	95.69	0.99	94.84
230	77	1.00	76.87	89.44	0.99	89.23
280	71	0.98	70.77	67.41	1.00	67.37
Total	591		398.3			387.4

counts were then smoothed using the *loess* algorithm with two modifications. First, the data at BAC = 0 was not used at all. Second, weights inversely proportional to the counts were employed, these being sensible if the counts are assumed Poisson. The smooth estimates in the last three cells were then scaled back up by the factor 5 and the results are listed in the column $\hat{\mu}_h(x)$. The last column, which is the product of the previous two, gives smooth estimates of attributable response and total 387, not too far from the pointwise estimate. A plot of $\hat{N}_h(x,0)$ against x is much smoother than $\hat{N}(x,0)$, however it is again appropriate to rescale the last three bins for this plot. When this is done the function is seen to begin turning down at the final bin. More refined plots using the exact x-values for all records confirm this downward trend which is due to there being small numbers of drivers at the extremely high BAC levels. The main peak of the graph is around 200 mg/100 ml, or four times the legal BAC limit.

5.4 SMOOTHING RESIDUALS

Residuals are used to diagnose model inadequancy primarily in two ways: (i) individual data points may be flagged as doubtful if their residual is too large. (ii) More global structural problems with the model will be revealed by global trends in the residuals. For detecting global trend, small n_i cause residual plots to display unwanted structure, even when the model is correct. The effect can be quite drastic and make informal interpretation of a residual plot virtually impossible. In the extreme case of $n_i = 1$, residuals each take only two possible values—conditional on the fitted value. For instance, the raw residual is $1 - \hat{\pi}_i$ or $-\hat{\pi}_i$.

Interpretation of large residuals clearly requires some care. While the mean and variance of residuals might still be approximately standard even when $n_i = 1$, the extreme discreteness means that "large" residuals cannot be calibrated by the normal distribution. Rather, cross-validation probabilities can be calculated; see p. 203.

Example 5.4: Difficulties with Binary Residual Plots Figure 5.6 shows logit residuals computed from binary data sets to which have been fitted the simple linear logistic model. In the first plot, this model is the correct one. In the second plot, the true model is $\text{logit}[\pi(x)] = x + x^2$ and so the fitted model is incorrect and it is the purpose of a residual plot to reveal this fact. Both plots display a high degree of structure which, if the data were normally or even continuously distributed, would immediately imply a serious defect in the model. The same unwanted structure will be apparent when the n_i equal 2 except there will be three "arms" in each plot. It is very difficult indeed to draw conclusions

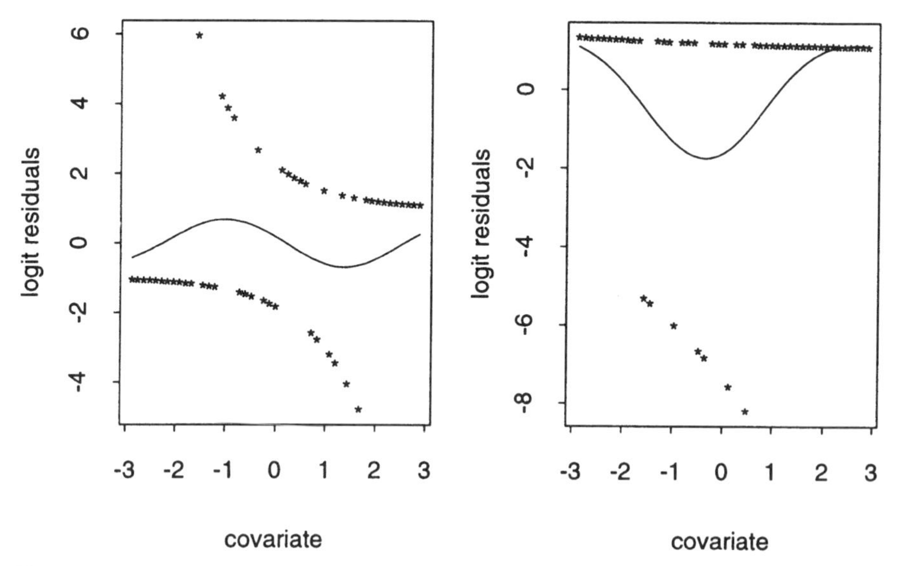

Figure 5.6. Logit residual plots for binary data. In both plots the fitted model is the simple linear logistic. In the left plot this model is correct but in the right plot the true logistic relationship is quadratic. Both plots display visual structure. Superimposed smooths can make interpretation of such plots easier.

about model adequacy from either of these plots without seeing the smooths that have also been displayed.

There are two ways that smoothing can be used to improve this situation. One is to simply smooth the residual plot—called *post-smoothing*. Indeed post-smoothing of residuals is useful for continuous data as well. A second approach is to build smoothing into the residuals so that the germane features of residuals from normal least squares models are recovered, called *pre-smoothing*. Of course, since binomial data may always be considered as binary data, these methods apply in general to binomial data but for moderately large n_i the residuals defined in the previous chapter are perfectly adequate.

5.4.1 Post Smoothing of Residuals

Let r_i be some kind of residual (for instance a Pearson residual) which is to be plotted against some other variables u_i (say an added variable residual) in the hope of revealing a defect in the model. Mostly one is hoping to detect some systematic departure of $\mathrm{E}(R|u)$ from zero. When the n_i are small the plot contains unwanted structure. One can simply smooth the residuals using any of the smoothers described in Section 5.1. The smooth $\hat{\mathrm{E}}(R|u)$ plotted against u estimates the underlying trend $\mathrm{E}(R|u)$ in a way much more easily interpreted visually.

Since under the null hypothesis, $\mathrm{E}(R|u) = 0$, the local mean smoother is as appropriate as any other for this purpose. The variance of the smoothed residuals is estimated using Eq. (5.6) and the variance matrix $\sum$ is the variance matrix of the residuals. For standardized residuals this may be taken as the identity, and for raw residuals $\mathrm{Var}(Y)(I - H)$ where H is the hat matrix of the fitted model. Choose the smoothing parameter by cross-validation, and then plot the smooth with bands. Significant departures from zero indicate possible deficiencies in the model and since the bias of the curve is towards zero, ignoring the bias in the bands will be conservative. There is no reason to stop at the cross-validation value of h, however. Smaller values of h will reveal deficiencies of the model on more local scales, so long as the smooth is calibrated with a reliable standard error.

While the unsmoothed residuals are very discrete in their distribution the smoothed residuals, being averages, will be more normally distributed. The effect sample size $\tilde{n}(x)$ determines how normal in distribution is the smoothed residual and as a rough guide one would not trust it unless the minimum of $\tilde{n}(x)\pi(x)$ and $\tilde{n}(x)(1-\pi(x))$ is at least 5. This is often a problem where the probability of success becomes extreme. For residual plots based on binary data, it is wise to suppress residuals for which the effective sample size is unsufficient to support the normal limit; most residuals, smoothed or unsmoothed, behave extremely erratically when the corresponding fitted value is too small or too large.

There is a further complication in interpreting smooths of residuals treated as a single curve—it is much more likely than 5% that *at least one point* of the smooth will violate a given 95% bound. Thus, we need simultaneous rather than pointwise calibration. A rough adjustment for this multiple comparison might be made by using the effective degrees of freedom p of the smooth, i.e., by making an adjustment for p effective comparisons.

Simple kernel smooths have been superimposed on the uninformative residual plots in Figure 5.6 and the second plot is hostile to the model and indeed identifies a quadratic discrepancy, although the smooth needs to be compared to its standard error. A further example of residual smoothing was already given in Figure 4.5 where the logit residuals are smoothed against the predictor age for the menstruation data. The dotted curve is a smooth (with four effective parameters) surrounded by two standard error bands. Recall that logit residuals have approximate variance the inverse of $n_i\pi_i(1-\pi_i)$ which has been estimated from the model. There is mild evidence that the residuals may deviate from zero in mean trend for the lower ages. Note that the extreme point at right has a strong effect on the smooth there, an effect which could be reduced by a robust smoother or simply by more smoothing. In the lower plot of logit residuals for the seed germination data, again four effective parameters have been used. The smooth, together with bands, contraindicate the simple linear model.

5.4.2 Presmoothing of Residuals

An alternative approach, suggested by Fowlkes (1987), removes the bizarre structure in binary residuals by incorporating smoothing directly into their construction. Let $\hat{\pi}_h^*(x)$ be an undersmooth of $\hat{\pi}_i$ against x_i. In the more general case we may smooth $\hat{\pi}_i$ against several predictors, say by generalized additive modeling. A smooth residual is defined by replacing the binary data Y_i by $Y_i^* = \hat{\pi}_h^*(x_i)$. An undersmooth is required so that the smoothed fitted values remain true to the shape of the data. For instance, the smooth Pearson residual is

$$r_{Pi}^* = \frac{Y_i^* - \hat{e}_i}{\text{s.e}(Y_i^*)} \tag{5.13}$$

where the standard error of the smooth is given in Eq. (5.10).

Replacing the data by a smoothed version eliminates the major problem with binary residuals, the extreme discreteness of their distribution; the distribution of the smoothed residuals depends on all data values near Y_i and so the distribution is much less discrete. Smooth residuals function quite well for most purposes of residual analysis; (i) they have approximate standard normal distributions and so large values relative to this distribution indicate an outlier, (ii) they can be plotted against various constructed variables to detect useful additions to the model, and (iii) they should have no mean trend if the model is correct.

Because of the presmoothing of the data, presmoothed residuals have fewer degrees of freedom than the data and are more highly correlated than unsmoothed residuals. If presmoothed residuals are to be further smoothed, then this correlation structure should be used as $\sum$ in the variance formulas. It is rather unclear how many effective degrees of freedom remain in these doubly smoothed residuals.

Example 5.2: (*Continued*) Revealing Quadratic Trend Using Presmoothed Residuals. Figure 5.7 follows on from Figure 5.6. The raw data are plotted after "jittering" in the left-hand plots. Superimposed are the fitted curve $\hat{\pi}(x)$ based on the linear logistic model as well as a (bias-corrected) kernel smooth $\hat{\pi}_h^*(x)$ of the data with smoothing constant $h = 0.2$, that corresponds to fitting around 12 parameters—an under-smooth in other words.

In the upper plots, the simple linear logistic model is the true one and the departures of the under-smooth Y_i^* from the parametric curve ($\hat{e}_i$) do not seem systematic. The plot at right is of the presmoothed residuals [see Eq. (5.13)] against x, and measures the standardized discrepancy between the two curves

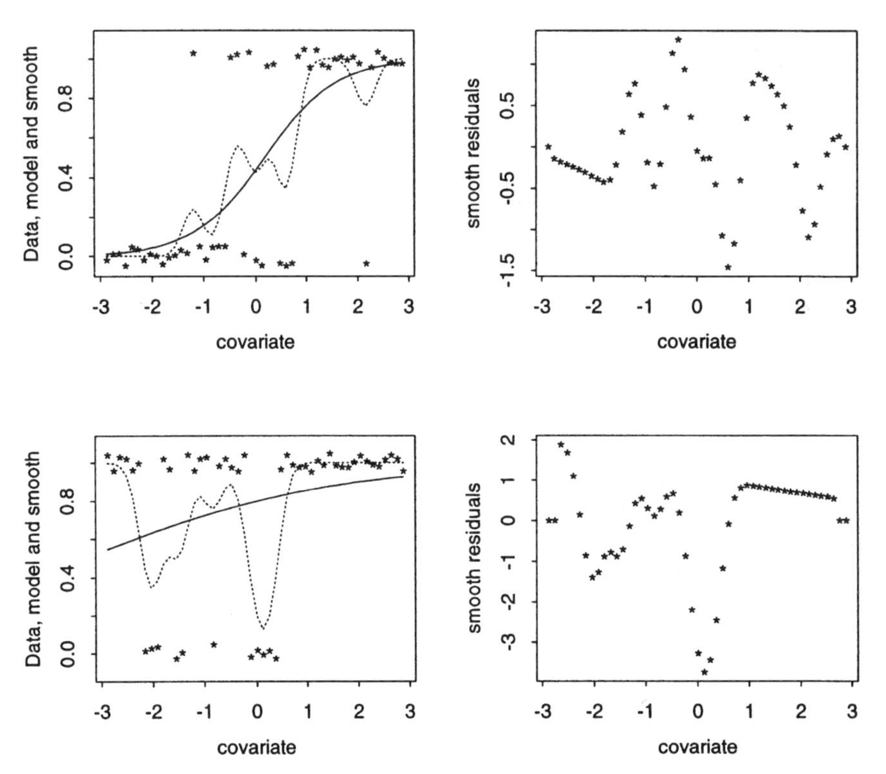

Figure 5.7. Presmooth residual plots for binary data. In the upper plots the true logistic relationship is linear. Comparison of the under-smooth with the linear fit gives the presmoothed residuals (right). In the lower plots the true logistic relationship is quadratic. Comparison of the under-smooth with the linear fit gives the presmoothed residuals (right) which contraindicate the linear model.

at left. The presmoothed residuals stay well within the usual normal bounds. In the lower plots the true model was quadratic. The under-smooth of the data disagrees with the linear fit and this disagreement is measured by the smooth residuals in the plot at right. The failure of the linear model to fit the (quadratic) data is clearly revealed.

How should h be chosen? There is no satisfactory answer—smaller values of h will reveal more local departures of the data from the parametric model, but both the variability and nonnormality of the residuals will be greater. The value $h = 0.2$ chosen leads to effective sample sizes $\tilde{n}(x)$ around 5, except at the endpoints. For a larger value of h the missing quadratic term in the lower plots is more clearly revealed in this particular example. Several values of h should be tried, corresponding to high, medium, and low frequency departures.

It is worth pointing out that the presmoothed residuals in Figure 5.7 already appear smooth, largely because there is only one covariate in the model. The presmoothed residuals are harder to implement with several predictors requiring multidimensional smoothing, but in this case plots of smooth residuals against one of the covariates look more like the random clouds we associate with residuals in normal linear models. This is because the variations of the unplotted covariates add "noise" to the underlying trend.

5.4.3 Smooth Test of Constant Scale

It is common to plot standardized residuals against fitted values to see if the scale of the errors in the data are correctly modeled by the error distribution. In Section 4.4 we saw how to model uniform over-dispersion of the data but the departure from the correct scale may not be uniform. Detecting a trend in mean for binomial data plots is hard enough. Detecting a trend in dispersion requires nothing short of clairvoyance.

If there is no mean trend in the residuals then the squared or absolute standardized residuals r_i^2 measure the dispersion and so a plot of these against fitted values reveals any trend in $\mathrm{E}(R^2 \mid \hat{Y})$. Smoothing this plot with confidence bands allows us to informally test various hypotheses about dispersion such as (i) is the error distribution correct?, i.e., is $\mathrm{E}(R^2 \mid \hat{Y})$ constantly equal to 1 and (ii) is there trend in the dispersion?, i.e., is $\mathrm{E}(R^2 \mid \hat{Y})$ constant or does it have a trend? To test the first hypothesis, we accept if the constant line at 1.0 fits within the confidence bands. To test (ii) we accept if any constant line fits within the confidence bands.

There are several complications. First, to compute standard errors of the smooth we need to know the variance of the squared residuals. If they are normal then this variance is 2 but it is more correct to use cumulants of the binomial distribution. Second, in comparing the whole curve against a constant line we must take some account of multiple comparisons as mentioned earlier. Thus, for an approximate 95% test the confidence bands should be larger than pointwise 95% bands and the degrees of freedom of the smooth might be used to make this adjustment. Third, since the sum of squares of the scaled residuals is approximately equal to Pearson chi-square whose expectation is $k - p$, it will be necessary to divide by $1 - p/k$ if scaled residuals are used to test (i). This is typically a small adjustment and is not necessary if standardized residuals (which are scaled residuals divided by $\sqrt{1 - H_{ii}}$) are used. It is better to use standardized residuals as these have variance closer to unity under the true model.

Example 5.5: Testing Constant Scale for Seed Data Refer to the seed germination data and the model fitted in Section 5.2.4. In the left plot of Figure 5.8 the squares of the Pearson residuals from this fit are plotted against fitted values for all 288 data points. It is difficult to see if there is any trend in this plot although there seems to be none. Note the erratic behavior of the residuals at the endpoints. In the right-hand plot a kernel smooth with $h = 3$ is plotted with as many of the residuals as the choice of scale on the y-axis allows. Since extreme expected values make residuals unreliable, residuals with expected values larger than 23 or smaller than 2 have been excluded leaving 160 of 288. It would be better, but more time consuming, to accumulate the cells so that their expected values are not too extreme. The effective number of parameters used in this smooth is 3.66 and the variance of the squared residuals is taken to be 2 (assuming normality) times 1.56 (which is the mean square of the remaining 160 residuals). Also displayed are 95% confidence bands accounting both for the multiple comparison [using $\Phi^{-1}(1-0.025/3.66) = 2.466$ instead of 1.96] and the inflated variance. The average fit of the data does seem to be slightly poorer in the center of the range. However, the evidence against uniform dispersion of the data is not particularly strong.

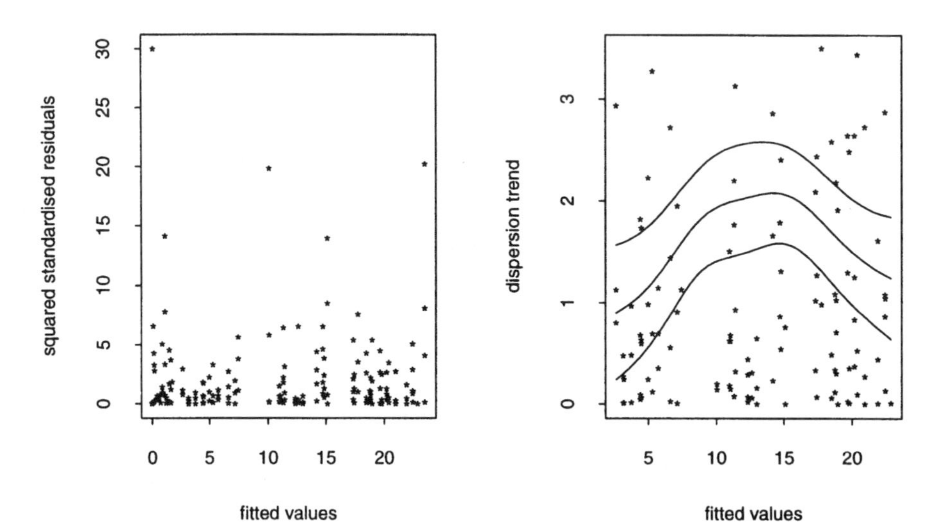

Figure 5.8. Graphical test of dispersion trend for seed germination data. Squared Pearson residuals show no clear trend in the left plot. In the right plot, a kernel smooth, applied only to data points with nonextreme fitted values, provides some evidence of dispersion trend.

5.5 GOODNESS-OF-FIT AGAINST A SMOOTH

The idea of replacing the data with an under-smooth leads to an alternative family of goodness-of-fit tests for binary or almost binary data. Consider a parametric model which specifies a relationship $\pi(x; \beta)$ for the probability of response as a function of the covariate x and parameters β. For instance, we might assume that the logit probability is quadratic in x. An alternative model that is smooth in x but with many more degrees of freedom (an under-smooth) might reveal interesting departures from the quadratic whose veracity we would like to test.

The deviance of the parametric model is not suitable for testing whether the parametrically modeled probability relationship is correct for several reasons. First, the distribution is not close to χ^2_{k-p} if the expected values are too small. Second, the data might be over-dispersed in which case a large deviance could be caused by the over-dispersion or the mean not being quadratic or both. Over-dispersion will also be difficult to detect for close to binary data.

Let $\mathcal{D}_0$ be the deviance of the parametric model and $\mathcal{D}_1$ the deviance of the under-smoothed model. The breakdown of the chi-square approximation of the deviance for sparse data is much less when the difference in the deviances of two nested models are compared. Thus, we might hope that the deviance difference $\mathcal{D}_0 - \mathcal{D}_1$ is closer to chi-square than the absolute deviances. Note however that $\mathcal{D}_0 - \mathcal{D}_1$ may actually be negative. This is because the parametric model is not a special case of the smooth model although this can be arranged by using a different kind of smoothing called *local likelihood*.

More importantly, $\mathcal{D}_0$ and $\mathcal{D}_1$ may both be inflated by over-dispersion which, by analogy with the classical F-statistic, suggests dividing by the deviance of the fuller model (the under-smooth) and using the test statistic

$$F = \frac{(\mathcal{D}_0 - \mathcal{D}_1)/(p_1 - p_0)}{\mathcal{D}_1/(k - p_1)}.$$

This statistic will not be very close to F in distribution, especially when the data are sparse binomial and it is probably necessary to simulate the distribution under the binomial model with fitted probabilities used as actual probabilities. While in principle one should simulate from the over-dispersed binomial distribution, the sensitivity of the distribution of F to over-dispersion is hopefully not very great since it is a *relative* change in deviance [see Glosup, Firth and Hinkley (1991) for further discussion and details.]

The choice of smoothing parameter is, as usual, problematic and will have two distinct effects here. If a small window width h is used then the test will be sensitive to very local departures from the parametric model, whereas if h is larger then the test will only pick up more global departures. In addition, the power of the test will break down as h decreases; in the extreme case as $h \rightarrow 0$ the above F-statistic will be equivalent to using the deviance of the parametric model which has very poor power for sparse data.

Example 5.5: Assessing Linearity for Remission Data For the cancer remission data there appears to be an increasing tendency in the probability of remission with the covariate. We earlier fitted a linear logistic model. A plot of the shape of the data is obtained by using the kernel smoother with $h = 2.77$ which corresponds to exactly five fitted parameters. The two model curves are displayed in Figure 5.9. The question is whether the interesting features of the nonparametric curve are to be believed or whether the simple linear logistic curve is a better sumary of the data.

Treating the data as binary and using Eq. (4.6) for the deviance of a model for binary data we obtain $\mathcal{D}_0 = 27.84$ on $27 - 2 = 25$ degrees of freedom and $\mathcal{D}_1 = 24.37$ and $27 - 5 = 22$ degrees. Even though these deviances are close to their degrees of freedom, recall that absolute deviances are not interpretable for binary data—the null means of the statistics are not their degrees of freedom. We measure the difference in deviances by

$$F = \frac{(27.84 - 24.37)/(25 - 22)}{24.37/22} = 1.044.$$

There is no real justification for using the $F_{3,22}$ distribution here which gives the P-value 0.585. A simulation using the linear logistic fitted probabilities as the true values produced 63 values as large as 1.044 out of 100 and so the approximate P-value is 0.63. In any case, there is no evidence against the linear logistic model in the direction of the nonparametric smooth and the features in this smooth are to be ascribed to random variation.

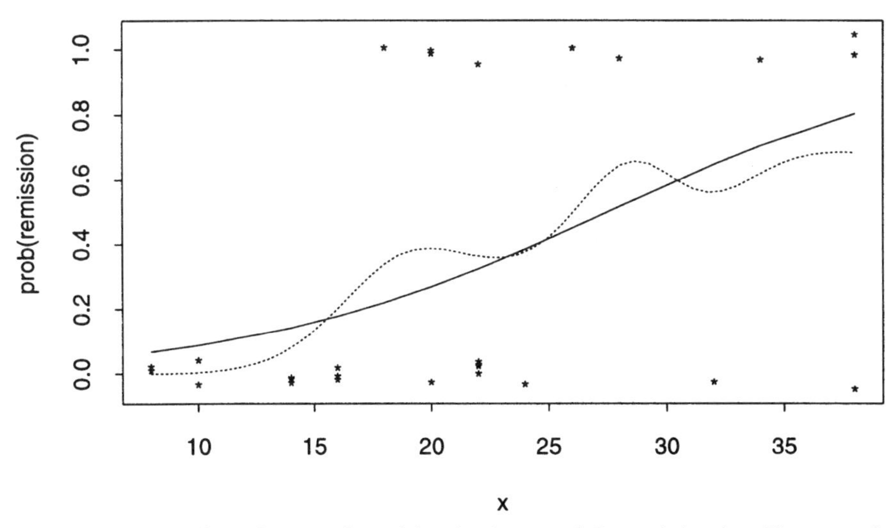

Figure 5.9. Comparison of parametric model and under-smooth for remission data. Three smooth estimates of the probability of remission as a function of x: (i) linear logistic (solid curve) and (ii) kernel smooth with bias correction (dotted curve).

FURTHER READING

Section 5.1: Smoothers in General

5.1.2. *For linear models, weighting by inverse variance is a good idea. For smoothing however, there is disagreement in the literature on whether such weighting is advantageous; compare, for instance, Silverman (1985) and Jones (1993).*

It can be shown that local polynomials of higher degree p *give asymptotically smaller MSE but for moderate sample sizes quadratics or cubics do not show any practical superiority over local lines. It can also be shown that for odd-degree polynomials the bias in the extreme of the* x*-space is the same as in the middle, whereas for even degrees it is an order of* h *larger. Thus, local lines would appear to best combine simplicity with good performance [see Wand and Jones (1995, Section 5.5)].*

5.1.5 *Cross-validation is an idea with wider application than choice of window width. Seminal references are Stone (1974) and Allen (1974). There are other methods based on more direct estimates of the average squared residual, mean prediction error, generalized Mallows* C_p*, and a likelihood based method called Akaike's criterion. These are described in Hastie and Tibshirani (1990, Section 3.4) who conclude that there is little to choose between the various methods.*

5.6 EXERCISES

5.1. Refer to the section on linear smoother.

a. Suppose we have data Y_i all equal to c. Show that a necessary and sufficient condition for the smoothing vector $H(x)$ to map these data to the constant function is that $\sum_{i=1}^{k} H_i(x) = 1$.

b. Suppose that H is symmetric and idempotent, i.e., $H(Hy) = Hy$ for all vectors y. The hat matrix of a parametric linear model is idempotent. Show that the row sums of H are equal to 1.

c. For the locally weighted mean smoother, find the limiting values of the smoothing matrix H as $h \to \infty$ and as $h \to 0$.

d. Can the entries of H for a local linear smoother ever be negative?

5.2. Let $\hat{Y} = HY$ be the vector of fitted values from a smoother H applied to data Y. Show that

a. The total of the fitted values and data values *necessarily* agree if, and only if, the column totals of H equal 1.

b. How does the property in (a) relate to the total of the residuals equaling zero?

c. Is smoothing useful for estimating the area under a curve?

5.3. Use the supplied *Splus* function `H.mean` and `H.line` to investigate the effective degrees of freedom measures $tr(H)$, $tr(HH^T)$, and $tr(2H - HH^T)$. In particular

a. For $x = (2,4,6,\ldots,20)$ plot the effective degrees of freedom as h ranges from 0 to ∞.

b. Repeat (a) for $x = (2,3,4,5,6,7,8,12,16,20)$.

c. For the x-values in (b), find the value h_m such that the degrees of freedom of the local mean equals 3 and the value h_l such that the degrees of freedom of the local linear smoother equals 3. These both do a similar amount of smoothing as fitting a quadratic. Generate some y data and compare these two fits, as well as the parametric quadratic fit.

5.4. The drinking and driving data are supplied as an *Splus* data set `bac.dat`.

a. Suppose that in age-group j, $\text{logit}\{\pi_j(x)\} = \alpha_j + f_j(x)$. Estimate the smooth curves $f_j(x)$ using 4 degrees of freedom.

b. Suppose that in age-group j, $\text{logit}\{\pi_j(x)\} = \alpha_j + f(x)$ i.e. $f_j(x)$ does not depend on j. Estimate the smooth curve $f(x)$ using 4 degrees of freedom.

c. Use the `gam.object` to test whether nonlinear terms are necessary in any of these fits. Also, test the model in (a) against the null model in (b).

d. The test in (b) suggests that the risk functions for the four age-groups are different. A different conclusion is reached when we fit linear logistic curves. Why do you think this is?

5.5. The data in the *Splus* object `kyphosis.dat` relates to 83 patients who underwent corrective spinal surgery. Kyphosis is defined by forward flexion of the spine exceeding 40 degrees from vertical, and presence or absence of this unfortunate condition is the response. The covariates are location of surgery on the spine and age of patients. The location is defined by two numbers, the starting number of the vertebrae, and the number of vertebrae affected.

a. Fit an additive model in age, start, and number using 3 degrees of freedom for each. Which of the effects are nonlinear at level 10%?

b. Fit the additive model that is parametric in number, but nonparametric in the other terms. Confirm the result in (a) using change in deviance.

c. Assess the significance of each effect by removing it from the additive model and recording the change in deviance.

d. In fact vertebrae 12 is special, numbers greater than 12 corresponding to *lumbar* vertebrae. Replace the term in `start` by a function `(start-12)I(start-12)`. This results in a constant for the nonlumbar group joined continuously to a linear fit for the lumbar group. Compare this with the earlier fit.

CHAPTER 6

Poisson Regression Models

In many instances, it is natural to assume that a set of counts are independent Poisson variables. For example, when large public records covering a fixed period of time are examined and records classified into various types—hospital records classified by gender, disease, prognosis, treatment, etc.—then the counts may be thought of as Poisson. On the other hand, if one is interested in describing the *pattern* of these counts, their associations, and relative sizes rather than their magnitude, then it might make more sense to consider the total number of records as fixed. In this case the distribution of the counts among the fixed total is multinomial.

In Chapter 2 a close connection between multinomial and independent Poisson models for a set of counts was established. A practical consequence of this duality is that models for binomial or multinomial data may always be recast as Poisson models and an illustration has been given in Section 3.3.3. However, Poisson models have advantages over and above their use as an alternative method of fitting multinomial models. Poisson models are more general than multinomial models—many Poisson models are not equivalent to a multinomial model—and these more general models have several useful features in their own right. Most important of these is that Poisson models, in particular multiplicative models for the means of the counts, are the most natural and flexible way to investigate the various relationships of marginal or conditional dependence that may exist between a set of factors.

Another even more fundamental advantage is that for a Poisson model no single factor is necessarily of particular importance; we do not seek solely to explain one factor in terms of the other factors. Rather than asking "conditional on all the other factors, which is the best description of the response factor?", Poisson models can address the question "which factors affect which and how?"

6.1 LOG-LINEAR MODELS

Let the data comprise a set of counts $Y^T = (Y_1, \ldots, Y_k)$, each component independently Poisson distributed whose respective means are $\mu_1, \ldots, \mu_k$. A log-linear model is defined by a set of relations

$$\eta_i(\beta) = \log(\mu_i) = x_{i1}\beta_1 + x_{i2}\beta_2 + \cdots + x_{ip}\beta_p; \qquad i = 1, \ldots, k \tag{6.1}$$

where $\beta_1, \ldots, \beta_p$ are unknown parameters and for each count Y_i the covariate values $x_i^T = (x_{i1}, \ldots, x_{ip})$ have been measured. This is the systematic component of the model, the random component being the assumed Poisson distribution. The $(k \times p)$ matrix X, with row vectors x_i^T, is called the design matrix.

Of course, an equally valid description of such models is that they are *multiplicative* models for the means, μ_i. Expressing the model on the log-scale reveals it to be a member of the class of generalized linear models (see Section 4.6) and brings to bear the theory and algorithms of this unified class of statistical procedures. Other link functions besides logarithm can be used, however only log-linear models correspond to multiplicative models for the mean. As mentioned in the preamble, such models are perfectly suited to the study and modeling of various forms of independence.

6.1.1 Likelihood Theory

For a single Poisson variable Y_i the contribution to the log-likelihood is

$$\log(e^{-\mu_i}\mu_i^{y_i}/y_i!) = c + y_i \log(\mu_i) - \mu_i,$$

[see also Eq. (2.11)]. The log-likelihood for the parameter vector β from the full data $Y_1, \ldots, Y_k$ obeying a log-linear Poisson model is

$$\begin{aligned} l(\beta) &= \sum_{i=1}^{k} y_i \log\ \mu_i(\beta) - \mu_i(\beta) \\ &= \sum_{i=1}^{k} y_i \sum_{j=1}^{p} x_{ij}\beta_j - \mu_i(\beta) \\ &= \sum_{j=1}^{p} \beta_j T_j - \mu_i(\beta) \end{aligned} \tag{6.2}$$

where $T_j = \sum_{i=1}^{k} x_{ij}Y_i$ is sufficient for the parameter β_j. Maximum likelihood estimates, their standard errors, and fitted values will all be functions of the data only through the p statistics $T_1, \ldots, T_p$.

General likelihood forms for Poisson models were given in Sections 2.3.1 and 2.5.1. The fitted values are simply the estimated means of each data point, i.e., $\hat{e}_i = \mu_i(\hat{\beta})$. The deviance Eq. (2.22), which is the likelihood ratio statistic for testing the model against the full model is

$$LR_P = 2\sum_{i=1}^{k} y_i \log\left(\frac{y_i}{\hat{e}_i}\right) - 2\sum_{i=1}^{k}(y_i - \hat{e}_i).$$

The second term will often vanish. Comparison of the deviance of two nested models is the basis for testing one model against a larger model. The Pearson goodness-of-fit statistic is

$$S = \sum_{i=1}^{k} \frac{(Y_i - \hat{e}_i)^2}{\hat{e}_i}$$

and could be used in a similar manner. Both deviance and Pearson goodness-of-fit statistics have approximate χ^2 distributions with $k - p$ degrees of freedom, provided that the fitted values $\hat{e}_i$ do not become too small or the number of parameters p in the model too large.

Example 6.1: Two-Way Classification Imagine a group of students entering the math department of a given university in a given year. Two variables are measured on these students. Factor C categorizes the student by course experience, specifically the mathematics level attained in high school. This might be rudimentary ($C = 1$), completion of year 11 ($C = 2$), or completion of year 12 ($C = 3$). Factor R categorizes students by their result when they finish, which might simply be the highest level year passed. Note that this is actually an ordinal variable but will be treated as categorical here.

The data could be presented as a two-dimensional contingency table with the say five levels of R defining rows and the three levels of C defining columns. Let Y_{ij} be the number of students who finish the course with $R = i$ after having entered with $C = j$, i.e., the count in cell (i,j) of the data table. Let π_{ij} be the probability an individual is one of these Y_{ij}, and T be the total number of students. The probability that a student has result $R = i$ is

$$\pi_{\bullet} = \pi_{i1} + \pi_{i2} + \pi_{i3}$$

and similarly $\pi_{\bullet j} = \pi_{1j} + \pi_{2j} + \pi_{3j} + \pi_{4j} + \pi_{5j}$ is the probability that a student enters with course experience $C = j$. When T is considered fixed the distribution of Y_{ij} is binomial with parameters T and π_{ij}. Consider the model that says that result R and course experience C are independent, which in mathematics specifies

$$E(Y_{ij}|T) = T\pi_{ij} = T\pi_{i\bullet}\pi_{\bullet j}. \tag{6.3}$$

Taking expectation with respect to T the left-hand side becomes unconditional expectation and T is replaced by $E(T) = \tau$. Taking logarithms gives

$$\eta_{ij} = \log(\mu_{ij}) = m + r_i + c_j \tag{6.4}$$

where $\mu_{ij} = \mathrm{E}(Y_{ij})$, $m = \log(\tau)$, $r_i = \log(\pi_{i\bullet})$, and $c_j = \log(\pi_{\bullet j})$. The additive model for the logarithm η_{ij} of the means corresponds to independence of the row factor R and column factor C. The "row effect" parameters r_i actually relate to logarithms of the probability of an individual being in row i. Similarly, the "column effect" parameters c_j are really logarithms of probabilities of being in column j.

The distribution of $Y_1, \ldots, Y_k$ will be independent Poisson if, and only if, the distribution of T is Poisson, with mean τ. However, inference on the parameters π_{ij}, and therefore on r_i and c_j, are not affected by this choice, so long as τ is a totally free parameter, unrelated to the π_{ij}. While it might not be plausible to assume that the total number of students entering a department in a given year is Poisson distributed (the number tends to be largely determined by quota), the actual distribution is irrelevant to inference on the probability parameters π_{ij} and the independence, or lack of independence, of the factors R and C.

The above example is about the simplest log-linear model encountered, both in conception and computation. The fact that r_i and c_j appear additively in the predictor η_{ij} immediately implies independence. This is a general feature—additive terms in a log-linear model formula imply independence of the corresponding factors. We will see later in Section 6.3 that log-linear model formulas are very easily built up and interpreted in terms of various kinds of independence.

6.1.2 Example 6.2: Marital Status of Danes

The data shown in Table 6.1 relate to a survey of 165 Danes who were classified by their age (categorized into one of eight groups) and their marital status. An individual is given marriage status `divorced` if he or she has been divorced at any time previously, regardless of whether or not he or she is now remarried. The legal age of marriage in Denmark at the time was 16. In addition,

Table 6.1. Marital Status Data[a]

Age	x_i	Single	Married	Divorced	Total
17–21	19	17 (94)	1 (6)	0 (0)	18
21–25	23	16 (67)	8 (33)	0 (0)	24
25–30	27.5	8 (31)	17 (65)	1 (4)	26
30–40	35	6 (19)	22 (69)	4 (12)	32
40–50	45	5 (16)	21 (66)	6 (19)	32
50–60	55	3 (11)	17 (61)	8 (29)	28
60–70	65	2 (12)	8 (50)	6 (38)	16
70+	75	1 (11)	3 (33)	5 (55)	9
All ages		58	97	30	185

[a]Survey of Danes classified by age-group and marital status; row percentages given in parentheses.

row percentages have been given in parentheses. For instance, among those 32 individuals surveyed in their forties, 16% were unmarried, 66% married, and 19% divorced.

The main interest is presumably in describing how the distribution of marital status changes with age, i.e., in the proportions of people in the three marital status categories in different age groups. This is why the row percentages have been given in parentheses. We observed that the proportion of unmarried individuals decreases with age and that the proportion of married individuals steadily increases up to the thirties age-group and then begins to decline as the proportion divorced begins to increase. Plots are given in Figure 6.1, together with various modeled curves to be described later. The row totals estimate the age distribution of Danes and the column totals the overall marriage distribution ignoring age. The actual size of the counts depends on the *sampling effort* and would seem to be of no interest.

Let us suppose that the sample was obtained by surveying individuals over a fixed period of time, perhaps at a shopping center. The counts in each age-group would then plausibly follow a Poisson distribution as would the counts in the body of the table. Let Y_{ij} be the number of individuals in age-group i with marital status j. The variable Y_{ij} is random because if we took another survey under similar conditions then different counts would be obtained. Assume that the Y_{ij} are independent Poisson variables with mean μ_{ij}. The independence model

$$\log(\mu_{ij}) = m + a_i + s_j \qquad i = 1, \ldots, 8 \qquad j = 1, 2, 3$$

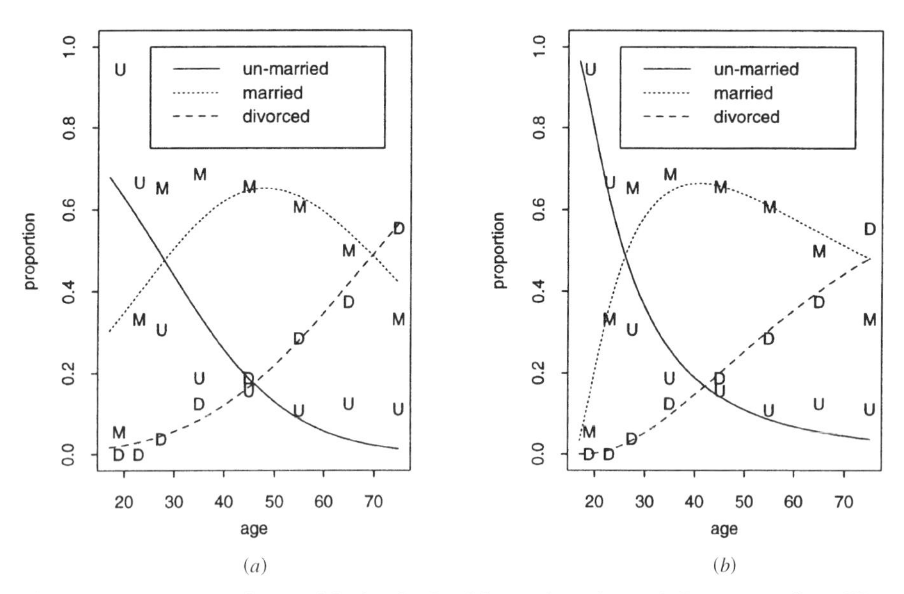

Figure 6.1. Two regression models for the Danish marriage data. (*a*) Data proportions (U = unmarried, M = married, D = divorced) and fitted model curves for simple linear model in age and (*b*) data proportions [as in (*a*)] and fitted model curves for linear model in log (age—16).

imposes the assumption that marital status and age-group are independent factors (see Example 6.1). This is a highly implausible hypothesis and the deviance of the model is 82.85 on $(8-1)(3-1) = 14$ degrees of freedom. In general, the degrees of freedom for an independence model of a row factor on r levels and a column factor on c levels is $(r-1)(c-1)$. Once the independence hypothesis is rejected, how can the dependence of marital status on age be described?

In Table 6.1, we have defined age covariate x_i as the center of the age-group. If we believe that the marital status distribution changes smoothly with age then age, or some function of it, should be used as a covariate rather than as a factor. Certainly, the plotted proportions in each column do vary quite smoothly with age and so some sort of regression on age x_i should provide a simple description of the data. Consider the following linear model

$$\log(\mu_{ij}) = m + a_i + s_j + \beta_j x_i \qquad i = 1, \ldots, 8 \qquad j = 1, 2, 3 \tag{6.5}$$

which is an extension of the independence model to allow the means μ_{ij} to depend regularly on the age x_i. Different slope parameters β_j are required for each column/marital status since we see in Figure 6.1 that the relationship of the counts with age are quite different for the three marital status groups.

While we are fitting the model using Poisson error, we are mainly interested in marital status and it helps to think of *status* as being a trinomial response, i.e., each individual may be in one of three possible marital status categories. The parameters s_j represent baseline means for these three marital status categories, on the log-scale. The parameters a_i represent overall log-mean counts in the eight age groups. The parameters β_j measure how fast the log-means of the counts for marital status j change with age x_i, over and above the demographic effects represented by the a_i.

Within each row, and conditional on the total of that row/age-group, the distribution of (Y_{i1}, Y_{i2}, Y_{i3}) is trinomial. Let $\pi_{j|i}$ be the probability that an individual in age-group i is in marital status j, describing this trinomial distribution. Recalling the relationship between Poisson means and multinomial probabilities we have

$$\pi_{j|i} = \frac{\mu_{ij}}{\mu_{i1} + \mu_{i2} + \mu_{i3}} = \frac{S_j e^{\beta_j x_i}}{S_1 e^{\beta_1 x_i} + S_2 e^{\beta_2 x_i} + S_3 e^{\beta_3 x_i}}$$

where $S_j = \exp(s_j)$. Notice that the parameters a_i are irrelevant to the multinomial model—these model the size of the counts in each row rather than the proportions within each row. More generally with $\pi_j(x)$ being the probability of being in marital status j as a function of age x

$$\pi_j(x) = \frac{S_j e^{\beta_j x}}{S_1 e^{\beta_1 x} + S_2 e^{\beta_2 x} + S_3 e^{\beta_3 x}}. \tag{6.6}$$

The model Eq. (6.5) uses this family of curves to describe the trinomial *proportions* given in the Table 6.1 and plotted as points in Figure 6.1.

In the **GLIM** analysis below I have defined a factor **STAT** for marital status on three levels, a factor **FAGE** for age-group on eight levels, and the covariate **AGE** containing the numerical age measurements x_i. The model Eq. (6.5) translates to

```
FAGE + STAT + STAT.AGE.
```

The deviance of the model is 24.324 on 12 degres of freedom, which is not a good fit. The estimated parameters are $\hat{s}_1 = 4.889$, $\hat{s}_2 = 2.846$, $\hat{s}_3 = -0.9994$, and $\hat{\beta}_1 = 0.1283$, $\hat{\beta}_2 = 0.0552$. The parameter β_3 is aliased to zero, simply because the three probabilities $\pi_1(x)$, $\pi_2(x)$, $\pi_3(x)$ must always sum to one.

These model curves are plotted together with the raw data proportions in the left part of Figure 6.1. The plot reveals why the model fits so poorly—at the younger age levels the fitted probabilities do not agree with the data. In fact at age $x = 16$, we must have $\pi(x) = (1, 0, 0)$ since nobody may marry before age 16. The model we have fitted never reaches this extreme distribution, even at age zero. How can we design a model that correctly predicts that all individuals are unmarried at age 16?

If we replace x by $\log(x - 16)$ in Eq. (6.6) then we obtain

$$\pi_j(x) = \frac{S_j(x - 16)^{\beta_j}}{S_1(x - 16)^{\beta_1} + S_2(x - 16)^{\beta_2} + S_3(x - 16)^{\beta_3}}, \tag{6.7}$$

and provided $\beta_1 > 0$, $\beta_2 \leq 0$, $\beta_3 \leq 0$ the equation implies that $\pi(16) = (1, 0, 0)$. Fitting this model (see **GLIM** output below), we obtain the deviance 7.825 on 12 degrees of freedom—a very good fit. Using the listed parameter estimates, the fitted model curves are displayed at right in Figure 6.1 and fit the observed proportions much better than do the fitted model curves of the previous model. The curves do not fit particularly well for the highest age-group; however, the counts here are small and the point estimates consequently unreliable.

```
[i]   ? $look y stat age fage$

[o]            Y      STAT      AGE     FAGE
[o] 1     17.000    1.000    19.00    1.000
[o] 2      1.000    2.000    19.00    1.000
[o] 3      0.000    3.000    19.00    1.000
[o] 4     16.000    1.000    23.00    2.000
[o] 5      8.000    2.000    23.00    2.000
[o] 6      0.000    3.000    23.00    2.000
[o] 7      8.000    1.000    27.50    3.000

  .
  .
```

```
! rest of data list omitted
[o]
[i]  ? $fact state 3 fage 8$
[i]  ? $yvar y$error p$
[i]  ? $fit stat+fage$                                    independence model
[o]  scaled deviance = 82.85 at cycle 5
[o]             d.f. = 14
[o]  ! First multinomial regression model of status on age
[i]  ? $fit stat+fage+stat.age-%gm$
[o]  scaled deviance = 24.324 at cycle 4
[o]             d.f. = 12
[o]
[i]  ? $disp e$
[o]     estimate      s.e.    parameter
[o] 1      4.889    0.5111    STAT(1)                                           s1
[o] 2      2.846    0.4400    STAT(2)                                           s2
[o] 3    -0.9994    0.5818    STAT(3)                                           s3
[o] 4     0.6809    0.3212    FAGE(2)
[o] 5      1.174    0.3454    FAGE(3)
[o] 6      1.995    0.4074    FAGE(4)
[o] 7      2.670    0.5131    FAGE(5)
[o] 8      3.063    0.6112    FAGE(6)
[o] 9      2.905    0.6984    FAGE(7)
[o] 10     2.627    0.7640    FAGE(8)
[o] 11   -0.1283   0.02067    STAT(1).AGE                                       b1
[o] 12  -0.05521   0.01514    STAT(2).AGE                                       b2
[o] 13     0.000   aliased    STAT(3).AGE                                       b3
[o]
[i]  ! The model still fits badly, particularly at low ages
[i]  ? $calc lage=%log(age-16)$
[i]  ? $fit stat+fage+stat.lage-%gm$
[o]  scaled deviance = 7.8249 at cycle 4
[o]             d.f. = 12
[o]
[i]  ? $disp e$
[o]     estimate      s.e.    parameter
[o] 1      6.166    0.6679    STAT(1)                                           s1
[o] 2      2.816    0.7286    STAT(2)                                           s2
[o] 3     -4.017     1.263    STAT(3)                                           s3
[o] 4      2.623    0.5638    FAGE(2)
[o] 5      3.901    0.7945    FAGE(3)
[o] 6      5.150     1.027    FAGE(4)
[o] 7      5.877     1.208    FAGE(5)
[o] 8      6.163     1.311    FAGE(6)
[o] 9      5.875     1.377    FAGE(7)
[o] 10     5.485     1.425    FAGE(8)
[o] 11    -3.127    0.5451    STAT(1).LAGE                                      b1
[o] 12    -1.677    0.4957    STAT(2).LAGE                                      b2
[o] 13     0.000   aliased    STAT(3).LAGE                                      b3
```

6.1.3 Residuals

Raw residuals for a Poisson model are simply the differences $y_i - \hat{e}_i$ between the observed and fitted values. However, since modeling is being done on the log-scale it is perhaps more useful to look at the differences

$$\log(y_i) - \log(\hat{e}_i) \approx \frac{y_i - \hat{e}_i}{\hat{e}_i} \tag{6.8}$$

where the right-hand side is a first-order Taylor expansion and will still be meaningful when $y_i = 0$. Notice that the $\hat{e}_i$ in the denominator does not have a square root. Plotting these raw residuals against predictors may reveal errors in the systematic part of the model and suggest terms to be added to the log-linear predictor.

Scaled residuals may be obtained as contributions to either the deviance or Pearson goodness-of-fit statistics. The Pearson measure induces residuals

$$r_{Pi} = \frac{y_i - \hat{e}_i}{\sqrt{\hat{e}_i}} \tag{6.9}$$

while the deviance measure induces the residuals

$$r_{Di} = \text{sign}(y_i - \hat{e}_i)\sqrt{2y_i \log(y_i/\hat{e}_i) - 2(y_i - \hat{e}_i)}. \tag{6.10}$$

Plotting these residuals against fitted values or ranked fitted values may reveal unusual points or outliers. However, when expected values become small, residuals become unstable and cannot be trusted. It will be pointed out in this chapter that binomial models may be fitted as log-linear Poisson models. In the binomial approach the squared scaled residual is

$$r_{Bi}^2 = \frac{(y_i - n_i\hat{\pi}_i)^2}{n_i\hat{\pi}_i(1 - \hat{\pi}_i)} = (y_i - \hat{e}_i)^2\left(\frac{1}{\hat{e}_i} + \frac{1}{n_i - \hat{e}_i}\right)$$

which is the sum of the squared Poisson residuals for the observed response y_i and observed nonresponse $n_i - y_i$.

Example 6.2: (Continued) **The Danish Marriage Data.** The first model fitted used the covariate `AGE` and gave large deviance 24.32(12). The left part of Figure 6.2 plots logit residuals against age with symbols 1, 2, 3 indicating the three marital status categories. There is a clear trend in these residuals that needs to be accounted for in the model. The center model used the covariate $\log(age - 16)$ giving deviance 7.825(12). While the model fits well overall, the

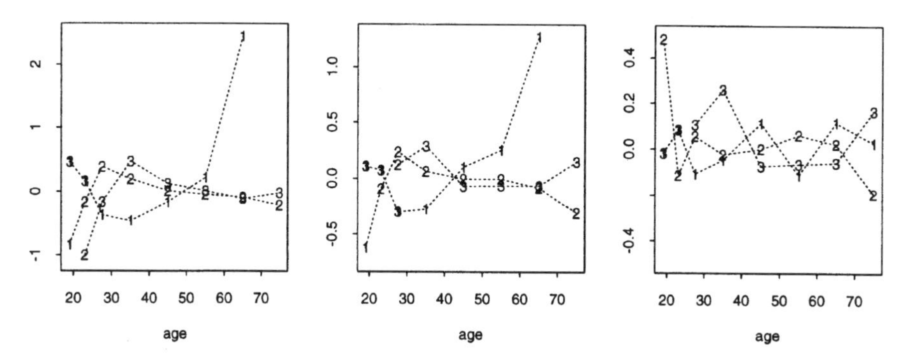

Figure 6.2. Logit residuals plotted against age. Left: Model in age. Center: Model in 1 = log(*age*−16). Right: Model with quadratic term in 1.

center plot indicates that a pattern is clearly present. The last model we try also includes a quadratic term in log(*age* − 16) giving the very small deviance 1.708(10) which is a significant reduction. Moreover, the trend in the residuals disappears.

6.1.4 Over-Dispersion

It will sometimes happen that no model will appear to fit a set of Poisson data. This can be recognized by the deviance or Pearson chi-square statistic being larger than the degrees of freedom. One will need to check that the fitted values are not too small and if they are then data points will need to be combined and goodness-of-fit statistics recalculated. One will also need to check that the systematic component of the model is not at fault by examining residual plots and adding sufficient terms to the linear predictor. Only then does the inflated deviance constitute evidence of *over-dispersion*.

If the systematic component of the model is not in error then the source of over-dispersion must be the error distribution, i.e., the assumption of a Poisson distribution itself. Over-dispersion occurs when the counts vary more widely about their mean than would be expected by Poisson random variables. The Poisson distribution with mean μ also has variance μ—this is a characteristic of Poisson variation.

The simplest way to explain the extra variation is to suppose that the mean μ of each count Y is not fixed but is itself a random variable with mean m. The Poisson distribution for Y now only holds conditional on the value of μ. Unconditionally the distribution is not Poisson. Using standard relations we find that

$$\begin{aligned} \mathrm{E}(Y) &= \mathrm{E}\{\mathrm{E}(Y|\mu)\} = E(\mu) = m \\ \mathrm{Var}(Y) &= \mathrm{E}\{\mathrm{Var}(Y|\mu)\} + \mathrm{Var}\{\mathrm{E}(Y|\mu)\} = m + \mathrm{Var}(\mu). \end{aligned}$$

The variance of Y is thus greater than its mean unless $\mathrm{Var}(\mu) = 0$ in which case Y has the Poisson distribution. Now μ is a positive random variable and a common model for positive random variables is the gamma distribution. A characteristic of the gamma(r, α) distribution is that for fixed rate parameter α, its variance $v = r/\alpha^2$ is proportional to its mean $m = r/\alpha$, i.e., $v = m/\alpha$. Substituting this into the expression for $\mathrm{Var}(Y)$ above we have

$$\mathrm{Var}(Y) = \mathrm{E}(Y)\phi^* \tag{6.11}$$

where $\phi^* = 1 + \alpha^{-1}$. The distribution of Y is actually *negative binomial.*

The log-linear model could still be fitted with the alternative assumption Eq. (6.11) on the error distribution. This is done simply by using weights equal to $1/\phi^*$. The deviance of all models will be a factor ϕ^* smaller and standard errors of parameter estimates will be a factor $\sqrt{\phi^*}$ larger. Since it is easy to show that the Pearson chi-square statistic, S, has mean $(k - p)\phi^*$ a sensible estimator of ϕ^* is

$$\hat{\phi}^* = \frac{S}{k - p}. \tag{6.12}$$

The Pearson statistic S employed for this should be from as full a model as necessary to ensure that the systematic component is correct. Call this model the *fullish* model. Counts with small expected values should be combined and the degrees of freedom adjusted accordingly. Call this adjusted degrees of freedom d_f.

As explained in Section 4.4.3 change in deviance tests using the chi-square distribution are now based on the F-distribution. Two hypotheses $\mathcal{H}_0$ and $\mathcal{H}_1$ both contained in the fullish model may be tested by recording the deviances of both models and computing

$$F = \frac{(\mathcal{D}_0 - \mathcal{D}_1)/(d_0 - d_1)}{\hat{\phi}^*}.$$

If the over-dispersion weights are being used then the deviances will automatically be a factor $\hat{\phi}^*$ smaller and so the denominator is not explicitly needed—we simply record the change in deviance per change in degrees of freedom. The null distribution of this test statistic is approximately F with degrees of freedom $d_0 - d_1$ and d_f.

Particular parameters in a given model may be tested by comparing estimates to their standard errors *with the estimated weights active*. Explicitly the test statistics are of the form

$$\frac{\hat{\theta}}{\text{s.e.}(\hat{\theta})\sqrt{\hat{\phi}^*}}$$

where s.e.($\hat{\theta}$) is the standard error computed without the over-dispersion weights. These ratios of estimates to standard errors are compared with the t distribution on d_f degrees of freedom.

The methods described may also be used for *under-dispersed* data although this is rather rare in practice. Under-dispersion may occur if the events of the underlying Poisson process generating the counts are negatively correlated. For such data, the deviance of the best fitting models will tend to be less than their degrees of freedom and $\hat{\phi}^*$ will be less than 1. Using the reciprocal as weights will increase deviances and decrease standard errors, making observed effects more statistically significant.

Example 6.2: (Continued) **Danish Marriage Data.** We saw from the earlier residual analysis that a regression on both $\log(x - 16)$ and its square were indicated and that the deviance of this model was 1.708(10). This deviance is unusually low and the probability of a χ^2_{10} variable being so small is 0.0019. The Pearson goodness-of-fit statistic is 1.525 which is even smaller. However, many fitted values are rather small and so these goodness-of-fit statistics cannot be taken at face value. Counts with fitted value smaller than 4 are accumulated. When this is done $S = 0.927$ and the degrees of freedom is reduced to $d_f = 4$. This is still reasonably strong evidence for under-dispersion.

Using the weights $1/\hat{\phi} = 4/0.927 = 4.513$ the output for the quadratic model is listed below. The deviance 7.707 is now a factor 4.513 larger. The estimated linear and quadratic coefficients are the last four listed and are all much larger than their standard errors, however these ratios are to be compared to critical values of the t_4 distribution. I have computed P-values for testing the hypothesis that each of these parameters is zero using the t_4 distribution and these are written opposite the parameter estimates. Removing the quadratic factor LAG2 increases the deviance from 7.7077 to 35.314. This change is $(35.31 - 7.71)/2 = 13.8$ which is much greater than the 95% quantile 6.94 of the $F_{2,4}$ distribution.

```
[i]  ? $calc w=0*y+4.513$weig w$
[i]  ? $fit stat*(lage+lag2)+fage$
[o]  scaled deviance = 7.707 at cycle 5
[o]             d.f. = 10
[o]
[i]  ? $disp e$
[o]     estimate      s.e.    parameter
[o] 1     0.1872     1.697    1
[o] 4      3.339     1.961    LAGE
[o] 5    -0.8318    0.3898    LAG2
!                                FAGE(j) parameters omitted from output
[o] 13     5.693    0.9369    STAT(2).LAGE                       P = .002
```

```
[o] 14     7.477     2.610     STAT(3).LAGE                    P = .023
[o] 15    -0.8024    0.1677    STAT(2).LAG2                    P = .004
[o] 16    -0.7925    0.4021    STAT(3).LAG2                    P = .060
[o]
[i]  ? $fit stat*lage+fage$
[o]  scaled deviance = 35.314 at cycle 4
[o]              d.f. = 12
```

Latent Variable Models An alternative approach to modeling over-dispersion is to assume that an unobserved or *latent* variable L is responsible for the extra variation around the mean. As a simple illustration, suppose that we have data classified by two factors A and B and that the additive independence model fits poorly. Suppose that there is an unmeasured latent variable L "causing" this association by which we mean that, conditional on the variable L, the cross-classification of A and B would be independent. The probabilities associated with the A–B cross-classification conditional on $L = k$ satisfy

$$\pi_{ij|k} = \pi_{i|k}\pi_{j|k}. \tag{6.13}$$

The log-linear model (LA, LB) describes this supposed L-conditional independence of A and B (see Section 6.3.2).

However, if the latent variable L is unmeasured then we cannot fit this model. Suppose that p_k gives the probability that $L = k$. For a survey of individuals, this would represent the proportion of the surveyed population with latent variable $L = k$. By multiplying Eq. (6.13) by p_k and summing, we find that the unconditional probabilities π_{ij} associated with the A–B cross-classification satisfy

$$\pi_{ij} = \sum_k p_k \pi_{i|k}\pi_{j|k} \tag{6.14}$$

and the parameters p_k, $\pi_{i|k}$, $\pi_{j|k}$ are unknown. The number of levels of the latent variable will need to be small if these parameters are to be identifiable. If A, B, and L have levels r, c, d, respectively, then the number of parameters to estimate is $d(r + c - 1)$ which will have to be less than the number of data values rc. For instance, if A and B are on five levels then d can be no more than 2.

In our earlier simple treatment of over-dispersion the mean μ_{ij} was assumed to be a negative binomial random variable with mean m_{ij}. Under the latent variable model the mean μ_{ij} of a cell is actually a random variable taking the value $\mu_{ij|k}$ with probability distribution p_k. This is a less restrictive model and latent variable models are more flexible and robust for this reason, subject to the limitation on the number of levels of the latent variable. More general latent variable models may have several latent variables, each affecting different subsets of the observable factors A, B, C, D, ... However, latent variable models are not expressible as log-linear models and different algorithms are required for their fitting. We will not cover this important topic in this book but refer the reader to the book by Lazarfeld and Henry (1968) and the seminal paper of Goodman (1974). Further references on latent variable models are Clogg and Goodman (1984, 1985), Gilula (1983, 1984), Henry (1983), McCutcheon (1987), and Palmgren and Ekholm (1987).

6.2 MULTINOMIAL LOGIT MODELS

In the example of the previous section, we were interested in a trinomial response `STATUS` and how its distribution could be explained in terms of age. However, we fitted a log-linear Poisson model to achieve this. Logistic models for binomial data readily generalize to multinomial logit models and such models may always be fitted through an appropriate Poisson model. Conversely, Poisson models can often be interpreted as multinomial logit models for one of the factors, or perhaps several factors jointly, in terms of the remaining factors. In this section, we investigate the general relationship between multinomial logit and log-linear Poisson models.

6.2.1 Simple Logistic Regression

Consider the following set of data found in Cornfield (1962). A sample of male residents of Framingham, Massachusetts, between the ages of 40 and 60 had their systolic blood pressure measured and over the six-year study period were classified by the absence or presence of coronary heart disease (Table 6.2). The variable cholesterol has been ignored here. A fuller version of the data are given in Table 4.25 on p. 261.

Table 6.2. Framingham Heart Study[a]

Blood Pressure	x_i	Disease (Y_i)	No Disease	Total (n_i)
<116	115	3 (5.6)	153 (150.4)	156
116–125	121	17 (10.5)	235 (241.5)	252
126–135	131	12 (14.9)	272 (269.1)	284
136–145	141	16 (18.0)	255 (253.0)	271
146–155	151	12 (11.6)	127 (127.4)	139
156–165	161	8 (8.9)	77 (76.1)	85
166–185	176	16 (14.3)	83 (74.7)	99
>185	190	8 (8.3)	35 (34.7)	43

[a]Cross-tabulation of males classified by blood pressure and presence or absence of heart disease with fitted values in parentheses [from Cornfield (1962)].

A straightforward approach to modeling these data would be to let the number of diseased men, Y be the response with probability of response π depending on blood pressure x. The simple logistic linear model assumes

$$\text{logit}(\pi_i) = \alpha + \beta x_i \tag{6.15}$$

and there are eight data points Y_i from eight fixed binomial denominators n_i. Blood pressure scores x_i have been assigned to each blood pressure group although it would be preferable to have the actual blood pressure measurements for each individual.

An alternative approach is to take all 16 counts in the table to be independent Poisson random variables. Let Y_{ij} be the count in cell (i,j) with mean μ_{ij} and consider the log-linear model

$$\begin{aligned}\log(\mu_{i1}) &= m + b_1 x_i + \tau_i \\ \log(\mu_{i2}) &= m + \delta + (b_1 + \beta)x_i + \tau_i.\end{aligned} \tag{6.16}$$

The parameter δ measures the difference in mean counts for diseased and nondiseased men, i.e., a main effect of disease status on the mean count. Now conditional on the number of individuals n_i with given blood pressure in row i, the distribution of the number with heart disease is binomial with parameters $n_i = y_{i1} + y_{i2}$ and $\pi_i = \mu_{i1}/(\mu_{i1} + \mu_{i2})$. Substituting the log-linear model formula for μ_{ij} into this, the parameters m, b_1, and τ_i cancel and we obtain

$$\text{logit}(\pi_i) = \log(\mu_{i1}) - \log(\mu_{i2}) = -\delta - \beta x_i \tag{6.17}$$

which is identical to Eq. (6.15) except for a sign change. The intercept α of the logistic regression becomes the main effect of disease δ. The slope parameter β in the logistic binomial model becomes the difference between the slope parameters $(b_1 + \beta)$ for nondiseased patients and b_1 for diseased patients, i.e., an interaction of the covariate x_i with the column/disease factor. This is a general feature—main effects in a logistic model become interactions (with the binomial response factor) in the equivalent Poisson model.

Why are the parameters τ_i required? You should reread Section 2.4 before continuing. We saw there that for inference on the probability parameters of a multinomial, a Poisson model gives the same results, provided that an extra free parameter τ is included to model the total count and that τ does not depend on the probability parameters. In the present case, we want this to be true for each blood pressure group/row and so we need a free parameter τ_i, not a function of α or β, for each row total n_i. These parameters may also be thought of as modeling the numbers in each blood pressure group, which does not tell us anything about the probability of being diseased *conditional on blood pressure*. If the parameters τ_i are not included in Eq. (6.16) the logistic relationship Eq. (6.17) *does* still hold conditional on the row totals n_i. However, the full likelihood from the Poisson model will not be the same as the logistic binomial likelihood as the terms in the Poisson likelihood modeling the n_i will depend on α and β. The parameter estimates will be different from the logistic binomial regression estimates and the fitted values in each row will not agree with the data totals.

Some **GLIM** output is given below. The model fitted is Eq. (6.16) with Poisson error which is equivalent to the logistic binomial regression Eq. (6.17). The parameter δ represents a difference between columns and so is simply included in the model formula as a column factor on two levels, called **SICK**. The parameters τ_i allow a different mean parameter for each row i and so we include an eight-level row factor, called **TAU**. The blood pressure score x_i is a covariate called **BLDP**. The slopes in Eq. (6.16) differ for the two columns. Thus, an interaction of the covariate **BLDP** with **SICK** is also included.

```
[i]   ? $LOOK Y SICK TAU BLDP$
[o]             Y      SICK      TAU     BLDP
[o]  1      3.000     1.000    1.000    115.0
[o]  2    153.000     2.000    1.000    115.0
[o]  3     17.000     1.000    2.000    121.0
[o]  4    235.000     2.000    2.000    121.0
[o]  5     12.000     1.000    3.000    131.0
[o]  6    272.000     2.000    3.000    131.0
[o]  7     16.000     1.000    4.000    141.0
[o]  8    255.000     2.000    4.000    141.0
[o]  9     12.000     1.000    5.000    151.0
[o] 10    127.000     2.000    5.000    151.0
[o] 11      8.000     1.000    6.000    161.0
[o] 12     77.000     2.000    6.000    161.0
[o] 13     16.000     1.000    7.000    176.0
[o] 14     83.000     2.000    7.000    176.0
[o] 15      8.000     1.000    8.000    190.0
[o] 16     35.000     2.000    8.000    190.0
[i]   ? $FACT TAU 8 SICK 2$YVAR Y$ERROR P$
[i]   ? $FIT SICK+TAU+BLDP+SICK$DISP E$
[o]   scaled deviance = 6.3649 at cycle 3
[o]             d.f. = 6
[o]
[o]      estimate       s.e.     parameter
[o]  1      1.726     0.2032     1
[o]  2      6.128     0.7367     SICK(2)                         deltahat
[o]  3     0.6221     0.1060     TAU(2)
[o]  4     0.9771     0.1263     TAU(3)
[o]  5      1.163     0.1603     TAU(4)
[o]  6     0.7237     0.2068     TAU(5)
[o]  7     0.4558     0.2535     TAU(6)
[o]  8     0.9332     0.3041     TAU(7)
[o]  9     0.3872     0.3677     TAU(8)
```

```
[o] 10      0.000   aliased     BLDP
[o] 11   -0.02471 0.004943     SICK(2).BLDP                          betahat
[o]
[i]  ? $PRIN %FV$
[o]  5.619  150.4  10.47  241.5  14.93  269.1  17.97  253.0
[o]  11.59  127.4  8.864  76.14  14.29  84.71  8.276  34.72
```

The default parametrization chosen by GLIM is exactly Eq. (6.16) except that τ_1 has been aliased to zero implicitly and b_1 explicitly. The estimates and standard errors of the interest parameters are $\hat{a} = -6.128(0.7367)$ and $\hat{\beta} = 0.0247(0.0049)$ and the deviance 6.36(6) indicates a good fit. Fitted values have been given in Table 6.2. Notice that within each row, fitted and observed values have the same totals. In other words, the model correctly predicts for instance 156 individuals in the lowest blood pressure group.

The alternative GLIM output below shows the results for fitting the model Eq. (6.16) but with the parameters τ_i removed. This model is *not* the same as the logistic binomial regression and it is difficult to see how it could be a sensible model. The total numbers in the different blood pressure groups are now parametrically modeled by

$$E(Y_{1i} + Y_{2i}) = E[n_i(x)] \propto e^{b_1 x}(1 + e^{\delta + \beta x}).$$

Unless we had some reason to believe this model for the total blood pressure counts we should not impose it, especially since we are not primarily interested in blood pressure at all but in how it affects the probability of disease. But let us see how it fits anyway.

```
[i]  ? $FIT SICK+BLDP*SICK$DISP E$PRIN %FV$
[o]  scaled deviance = 172.19 at cycle 3
[o]              d.f. = 12
[o]        estimate         s.e.    parameter
[o]  1        2.371       0.6329    1
[o]  2        5.387       0.6585    SICK(2)            ahat
[o]  3    0.0004794     0.004206    BLDP
[o]  4     -0.01951     0.004403    SICK(2).BLDP       bhat
[o]  !                Fitted values below
[o]  11.32  262.4  11.35  234.1  11.40  193.6  11.46  160.0
[o]  11.51  132.3  11.57  109.4  11.65  82.21  11.73  62.98
```

The model fits extremely poorly with the deviance 172.19(12). The parameter estimates are now $\hat{a} = -5.387(0.6585)$ and $\hat{\beta} = 0.0195(0.0044)$ being moderately different to the earlier estimates. However, the fitted values no longer add up to the correct totals within each blood pressure group. For instance, in the lowest group, the model predicts 11.32 + 262.45 = 273.77 individuals instead of 156. The model is plainly wrong.

The General Case Moving to the more general problem, imagine a set of binomial data $Y_1, \ldots, Y_k$ from totals $n_1, \ldots, n_k$ and the general logistic regression model

$$\log(\mu_{ij}) = m + \tau_i + \delta_j + x_i^T \beta_j; \qquad i = 1, \ldots, k \qquad j = 1, 2 \tag{6.18}$$

For identifiability, we set $\delta_1 = 0$ and $\beta_1 = 0$ and rename δ_2, β_2 by δ and β. We could have set $b_1 = 0$ in Eq. (6.16) from the beginning; it is automatically aliased to zero because the probabilities in each row add up to one. However, it would then have been less clear that β was an interaction of the covariate x_i with the column/disease factor. Anyway, with this restriction Eq. (6.18) becomes

$$\begin{aligned}\log(\mu_{i1}) &= m + \tau_i \\ \log(\mu_{i2}) &= m + \delta + x_i^T \beta + \tau_i.\end{aligned}$$

The parameter δ measures the difference in the mean counts of responders and nonresponders and is included in a `GLIM` model through a two-level factor indicating whether a count is for responders or nonresponders. Conditional on each row total n_i, the distribution of Y_{i1} is binomial with probability parameter $\pi_i = \mu_{i1}/(\mu_{i1} + \mu_{i2})$ and so

$$\operatorname{logit}(\pi_i) = \log\left(\frac{\mu_{i1}}{\mu_{i2}}\right) = -\delta - x_i^T \beta$$

which is identical to the original logistic model, except for a sign change. In the log-linear Poisson formulation, the intercept parameter α becomes the main effect δ of the two-level response factor while the logistic regression parameter β become an interaction of the covariate with the response.

6.2.2 Multinomial Logit Models

The model fitted to the Danish marriage data was a multinomial logit model which is the direct extension of logistic binomial models to responses on more than two categories. Let R be a response factor on c levels. For instance, marriage status is a response on three levels. Let $\pi = (\pi_1, \ldots, \pi_c)$ be the probability vector listing the probabilities that an individual responds at various levels.

Suppose we have a set of independent observations on the multinomial variable R, i.e., $Y_i^T = (Y_{i1}, \ldots, Y_{ic})$ gives the numbers of individuals in the c categories under the ith set of experimental conditions. These experimental conditions are summarized by a set of covariate measurements x_i. The probability vector describing the ith multinomial data vector Y_i^T is

$$\pi_i = (\pi_{1|i}, \ldots, \pi_{c|i}).$$

A multinomial logit model is one that restricts the probability vector π_i to have components of the form

$$\Pr(R = j|i) = \pi_{j|i} = \frac{e^{x_i^T \beta_j}}{e^{x_i^T \beta_1} + \cdots + e^{x_i^T \beta_c}} \tag{6.19}$$

or equivalently satisfying

$$\psi_{jl|i} = \log\left(\frac{\pi_{j|i}}{\pi_{l|i}}\right) = x_i^T(\beta_j - \beta_l). \tag{6.20}$$

Equation (6.6) is clearly of this form and the model fitted to the Danish marriage data was a trinomial logit model. The parameter $\psi_{jl|i}$ is the log-odds for category j compared to l (under conditions i). Only $(c - 1)$ of these are needed to describe the probability vector π_i. In the special case of $c = 2$ categories there is really only one such logit. Renaming $\pi_1 = \pi$ and $\pi_2 = 1 - \pi$ this sole logit is $\psi_{12} = \log(\pi_1/\pi_2) = \text{logit}(\pi)$.

Because the $\pi_{j|i}$ in Eq. (6.19) add up to one, it follows that one of the parameter vectors β_j must be aliased to zero. It makes no real difference which is chosen, though **GLIM** automatically chooses the last category. The regression parameters β_j then measure how quickly the mean counts change relative to that category whose parameter has been aliased. We think of this category as the baseline category. For instance, with the **GLIM** parametrization $\beta_c = 0$, we may describe the probability vector π_i by the logits of the first $c - 1$ categories relative to the last. Renaming $\psi_{jc|i}$ by $\psi_{j|i}$ and using Eq. (6.20) we have

$$\psi_{j|i} = \log(\pi_{j|i}/\pi_{c|i}) = x_i^T \beta_j. \tag{6.21}$$

This is the defining property of multinomial logit models—the jth logit (relative to some baseline category) is modeled linearly in terms of a set of parameters β_j and a set of covariates x. Note that there is a slope parameter β_j for each column/category.

Parametric estimates of $\psi_{j|i}$ follow upon replacing β_j with estimates $\hat{\beta}_j$. Empirical logits can be defined by replacing the $\pi_{j|i}$ by their empirical estimates $Y_{ij}/Y_{i\bullet}$ giving

$$\hat{\psi}_{jl|i} = \log(Y_{ij}) - \log(Y_{il})$$

and plotting these against various covariates at the data exploration stage might suggest suitable covariates to include in the vector x of explanatory variables.

The standard error of the point estimator $\hat{\psi}_{jl|i}$ is

$$\text{s.e.}(\hat{\psi}_{jl|i}) = \frac{Y_{ij} + Y_{il}}{Y_{ij}Y_{il}}.$$

Example 6.2: (Continued) **The Danish Marriage Data.** The response of marital status is on $c = 3$ categories and so the probability vector (π_1, π_2, π_3) may be described by two logits, which are taken to be ψ_{13} and ψ_{23}, and the divorced category three is the baseline. The first logit compares numbers of unmarried to divorced individuals while the second compares married to divorced. Since there are zeros in the data, the ordinary empirical logits are infinite. This can be avoided by adding 1/2 to all the counts. This actually minimizes the bias of $\hat{\psi}_{jl}$ as an estimator of ψ_{jl}, and these are called *modified* empirical logits. Figure 6.3 shows plots of $\hat{\psi}_{13}(x)$ against age x, in the left plot and against $\log(x - 16)$ in the right plot. It is apparent that a logit-linear model in $\log(x - 16)$ rather than x is appropriate. Plots of $\hat{\psi}_{23}(x)$ lead to a similar conclusion. Earlier we fitted the logistic linear model in $\log(x-16)$. Using notations defined there, Eq. (6.7) implies that the logits satisfy

$$\psi_{jl} = s_j - s_l + (\beta_j - \beta_l)\log(x - 16).$$

The particular parametrization used by `GLIM` sets $\beta_3 = 0$ which implicitly uses the divorced category as the baseline. We then have

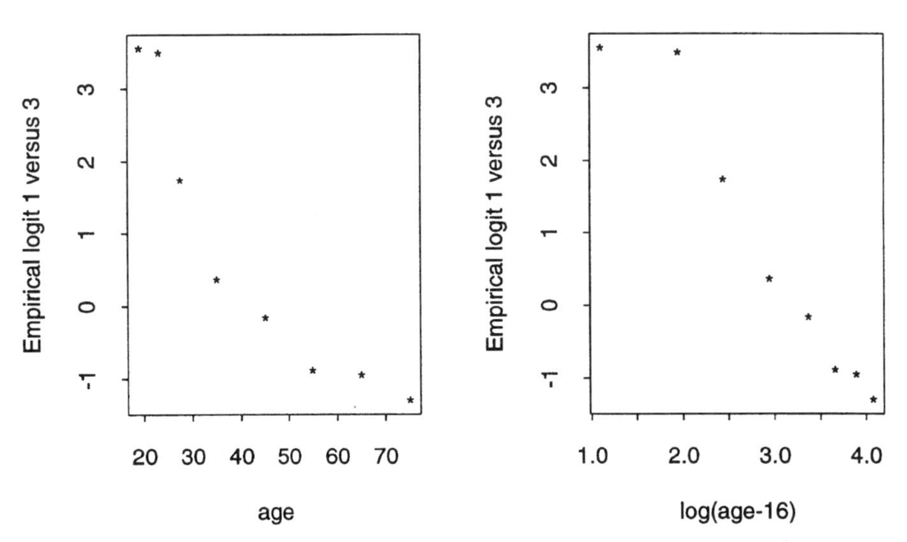

Figure 6.3. Plots of empirical logits for Danish marriage data, comparing unmarried to divorced. Left against age and right against log(age – 16).

$$\psi_{13} = s_1 - s_3 + \beta_1 \log(x - 16), \qquad \psi_{23} = s_2 - s_3 + \beta_2 \log(x - 16).$$

The first equation is a logistic regression of the proportion of unmarried among unmarried or divorced individuals; the second equation is a logistic regression of the proportion of married among married or divorced individuals. The parameter estimate $\hat{\beta}_1 = -3.127$ says that $\log(\pi_1/\pi_3)$ decreases with $\log(x - 16)$ at the rate 3.127 or, equivalently that the ratio π_1/π_3 of unmarried to divorced individuals decreases like $(x - 16)^{-3.127}$. Similarly $\hat{\beta}_2 = -1.677$ gives the rate at which the ratio of married to divorced individuals decreases.

Generalized Multinomial Logit Models The defining property Eq. (6.19) of a multinomial logit model can actually be further extended to allow for covariates that differ with the different response categories. This has many applications. For instance, one might be investigating consumer preferences among c competing products. If the competing products have different prices then we might wish to separate the price effect from the underlying customer satisfaction with the product. There might also be further covariates that describe the experimental conditions and are not category specific.

Again let $(Y_{i1}, \ldots, Y_{ic})$ be the numbers responding in the various categories under experimental conditions i. Let x_{ij} represent a set of covariates that cannot only vary with experimental conditions i but also vary across the different categories j for a given set of experimental conditions. The generalized multinomial logit model imposes the curves

$$\pi_{j|i} = \frac{e^{x_{ij}^T \beta_j}}{e^{x_{i1}^T \beta_1} + \cdots + e^{x_{ic}^T \beta_c}}. \tag{6.22}$$

Covariates describing characteristics of the individual choosers can be included without further extension of the theory. We simply take the response to be that of an individual, i.e. $(Y_{i1}, \ldots, Y_{ic})$ is a vector containing a single 1 and the rest zeros, and i now indexes individuals in the study rather than sets of individuals deemed to have been subjected to identical conditions. An example of this nature has been given by McFadden (1974) and is discussed in Agresti (1990, p. 317). Quite a different model for analyzing consumer preference data will be presented in Section 6.5.5. Generalized logit models are discussed by Amemiya (1981), Theil (1969, 1970), Small (1987, 1988), and McFadden (1974, 1981, 1982, 1984).

6.2.3 General Conversion Formula

In this section, we show that multinomial logit models can generally be fitted as log-linear Poisson models, explain the relationships between the parameters of the models, and give a general scheme for converting a multinomial model formula into an equivalent Poisson model formula.

Imagine a set of data organized as an array of independent Poisson variables with means μ_{ij}. We think of i as indexing experimental conditions and j as indexing categories of a factor of interest. Conditional on the experimental conditions, the ith row of data is multinomial with probability vector

$$\pi_{j|i} = \mu_{ij}/(\mu_{i1} + \mu_{i2} + \cdots + \mu_{ic}).$$

Consider the log-linear model

$$\log(\mu_{ij}) = m + \tau_i + \delta_j + x_i^T \beta_j \qquad i = 1, \ldots, k \qquad j = 1, \ldots, c \tag{6.23}$$

which is identical to Eq. (6.18) except that j now has c levels rather than 2. It then follows that

$$\psi_{jl|i} = \log(\mu_{ij}) - \log(\mu_{il}) = \delta_j - \delta_l + x_i^T(\beta_j - \beta_l)$$

and setting $\delta_c = 0$ and $\beta_c = 0$ for identifiability, the logits $\psi_{j|i}$ relative to the baseline category c are given by

$$\psi_{j|i} = \delta_j + x_i^T \beta_j \tag{6.24}$$

which is identical to Eq. (6.21) except that the intercept parameters δ_j have been displayed separately. In the Poisson model these parameters are associated with a c level factor R identifying the column or response category of each count whereas in the multinomial model they represent intercepts in the logistic regression of category j against the baseline category. The parameters β_j, being possibly different for each j, are associated in the Poisson model, with an interaction of the covariate x and the response factor R, but in the multinomial model are the slope coefficients of the logistic regression of category j against the baseline category.

Role of the Parameters $\boldsymbol{\tau}_i$ The parameters τ_i do not have any bearing on the multinomial model formula. They are included by defining a factor which identifies each set of experimental conditions. Only when these free parameters are included does the Poisson likelihood for each row give the same results as the conditional multinomial likelihood for each row. One may think of the τ_i as modeling the totals $n_i = Y_{i1} + \cdots + Y_{ic}$. These totals are not of primary interest as they depend on how much data was collected for each set of experimental conditions/row and this may well be entirely nonrandom for a planned experiment. The fitted values in each row will total exactly n_i. In this sense, we have not modeled the n_i at all and so difficulties in fitting the data to these observed totals cannot upset our efforts to model the *pattern of the proportions* within each row.

General Scheme A general scheme is now given for converting a multinomial logit model into the `GLIM` model formula of an equivalent log-linear Poisson model. Consider a response variable R on c levels that is to be explained in terms of factors `A1`, `A2`, ... and covariates `X1`, `X2`, ... A set of observations on R are taken under k sets of conditions over which these factors and covariates vary. Let `M(A1, A2, ..., X1, X2, ...)` represent the multinomial model formula. For instance, for the Danish marriage data the response was `STAT` and there was a single covariate `AGE` and the multinomial logit model formula for the first model fitted was `M(AGE)=AGE`, which says that the probabilities of being in each category vary logistically with the covariate `AGE`.

Alternatively, consider the set of multinomial counts as a $k \times c$ array with Y_{ij} Poisson distributed. Let `N` be a factor that identifies each set of experimental conditions applying to Y_{ij}, i.e., a row factor. Let `R` be a factor that identifies the response level of each count Y_{ij}, i.e., a column factor. Then the log-linear Poisson model with model formula

$$\mathrm{R} * \mathrm{M} + \mathrm{N} \tag{6.25}$$

is equivalent to the multinomial logit model with model formula `M`. The parameter estimates associated with the multinomial model will appear as interactions with `R`. For instance, in the package `GLIM` the estimates associated with the various levels of factor `A1` will carry the labels `R(2).A1(i)` and estimates of the slope parameter for the covariate `X1` carry the label `R(2).X1`.

The parameter estimates associated with `N` are entirely irrelevant to inference on the multinomial model and are typically numerous. For example, a simple logistic regression on 100 binomial counts will have only two parameters, the slope and intercept, of interest, but a further 100 parameters associated with `N`. There is a simple and often convenient method of constructing the factor `N`. Let `FX1`, `FX2`, ... be factors indicating the different distinct values of the corresponding covariates `X1`, `X2`, ... For example, the covariate values 19, 23, 27.5, 35, 45, 55, 65, 75 of `AGE` for the Danish marriage data are replaced with integer labels 1–8. Then the interaction term

$$\mathrm{FX1} * \mathrm{FX2} * \cdots * \mathrm{A1} * \mathrm{A2} * \cdots \tag{6.26}$$

adds to the model a separate free parameter for each possible combination of experimental conditions. This factor can be used in the general scheme [see Eq. (6.25)] instead of `N`. Notice that `N` is the interaction of all factors and *factored covariates* excluding the response factor `R`. If `R` is mistakenly included then the model will be full.

Earlier Examples For the Danish marriage data we fitted two models whose `GLIM` model formulas with Poisson error and log-link were

$$\text{STAT} + \text{STAT.AGE} + \text{FAGE}$$
$$\text{STAT} + \text{STAT.LAGE} + \text{FAGE}$$

We have not included the `-%GM` term which merely changes the parametrization by setting the mean term m in Eq. (6.5) equal to zero so that the s_j are all unaliased. These formulas may be interpreted as multinomial logit models for the response `R = STAT` with model formulas `M = AGE` and `M = LAGE`, respectively. The term `FAGE` corresponds to the general term `N`; it includes an extra free parameter for each set of conditions/age. According to Eq. (6.26), with the single covariate `X = AGE` and no other factors present, the term `N` is simply `FX = FAGE`. For the Framingham Heart Data, the response was `R = SICK` and the binomial logit model was again in terms of a single covariate `BLDP`. The term `N` identifies each set of conditions/blood pressure levels and was called `TAU`. Thus, the required Poisson model formula is

$$\text{SICK} * \text{BLDP} + \text{TAU}$$

which agrees with the model formula used in the earlier `GLIM` listing. The parameters of the binomial logit model appear as interactions with `SICK(2)`.

Example 6.3: Sexual Behavior and Divorce A confidential survey of 494 divorcees and 542 married individuals was conducted. Each respondent was asked to confidentially reveal whether or not he/she had had (i) premarital sex (PMS) with any partner besides his/her spouse or (ii) extramarital sex (EMA). The gender of each respondent was also recorded. There is interest in trying to explain divorce rates in terms of PMS, EMS, and gender. The data are shown in Table 6.3. For each of the eight combinations of PMS, EMS, and gender, the table lists the observed number divorced and the total number in parentheses. Recall from Chapter 3 that even though this is a cross-sectional rather than a prospective study, we may code the data as if it were prospective. For instance, in the first cell we may imagine that the 198 males who either did not engage in, or do not admit to, premarital or extramarital sex are a fixed random sample of individuals of whom, upon follow up, 68 are found to be divorced.

We initially model the data as binomial and some `GLIM` output is given below. The factors were input with the convention 1 = no, 2 = yes. Let π_{ijk} be the probability of divorce for gender i, PMS status j and EMS status k. The

Table 6.3. Divorce and Sexual Behavior [Gilbert (1981)]

PMS EMS	No No	No Yes	Yes No	Yes Yes
Males	68 (198)	17 (21)	60 (102)	28 (39)
Females	214 (536)	36 (40)	54 (79)	17 (21)

first fitted model assumes

$$\text{logit}(\pi_{ijk}) = (GP)_{ij} + (EP)_{jk}$$

where E, P, G represent parameters for EMS, PMS, and gender. The deviance is 0.4396(2) and so the model fits well. Examination of the `GEND(2).PMS(2)` parameter estimate suggests that this interaction is not necessary and we next fit the model

$$\text{logit}(\pi_{ijk}) = G_i + P_j + (EP)_{jk}$$

which has deviance 0.6978(3). The estimate 1.016 labeled `PMS(2)` suggests that for those not involved in extramarital sex [EMS(1)] the odds of divorce is about 2.76 times higher for those engaging in PMS than those who do not. What is the effect of EMS over and above this? The estimate 2.396 labeled `PMS(1).EMS(2)` suggests that for those who did not engage in PMS, the odds of divorce is about 11 times higher for those engaging in EMS than for those who do not. The estimate 0.609 labeled `PMS(2).EMS(2)` suggests that for those who engaged in PMS, the odds of divorce is just under two times higher for those engaging in EMS than for those who do not.

```
[i]   ! ANALYSIS USING BINOMIAL MODELING
[i]   ? $fit gend*pms+ems.pms$disp e$
[o]   scaled deviance = 0.4396 at cycle 4
[o]              d.f. = 2
[o]      estimate      s.e.     parameter
[o] 1    -0.6664     0.1474     1
[o] 2     0.2642     0.1699     GEND(2)
[o] 3      1.016     0.2430     PMS(2)
[o] 4     0.1677     0.3306     GEND(2).PMS(2)
[o] 5      2.390     0.3878     PMS(1).EMS(2)
[o] 6     0.6090     0.3383     PMS(2).EMS(2)
[o]
[i]   ? $fit gend+pms+pms.ems$disp e$
[o]   scaled deviance = 0.6978 at cycle 4
[o]              d.f. = 3
[o]      estimate      s.e.     parameter
[o] 1    -0.6997     0.1327     1
[o] 2     0.3089     0.1458     GEND(2)
[o] 3      1.099     0.1787     PMS(2)
[o] 4      2.396     0.3879     PMS(1).EMS(2)
[o] 5     0.5962     0.3366     PMS(2).EMS(2)
```

The data have also been analyzed using log-linear Poisson models. Binomial data points such as 68(198) are instead coded as two Poisson counts (68 and 130) for divorcees/nondivorcees, respectively. The factor `DIV` is defined on two levels and we think of this as the response factor R. The factor N takes levels one through eight for the eight different combinations of PMS, EMS, and gender. Alternatively, according to Eq. (6.26), we could replace N in all model formulas by the interaction of all factors to be used in the modeling of DIV, i.e., N=EMS*PMS*GEND.

The model fitted below is identical to the second binomial model above and the model formula was obtained by substituting the GLIM binomial model formula directly into Eq. (6.25). The reader should note that the deviances of the models are identical. Except for a sign change, the parameter estimates listed for the binomial model appear in the Poisson model with virtually identical labels except preceded by DIV(2), and have been marked by asterisks. Some aliased parameters have been deleted from the parameter estimate lists but are in any case irrelevant to relating the Poisson models to the equivalent binomial models. The parameter estimates associated with N=EMS*PMS*GEND relate to the relationships between EMS status, PMS status, and gender. While these might also be of interest in their own right, they are not relevant to the main issue of how EMS status affects divorce rates.

```
[i]   ! ANALYSIS USING POISSON MODELING
[i]   ? $fit div*(gend+pms+ems.pms)+N$disp e$
[o]   scaled deviance = 0.69784 at cycle 3
[o]                d.f. = 3
[o]
[o]      estimate      s.e.     parameter
[o]  1      4.185     0.1136    1
[o]  2     0.6997     0.1327    DIV(2)*
[o]  3    -0.6227     0.2025    GEND(2)
[o]  4    -0.7580     0.2809    PMS(2)
[o]  5     -1.309     0.2494    N(2)
[o]  6     0.6846     0.3180    N(3)
[o]  7   -0.07805     0.3321    N(4)
[o]  8      1.814     0.1766    N(5)
[o] 10      1.165     0.2588    N(7)
[o] 12    -0.3089     0.1458    DIV(2).GEND(2)*
[o] 13     -1.099     0.1787    DIV(2).PMS(2)*
[o] 16     -2.396     0.3878    DIV(2).PMS(1).EMS(2)*
[o] 17    -0.5962     0.3366    DIV(2).PMS(2).EMS(2)*
```

6.2.4 Interaction of Factors as Response

We have seen how Poisson models can be used to implicitly fit a logistic model to one of the measured factors treated as a multinomial response. Such a model explains the response factor in terms of the other factors and covariates. Sometimes, however, we may be interested in explaining two factors simultaneously.

Let A and B be two such factors. We wish to explain both A and B in terms of the other factors and covariates. What do we mean by "both A and B?" Presumably we mean the pattern of counts across all the different levels of A and B or, in other words, the cross-classification of A and B. If A is on r levels and B on c levels then there are rc possible combinations of A and B. The interaction term A*B assigns different labels to these rc combinations. We thus want the interaction A*B to be the multinomial response on rc levels.

The general Eq. (6.25) still applies except that the response R is now A*B and the factor N now distinguishes all the different experimental conditions under which R=A*B has been observed. This factor is identical to an interaction of all measured factors, excluding response factors A and B.

Example 6.3: (Continued) **Modeling Divorce and EMS Jointly.** If one thinks not only of divorce but also of extramarital sex as a sign of marital breakdown then we might seek to explain the cross-classification of DIV and EMS in terms of the other factors GEND and PMS. The response R=DIV*EMS is a multinomial response on four levels, describing four different levels of marital status—for example, among males who engaged in no PMS there are 68 divorcees and 130 nondivorcees who did not engage in EMS, and 17 divorcees and 4 nondivorcees who did engage in EMS. Why are these four counts (68, 130, 17, 4) in the proportions we observe and how do the gender and PMS factors affect the pattern of these four counts?

The term N comprises all factors besides the response factor(s) so in this case N=GEND*PMS and a multinomial model M for DIV*EMS can be fitted as a log-linear model using the general formula

$$\mathrm{DIV} * \mathrm{EMS} * \mathrm{M} + \mathrm{GEND} * \mathrm{PMS}.$$

In the GLIM output below we have fitted the model that says that the response DIV*EMS depends additively on gender and PMS. The deviance is 0.367(3) and so the model fits well and there is no evidence for an interaction of the effect of gender and PMS on DIV*EMS (that would be a four-way interaction and therefore the full model). Examination of the parameter DIV(2).EMS(2).GEND(2) suggests that gender might not have a significant effect. However, on fitting the model with M=PMS we find that the deviance increases to 8.153(6) and we reject the hypothesis of no gender effect.

The reason for this apparent contradiction is that the response R here has four levels and so even a simple gender effect has three degrees of freedom, not one. Consequently, we cannot simply look at the single parameter estimate DIV(2).EMS(2).GEND(2) to decide if GEND is required in the model. This only measures one aspect of the effect of gender on the response. The estimates DIV(2).GEND(2) and EMS(2).GEND(2) describe the other two degrees of freedom of the gender effect and the last of these is highly significant.

```
[i]  ? $fit div*ems*(gend+pms)+gend*pms$disp e$
[o]  scaled deviance = 0.36740 at cycle 3
[o]            d.f. = 3
```

```
[o]      estimate      s.e.    parameter
[o]  1      4.205    0.1131    1
[o]  2     0.6750    0.1356    DIV(2)
[o]  3     -1.390    0.2263    EMS(2)
[o]  4      1.166    0.1261    GEND(2)
[o]  5   -0.09448    0.1480    PMS(2)
[o]  6     -2.194    0.4956    DIV(2).EMS(2)
[o]  7    -0.2761    0.1513    DIV(2).GEND(2)
[o]  8    -0.3884    0.2420    EMS(2).GEND(2)
[o]  9     -1.088    0.1785    DIV(2).PMS(2)
[o] 10     0.6227    0.2431    EMS(2).PMS(2)
[o] 11     -1.305    0.1610    GEND(2).PMS(2)
[o] 12    -0.3283    0.5295    DIV(2).EMS(2).GEND(2)
[o] 13      1.692    0.5363    DIV(2).EMS(2).PMS(2)
[o]
[i]   ? $fit div*ems*pms+gend*pms$
[o]   scaled deviance = 8.1535 at cycle 3
[o]             d.f. = 6
[i]   ! THIS LAST MODEL IS REJECTED COMPARED TO THE FIRST.
```

To make this a little clearer, instead of using `GEND*PMS` define the factor `N` explicitly as a four-level factor, and also the response `R=DIV.EMS`. The parameter estimates labeled `R(j).GEND(2)` now describe the effects of gender on the mean counts at response level j, compared to response level 1 (which has been aliased). All these parameter estimates appear significant.

```
[i]   ! DEFINE 4 LEVEL FACTORS R AND N.
[i]   ? $fit r*(gend+pms)+n$disp e$
[o]   scaled deviance = 0.36740 at cycle 3
[o]             d.f. = 3
[o]      estimate      s.e.    parameter
[o]  1      4.205    0.1131    1
[o]  2     0.6750    0.1356    R(2)
[o]  3     -1.390    0.2263    R(3)
[o]  4     -2.909    0.4450    R(4)
[o]  5      1.166    0.1261    GEND(2)
[o]  6     -1.400    0.1349    PMS(2)
[o]  7      1.305    0.1610    N(2)
[o]  8      0.000   aliased    N(3)
[o]  9      0.000   aliased    N(4)
[o] 10    -0.2761    0.1513    R(2).GEND(2)
[o] 11    -0.3884    0.2420    R(3).GEND(2)
[o] 12    -0.9927    0.4737    R(4).GEND(2)
[o] 13     -1.088    0.1785    R(2).PMS(2)
[o] 14     0.6227    0.2431    R(3).PMS(2)
[o] 15      1.227    0.4739    R(4).PMS(2)
```

6.3 MODELING INDEPENDENCE

In the previous section, the models fitted sought to explain one factor, the response, in terms of other factors and covariates. The relationship between the explanatory factors was not of interest and indeed the term `N` explicitly fits the full model to these explanatory factors.

In many cases, no single classifying factor is the natural response. Rather, we have a set of counts cross-classified by various factors and we are interested in describing how these factors are related. Log-linear models are the most natural tool for investigating various types of independence between such factors.

6.3.1 Types of Dependence

There are several types of independence that are described in this section. In order to present the main ideas, it is enough to consider three factors A, B, and C. Individuals are sampled in some manner and each is then classified according to these factors. Let π_{ijk} represent the probability that an individual is classified as $A = i$, $B = j$, and $C = k$. We can think of the labels (A, B, C) as being a 3-dimensional discrete random variable with probability distribution π_{ijk}. All types of independence are defined in terms of these probabilities.

From elementary probability theory, the marginal distribution of any two variables is obtained by summing the π_{ijk} over the ignored variable. For instance, the probability distribution of (A, B) is given by

$$\Pr(A = i, B = j) = \pi_{ij\bullet}$$

where we will use $\cdot$ as usual to denote summation over the index it replaces. Similarly, the marginal distribution of C is given by

$$\Pr(C = k) = \pi_{\bullet\bullet k}.$$

Conditional distributions now follow simply. Dividing $\Pr(A = i, B = j)$ by $\Pr(B = j)$ for instance, the probability distribution of A conditional on B is

$$\Pr(A = i | B = j) := \pi_{i|j} = \frac{\pi_{ij\bullet}}{\pi_{\bullet j\bullet}}.$$

The various types of independence to be described are all defined in terms of various marginal or conditional distributions *factorizing*.

Definition 1. *Mutual independence* Three variables A, B, C are mutually independent if the joint distribution of (A, B, C) is the product of the marginal distributions of A, B, and C. In terms of conditions on the π_{ijk} this requires

$$\pi_{ijk} = \pi_{i\bullet\bullet}\pi_{\bullet j\bullet}\pi_{\bullet\bullet k}. \tag{6.27}$$

Definition 2. *Joint independence* Two variables A and B are jointly independent of C if the distributions of (A, B, C) is the product of the joint distribution of (A, B) with the distribution of C. In terms of the π_{ijk}

$$\pi_{ijk} = \pi_{ij\bullet}\pi_{\bullet\bullet k}. \tag{6.28}$$

Definition 3. *Marginal independence* Two variables A, B are marginally independent if the marginal distribution of (A, B) is the product of the marginal distributions of A and B. In terms of the π_{ijk}

$$\pi_{ij\bullet} = \pi_{i\bullet\bullet}\pi_{\bullet j\bullet}. \tag{6.29}$$

Definition 4. *Pairwise independence* Three variables A, B, C are pairwise independent if (A, B), (A, C), and (B, C) are each marginally independent.

Definition 5. *Conditional independence* Two variables A and B are independent given C if the distribution of (A, B) given C is the product of the distributions of A given C, and B given C. In terms of the π_{ijk}

$$\pi_{ij|k} = \pi_{i|k}\pi_{j|k} \tag{6.30}$$

which, upon multiplying by $\pi^2_{\bullet\bullet k}$, is equivalent to

$$\pi_{\bullet\bullet k}\pi_{ijk} = \pi_{i\bullet k}\pi_{\bullet jk}. \tag{6.31}$$

Roughly speaking, two variables are independent if values of one do not affect likely values of the other. This sounds rather simple however, the relationships between the various kinds of independence above require a little careful thought. For instance, independence does not imply mutual independence. In the log-linear models described later, pairwise independence corresponds to no two-way interactions. However, a model with no two way interactions may still have a three-way interaction. Mutual independence is the strongest (and least interesting) type of independence and implies all the others.

Marginal and Conditional (Partial) Associations A fact of some practical importance is that conditional independence does not imply marginal independence, nor vice versa. It is easy to think of examples illustrating this.

Consider the three factors—gender, height, and income. Considering only males, it seems unlikely that income and height are related. The same is presumably true of females. In terms of independence, we have asserted that income and height are independent conditional on gender. However, in most countries males have higher incomes than females. Moreover, males are on average taller than females. It follows that if we collected data on the heights and incomes of individuals we would find that these variables have a positive relationship, i.e., income and height are *not* marginally independent. The apparent association is "caused by" the association of the hidden variable gender with both income and height. Indeed, one of the purposes of statistical modeling is to explain associations of variables in terms of other underlying variables in precisely this way.

We have concluded that marginal independence is not implied by conditional independence. Though less common in practice, it is also possible for two factors to appear independent until broken down by another factor when they appear dependent. That is, conditional independence is not implied by marginal independence. The fact that a pair of variables can have a marginal association of a different sign to their conditional (also known as *partial*) association is known as Simpson's paradox and was already discussed in Section 3.6.1. Conditional associations are usually of more interest and importance than marginal associations since they are controlled for the conditioning variable. Look back over the above example and convince yourself that the conditional association is the better index of relationship.

Joint Implies Marginal and Conditional Independence If (A, C) are jointly independent of B then (i) A and B are marginally independent, ignoring C and, (ii) A and B are conditionally independent given C. This can be shown directly from the definitions with straightforward algebra. It can be shown more simply, however, by recalling two facts about independence: (1) Two variables are independent if, and only if, conditioning on one variable does not alter the distribution of the other. (2) If A and B are independent then any functions of A and B are independent.

If (A, C) are jointly independent of B then B is independent of any function of (A, C), in particular it is (marginally) independent of A. Next, to show that A and B are independent conditional on C, it is enough to show that after conditioning on C, conditioning on A does not further change the distribution of B. In fact, the distribution of B conditional on C is just the marginal distribution of B since B and C are independent. Further conditioning on A gives the distribution of B conditional on (A, C) which is again the marginal distribution of B.

Collapsibility: Equal Marginal and Partial Associations When can Simpson's paradox *not* occur i.e. when are the partial and marginal associations necessarily equal? It has been shown by Bishop (1971) that the partial (i.e. conditional) and marginal associations of A and B will be identical provided that C is conditionally independent of either A or B i.e. if C is independent of A given B or if C is independent of B given A. This condition is sometimes called the condition for *collapsibility* over the factor C. This result generalizes to more than three factors (see next section).

Collapsibility, in implying that associations are the same marginally and conditionally, consequently implies that the partial associations are the same for each value of the conditioning variable. The reverse is not true. For instance, we could have a situation where there is the same positive association of A and

B conditional on all values of C and yet the marginal association is different, and even of the opposite sign.

6.3.2 Interpretation of Model Formulas

Log-linear models are simply interpreted in terms of various kinds of independence. The algebra underlying such interpretations is conceptually simple if often tedious in detail. However, in practice, the interpretation of a given log-linear model formula can be given without going through the algebraic details. In this section, we will explain the basis of the algebra, go through one example in detail, and then illustrate how to interpret log-linear model formulas in general.

For simplicity and definiteness, consider three factors A, B, C. Log-linear model formulas specify restrictions on the means μ_{ijk} of the cross-classification of these factors. Probabilities and conditional probabilities are simple ratios of various partial totals of the μ_{ijk}. For example

$$\begin{aligned}
\pi_{ijk} &= \Pr(A = i, B = j, C = k) = \mu_{ijk}/\mu_{\bullet\bullet\bullet} \\
\pi_{ij} &= \Pr(A = i, B = j) = \mu_{ij\bullet}/\mu_{\bullet\bullet\bullet} \\
\pi_{i} &= \Pr(A = i) = \mu_{i\bullet\bullet}/\mu_{\bullet\bullet\bullet} \\
\pi_{ij|k} &= \Pr(A = i, B = j | C = k) = \mu_{ijk}/\mu_{\bullet\bullet k} \\
\pi_{i|j} &= \Pr(A = i | B = j) = \mu_{ij\bullet}/\mu_{\bullet j\bullet}.
\end{aligned}$$

A log-linear model formula specifies a *multiplicative* formula for μ_{ijk}. Substituting the formula into these conditional probabilities we can deduce which factors are independent of which by seeing which conditional probabilities are the same as which unconditional probabilities. For instance, A and B are independent if $\pi_{ij\bullet} = \pi_{i\bullet\bullet}\pi_{\bullet j\bullet}$, as stated in Eq. (6.29).

As an illustration consider the log-linear model with formula

$$\log(\mu_{ijk}) = m + a_i + b_j + c_k + (ac)_{ik} + (bc)_{jk}. \tag{6.32}$$

The parameters a_i, b_j, c_k represent main effects of the factors and $(ac)_{ik}$ and $(bc)_{jk}$ interactions of C with A and B, respectively. How is this model interpreted in terms of independence, rather than main effects and interactions?

First, exponentiate the model formula and use notation such as $A_i = \exp(a_i)$ and $(AC)_{ik} = \exp(ac)_{ik}$, etc., to obtain

$$\mu_{ijk} = MA_iB_jC_k(AC)_{ik}(BC)_{jk}.$$

All conditional probabilities may be written in terms of ratios of partial totals of these μ_{ijk}. For instance

$$\pi_{ijk} = \Pr(A = i, B = j, C = k)\& = \frac{A_i B_j C_k (AC)_{ik}((BC)_{jk}}{A_\bullet B_\bullet C_\bullet (AC)_{\bullet\bullet}(BC)_{\bullet\bullet}}$$

$$\pi_{ik} = \Pr(A = i, C = k)\& = \frac{A_i B_\bullet C_k (AC)_{ik}(BC)_{\bullet k}}{A_\bullet B_\bullet C_\bullet (AC)_{\bullet\bullet}(BC)_{\bullet\bullet}}$$

$$\pi_{jk} = \Pr(B = j, C = k)\& = \frac{A_\bullet B_j C_k (AC)_{\bullet k}(BC)_{jk}}{A_\bullet B_\bullet C_\bullet (AC)_{\bullet\bullet}(BC)_{\bullet\bullet}}$$

$$\pi_k = \Pr(C = k)\& = \frac{A_\bullet B_\bullet C_k (AC)_{\bullet k}(BC)_{\bullet k}}{A_\bullet B_\bullet C_\bullet (AC)_{\bullet\bullet}(BC)_{\bullet\bullet}}.$$

Consequently, we have expressions for the following conditional probabilities:

$$\pi_{ij|k} = \frac{\pi_{ijk}}{\pi_k} = \frac{A_i B_j (AC)_{ik}(BC)_{jk}}{A_\bullet B_\bullet (AC)_{\bullet k}(BC)_{\bullet k}}$$

$$\pi_{i|k} = \frac{\pi_{ik}}{\pi_k} = \frac{A_i (AC)_{ik}}{A_\bullet (AC)_{\bullet k}}$$

$$\pi_{j|k} = \frac{\pi_{jk}}{\pi_k} = \frac{B_j (BC)_{jk}}{B_\bullet (BC)_{\bullet k}}.$$

By inspection we see that

$$\pi_{ij|k} = \pi_{i|k}\pi_{j|k}$$

which is to say that A and B are independent conditional on C.

Because the term $(ac)_{ik}$ does not contain the subscript j, the partial association of A and C when B is held fixed at j does not depend on j. Is this single partial association the same as the marginal association? From the collapsibility condition of the previous section the marginal association of A and C *is* the same as the partial association conditional on B. Similarly, the marginal and partial association of A and C are the same.

You should go through the details of the above algebra. The reason that the conditional independence holds is that various factors in numerator and denominator cancel and the reason for this is that there are no terms present that involve both i and j. Using similar methods we may show that all other variables are dependent, both conditionally or unconditionally. There are no other statements about independence to be made about this model and it is defined by this single relation of conditional independence.

It is not necessary to go through such a tedious calculation for every different model. The various kinds of independence can be identified using the following ideas.

1. Factors can only be independent if they do not occur in a single term of the log-linear model.
2. Factors will be independent if they occur additively.
3. Factors will be conditionally independent if, after crossing out the conditioning factor, they occur additively.

It is also convenient to have a simpler notation for the log-linear model formula than writing it out in full. *Hierarchical* models are models that contain all main effects of lower order than their interactions. For instance, the model

$$\log(\mu_{ijk}) = m + a_i + b_j + (ab)_{ij}$$

is hierarchical because main effects for A and B are present. The model

$$\log(\mu_{ijk}) = m + a_i + (ab)_{ij}$$

is not hierarchical because the main effect for B is not present. This model allows B to modify the effect of A through the interaction parameters $(ab)_{ij}$ but assumes that B has no direct effect itself. The main use of nonhierarchical models is for so-called nested designs. Hierarchical models are easier to talk about because they are determined by their highest order interactions—there is no need to detail the main effects and lower order interactions as these are automatically implied by the highest order terms.

We use the following notation. *Identify the highest order interactions in the model formula. List these interactions as symbolic products of the corresponding factors*. Thus, we use the notation (AC, BC) for the model Eq. (6.32). Using the three ideas listed above, we see that none of the factors A, B, C are independent because they do not occur separately. However, crossing out the factor C we are left with (A, B) and A and B are now entirely separate. Thus, A and B are independent conditional on C, which is the same conclusion we reached earlier from a long algebraic argument.

I now list the types of models available for three factors and what these models assume about independence.

Mutual Independence of A, B, and C The model formula (A, B, C) has the three factors in separate terms and so this model assumes mutual independence of these factors. Mutual independence also implies conditional independence. For instance, crossing out C we have the model (A, B) and so A and B are also conditionally independent. This model is unlikely to be of much ultimate use as experiments are usually performed and data collected with the aim of establishing some form of dependence between factors.

A and B Jointly Independent of C The model formula (AB, C) assumes that A and B are jointly independent of C. Crossing out A we see that this implies that C is independent of B conditional on A and, by symmetry, of A conditional on B. This model assumes *two* pairs of variables are conditionally independent. We also have from the collapsibility condition that the marginal association of A and B is the same as the partial association conditional on C.

This model is equivalent to a multinomial model for the response C since (AB, C) is of the form

$$R * M + N$$

where the response `R` is C, the factor `N` is AB and the model `M` is the null model, i.e., this model says that the multinomial distribution of C is unaffected by the covariates A or B.

Independence of A and B Conditional on C The model formula (AC, BC) was already treated in detail earlier. There is only a single pair of variables that display independence, namely A and B conditional on C. By the collapsibility condition, the marginal association of A and C is the same as their partial association conditional on B. By symmetry, the marginal association of B and C is the same as their partial association conditional on A.

This model is equivalent to two multinomial models of the form `R*M+N`. First, it is equivalent to the model `M=C` for the response A so that the factor `N` is BC, i.e., multinomial response A is explained in terms of C only. Second, it is equivalent to the model `M=C` for the response B so that the factor `N` is AC, i.e., multinomial response B is explained in terms of C only.

A model of this type would be of interest if a marginal association of A and B had been observed. This model explains the association in terms of C since after C is known, A and B are independent. An example was given earlier where a marginal association of income and height was explained through their association with a third-factor gender.

No Three-Factor Interaction The model (AB, BC, AC) has no three-factor interaction. No variables are independent conditionally or otherwise. The model assumes that the *partial* associations of any pair of variables conditional on the third variable are the same regardless of the value of that third variable. However, collapsibility conditions are not satisfied and so the partial associations of variables, while constant, may be quite different to their marginal associations.

The model is equivalent to various multinomial logit models. For instance, let C be the response `R`, which implies that the factor `N` is AB. Let the model `M` be `A+B`. Then we see that

$$R * M + N = C * (A + B) + A * B = A * B + B * C + A * C.$$

The model is thus equivalent to a simple additive model for the effects of A

and B on the multinomial distribution of C. This does not strictly imply independence of any variables. However, we might think of this as saying that the effects of A and B do not interact. This is just another way of saying that there is no three factor interaction. However, it is worth noting that this notion of 'no interaction' is specific to the logit scale on which effects are measured. A model which assumed say multiplicative effects of A and B on the distribution of C, i.e. additive on the log rather than the logit scale, would not correspond to the no three-factor interaction model.

Finally, we could equally have taken B as the response and shown that the model is equivalent to the logit model with formula $A + C$.

6.3.3 Models with More than Three Factors

The main ideas of independence and collapsibility discussed in the previous section readily generalize to more than three factors.

Suppose that there are four factors (A, B, C, D) and consider the model formula (AB, BC, AD). By crossing out various factors in this formula we conclude (i) conditional on A, (B, C) are jointly independent of D, (ii) conditional on B, (D, A) are jointly independent of C, and (iii) conditional on A and C, B is independent of D.

Collapsibility Collapsibility conditions are very easy to state in general, now that we have a simple notation for the model formula. The marginal associations of a set of factors are the same as the partial associations conditional on some variable C, provided that these variables do not occur together with C in the same term. Only in this case can one collapse over the factor C without the associations changing. Collapsibility, if permissible, is worth doing as the analysis of the dependence of the other factors is simplified.

For instance, in the model (AB, BC, AD) note that AB and BC do not occur together with D. Thus, we could collapse over D without altering the associations of A, B and C. If we were primarily interested in only these associations then we could completely ignore the D measurements by simply fitting the model (A, AB, BC) which is the same as the model (AB, BC). The association of D with A is not relevant to the A–B–C associations. The marginal effect of A *would* be affected by collapsing over D and so collapsing would not be indicated if this marginal effect were of interest.

An immediate consequence of the collapsibility condition is that if the factor C is independent of all the other factors jointly then it can be ignored. For instance, the model (ABC, D) says that the associations of A, B, and C are the same marginally as conditionally on D. The data can be thought of as a separate A–B–C cross-classifications for different levels of D. The model says that the associations are the same for each level of D and this factor is only present to model the different total of counts for each of these cross-classifications. Nothing is lost at all by collapsing the data over D.

Pictorial Representation of Models Models that contain only two- and three-way interactions, and not too many of them, may be usefully represented as a diagram. For each factor draw a circle with the factor label inside. If two factors share a two-way interaction they are joined by a line or curve. If three factors share a three-way interaction then they are joined by three spokes from a common axle. An example is shown in Figure 6.4.

Ideally, the factors can be moved around in such a configuration that the number of crossing curves is minimized or avoided altogether. Higher order interactions are harder to effectively represent and are in any case difficult to explain. Such pictures are useful for conveying the meaning of models to non-specialists.

6.3.4 A Worked Example with Five Factors

The data shown in Table 6.4 are the results of a survey of 4502 individuals cross-classified by five measured factors and are taken from Sewell and Orenstein (1965). The factor IQ was categorized from an original numerical score as was the factor ASP which stands for Occupational Aspiration. The factor EC stands for socioeconomic status and was reduced to a two-points scale of high or low. The place of residence was also classified as rural, small urban, and large urban; the last two are called *town* and *city* in the table.

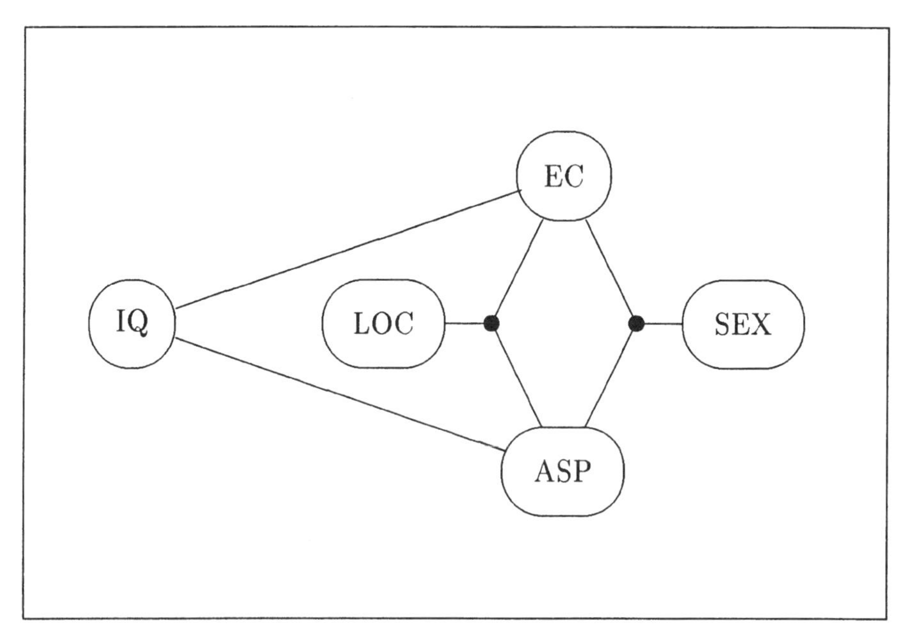

Figure 6.4. Pictorial representation of the log-linear Poisson model fitted to the Occupational Survey Data.

	Rural	Town	City	Total
ASP (high)	445	1004	412	1861
ASP (low)	1075	1196	370	2641
Total	1520	2200	782	4502
% high	29	45	53	

There are 48 counts cross-classified by five factors. If no single variable is to be considered the response, then we might investigate the pairwise dependence by tabling the data by each pair of factors. For instance, the cross-classification of the factors ASP and Residence is shown in the table above. It is apparent that the proportion with high occupational aspirations increases from the lowest value 29% for rural residents to the highest value 53% for city residents. We could construct nine other similar tables to look at the other pairwise associa-

Table 6.4. Occupational Survey Data

Sex	Residence	IQ	EC	ASP (High)	ASP (Low)
Male	Rural	High	High	117	47
			Low	54	87
		Low	High	29	78
			Low	31	262
	Town	High	High	350	80
			Low	70	85
		Low	High	71	120
			Low	33	265
	City	High	High	151	31
			Low	27	23
		Low	High	30	27
			Low	12	52
Female	Rural	High	High	102	69
			Low	52	119
		Low	High	32	73
			Low	28	349
	Town	High	High	338	96
			Low	44	99
		Low	High	76	107
			Low	22	344
	City	High	High	148	35
			Low	17	39
		Low	High	21	47
			Low	6	116

tions. There are two things wrong with this simple minded approach. First, it is very long and complicated. Second, and more importantly, all these tables display marginal associations of two variables and may give quite the wrong picture of the conditional associations—there is no reason to assume that we can collapse over any factors.

I will first consider models that explain the occupational aspirations of individuals in terms of the other four factors. The data are input into a package such as `GLIM` or *Splus* as 48 Poisson counts with the corresponding labels of the five classifying factors. Logistic models `M` for the binary response ASP will, in the Poisson formulation, have the form

$$\text{ASP} * \text{M} + \text{IQ} * \text{LOC} * \text{EC} * \text{SEX}.$$

The last term `IQ*LOC*EC*SEX` is the term `N` in the general multinomial-Poisson conversion scheme Eq. (6.25). It is worth pausing to ask why this term is required. Inclusion of this term adds $2 \times 3 \times 2 \times 2 = 24$ extra free parameters to the model, exactly one for each row of the data in Table 6.4. For example, in the first row there are $117 + 47 = 164$ individuals, 117 of whom have high ASP. In modeling ASP as a binary response, we ask "Why did 117 of these 164 individuals have high response." The fact that there are 164 individuals in this group is of no relevance to this question. Inclusion of the term `N` takes each row total as fixed, and does not model these totals at all. There may well be relations between these totals, i.e., between the factors `IQ`, `LOC`, `EC`, and `SEX`, and these will be modeled later, but we are not interested in them now. Rather, we are using these factors to explain ASP.

I have fitted models `M` with various orders of interactions in the explanatory factors `IQ`, `LOC`, `EC`, and `SEX`. The P-value for testing each model against the next most general one is given in the last column. We reject the main effects model for the two-way model. There is little evidence that three-way interactions are required ($P = 0.177$) although the three-way model fits well.

Model	Deviance	df	P-Value
Main effects	37.66	16	0.0004
All two-way	11.259	9	0.1773
All three-way	1.0378	2	0.9971
Four-way	0	0	

For the two-way model, one finds that all three-way interaction parameters involving `ASP` and `IQ` are not significant. For instance, the parameter labeled `ASP(2).IQ(2).LOC(2)` is the estimate `IQ(2).LOC(2)` in the logistic binomial model for ASP. This parameter is small compared to its standard error and is not required. On removing all the interactions involving `IQ` in the model `M` we obtain the logistic binomial model

$$\mathrm{LOC} * \mathrm{EC} + \mathrm{LOC} * \mathrm{SEX} + \mathrm{EC} * \mathrm{SEX} + \mathrm{IQ}$$

that has deviance 14.188 on 13 degrees of freedom and has all terms significant. This is the best model for `ASP` using categorical methods. Of course, the factor `LOC` is ordinal and we may well obtain a simpler model by using the methods described later in Section 6.4.

Next we look for models for the binary response `IQ`. These will have the form `IQ*M+N` where the factor `N` is now

$$\mathrm{ASP} * \mathrm{LOC} * \mathrm{EC} * \mathrm{SEX}.$$

Consider male individuals with rural place of residence, high socioeconomic status, and high occupational aspirations. There are 146 of these, 117 of whom have high IQ's and 29 low IQ's. If we are interested in explaining why a proportion 117/146 have high IQ's then the number 146 is of no consequence. Including the factor `N` in the model treats such totals as fixed.

If you wanted to test whether or not `LOC` was an important factor in predicting `IQ` you would fit the null model with `M=ASP*EC*SEX` which has deviance 16.528 on 16 degrees of freedom. This model fits well and so there is no evidence that `LOC` is required to predict IQ. A more specific test would compare this model to the alternative model `M=LOC+ASP*EC*SEC` which assumes a simple additive effect of `LOC`. This model has deviance of 11.914, a reduction of 4.614 on two degrees of freedom. The associated P-value is 0.0995 and so there is mild evidence that place of residence might predict IQ. By narrowing the focus of the test, rather than just performing a goodness-of-fit test, a slight effect for place of residence has been revealed.

Various models `M` for `IQ` are tabulated below. The most simple and parsimonious appears to be the model `ASP+LOC+EC` that has deviance 14.906 on 19 degrees of freedom. This model can be obtained formally using the backwards elimination procedure for model selection.

Model for IQ	Deviance	df
ASP * EC	19.468	20
ASP * SEX	160.95	20
ASP * SEX + EC	19.193	19
EC * SEX	609.66	20
EC * SEX + ASP	18.525	19
ASP + LOC + EC + SEX	14.855	18
ASP + LOC + EC	14.906	19
ASP + EC	19.548	20

Finally, let us abandon the idea of any single factor being the response and simply ask "What are the associations of these five factors?" There is no longer any need for a factor N since we do not require the model to be equivalent to any multinomial logit model for a response factor. One could start the search for a simple model by beginning with the full five-way interaction model and proceeding to reduce by backwards elimination, however, we have already found a good fitting model above of the form

$$IQ * (ASP + LOC + EC) + F$$

where F is currently the 4-way interaction ASP*LOC*EC*SEX. The table below shows the results of replacing F with all three-way interactions of the factors ASP, LOC, EC, and SEX that increases the deviance by 3.592 for two degrees of freedom. This leads us to accept the all three-ways factor model for F($P = 0.1760$). The table also shows the deviances upon removing each of these three-way interactions singly. It appears that the LOC*EC*SEX and ASP*LOC*SEX interactions are the least significant. Upon removing these interactions the model has deviance 27.900 on 25 degrees of freedom. Examination of the parameter estimate reveals that the LOC*SEX interaction is not required.

Model Term F	Deviance	df
Four-way	14.906	19
Three-way	18.498	21
–ASP.LOC.EC	35.679	23
–ASP.EC.SEX	30.289	22
–ASP.LOC.SEX	24.902	23
–LOC.EC.SEX	19.124	23

Our final model, with deviance 29.17 on 28 degrees of freedom, has two of the 10 possible three-way interactions, namely ASP*LOC*EC and ASP*EC*SEX and nine of the 10 two-way interactions. The interaction IQ*LOC is of borderline significance and when removed the deviance is 32.26 on 29 degrees of freedom. This model is shown in Figure 6.4.

6.4 SOME MODELS FOR ORDINAL DATA

The models discussed so far treat all factors as categorical. When some of these factors are *ordinal* then potentially, associations of the factors may be modeled more simply and estimated more precisely. A simple and common approach is to replace the ordinal labels by numerical scores, essentially converting the ordinal measurement artificially into a numerical measurement. This method would seem to be natural and appropriate provided that the ordinal measurement really derives from some underlying numerical measurement which has later been categorized. Classifying the income of an individual into ranges would be a typical example.

The main advantages of the models described in this chapter over earlier models are twofold. First, the associations of ordinal variables can sometimes be more simply described by a small number of regression or slope parameters, whereas for the general nonindependence model there is a large number of interaction parameters required. Second, the narrower focus afforded by the simpler ordinal models of association increases the residual degrees of freedom and consequently the power of any tests of the association.

We will give only a brief account of some models that fit naturally into the framework of this chapter. Techniques for the analysis of ordinal data could easily fill a book on their own and a good starting point is Agresti (1984).

For simplicity we begin with a response factor B on c levels which is to be explained in terms of a classifying factor A on r levels. Imagine the data set out as an $r \times c$ contingency table, with counts Y_{ij} and means μ_{ij} in each cell. Treating the counts Y_{ij} as independent Poisson random variables, the most general model of association is

$$\log(\mu_{ij}) = m + a_i + b_j + (ab)_{ij} \; i = 1, \ldots, r \qquad j = 1, \ldots, c \tag{6.33}$$

and there are $(r-1)(c-1)$ free parameters $(ab)_{ij}$ describing the way in which the factor A affects the response B. These parameters are essentially log-odds ratios for 2×2 sub-tables comparing $A = i$ and $B = j$ with baseline levels (that depend on the precise parametrization). The various models for ordinal data to be described in this section are obtained by replacing the $(ab)_{ij}$ term by a simpler term involving *scores* for the ordinal factors.

6.4.1 Local Logits, Local Odds-Ratios, and Scores

Suppose now that B were an ordinal response; i.e., the c levels of B can be arranged in a definite order. Testing the independence model gives the same change in deviance regardless of the ordering of columns and in this since the information in the ordering of the responses has not been utilized. What sort of simple model might describe the effect of the row factor on the ordinal response B?

Let $\pi_i = (\pi_{1|i}, \ldots, \pi_{c|i})$ be the probability vector conditional on row i, i.e., $\pi_{j|i} = \Pr(B = j|A = i)$. The ordering of the responses immediately suggests describing the multinomial probability vector $\pi_{j|i}$ by relationships of *adjacent* probabilities. The parameters

$$l_{j|i} = \log\left(\frac{\pi_{j+1|i}}{\pi_{j|i}}\right) \; j = 1, \ldots, c-1 \tag{6.34}$$

measure the chance of responding at level $j + 1$ rather than j when the row factor A is at level i. These are called *local* or adjacent logits.

The $c - 1$ local logits can be easily converted back into a probability vector. From the definition of local logit we have

$$\log\left(\frac{\pi_{j+1|i}}{\pi_{1|i}}\right) = \log\left(\frac{\pi_{j+1|i}}{\pi_{j|i}} \frac{\pi_{j|i}}{\pi_{j-1|i}} \cdots \frac{\pi_{2|i}}{\pi_{1|i}}\right) = l_{1|i} + l_{2|i} + \cdots + l_{j|i}$$

and so the log-odds of response at level $j + 1$ compared to level 1 is the partial sum of the first j local logits. Rearranging and exponentiating we find that

$$\pi_{j|i} = \pi_{1|i} L_{1|i} L_{2|i} \ldots L_{j-1|i} \tag{6.35}$$

where $L_{j|i} = \exp(l_{j|i})$.

Suppose that A is also an ordinal variable. Then the ordering again suggests expressing the means μ_{ij} in terms of *adjacent* log-odds ratios. Specifically we define

$$\theta_{ij} = l_{j|i+1} - l_{j|i} = \log\left(\frac{\mu_{i+1,j+1}\mu_{ij}}{\mu_{i+1,j}\mu_{i,j+1}}\right) \tag{6.36}$$

that measures the association between A and B in the local 2×2 sub-table with cell (i, j) in the upper-left corner.

So far, we have just used the ordering to choose a particular parametrization for the logits of the multinomial model or the log-odds ratios of the Poisson model. The ordinal nature of the data has not been utilized. One way to do so is to let the c levels of B be associated with c numerical scores $v_1 < v_2 < \ldots < v_c$. Similarly, when A is ordinal as well, let the r levels of A be associated with numerical scores $u_1 < u_2 < \ldots < u_r$.

Ideally, these scores should relate directly to an underlying numerical measurement. For instance, if B categorizes individuals by their income group then the midpoint of each group might be used as scores. A less transparent example might be the ordinal variable measuring the status of an individual's chosen job. A classification of (low, middle, high) status might be based on several numerical factors such as the individual's income, degree of training, and degree of responsibility. Some numerical combination of these factors presumably underlies the classification into the three status categories and might be used to construct appropriate scores. Certainly, the choice of scores may, in some cases, have a large effect on the final results [see, for instance, Graubard and Korn (1987)] and when there are no natural scores it is worth doing a sensitivity analysis.

6.4.2 Local Logit Models for a Categorical Row Effect

The model described in this section provides a simple way of describing the effect of a categorical factor A on the distribution of an ordinal response B. It is the ordinal data analogue of the one-way analysis of variance model.

The quantity $v_{j+1} - v_j$ measures how different are response levels $j + 1$ and j. If v_{j+1} and v_j are very close then we might expect the row effect, if present at all, to be small. On the other hand, if v_{j+1} and v_j are very different, then the row effect, if present, might tend to be larger. A model that is *multiplicative* in the score differences $v_{j+1} - v_j$ is suggested. The simplest such model is

$$l_{j|i} = \alpha_j + \beta_i(v_{j+1} - v_j) \; i = 1, \ldots, r \qquad j = 1, \ldots, c. \tag{6.37}$$

describing the effect of the row factor A on the local logits $l_{j|i}$. The model is called the *row effects model* although it is quite a special and specific model for the row effect.

When $\beta_i = 0$, the local logits are just α_j. These parameters represent the underlying probability distribution "without any row effect." Denote this probability vector by π_0. Using Eq. (6.35) to express the local logits in terms of probabilities we find

$$\pi_{j|0} = \pi_{1|0} A_1 A_2 \ldots A_{j-1} \tag{6.38}$$

where $A_j = \exp(\alpha_j)$. As a simple special case, when $\alpha_j = \alpha$ does not depend on j it follows that $\pi_{j|0}$ is proportional to $\exp\{\alpha(j - 1)\}$ which explicitly gives

$$\pi_{j|0} = c^{-1} e^{\alpha(j-1)} \frac{e^{\alpha c} - 1}{e^{\alpha} - 1}, \qquad j = 1, \ldots, c$$

and when $\alpha = 0$ this is the uniform distribution on the integers $1, \ldots, c$.

The parameters β_i modify the underlying distribution summarized by the α_j. When $\beta_i > 0$, the distribution is systematically pushed to the right and the response tends to be at higher levels than for the unmodified distribution. For instance, when the scores $v_j = j$, the relative odds of responding at level $j + 1$ compared to j is increased by β_i, right across the range of possible responses. When $\beta_i < 0$, the distribution is pushed to the left and responses tend to be at lower levels than for the unmodified distribution.

Some numerical examples of how a row effect satisfying Eq. (6.37) modifies a given underlying distribution are given in Table 6.5. The first block relates to an underlying uniform distribution (i.e., to $\alpha_j = 0$ for all j). Notice that when $\beta = -0.1$ the row effect is precisely the opposite of when $\beta = 0.1$. The second block relates to the underlying distribution given by the values (0, 0.539, −0.442, −1.253, −2.234) for the α_j. These local logits correspond to the probability vector $\pi = (0.240, 0.411, 0.265, 0.076, 0.008)$ which is actually the $Bi(4, 0.3)$

Table 6.5. Illustration of Various Row Effects on Underlying Uniform and Underlying Bi(4, 0.3) Distributions

β	0	1	2	3	4	Mean
−0.2	0.287	0.235	0.192	0.157	0.129	1.606
−0.1	0.212	0.219	0.198	0.179	0.162	1.800
0.0	0.200	0.200	0.200	0.200	0.200	2.000
0.1	0.162	0.179	0.198	0.219	0.242	2.200
0.2	0.129	0.157	0.192	0.235	0.287	2.394
0	0.240	0.411	0.265	0.076	0.008	1.200
0.1	0.212	0.402	0.285	0.091	0.011	1.285
0.2	0.185	0.389	0.305	0.106	0.014	1.374
0.3	0.161	0.373	0.323	0.125	0.018	1.466
0.4	0.138	0.354	0.339	0.145	0.023	1.560
0.5	0.117	0.333	0.353	0.166	0.029	1.656
1.0	0.045	0.212	0.371	0.288	0.083	2.152
∞	0.000	0.000	0.000	0.000	1.000	4.000

distribution. The table shows how this distribution changes when modified by a row effect β. The mean of each distribution is given in the last column. As β increases the distribution is shifted more and more to the right. When $\beta = \pm\infty$ the distribution is concentrated at the highest/lowest response level.

The row effects model imposes the restriction that the distribution of responses across the ordinal factor B is made *stochastically larger or smaller* by the row factor. In particular, the row effect leaves a unimodal distribution unimodal. It would not be possible to model a row factor that (i) modified the probability of only one or some of the response levels or (ii) made the distribution more or less concentrated. Nevertheless, the model is useful in describing any primary tendency of the row factor to systematically move the response distribution towards higher or lower levels.

The row effects model, being a linear model for the local logits, can also be written as a log-linear Poisson model. Consider the model

$$\log(\mu_{ij}) = m + a_i + b_j + \beta_i v_j \qquad i = 1, \ldots, r \qquad j = 1, \ldots, c. \tag{6.39}$$

We note that there is a free parameter a_i for each row and so the Poisson model is equivalent to the multinomial model with systematic component

$$l_{j|i} = \log\left(\frac{\mu_{i,j+1}}{\mu_{ij}}\right) = b_{j+1} - b_j + \beta_i(v_{j+1} - v_j) \tag{6.40}$$

that is identical to Eq. (6.37) except that α_j has been relabeled $b_{j+1} - b_j$. The local log-odds ratios are given by

$$\theta_{ij} = l_{j|i+1} - l_{j|i} = (\beta_{i+1} - \beta_i)(v_{j+1} - v_j). \tag{6.41}$$

Examining the model Eq. (6.39) we see that it is a special case of the general nonindependence model Eq. (6.33). The term $(ab)_{ij}$ has been replaced by the term $\beta_i v_j$ that models the dependence in terms of the $r-1$ parameters β_i instead of the full $(r-1)(c-1)$ parameters in the full model. For each row, the effect of moving across the columns is linear in the scores with slope β_i, whereas in the full independence model the means can change across columns in an arbitrary manner. For a given column however, the pattern of the log means follows the α_i's and is arbitrary.

The row effects model replaces $(r-1)(c-1)$ association parameters with $r-1$ and so the degrees of freedom for testing goodness-of-fit is

$$df = (r-1)(c-1) - (r-1) = (r-1)(c-2).$$

Fitting the model is extremely simple. If `A`, `B` contain the row and column labels respectively of each count and if `V` contains the row scores of each count then the model

$$\text{A} + \text{B} + \text{A} * \text{V}$$

fits the row effects model. The parameter estimates `A(i).V` give the row effects β_i on the distribution of the multinomial response B.

Example 6.4: Mental Health and Socioeconomic Status The data in Table 6.6 are a cross-classification of the mental health (MH) and socioeconomic status (EC) of the parents of a sample of individuals. The sample was not random but ensured that the numbers in each mental health category were of similar magnitude. The categories of socioeconomic status have been reduced to three levels (from the original six). It is suspected that low status is associated with poorer levels of mental health.

Table 6.6. Mental Health and Socioeconomic Status of Parents

Status	Well	Mild	Moderate	Impaired
High	121	188	112	86
	117.5, 120.0	198.7, 199.7	101.0, 99.9	89.8, 87.4
Medium	129	246	142	154
	126.4, 122.1	247.9, 245.1	146.1, 147.8	150.6, 156.0
Low	57	168	108	149
	63.1, 64.9	155.4, 157.2	114.9, 114.3	148.6, 145.6

Note: Fitted values are printed below each count, (i) for the row effects model at left and (ii) for bilinear model at right [from Srole et al. (1978, p. 289)].

The GLIM output below shows the results of fitting the row effects model with unit space scores for the mental health factor. This factor is labeled **FMH** and the scores are taken to equal the values 1, 2, 3, 4 and are labeled **MH**. The model fits well with deviance 4.21 on $(3-1)(4-2)=4$ degrees of freedom. The fitted values are listed and have been tabulated in Table 6.6. The parameters labeled **FEC(i).MH** are the β_i. The parameter β_1 has been aliased and so the distribution for the first (i.e., highest) socioeconomic status is implicitly taken as the baseline. The parameters labeled **FEC(j)** give this fitted distribution. They estimate that the log-means for the first row have the pattern (0, 0.5255, −0.1515, −0.2619). Exponentiating and normalizing to add to 1.0 gives the baseline probability vector

$$\pi = (0.232, 0.392, 0.199, 0.177)$$

and multiplying by the first row total 507 gives the fitted values in this row.

This baseline distribution is modified in the second row through the parameter $\beta_2 = 0.1483$. This parameter is significantly greater than zero and indicates that for medium socioeconomic status, more individuals tend to have higher (i.e., worse) levels of mental health than for those with high socioeconomic status. The parameter β_3 is 0.3752 which suggests that this effect is even greater for those of low socioeconomic status. The difference between the distribution of mental health for those with medium compared to low socioeconomic status is measured by $\beta_3 - \beta_2 = 0.2269$.

Parameters labeled 1, FEC(2), FEC(3) are present so that fitted value totals agree with the data totals in each row. It is these terms that make the Poisson model equivalent to the multinomial logit model.

```
[i]  ? $look y fec fmh
[o]            Y       FEC      FMH
[o]  1    121.00    1.000    1.000
[o]  2    188.00    1.000    2.000
[o]  3    112.00    1.000    3.000
[o]  4     86.00    1.000    4.000
[o]  5    129.00    2.000    1.000
[o]  6    246.00    2.000    2.000
[o]  7    142.00    2.000    3.000
[o]  8    154.00    2.000    4.000
[o]  9     57.00    3.000    1.000
[o] 10    168.00    3.000    2.000
[o] 11    108.00    3.000    4.000
[o] 12    149.00    3.000    4.000
[o]
[i]  ? $calc mh=fmh$fact fmh fec 3$
[i]  ? $yvar y$error p$
[i]  ? $fit fec+fmh+fec*mh$! fec*mh is the factor N
[o]  scaled deviance = 4.3113 at cycle 3
```

```
[o]             d.f. = 4
[i]   ? $disp e$
[o]       estimate       s.e.     parameter
[o] 1       4.767    0.07995     1
[o] 2    -0.07567     0.1494     FEC(2)
[o] 3     -0.9966     0.1704     FEC(3)
[o] 4      0.5255    0.07690     FMH(2)
[o] 5     -0.1515     0.1032     FMH(3)
[o] 6     -0.2694     0.1337     FMH(4)
[o] 7       0.000    aliased     MH
[o] 8      0.1483    0.05746     FEC(2).MH
[o] 9      0.3752    0.06243     FEC(3).MH
[o]
[i]   ? $prin y %fv$
[o]   117.5   198.7   101.0   89.76   126.4   247.9
[o]   146.1   150.6   63.12   155.4   114.9   148.6
```

6.4.3 Bilinear Model for an Ordinal Row Effect

What if the row factor A is also ordinal? The levels of the row factor are associated with scores $u_1 < u_2 < \ldots < u_r$ as described earlier. The simplest way to utilize these numerical scores is to replace the unrestricted slope parameters β_i of the row effects model by the term βu_i. This means that the size of the row effect changes systematically with the level i of the row factor through the scores u_i. The local logit formulation of the row effects model now becomes

$$l_{j|i} = \alpha_j + \beta u_i(v_{j+1} - v_j) \qquad i = 1, \ldots, r \qquad j = 1, \ldots, c-1 \tag{6.42}$$

while the Poisson formulation becomes

$$\log(\mu_{ij}) = m + a_i + b_j + \beta u_i v_j \qquad i = 1, \ldots, r \qquad j = 1, \ldots, c. \tag{6.43}$$

For each row i, the log-means are a linear function of the column scores v_j with slope β. For each column j, the log-means are a linear function of the row scores u_i with the same slope β. The model is then often called the *bilinear* model. There is a single parameter β describing the association of the ordinal variables A and B. If $\beta > 0$ then when A takes high values B tends to take high values and when A takes low values B tends to take low values. This model tends to fit well when the ordinal variables A and B come from underlying continuous measurements with bivariate normal distribution. If the correlation of these measurements is ρ then β will be close to $\rho/(1-\rho^2)$. Even when this is not true, the parameter β still has the simple interpretation as a slope parameter for both rows and columns in their respective scores. For each unit increase in the row or column scores, the mean count changes by a factor e^{β}.

The parameters b_j relate to the underlying multinomial distribution when the row score if $u_i = 0$. If one particular level of the row factor is of interest or is the natural baseline, then the u_i-scores may be translated so that this row has score zero. The bilinear model replaces $(r - 1)(c - 1)$ association parameters with a single parameter β. Thus, $df = rc - r - c$.

6.4.4 Equal or Unit Spacing of Scores

A special and common case is that the scores describing an ordinal response are *equally spaced*. Without loss of generality, we may take the scores to have unit spacing. In this case Eq. (6.37) becomes

$$l_{j|i} = \alpha_j + \beta_i \; i = 1, \ldots, r \qquad j = 1, \ldots, c - 1. \tag{6.44}$$

If we plot the local logits against i for various j then the plots will all be parallel and the model is often called the *parallel odds model* [see Goodman (1983)]. The bilinear model Eq. (6.43) can be expressed in terms of local log-odds ratios as

$$\theta_{ij} = l_{j|i+1} - l_{j|i} = \beta(u_{i+1} - u_i)(v_{j+1} - v_j) \tag{6.45}$$

and when the scores have unit spacing, all these simply equal β. This model imposes the restriction that within any 2×2 local sub-table of the full $r \times c$ contingency table, the log-odds ratio is β. For the 2×2 sub-table comprising rows one and three and columns two and five, the log-odds ratio would be 4β.

Equal or unit spacing of the scores certainly simplifies the interpretation of the association parameter β. If the unit spaced scores are inappropriate however, then the model will presumably fit poorly. More generally, whichever scores we use, they may not be the best scores and the models described may not in fact be true for any scores. Nevertheless, the models provide a powerful method of picking up a specific and important component of the association of A and B. Put more strongly, fitting a linear model does not require that the trend in log-means really be linear. Regardless of the truth of the model, the estimated slope parameter measures a linear component of the association. This provides both a simple description of the association as well as enhanced power for detecting association even if the model is only roughly true.

Example 6.4: (Continued) **Mental Health and Status.** The ordinal nature of the socioeconomic status row factor can be exploited using the bilinear model. Using unit space scores 0, 1, 2 treats the first (high) socioeconomic status as the baseline. The b_j parameters will describe this baseline distribution. The variate `INT=MH*EC` contains the product of the scores $u_i v_j$ [see Eq. (6.43)].

On fitting the model, the deviance is 4.96 on five degrees of freedom and the estimate of β is 0.1875 with standard error 0.0312. This estimates the assumed constant log-odds ratio in the four local 2×2 sub-tables of the 3×4 data table. The mean count increases by a factor $\exp(0.1875) = 1.206$ for each category we move to the right or down. The parameters FMH(j) estimate the pattern of log-means in the baseline (i.e., first) row. The pattern of log-means (0, 0.5094, −0.1893, −0.3173) translates into the probability vector

$$\pi_1 = (0.237, 0.394, 0.197, 0.172).$$

Multiplying this by the first row total 507 gives the fitted values for those with high socioeconomic status, shown in Table 6.6 on p. 343.

```
[i]  ? $calc ec=fec-1!              score for row 1 is zero
[i]  ? $calc int=mh*ec!             interaction covariate
[i]  ? $fit fec+fmh+int$disp e$
[o]  scaled deviance = 4.9697 at cycle 3
[o]             d.f. = 5
[o]
[o]      estimate       s.e.     parameter
[o]  1      4.788    0.07509     1
[o]  2    -0.3580     0.1190     FEC(2)
[o]  3     -1.365     0.2283     FEC(3)
[o]  4     0.5094    0.07420     FMH(2)
[o]  5    -0.1839    0.09521     FMH(3)
[o]  6    -0.3173     0.1202     FMH(4)
[o]  7     0.1875    0.03124     INT
[o]
[i]  ? $prin %fv$
[o] 120.02  199.74   99.86   87.39  122.08  245.09
[o] 147.81  156.02   64.90  157.17  114.34  145.59
```

6.4.5 Extensions of These Models

There are several models of the same flavor as those described available for modeling the association of two ordinal variables A and B. For instance, in the row effects model, we saw that in the Poisson formulation the log-means

$$\log(\mu_{ij}) = m + a_i + b_j + \beta_i v_j \tag{6.46}$$

were a linear function of the column scores v_j. We may not wish to impose such a linear assumption in which case we could simply replace the scores v_j by a free parameter for each column. This gives the model

$$\log(\mu_{ij}) = m + a_i + b_j + \alpha_i \beta_j$$

where the row effects α_i and column effects β_j are all to be estimated. The model is called the *row and column effects* model. In fact, this is not a log-linear model at all since the parameters are multiplied in the log-linear predictor. There are other statistical difficulties. For instance, the log-likelihood may have local maxima and testing the hypothesis of no row or column effects (i.e., of independence) cannot be based on the chi-square distribution.

The bilinear model is essentially a linear regression on the covariate $u_i v_j$. One could consider other regression models involving the scores u_i and v_j and various functions of these. One might also consider modeling the dependence on these scores by a smooth function. For instance, the row effects model Eq. (6.46) could be modified to

$$\log(\mu_{ij}) = m + a_i + b_j + f_i(v_j) \tag{6.47}$$

where f_i is a smooth function. This is a generalized additive model as described in Chapter 5. When the log-means do not vary linearly in the scores v_j, such a model may fit better but the row effects themselves will be summarized by the estimated smooth functions $\hat{f}_i$—the simplicity of the linear interpretation will be lost.

The earlier models extend in a straightforward manner to the investigation of the association of more than two variables. Two simple examples should serve to illustrate. Suppose there were three factors A, B, and C on respectively r, c, and d levels. Suppose that B and C are ordinal and are associated with scores v_j and w_k, respectively. Consider the model formula for the means μ_{ijk} of the cross-classification table given by

$$\log(\mu_{ijk}) = m + a_i + b_j + c_k + (ab)_{ij} + \beta_i^A w_k + \beta v_j w_k.$$

Because of the term $(ab)_{ij}$ we can think of this model as a multinomial model for the ordinal response factor C in terms of the categorical factor A and the ordinal factor B. The parameters β_i^A describe a row effect of the categorical factor A on the multinomial distribution of C, as described in Section 6.4.2. This effect is the same regardless of the level of B. The parameter β describes a bilinear association of the ordinal variables B and C. This association is the same regardless of the level of A. There are a total of r association parameters in this model, compared to $(r-1)(c-1)(d-1)$ in the general dependence model and so the degrees of freedom is $(r-1)(c-1)(d-1) - r$.

As another example suppose that the row factor A is ordinal with associated scores u_i and there are two binary factors B and C whose association is of interest. A plot of the log-odds ratio measuring the association of B and C against u_i reveals a linear trend. Thus, some sort of three-way interaction of the factors is required. The log-linear model

$$\log(\mu_{ijk}) = (ab)_{ij} + (ac)_{ik} + \beta_{jk} u_i$$

replaces the general three-way interaction term $(abc)_{ijk}$ in the full model by the term $\beta_{jk} u_i$. Only $(c-1)(d-1) = 1$ of the β_{jk} will be unaliased. For instance, `GLIM` takes the unaliased parameter to be β_{22}. The single log-odds ratio measuring the association of B and C at age-group $A = i$ is $\beta_{22}u_i$ and increases linearly with u_i. We have replaced $(r-1)(c-1)(d-1) = r-1$ association parameters in the full model by a single association parameter β_{22} and the degrees of freedom is

$$(r-1)(c-1)(d-1) - (c-1)(d-1) = (r-2)(c-1)(d-1)$$

which equals $r-2$ for binary factors B and C. An example involving the association of breathlessness and wheezing of miners, for various age groupings, is given in exercise 6.4.

6.4.6 Methods Based on Other Ordinal Logits

Local logits are made possible by the ordering of the response categories. However, these are not the only natural parameters for describing the probability distribution of an ordinal response. One alternative to consider is

$$l_j = \log\left(\frac{\pi_j}{\pi_{j+1} + \pi_{j+1} + \cdots + \pi_c}\right). \tag{6.48}$$

This is just the logit of the conditional probability

$$p_j = \pi_j/(\pi_j + \cdots + \pi_c),$$

of response at level j given a response at level j or higher. The logits l_j are called *log-hazard* or *continuation ratio* logits. Some linear models for these parameters can be fitted using standard procedures for binary responses.

Log-Hazard Models Consider a multinomial response $Y_1, \ldots, Y_c$ on the ordinal factor B. The joint density of this response can be factorized as

$$f(y_1, \ldots, y_c) = f(y_1)f(y_2|y_1)f(y_3|y_2, y_1)\ldots f(y_c|y_{c-1}, \ldots, y_1)$$

with obvious notation. What are these conditional distributions? When $Y_1, \ldots, Y_{j-1}$ is given their total is also given and so

$$n_j = n - \sum_{i=1}^{j-1} Y_i = \sum_{i=j}^{c} Y_i$$

is given. Thus, the distribution of Y_j conditional on $Y_1, \ldots, Y_{j-1}$ is that of a binomial with denominator n_j and probability parameter p_j. The likelihood for a set of data on the ordinal multinomial response B across a range of k

conditions would factorise into

$$L(Y_1, \ldots, Y_c) = L(Y_1)L(Y_2|Y_1)\ldots L(Y_c|Y_{c-1}, \ldots, Y_1)$$

where

$$L(Y_j|Y_{j-1}, \ldots, Y_1) = \prod_{i=1}^{k} f(y_{ij}|y_{i1}, \ldots, y_{i,j-1})$$

using an obvious notation. Each of these component likelihoods is a binomial likelihood from the conditional binomial data y_{ij} with independent binomial distributions and parameters n_j, p_j depending on i. Provided that *different parameters* are used to model the logits of the $p_{j|i}$ for each component j of the ordinal response, the full likelihood can be built up by fitting separate linear logistic models and adding the deviances. If there are parameters in common then fitting cannot be achieved using logistic binomial procedures.

When the ordinal factor B relates to a *progressive* phenomenon, where individuals must pass through all earlier levels to reach their given level, then continuation ratio logits give the probability that an individual does not progress past level j, conditional upon reaching this level. These probabilities are of intrinsic interest.

Example 6.4: (Continued) **Mental Health and Status.** The ordinal factor of mental health is progressive in that we would expect impairment to follow moderate symptoms, mild symptoms, or no symptoms. If there is interest in modeling the *progression* of the disease then quantities such as

$$p_2 = \Pr(\text{mild symptoms}|\text{not well})$$

which measures the probability that an individual who gets mild symptoms of mental illness does not deteriorate further, are obviously of interest.

The data from Table 6.6 have been reorganized in Table 6.7 to concentrate on these conditional probabilities. The numbers in parentheses give the numbers in that mental health category or higher. For instance, in the first row, of the 386 individual who were not well, 188 had only mild symptoms.

Let us fit a linear logistic regression in the socioeconomic scores u_i, separately for (i) p_1, the probability of being well, (ii) p_2, the probability of having mild symptoms conditional on being not well, and (iii) p_3, the probability of having moderate symptoms conditional on having more than mild symptoms. Since mental health is on four levels, there are only three degrees of freedom and so these three models define the multinomial model. Explicitly

$$l_{j|i} = \alpha_j + \beta_j u_i \qquad j = 1, 2, 3$$

Table 6.7. Fit of Log-Hazard Model to Mental Health Data[a]

Status	Well	Mild	Moderate
High	121 (507)[a]	188 (386)[a]	112 (198)[a]
	125.3[b]	184.9[b]	110.8[b]
Medium	129 (671)[a]	246 (542)[a]	142 (296)
	120.4[b]	252.2[b]	144.3[b]
Low	57 (482)[a]	168 (365)[a]	108 (257)[a]
	61.3[b]	164.8[b]	106.8[b]

[a]Conditional binomial denominators are in parentheses.
[b]Fitted values.

which allows different intercepts α_j and slopes β_j for each of the three conditional probabilities.

Some GLIM output is given below. The slope parameters for the three regressions are -0.4061, -0.0547, -0.2904. These model the probabilities of *not progressing*. To express results in terms of the probabilities of progressing we simply change their signs. Thus, for each drop in socioeconomic status score, the odds of not being well increases by 50%, the odds of progressing from mild symptoms increases by 5.6%, while the odds of progressing from moderate symptoms to actual impairment increases by 34%. The deviance of this model is $1.2828 + 0.4942 + 0.1207 = 1.898$ on three degrees of freedom. Fitted values are given in the table and the model clearly fits extremely well. Moreover, it is very easy to interpret. It is not directly comparable with the earlier row effects and bilinear models.

```
[i]  ? $look y1 n1 y2 n2 y3 n3 ec$
[o]          Y1        N1       Y2        N2        Y3        N3        EC
[o]  1    121.00     507.0     188.0     386.0     112.0     198.0     0.000
[o[  2    129.00     671.0     246.0     542.0     142.0     296.0     1.000
[o]  3     57.00     482.0     168.0     365.0     108.0     257.0     2.000
[i]  ? $yvar y1$error bino n1$fit ses$disp e$prin %fv$
[o]  scaled deviance = 1.2828 at cycle 3
[o]              d.f. = 1
[o]
[o]     estimate      s.e.     parameter
[o]  1    -1.114   0.09482     1
[o]  2   -0.4061   0.08419     EC
[o]
[o]  125.3    120.4    61.30
[i]  ? $yvar y2$error bino n2$fit ses$disp e$prin %fv$
[o]  scaled deviance = 0.49423 at cycle 3
[o]              d.f. = 1
[o]
```

```
[o]      estimate      s.e.    parameter
[o] 1  -0.08408   0.09095    1
[o] 2  -0.05475   0.07320    EC
[o]
[o]  184.9     252.2     164.9
[i]  ? $yvar y3$error bino n3$fit ses$disp e$prin %fv$
[o]  scaled deviance = 0.12068 at cycle 3
[o]              d.f. = 1
[o]
[o]      estimate      s.e.    parameter
[o] 1     0.2405    0.1259    1
[o] 2    -0.2904   0.09525    EC
[o]
[o]  110.8     144.3     106.8
```

Cumulative Logits Another natural way of exploiting the ordering of the categories is to simply use the logits of the cumulative probabilities

$$F_j = \pi_1 + \pi_2 + \cdots + \pi_j.$$

The *cumulative logits* are defined by

$$l_j = \log\left(\frac{F_j}{1 - F_j}\right) = \log\left(\frac{\pi_1 + \cdots + \pi_j}{\pi_{j+1} + \cdots + \pi_c}\right). \tag{6.49}$$

These logits, and indeed the continuation ratio logits, do not make use of scores for the ordinal response. The ordering is contained within the logits. For example, the cumulative logits necessarily increase with j. A very simple model exploiting this parametrization would be the linear model

$$l_{j|i} = \alpha_j + \beta^T x_i$$

where x is a vector of covariates and β a vector of unknown parameters. Unfortunately, cumulative logit models cannot be fitted using simple binomial procedures. An iterative algorithm that can be implemented within GLIM, is given by McCullagh (1980). An interactive package for ordinal data called PLUM has been developed by McCullagh at the University of Chicago. The SAS procedure LOGISTIC may also be used.

6.5 MODELS FOR INCOMPLETE TABLES

We have been looking at models for count data cross-classified by several factors. We may think of such data as a (multidimensional) contingency table. While not stated explicitly, in all of the models encountered so far the mean of each cell is strictly positive. Tables where some cells have zero mean are called *incomplete*.

The cells of a contingency table may contain two distinct kinds of zero counts. A *sampling* zero occurs whenever the number of individuals classified into a cell just happens to be zero. There is nothing particularly special about sampling zeros although too many sampling zeros can cause technical problems in maximizing the likelihood. Sampling zeros contain information, in particular, about the mean of the corresponding cell (it is probably small) and more generally, about the entire model (sampling zeros contribute to the likelihood equations like any other data point). From a technical point of view, on collecting more data all sampling zeros will presumably disappear and so it can be assumed in certain asymptotic analyses that all counts in the table are positive.

On the other hand, if the mean of a cell is zero then it will necessarily contain a *structural* zero count no matter how much data is collected. If it is assumed that the cell mean is zero then such a count gives no further information about the fitted model. Structural zeros occur in a variety of contexts. Some observations from a cross-classification are sometimes not reported, or certain combinations may be impossible *a priori* and zero probability attached to the corresponding cell. For instance, in certain biological populations certain types of individuals will only mate with certain other types. Another natural application occurs when we want to fit two different models to different parts of a contingency table. To do this we may simply ignore the part of the table not being modeled by treating all cells in that section as structural zeros.

Most of the ideas of independence, and the associated log-linear models, apply to incomplete tables. In particular, additive terms in the linear predictor correspond to independence or conditional independence of the underlying categories.

6.5.1 Existence of ML Estimates

For log-linear models, the log-likelihood function is a strictly concave function of $\log \mu_i$ and so ML estimates, if they exist, will certainly be unique [see Birch (1963)]. We also have the following results concerning existence of ML estimates [see Birch (1963) and Fienberg (1970)].

1. ML estimates, if they exist, satisfy the matrix equation $X^T y = X^T \mu$.
2. A sufficient condition for existence is that all counts $y_i > 0$.
3. If $X^T y$ contains any zeros then ML estimates do not exist. For instance, in a model involving factors only, the ML equations equate observed and fitted counts of certain marginal tables. These marginal tables must not contain any zeros.

Conditions 2 and 3 still leave plenty of room for uncertainty; data sets very commonly contain zero counts but these zeros are often sufficiently rare and evenly distributed that for most models the sufficient statistic $S = X^T y$ contains no zeros. In this case the existence of ML estimates is not determined by 2 and 3, but in practice, nonexistence of ML estimates is soon discovered by fitting the model and receiving an appropriate error message. In `GLIM` one may receive a message such as *unit held at limit* which means that the fitted value for a data unit is converging to zero. This implies that the linear predictor is converging to $-\infty$ and that therefore one or more of the $\hat{\beta}_j$ are diverging. `Glim` prevents the fitted value falling below some small predefined tolerance limit. Estimates of other parameters need not be invalidated by certain other parameter estimates diverging. Indeed, if one defines the parameters of the model to be $\exp(\beta)$ rather than β, then vanishing fitted values are quite acceptable.

The exact pattern of the structural zeros has a bearing on the existence and uniqueness of maximum likelihood estimators for the cell means [see Haberman (1977) or Bishop, Fienberg, and Holland (1975, Chapter 5)]. Technical details will not be given in this book. Instead, we look at how the notion of independence can be extended to incomplete tables and give some common and interesting applications.

6.5.2 Quasi-Independence and Log-Linear Models

The concept of quasi-independence generalizes the notion of independence to incomplete contingency tables that could not possibly exhibit independence because of structural zeros. Quasi-independence, roughly speaking, means that the factors are independent apart from the structural zeros.

Consider a set of individuals cross-classified by three factors A, B, and C and let π_{ijk} be the probability that an individual is classified into cell (i, j, k). Then we say that mutual independence of A, B, and C holds if

$$\pi_{ijk} = A_i B_j C_k \tag{6.50}$$

where A_i, B_j, C_k are sequences of positive parameters. This is in fact equivalent to the earlier definition [see Eq. (6.27)] on p. 326. The salient feature is that the joint probability π_{ijk} *factorizes*.

What is Quasi-Independence? Suppose now that the cross-classification contains one or more structural zeros, i.e., some of the π_{ijk} are known to be zero. Let S denote the set of indices (i, j, k) that are *not* structural zeros. We say that the cross-classification exhibits *quasi-independence* (QI) if

$$\pi_{ijk} = A_i B_j C_k \qquad \forall (i, j, k) \in S. \tag{6.51}$$

Quasi-independence is just independence applied to the nonstructural zeros of the contingency table. There is some redundancy in the parametrization since the parameters A_i may be multiplied and the B_j divided by the same constant.

If the parameters A_i, B_j, C_k are assumed positive, then the π_{ijk} are all positive within the set S, i.e., S is as small a set as possible. Without loss of generality, we make this assumption from here on.

When the data are taken to be independent Poisson rather than multinomial an equivalent condition to Eq. (6.51) on the cell means μ_{ijk} is

$$\mu_{ijk} = MA_iB_jC_k \qquad \forall(i,j,k) \in S \tag{6.52}$$

where M is the mean value of the total of the data. Clearly, this second definition implies the first and, by the connection between multinomial and Poisson distributions, the first implies the second.

What Does QI Mean? First, incomplete contingency tables often contain complete sub-tables. For instance, in a two-dimensional table there may be rectangular blocks free of any structural zeros. From the Poisson formulation Eq. (6.52) it follows that the row and column factors are independent when restricted to the levels of this sub-table. *Under quasi-independence, all complete sub-tables satisfy the mutual independence model.*

What if there are no complete sub-tables? Imagine that the incomplete contingency table was obtained from a complete contingency table by discarding or ignoring counts in the structural zero cells. If the classification factors were independent in the complete cross-classification then Eq. (6.50) holds. Ignoring all structural zeros it immediately follows that the QI relation Eq. (6.51) holds for the nonstructural zeros. Thus, a natural interpretation of the QI model is that of ordinary independence of the underlying factors. In other words, we might say that *under the quasi-independence model, the factors are mutually independent apart from the fact that certain combinations of states are prohibited.*

We also note that assuming quasi-independence for any incomplete table, each and every sub-table is quasi-independent, whether complete or not. This generalizes the result that every sub-table of an independent complete contingency table is independent.

Quasi Log-Linear Models The quasi-independence model Eq. (6.52) may be expressed as a log-linear model with systematic component

$$\eta_{ijk} = \log(\mu_{ijk}) = m + a_i + b_j + c_k \ \forall(i,j,k) \in S \tag{6.53}$$

that is identical to the usual log-linear model for independence except that only cells in S are included. Provided that none of the marginal totals are zero, under QI the maximum likelihood estimates of the cell means satisfy the $r + c + d$ equations

$$\hat{\mu}_{i\bullet\bullet} = y_{i\bullet\bullet}, \hat{\mu}_{\bullet j\bullet} = y_{\bullet j\bullet}, \hat{\mu}_{\bullet\bullet k} = y_{\bullet\bullet y}, \tag{6.54}$$

i.e., estimates are obtained by equating marginal totals of cell means μ_{ijk} and cell counts y_{ijk}. These equations may be solved using the iterative proportional fitting algorithm of Deming and Stephan (1940) or using the more general algorithm of generalized linear models.

The degrees of freedom of the QI model is sometimes difficult to determine. However, suppose that all marginal totals of the data such as

$$Y_{i\bullet\bullet}, Y_{\bullet j\bullet}, Y_{\bullet\bullet k}, Y_{ij\bullet}, Y_{i\bullet j}, Y_{\bullet jk}$$

are positive. Let z be the number of structural zeros in the table. Then the degrees of freedom is

$$(r-1)(c-1)(d-1)-z.$$

If some of the marginal totals are zero then all the cells contributing to that total are necessarily zero as are the fitted values. These cells should be removed entirely, producing a corresponding decrease in the degrees of freedom. Indeed, the same is true for complete tables—a point not mentioned in earlier sections on log-linear models for independence. In any case, the degrees of freedom is returned automatically by most packages.

Example 6.5: Discharge of Stroke Patients Table 6.8 is a tabulation of degree of disability due to stroke for patients measured at admission and then at discharge at a certain hospital. These data are analyzed in Bishop and Fienberg (1969) and Bishop, Fienberg, and Holland (1975). Rating 1 is the least serious disability and rating 5 the worst. Since patients are never discharged if they are worse (deaths are ignored) the upper-right portion of the table contains structural zeros.

Under the quasi-independence model the mean in cell (i,j) is given by

Table 6.8. Disability Ratings of Stroke Patients at Admission and Discharge

Admission	Discharge Status					
Status	1	2	3	4	5	Total
1	5 (5.00)	—	—	—	—	5
2	4 (3.75)	5 (5.25)	—	—	—	9
3	6 (4.43)	4 (6.20)	4 (3.37)	—	—	14
4	9 (6.16)	10 (6.63)	4 (4.69)	1 (4.52)	—	24
5	11 (15.66)	23 (21.92)	12 (11.94)	15 (11.48)	8 (8.00)	69
Total	35	42	20	16	8	121

Note: Fitted values are given in parentheses.

$$\mu_{ij} = MA_iB_j \qquad (i,j) \in S$$

where $S = \{(i,j) : i \geq j\}$. Now look at the line for those admitted at status 5. Conditional on this admission status, the discharge distribution of these 69 individuals is multinomial. Under the quasi-independence model the response probabilities are

$$\pi_{j|5} = \frac{\mu_{5j}}{\mu_{5\bullet}} = \frac{B_j}{\sum_{i=1}^{5} B_k}.$$

Note that M and the A_i disappear. The B_j, appropriately rescaled, represent underlying probabilities of being in column j. For the other rows, the conditional multinomial probabilities $\pi_{j|i}$ are simply B_j divided by the (partial) row total of the B_j.

What does the quasi-independence model mean here? We may interpret the probability vector $\pi_{j|5}$ above as the probability that a patient eventually stabilizes at rating j. The QI model says that, given that a discharged patient's state cannot be worse, the state at discharge is independent of the state at admission. This would be the case if treatment was ineffective and patients' states changed randomly, a rather unlikely hypothesis.

The model was fitted in `GLIM` by loading the 15 counts into a column vector, defining row and column factors each on five levels and fitting the additive model `ROW+COL` with Poisson link. The degrees of freedom is

$$(r-1)(c-1) - z = 16 - 10 = 6$$

with deviance 9.60. This is not a particularly good fit but the model is accepted at level 5% (P = 0.142). Nor is there any systematic pattern in the deviations of the fitted values from the observed values (see Table 6.8). It seems that we can accept the hypothesis that treatment is ineffective. The estimated probability vector of $\pi_{j|5}$ is

$$\hat{\pi}_{j|5} = \hat{e}_{5j}/69 = (0.227, 0.318, 0.173, 0.166, 116).$$

These represent underlying probabilities of the five status classes. Under the model, patients move randomly between these classes. Apart from the fact that they are never released if they are worse, they do not benefit from treatment.

6.5.3 Mobility Models

One application of quasi-independence among many, is to the study of how and why individuals change from one category to another. Two applications are given in this section. We also demonstrate how more complicated models than the straight QI model can be built up by dividing the table into sections, within which quasi-independence holds but between which it does not. This simply requires the construction of another factor indexing these sections of the table and appropriate interaction terms in the model formula.

Example 6.6: British Social Mobility Data The data in Table 6.9 were collected by Glass (1954) and his co-workers and have been analyzed in many other textbooks. About 3500 fathers and sons were sampled. We will assume that the sample was a simple random one even though it was in fact stratified and take the counts of the table to be Poisson. The career status of father and son were measured on a five-point scale. Note that the zero count in the bottom left corner is a sampling zero, not a structural zero. There is interest in understanding the degree to which the career status of the father determines the career status of the son.

Fitting the independence model gives deviance 811.0 on 16 degrees of freedom. We note first that the counts on the diagonal are larger and so there is apparently a tendency of sons to have the same status as their father. This suggests ignoring the diagonal data and fitting the quasi-independence model to the remaining incomplete data. This will determine whether, among those sons who are *socially mobile*, their status is affected by the status of their father over and above their not having the same status as their father. To do this we can either (i) simply remove the diagonal data or, (ii) include five free parameters for the five diagonal elements in which case the fitted values and observed values on the diagonal will agree and contribute nothing to the overall deviance. This can simply be done by defining a factor indicating the diagonal and the interaction of this with the row factor gives the required five parameters. The model formula in the `GLIM` output below is `DIAG*FATH+SON`. The fit of this model is still very poor having deviance 249.4 and 11 degrees of freedom. We can see why—the counts *near* the main diagonal are inflated, but higher just above than just below.

Table 6.9. Career Status of Fathers and Sons [from Glass (1954)]

	Son's Status					
Father	1	2	3	4	5	Total
1	50	45	8	18	8	129
2	28	174	84	154	55	495
3	11	78	110	223	96	518
4	14	150	185	714	447	1510
5	0	42	72	320	411	845
Total	103	489	459	1429	1017	3507

How can we further model these data? One imagines that the dynamics of advancing social status are quite different to the dynamics of descending social status. The quasi-independence model does not recognize this. A model taking this into account is easily developed. We simply assume that the parameters A_i, B_j depend on whether $i < j$ or $i > j$. This is entirely equivalent to fitting separate quasi-independence models separately to the upper and lower triangular arrays of the data. The model is

$$\pi_{ijk} = \begin{cases} A_i B_j & i > j \\ A_i^* B_j^* & i < j \end{cases},$$

and may be fitted very simply. Let `T` be a three-level factor indicating the section to which a data point belongs, i.e., the upper triangle, the lower triangle, or the main diagonal. Then the `GLIM` model formula `T.(FATH+SON)` fits the required model.

In fact, this model is an ordinary quasi-independence model if the data is suitably rearranged. Divide the fathers up into three groups indexed by `T` above. Then the cross-classification of the father's status with this three-level factor gives 15 classes. We then retabulate the data with respect to this 15-level factor and the five-level factor of the son's status. For two of the rows, there are all structural zeros and so we may omit these lines. The result is given in the 13 × 5 array displayed in Table 6.10. The model being fitted is just the ordinary quasi-independence model for this contingency table.

Table 6.10. Rearrangement of Social Mobility Data as a 13 × 5 Array

		Son's Status				
T	Father	1	2	3	4	5
Lower	2	28	—	—	—	—
	3	11	78	—	—	—
	4	14	150	185	—	—
	5	0	42	72	320	—
Diagonal	1	50	—	—	—	—
	2	—	174	—	—	—
	3	—	—	110	—	—
	4	—	—	—	714	—
	5	—	—	—	—	411
Upper	1	—	45	8	18	8
	2	—	—	84	154	55
	3	—	—	—	223	96
	4	—	—	—	—	445

```
[i]  ? $look y son fath diag t$
[o]        Y       SON     FATH     DIAG       T
[o]   1 50.000   1.000   1.000   1.000   2.000
[o]   2 45.000   2.000   1.000   2.000   3.000
[o]   3  8.000   3.000   1.000   2.000   3.000
[o]   4 18.000   4.000   1.000   2.000   3.000
[o]   5  8.000   5.000   1.000   2.000   3.000
[o]   6 28.000   1.000   2.000   2.000   1.000
.
.
!              rest of data list is omitted
[i]  $fact fath 5 son 5 diag 2 t 3$yvar y$error p$
[i]  ? $fit fath+son$
[o]  scaled deviance = 811.0 at cycle 5
[o]            d.f. = 16
[o]
[i]  ? $fit fath*diag+son$
[o]  scaled deviance = 249.43 at cycle 4
[o]            d.f. = 11
[o]
[i]  ? $fit t*(fath+son)$
[o]  scaled deviance = 13.954 at cycle 4
[o]            d.f. = 6
```

The deviance of this model is 13.95 on six degrees of freedom. This comprises deviance 1.4 on three degrees of freedom for the lower triangle of data corresponding to sons with lower status than their fathers and 12.6 on three degrees of freedom for the upper triangle of data corresponding to sons with higher status than their fathers. There is certainly no evidence against quasi-independence for downwardly mobile sons—their status does not appear to be dependent upon their father's status apart from the fact that it is different (by definition of being downwardly mobile). For upwardly mobile sons, the quasi-independence model is not adequate.

This does not exhaust the possible models for this data set. One obvious model could be based on the idea of distance between status categories. For instance, consider the multiplicative model

$$\mu_{ij} = A_i B_j D_k$$

where D_k indicates the different diagonals, e.g., $D = k$ corresponds to cells with $i - j = k$. This is nothing but the quasi-independence model for the three-way classification with respect to A, B, and D. This model actually fits slightly worse than the previous model (deviance 19.1 on eight degrees of freedom). Models exploiting the ordinal nature of the status classification could also be tried (see Section 6.4), but we leave it as an exercise to show that the models discussed there do not fit well.

Table 6.11. Faculty of Enrollment and Final Degree

	Final Degree						
Enrollment	S	C	E	H	FA	M	Total
S	—	9	8	3	3	4	27
C	18+	—	15+	7	4	3−	47
E	12	15+	—	2−	2−	7	38
H	5−	8	2−	—	21+	15+	51
FA	1	1	0	7+	—	3	12
M	3	2	1	11+	4	—	21
Total	38	35	26	30	34	32	195

Note: For those students who changed faculty only. Faculties are arranged into two groups.

Example 6.7: Mobility of Students Across Faculties Table 6.11 relates to students at La Trobe University in 1990. Students had two categorical variables measured. The first was their faculty of enrollment at entry to the university and the second the faculty in which they received their final degree. Only data for those students who changed faculty is recorded. We are interested in understanding the movement of students from one faculty to another.

The six faculties have been divided into two groups of three, Science (S), Commerce (C), and Engineering (E) in the first group and Humanities (H), Fine Arts (FA), and Music (M) in the second group. We first fit the quasi-independence model which says that the faculty of final degree is not affected by the faculty of enrollment, over and above these two faculties being different. The deviance is 91.94 with degrees of freedom $5 \times 5 - 6 = 19$. Clearly there is no quasi-independence.

I have added + and − signs to the table to indicate those data points that have significant positive or negative residuals in this model. Apparently, students tend to remain within the faculty blocks. A model to this effect is constructed by dividing the table up into four contingency tables. In the upper left and lower right 3×3 tables we assume separate quasi-independence models. In other words, we assume that, within each block of faculties, the faculty of enrollment does not affect which faculty students move to. In the upper-right and lower-left tables, which are complete, we fit the full independence model.

This can simply be done by defining a block factor `BLOCK` on four levels indicating the section of the table to which each data point belongs. The model formula `BLOCK.(ENTER+LEAVE)` fits the required model. The deviance is 6.83 on 10 degrees of freedom. There does seem to be quasi-independence within the blocks of faculties.

```
[i]  ? $units 30
[i]  ? $data y$read
[i]  $REA? 9 8 3 3 4 18 15 7 4 3 12 15 2 2 7 5 8 2 21 15 1
[o]  1 0 7 3 3 2 1 11 4
```

```
[i]  ? $data enter$read
[i]  $REA? 1 1 1 1 1 2 2 2 2 2 3 3 3 3 3 4 4 4 4 4 5 5 5 5
[o]  5 6 6 6 6 6
[i]  ? $data leave$read
[i]  $REA? 2 3 4 5 6 1 3 4 5 6 1 2 4 5 6 1 2 3 5 6 1 2 3 4
[o]  6 1 2 3 4 5
[i]  ? $data block$read
[i]  $REA? 1 1 2 2 2 1 1 2 2 2 1 1 2 2 2 3 3 3 4 4 3 3 3 4
[o]  4 3 3 3 4 4
[i]  ? $fact enter 6 leave 6 block 4$
[i]  ? $yvar y$error p$
[i]  ? $fit enter+leave$
[o]  scaled deviance = 91.94 at cycle 4
[o]                d.f = 19
[i]  ? $fit block*(enter+leave)$
[o]  scaled deviance = 6.8342 at cycle 3
[o]               d.f. = 10
```

6.5.4 Symmetry and Quasi-Symmetry

Models for symmetry are useful for analyzing matched pairs data. Imagine that individuals are classified according to an m-level categorical factor and at some later date are again classified according to the same factor. Let Y_{ij} be the number of individuals who respond (i, j) for the two measurements. These data can be arranged as a square $m \times m$ contingency table.

Symmetry Models The independence model is not likely to be of interest for such data since individuals' measurements are probably positively correlated and so counts along or near the main diagonal will likely be inflated. One issue of obvious interest is to describe the patterns among those who record *different* measurements. For instance, if the classifying factor was the political party of first preference and individuals were surveyed and then followed up at a later date, then the hypothesis of *no swing* would be of primary interest. Mathematically, this hypothesis assumes that the probability π_{ij} of an individual being recorded in cell (i, j) satisfies

$$\pi_{ij} = \pi_{ji}.$$

Symmetry can be formulated as a log-linear Poisson model. Suppose that the Y_{ij} have independent Poisson distributions with mean μ_{ij} and let S be the set of indices for which $\mu_{ij} > 0$. The symmetry model is the log-linear model

$$\log(\mu_{ij}) = m + a_i + a_j + (ab)_{ij} \qquad (i, j) \in S \tag{6.55}$$

with the additional restriction $(ab)_{ij} = (ab)_{ji}$. This immediately implies that the set S will be a symmetric set. Note that the main effects a_i and a_j for the row and column factors of the two-way table are assumed identical. This is necessary if we are to have $\mu_{ij} = \mu_{ji}$. Under the symmetry model the marginal distributions of the row and column measurements are identical. This is called *marginal homogeneity*. For the aforementioned political poll this means that the support for the different parties is the same on the two sampling occasions.

Fitting the symmetry model may be done by hand by noting that, conditional on an individual responding (i,j) or (j,i), the number Y_{ij} has binomial distribution with parameters $Y_{ij} + Y_{ji}$ and $1/2$. Thus,

$$\hat{e}_{ij} = (Y_{ij} + Y_{ji})/2.$$

When there are no structural zeros, the degrees of freedom is $m(m-1)/2$ where m is the number of levels of the classifying factor.

Quasi-Symmetry Models The symmetry model is quite restrictive since it imposes marginal homogeneity. This can be relaxed by removing the assumption that the main effects for the row and column classification are identical. This gives the *quasi-symmetry* (QS) model

$$\log(\mu_{ij}) = m + a_i + b_j + (ab)_{ij} \qquad (i,j) \in S \tag{6.56}$$

where, as before, $(ab)_{ij} = (ab)_{ji}$. The cell probabilities π_{ij} satisfy

$$\pi_{ij} = A_i B_j S_{ij}$$

where $S_{ij} = S_{ji}$ describes a symmetric disturbance to an underlying independence model. The degrees of freedom is $(m-1)(m-2)/2$. When the parameters $(ab)_{ij}$ are constant, they may be absorbed into the intercept term m and the model becomes that of quasi-independence. Quasi-symmetry is thus an extension of the quasi-independence model. Note in particular that the QS model will fit the data perfectly along the main diagonal.

The QS model is best understood in terms of odds ratios. Let

$$\theta_{ij} = \log\left(\frac{\mu_{ij}\mu_{11}}{\mu_{i1}\mu_{1j}}\right)$$

be the log-odds ratio of the 2×2 sub-table comprising rows $(1, i)$ and columns $(1, j)$. Then, under the quasi-symmetry model, $\theta_{ij} = \theta_{ji}$ and for this reason is sometimes called the *symmetric association* model.

Fitting the QS Model There are two ways to fit the quasi-symmetry model Eq. (6.56). The most straightforward is to define a factor `SYM` on $m(m+1)/2$ levels which groups cells (i,j) without regard to order. For instance cells (1, 2) and (2, 1) form a group. The model

$$\text{ROW} + \text{COL} + \text{SYM}$$

fits the quasi-symmetry model. However, either all the row effects or all the column effects will be aliased in the standard parametrization and the models `ROW+SYM` or `COL+SYM` produce identical fits. The full symmetry model has the simple model formula `SYM`.

A second method is based on constructing a second data table being the *transpose* of the original table. Let the two-level factor `TABLE` indicate whether a data point comes from the true or transposed data table. Fitting the model

$$\text{ROW.COL} + \text{ROW.TABLE} + \text{COL.TABLE}$$

the fitted values in the original table are just the fitted values under the quasi-symmetry model. This may be proven by showing that the likelihood equations for the above model are identical to those for the quasi-symmetry model. The deviance will be a factor of two too large since it is calculated across both tables. The simpler model `ROW.COL` corresponds to the full symmetry model and so marginal homogeneity is easily tested in either approach.

Example 6.6: (Continued) **British Social Mobility Data.** The symmetry model has deviance 46.20 on $(5 \times 4)/2 = 10$ degrees of freedom. The quasi-symmetry model has deviance 10.95 on $(4 \times 3)/2 = 6$ degrees of freedom. Fitted values are displayed for both of these models in Table 6.12, the QS fitted values at right. While not a great fit, the quasi-symmetry model is an improvement on the blocked quasi-independence model fitted in Section 6.5.3 and a huge improvement on the full quasi-independence model.

The difference between the symmetry and quasi-symmetry models is mar-

Table 6.12. Symmetry and QS Models Fitted to the British Mobility Data

	Son's Status				
Father	1	2	3	4	5
1	50	45	8	18	8
		36.5, 43.3	9.5, 11.0	16.0, 19.3	4.0, 5.4
2	28	174	84	154	55
	36.5, 29.7		81.0, 78.5	152.0, 155.7	48.5, 57.0
3	11	78	110	233	96
	9.5, 8.0	81.0, 83.5		204.0, 215.2	84.0, 101.3
4	14	150	185	714	447
	16.0, 12.7	152.0, 148.3	204.0, 192.8		383.5, 442.2
5	0	42	72	320	411
	4.0, 2.6	48.6, 39.9	84.0, 66.7	383.5, 324.8	

Fitted values displayed below counts.

ginal homogeneity. Thus, the increase of deviance of 46.20 – 10.95 = 35.25 on four degrees of freedom constitutes massive evidence against this hypothesis. The underlying distributions of job status are different for fathers and sons. From examining the row and column totals, the difference is that more sons have jobs in the highest status and less in the middle status. The rather poor fit of the QS models is mainly due to the fact that counts are inflated more above than below the diagonal, even after taking account of the generally increasing status of sons over fathers.

6.5.5 A Ranking Model for Preference Data

Imagine m products that are to be ranked according to consumer preferences. A sample of consumers are surveyed. It may not be feasible to have each consumer rank the m products and, even if such a ranking were available, it may be a complicated task to combine these rankings into a single ranking summarizing the relative merits of the m products.

Suppose instead that each consumer is asked to state a preference among only two of the m products. Suppose that there are N_{ij} consumers asked to compare products i and j and that Y_{ij} of them prefer i to j while Y_{ji} prefer j to i. These *paired preference* data may be displayed as a square $m \times m$ contingency table with missing diagonal, since we do not compare a product with itself. If we assume that all consumers respond independently then

$$Y_{ij} \stackrel{d}{=} Bi(N_{ij}, \pi_{ij}) \qquad i \neq j$$

where π_{ij} is the probability that a randomly chosen consumer, asked to choose between products i and j, will prefer i.

For individual consumers, rankings should be *transitive*, i.e., prefering i over j and j over k should imply a preference for i over k. For a population of individuals, we might define i to be preferred to j if the majority prefer i over j, i.e., if $\pi_{ij} > 0.5$. This preference ranking is *not* necessarily transitive and such anomolies may well appear in the table. To arrive at a ranking of the m products, we require a less simple minded method than simply saying i is ranked higher than j if $y_{ij} > y_{ji}$.

Bradley–Terry Model A very simple summary of the data would associate with each product a single parameter p_i, measuring how attractive product i is. If $p_i > p_j$ then it should follow that $\pi_{ij} > 0.5$. The simple model

$$\pi_{ij} = \frac{p_i}{p_i + p_j} \tag{6.57}$$

due to Bradley and Terry (1952), ensures this though it is not the only such model. The parameters p_i not only imply a ranking of the products but give a *quantitative* ordering. Notice that the parameters p_i may be multiplied by an arbitrary scaling constant without affecting the above relation. It is usually assumed that the p_j sum to m although any other value will do just as well.

An interesting application of this model is to the analysis of sporting records. Suppose that m players are to be compared. Records of matches between these players over a period of time are collected and displayed as a square contingency table where Y_{ij} is the number of times player i beat player j. The parameter p_i describes the overall strength of player i and can be used to predict the outcome of a future match according to Eq. (6.57).

Fitting the Bradley–Terry Model The Bradley–Terry model is in fact identical to the quasi-symmetry model with the identification

$$\log(p_i) = a_i - b_i,$$

see Fienberg and Larntz (1976). If the second method on p. 364 is used to fit the quasi-symmetry model, then the estimates of $\log(p_i)$ will appear as simple interactions of the column and table effects. An *Splus* function `preference` that uses quite a different fitting algorithm is at the Wiley website.

Example 6.9: Tourist Destinations A tourism firm surveyed Australian vacationers on their preferences among six Pacific Island destinations. Each was asked to choose between two destinations only, after examining descriptions of their respective vacation packages. Table 6.13 gives the results.

The preferences suggested by this table are intransitive. For instance, Vanuatu is strongly preferred to Fiji (12 > 7) and Fiji is preferred to the Cook islands (14 > 12), however the Cook Islands are preferred to Vanuatu (14 > 12). The `GLIM` outputs fit the model as a quasi-symmetry model after having input the table twice, one in reflected form. The factor `TABLE` is on two levels. Most of the parameter estimates have been deleted from the list. The estimates of $\log(p_i)$ appear as interactions of `ROW(i).TABLE(2)`. Exponentiating these and normalizing to sum to six gives

Table 6.13. Preferences for Pacific Island Vacation Destinations

Place	Bali	Cook	Fiji	Norfolk	Tahiti	Vanuatu
Bali	—	14	16	24	14	7
Cook	26	—	12	30	17	14
Fiji	20	14	—	10	17	7
Norfolk	38	12	24	—	10	11
Tahiti	16	26	18	16	—	10
Vanuatu	15	12	12	18	9	—

$$\hat{p} = (0.66, 1.18, 0.81, 0.90, 1.19, 1.25)$$

for the six respective destinations. From a future sample we might expect that a proportion

$$\pi_{61} = \frac{1.25}{1.25 + 0.66} = 0.654$$

prefer Vanuatu to Bali. The model does not fit particularly well, the main deficiency being in its prediction of the counts relating to preferences between Norfolk and Cook Islands. But it does give a single numerical measure of the attractiveness of the six destinations.

```
[i]   ? $fit row*col+row*table+col*table$
[o]   scaled deviance = 31.608 at cycle 3
[o]               d.f. = 19
[o]
[i]   ? $fit -row-col-table-%gm$
[o]   scaled deviance = 31.608 (change = 0.0000) at cycle 3
[o]               d.f. = 19     (change = 0      )
[o]
[i]   ? $disp e$
[o]      estimate       s.e.     parameter
[o] 37     0.0000     0.2287     ROW(1).TABLE(2)
[o] 38     0.5842     0.2066     ROW(2).TABLE(2)
[o] 39     0.2023     0.2123     ROW(3).TABLE(2)
[o] 40     0.3083     0.1881     ROW(4).TABLE(2)
[o] 41     0.5959     0.2185     ROW(5).TABLE(2)
[o] 42     0.6434     0.2383     ROW(6).TABLE(2)
[o] 43    -0.5842     0.2197     COL(2).TABLE(2)
[o] 44    -0.2023     0.2292     COL(3).TABLE(2)
[o] 45    -0.3083     0.2205     COL(4).TABLE(2)
[o] 46    -0.5959     0.2269     COL(5).TABLE(2)
[o] 47    -0.6434     0.2441     COL(6).TABLE(2)
```

6.6 MULTIPLE RECORD SYSTEMS

Imagine a population of unknown size N. There are t lists each of which contain a subset of the population. The lists are produced by randomly sampling the population but the intensity of sampling might differ from list to list. This experimental set up is called a *multiple record system*. There is close link between multiple record systems and so-called capture–recapture experiments, the definitive reference on the latter being Seber (1982).

One important example is the United States census. One of the lists, the

United States census itself, contains almost but not quite, every individual in the population. In addition there are one or two other lists with much smaller sampling intensities. The aim is to estimate the actual population size from the three lists, and then to estimate how many people are missed by the census. There is special interest in the extent to which the undercount might be worse for minority groups.

Dependence between lists is allowed. Later we will look at lists of drug addicts collected by different social service agencies and not surprisingly find that individuals tend to appear on one list if they appear on the other lists. Higher order dependencies, for instance a tendency to appear on list three if one appears on list one *and* list two, can be included, but in practice such effects are often small.

Heterogeneity of individuals can be accounted for by marrying the Rasch model, popular among educational researchers, with an assumption that high-order interactions are negligible. We look at this in Section 6.6.3, but until then homogeneity of the population is assumed.

6.6.1 Matched Lists and Incomplete Tables

We assume that the identity of individuals is sufficiently well recorded that we can identify those appearing on more than one list. For instance, recording the age, sex, middle initial, and height of an individual would probably be sufficient for human populations, while preserving individual anonymity. With this assumption, we can identify r distinct individuals who appeared on at least one of the lists and also give a full list history of these individuals.

Let us describe an individual's list history by a sequence of ones and zeros. For instance, $\omega = 1001$ describes an individual on the first and fourth list only. Those individuals who were never seen have list history $\omega = 00..0$. The experiment assigns each individual to one of the 2^t histories, with probabilities denoted p_ω. The sufficient statistics are the frequencies, r_ω, of all the $2^t - 1$ observable histories. Let L1 denote the two-level factor measuring whether or not an individual is on list 1, L2 on list 2, etc. The data $\{r_\omega\}$ are a 2^t cross-classification of the population with respect to these t two-level factors. There is one missing frequency $r_{00..0}$. By fitting models to the p_ω we aim to estimate $p_{00..0}$ and then the missing frequency. The data being a simple $2 \times 2 \times \ldots \times 2$ table, log-linear models are an obvious and flexible family of models.

Simplest Case: Two Lists There are three observable list histories, namely 01, 10, 11. The data r_{01}, r_{10}, r_{11} may be arranged as a 2×2 contingency table with respect to the indicating factors L1, L2, as shown in Table 6.14. There are n_i individuals on list i, and r_{11} individuals on both lists.

The count r_{00} is unknown which prevents us from knowing the population

Table 6.14. Multiple Record Data for $t = 2$ Lists as 2×2 Table

	L2 = 1	L2 = 0	Total
L1 = 1	r_{11}	r_{10}	n_1
L1 = 0	r_{01}	$r_{00} = ?$	?
Total	n_2	?	$N = ?$

size N. An estimate $\hat{\mu}_{00}$ provides the estimate

$$\hat{N} = r_{11} + r_{10} + r_{01} + \hat{\mu}_{00}.$$

Clearly no estimate of r_{00} can be made without some assumptions about the relationships of the cell means, $\mu_{ij} = \mathrm{E}(r_{ij})$. There may or may not be dependence between the lists. Let us measure the dependence by the odds ratio ϕ_{12} defined as usual as $\phi_{12} = (\mu_{11}\mu_{00})/(\mu_{10}\mu_{01})$. Rearranging we have $\mu_{00} = \phi_{12}\mu_{10}\mu_{01}/\mu_{11}$. Replacing the μ_{ij} by their estimates r_{ij} we have the estimator

$$\hat{\mu}_{00} = \phi_{12}\,\frac{r_{01}r_{10}}{r_{11}}. \tag{6.58}$$

Now ϕ_{12} is not estimable from three counts. One might assume that $\phi_{12} = 1$, which is to say that being on list 1 has no effect on the probability of being on list 2. This is the independence model for the two-way classification by the factors L1, L2. The goodness-of-fit of this model is also untestable from three counts. We then obtain

$$\hat{\mu}_{00} = \frac{r_{01}r_{10}}{r_{11}} \tag{6.59}$$

and so

$$\hat{N} = r_{11} + r_{10} + r_{01} + \frac{r_{01}r_{10}}{r_{11}} = \frac{n_1 n_2}{r_{11}}.$$

This is the well-known Petersen estimator [see Seber (1982)].

Three Lists With $t = 3$ lists, we do not need to assume a value for ϕ_{12}. We can estimate it directly from observable frequencies provided that we make an assumption about *higher order* dependencies. To elaborate, there are seven observable capture histories, that we arrange as a $2 \times 2 \times 2$ table with a single missing cell, as displayed in Table 6.15. There is a complete 2×2 table for those on list 3, and an incomplete 2×2 table for those not on list 3. From the

Table 6.15. Multiple Record Data for $t = 3$ Lists as $2 \times 2 \times 2$ Table

	L3—Yes		L3—No	
	L2—Yes	L2—No	L2—Yes	L2—No
L1—Yes	r_{111}	r_{101}	r_{011}	r_{100}
L1—No	r_{011}	r_{001}	r_{010}	$r_{000} = ?$

incomplete table, we have the 2-list estimator

$$\hat{\mu}_{000} = \phi_{12} \frac{r_{010} r_{100}}{r_{110}}$$

where ϕ_{12} measures the dependence of lists 1 and 2. However, we have an estimate

$$\hat{\phi}_{12} = \frac{r_{111} r_{001}}{r_{101} r_{011}}$$

from the complete table, provided that we assume ϕ_{12} is the same for both tables, in other words there is *no three-way interaction*. This gives the estimator

$$\hat{\mu}_{000} = \frac{r_{111} r_{100} r_{010} r_{001}}{r_{110} r_{101} r_{011}} \tag{6.60}$$

and hence an estimate of N.

Log-Linear Models Other assumptions lead to alternative estimates. For instance, there might be reason to assume that list 3 is independent of lists 1 and 2, for instance, if list 3 were a police list and lists 1 and 2 were drug therapy lists with a policy of not reporting users to police. There might be quite a high, but unknown, degree of dependence between lists 1 and 2, however. In this case it seems that there is no loss in combining lists 1 and 2 into a joint list, containing individuals on list 1 or list 2 or both. The data are then arranged as in Table 6.16 and from the two-list estimate Eq. (6.59) we obtain the estimator of r_{000} as

$$\hat{\mu}_{000} = \frac{(r_{110} + r_{100} + r_{010}) r_{001}}{r_{111} + r_{101} + r_{011}}. \tag{6.61}$$

We can obtain this estimator using an appropriate log-linear model. Arranging the data as a 2^3 contingency table, one parametrization of the full log-linear

Table 6.16. Multiple Record Data for $t = 3$ After Joining Lists 1 and 2

	L3—Yes	L3—No
L12—Yes	$r_{111} + r_{101} + r_{011}$	$r_{110} + r_{100} + r_{010}$
L12—No	r_{001}	$r_{000} = ?$

model for the complete table is

$$
\begin{aligned}
\log(\mu_{111}) &= m \\
\log(\mu_{011}) &= m + u_1 \\
\log(\mu_{101}) &= m + u_2 \\
\log(\mu_{001}) &= m + u_1 + u_2 + u_{12} \\
\log(\mu_{110}) &= m + u_3 \\
\log(\mu_{010}) &= m + u_1 + u_3 + u_{13} \\
\log(\mu_{100}) &= m + u_2 + u_3 + u_{12} \\
\log(\mu_{000}) &= m + u_1 + u_2 + u_3 + u_{12} + u_{23} + u_{13} + u_{123}.
\end{aligned}
\tag{6.62}
$$

For the incomplete table, the last relation is not required. The parameter μ_i is a main effect for occasion i giving the log-ratio of the means counts of those not on list i to those not on this list or equivalently the logit probability of *not* being on list i. The parameter u_{12} is an interaction of list 1 and 2 that measures the extent to which being on list 1 affects the chance of being on list 2. How do we get the estimator in Eq. (6.61)?

Set the highest order parameter u_{123} to zero. Then set u_{13}, u_{23} to zero to model lists 1 and 2 being independent of list 3. The nonzero parameters $(m, u_1, u_2, u_3, u_{12})$ are estimated from the seven observable frequencies and then μ_{000} estimated from the last line of these equations.

Four or More Lists Obtaining data for a large number of lists may become impractical because of the large effort needed to match individuals across all the lists, i.e., to identify the r distinct individuals and to construct their histories. Nevertheless, the statistical advantages of having an extra list was apparent above, where we were able to replace the assumption of list independence for the two-list system, with a more plausible assumption of no three-way dependence in the three-list system.

For t lists the data comprise a 2^t factorial design. Since there are only $2^t - 1$ observable counts, estimation of N requires the elimination of at least one of the 2^t factorial parameters. The weakest assumption is that the highest order dependence of the lists is negligible. Since the residual degrees of freedom is zero, it will in general not be possible to test this assumption, although if the next highest interactions appear to be negligible then the assumption will be more plausible. These next higher interactions, and all other terms in the model, can be tested after the assumption of no highest order interaction. Through a process of model selection we aim to find a simple well-fitting model from which an estimate of N is obtained.

For instance, with $t = 4$ we assume $u_{1234} = 0$. If many of the three-way interactions u_{ijk} were significantly different from zero then we might question this assumption, and it would be difficult to fit any model to the data with much confidence. If the three-way interactions are all insignificant then we might assume they are zero. Examination of the estimated parameters may suggest that further parameters can be set to equal to each other, or to zero. Provided these model reductions are close to true, a more accurate estimate of N is obtained.

A simple device for obtaining the estimate of the missing cell is to include a value for this cell but to give it zero weight in the fitting procedure. The fitted value for this cell is then the required estimated mean $\hat{\mu}_{00..0}$ and the variance of $\log \hat{\mu}_{00..0}$ is the variance of the linear predictor for this cell.

Adjusting for N Being Fixed In the log-linear model treatment of multiple lists, the unseen count is considered Poisson distributed and therefore N itself is considered Poisson. It usually makes more sense to consider N to be a fixed unknown value that we wish to estimate, even though in hypothetic repetitions of the multiple listing experiment it may well be different. When N is treated as fixed, i.e., when we condition on N, the distribution of the r_ω is multinomial rather than Poisson.

In Section 2.4 we argued that it makes no difference to inference on the probability parameters of a multinomial whether or not the total is considered fixed or random. However, here the parameter of interest is the total N itself and it certainly does make a difference to inference whether we consider it fixed or random.

It can be shown however that the ML estimator of N under multinomial sampling is hardly different from the estimator under Poisson sampling. Letting P and M subscripts denote Poisson and multinomial sampling, respectively, it can also be shown that

$$\text{Var}_M(\hat{N}) = \text{Var}_P(\hat{N}) - N. \tag{6.63}$$

Thus, the width of intervals will be less under multinomial than under Poisson sampling [see Sandland and Cormack (1984) for details].

There are several reasonable methods of computing confidence intervals for N. Let R be the total number of *distinct* individuals sampled. Logically, N can be no smaller than R. The method of Chao (1989) is based on the approximate result

$$\log\left(\frac{N - R}{\hat{N} - R}\right) \xrightarrow{d} \mathcal{N}(0, \sigma^2) \tag{6.64}$$

where $\hat{\sigma} = \text{s.d}(\hat{N})/[\hat{N} - R]$. This leads to confidence intervals of the form

$$(R + (\hat{N} - R)e^{-c\hat{\sigma}}, R + (\hat{N} - R)e^{c\hat{\sigma}}). \tag{6.65}$$

Among interval constructions with closed form, this is the only one that can be recommended. It is easily applied so long as a standard error for $\hat{N}$ is available. The lower limit is necessarily larger than the logical lower bound R.

Profile likelihood intervals can also be constructed from the multinomial profile likelihood function. It is shown in Cormack (1992) that

$$\hat{\ell}_M(N) = \hat{\ell}_P(N) + \log(1 - r/N).$$

The profile log-likelihood under Poisson sampling can be obtained by setting the mean of the unobserved cell equal to $N - R$ (using the `offset` command in `GLIM`) and recording the deviance, for various choices of N. Subtracting the function $\log(1 - r/N)$ gives a picture of the multinomial deviance. The confidence interval for N comprises those values of N for which this is within a χ_1^2 quantile of its minimum. Again the lower limit of this interval is necessarily greater than R.

A third method is to compute the variance of the linear predictor for the missing cell, $\log \hat{\mu}_{00..0}$. In the standard parametrization of Eq. (6.62) this variance is just $1^T V 1$ where $V = (X^T W X)^{-1}$ is the variance matrix of the parameter estimates. From this we compute a confidence interval for $\log \mu_{00.00}$, then $\mu_{00..0}$ and finally N.

6.6.2 Illustration: Addicts in Edinburgh

The methods are illustrated on data listed in Frischer et al. (1995). One aim of that study was to estimate the number of intravenous drug users in Glasgow, Scotland in 1990 (as a step in estimating the number of HIV infected drug users). There were four lists chosen on which drug injectors might appear. List 1 was the HIV test register. List 2 was a combination of various community treatment agencies for problem drug users. List 3 was a police list of arrests for nonmarijuana related drug offences. List 4 was a needle and syringe exchange. We note that list 3 is qualitatively different to lists 1, 2, and 4 which are social services.

The identity of individuals on the police list was known completely. Individuals on the social lists 1, 2, and 4 were identified by their date of birth, gender, initials, and residential zip code. Two identical individuals in these respects were deemed to be the same individual; see the above cited paper for details. The 1990 data are listed in Table 6.17, broken down into four age-groups, 15–19, 20–24, 25–29, and 30–55, respectively. The data are input into `GLIM` as 64 counts with four two-level factors `L1`, `L2`, `L3`, `L4` and a factor `age` on four

Table 6.17. List Histories of Glaswegian IVD Users for Four Age-Groups

	L1 (HIV)	Yes	Yes	No	No	Yes	Yes	No	No
	L2 (TRE)	Yes	No	Yes	No	Yes	No	Yes	No
L3 (POL)	L4 (NES)			Age 15–19				Age 20–24	
Yes	Yes	0	0	0	7	1	1	7	26
Yes	No	1	3	7	63	3	8	25	121
No	Yes	1	0	14	76	23	27	74	392
No	No	6	21	65	?	61	98	270	?
L3 (POL)	L4 (NES)			Age 25–29				Age 30–55	
Yes	Yes	1	1	4	16	2	0	2	7
Yes	No	2	3	11	87	2	3	7	87
No	Yes	13	20	36	282	4	5	23	114
No	No	36	88	278	?	13	60	258	?

levels. The missing values are coded as an arbitrary nonnegative integer, with associated weight zero.

We first estimate the four missing counts by the additive model, separately for each age-group. The model formula in `GLIM` is

$$(L1 + L2 + L3 + L4) * AGE.$$

The fitted values for cells 16, 32, 48, 64 estimate the numbers $\hat{\mu}_{0000}$ unsampled. In this and later models, the standard error of $\hat{\mu}_{0000}$ is computed from

$$\text{s.e}(\hat{\mu}_{00..0}) = \text{s.e}\{\log(\hat{\mu}_{00..0})\} \times \hat{\mu}_{00..0}$$

and the variances of log $\hat{\mu}_{0000}$ for the three groups are in the positions 16, 32, 48, 64 of the system vector `%v1`. Estimates and standard errors are given in Table 6.19 under the heading "Simplest Model." The model fits extremely poorly, the total deviance being 135.63(40). This is not surprising as the lists are surely not all independent.

Let us find a more appropriate model, initially just for the youngest age-group. Starting with the model with all three-way interactions (the full model here since one count is missing) we proceed by removing the least significant three-way interactions, then the least significant two-way interactions until we arrive at a hierarchical model with all nonsignificant terms removed. Table 6.18 lists the results of this procedure for the youngest age-group. For instance, the model on the second last line has the police list 3 independent of the treatment lists 1, 2, and 4, with the model formula `L1*L2*L4+L3`. If the three-way interaction `L1.L2.L4` is removed the deviance increases by 2.104 and the esti-

Table 6.18. Model Selection and Estimates for 15–19 Age-Group

Model	Deviance (df)	$\hat{\mu}_{00.0}$	SE($\log(\hat{\mu})$)
Full	0(0)	529.3	255
−L1.L2.L3	0.0001818(1)	682.5	1.312
−L1.L3.L4	0.0002206(2)	682.5	1.312
−L2.L3.L4	1.655(3)	245.4	1.133
−L1.L3	1.797(4)	349.0	0.539
−L2.L3	2.659(5)	526.9	0.343
−L3.L4	3.437(6)	640.5	0.277
−L1.L2.L4	5.540(7)	670.2	0.278

mate changes to 670.2. It is probably safer to retain this three-way interaction, since we would expect association between the three treatment lists.

Table 6.19 shows results for this same model, for all four age-groups. Estimates of missing cells are rounded to the nearest integer. A confidence interval for the number in each age-group is easily constructed. For the eldest age-group for instance, a 95% confidence interval for $\log(\mu_{00..0})$ is

$$\log(1804) \pm 1.96 \times 0.238 = (7.031, 7.964)$$

where $0.238 = \text{se}(\log \hat{\mu}_{0000})$. Exponentiating this and adding the 587 known individuals gives the interval (1718, 3463). Such intervals should accompany the estimates of N in the last column, each obtained by adding $\hat{\mu}_{00..0}$ to the number of distinct individuals given in column 2. In all, we estimate 5780 unseen individuals to be added to the 2866 seen individuals.

Since the four age-groups are modeled separately, the variances may be added to obtain a standard error 628.2 for the estimate 5780. Again computing an interval on the log-scale first we obtain the 95% interval

$$\log(5780) \pm 1.96 \times 628.2/5780 = (8.449, 8.875)$$

Table 6.19. Estimates of Missing Cells, All Age-Groups

		Simplest Model		L1*L2*L4+L3		
Age Group	Seen	$\mathcal{D}$ (0)	$\hat{\mu}_{00.0}$ (se)	$\mathcal{D}$ (6)	$\hat{\mu}_{00.0}$ (se)	$\hat{N}$
15–19	264	15.73	505 (4.36)	3.44 (6)	640 (177.3)	904
20–24	1137	59.82	1103 (2.74)	3.24 (6)	1612 (246.6)	2749
25–29	878	36.71	1273 (3.60)	3.37 (6)	1724 (340.9)	2602
30–55	587	23.37	1212 (0.14)	10.91 (6)	1804 (431.1)	2391
All ages	2866	135.63	4093	20.96 (24)	5780 (628.2)	8646

and exponentiating and adding 2866 gives the interval (7537, 10018) for the total number of injecting drug users.

Finally, some GLIM output is displayed for computing the four estimates in Table 6.19. The missing cells are in positions 16, 32, 48, 64 and these values are stored in the vector index. The vector sampled contains the value 0 for the missing cells and 1 for the cells with data and is used to weight the fit. The fitted model has lists 1, 2, 4 independent of list 3, with the dependencies and marginal effects possibly differing across age-groups. The standard errors have been computed at the last line, with the multinomial adjustment given in Eq. (6.63).

In the second half, the data have been input as 16 counts ignoring age-group. In other words, the four sets of data have been aggregated. The model with treatment and police lists independent has a very small deviance of 2.93(6) and the estimate of the missing cell is 5628 giving the estimate 8494. The standard error for log $\hat{\mu}_{00.000}$ is 0.0985 and gives the interval (7506, 9692) for the total number of injecting drug users. By comparison, the likelihood based interval is (7491, 9721).

```
[i]  ? $assign index=16,32,48,64$
[i]  ? $yvar y$error poisson$weig sampled$
[i]  ? $fit age*(l1*l2*l4+l3)$
[o]  scaled deviance = 20.973 at cycle 8
[o]       residual df = 24     from 60 observations
[i]  ? $extract %vl$
[i]  ? $calc nest=%fv(index):sd=%sqrt(%vl(index))$
[i]  ? $prin nest sd$
[o]  640.5  1612.  1724.  1804.  0.2773  0.1530  0.1978  0.2389
[i]  ? $calc sd=%sqrt(%vl(index)*nest**2-nest)$print sd$
[o]  175.7  246.6  338.4  428.9
[o]  ? $comm Data reinput as 16 counts ignoring age . . . .
[i]  $print y$
[o]  4.00  2.00  13.00  56.00  8.00  17.00  50.00  358.00
[o]  41.00  52.00  147.00  864.00  116.00  267.00  871.00  0.00
[i]  ? $data sampled$read
[i]  $REA? 1 1 1 1 1 1 1 1 1 1 1 1 1 1 1 0$
[i]  ? $yvar y$error poisson$weig sampled$
[i]  ? $fit l1*l2*l4+l3$
[o]  scaled deviance = 2.9292 at cycle 3
[o]       residual df = 6   from 15 observations
[i]  ? $look 16 %fv$
[o]          %FV
[o]  16  5628.
[i]  ? $extract %vl$calc sd=%sqrt(%vl(16))$prin sd$
[o]  0.0985
```

6.6.3 The Rasch Model for Heterogeneity*

Lists collected from a heterogeneous population may have the appearance of dependence. For example, suppose that there are three lists. Imagine that half the population avoids the lists and half are attracted to the lists. Those individuals who are attracted to the list would tend to appear on all three lists, or two out of three lists. On average the numbers of individuals appearing on more than one list would exceed expectations under the independence hypothesis.

As a numerical example, Table 6.20 gives expected values of the eight capture histories assuming that the probability of appearing on any list is 0.3 for the avoiding group and 0.7 for the attracting group. The odds ratios ϕ_{12} measuring dependence between lists 1 and 2 is 1.76 for both of these tables. Since these two odds ratios are equal, there is no three-way interaction. The other odds ratios ϕ_{13} and ϕ_{23} also equal 1.76. Thus, three (equal intensity) independent lists on this type of heterogeneous population are mathematically indistinguishable from three dependent lists on a homogeneous population. The reader might like to explore the pattern of cell means for other parameter values. You will find that there are always dependencies.

Bearing in mind the above example, it is doubtful whether heterogeneity and dependence can be separated. For the remainder of this section we assume that the lists are independent at the level of the individual, i.e., the probability of an individual appearing on one list is independent of appearance on other lists. Let p_{ih} be the probability that individual h appears on list i. Then the probability of list history ω is

$$p_h(\omega) = \prod_{i=1}^{t} p_{ih}^{x_i}$$

where $x_i = 1$ if the individual is on list i. If we aggregate over the population then the mean values of the list history counts are

$$\mathrm{E}(R_\omega) = \sum_{h=1}^{N} \prod_{i=1}^{t} p_{ih}^{x_i}. \tag{6.66}$$

If, for each list i, the p_{ih} are independent random variables from common distributions P_i, then the joint distribution of $\{R_\omega\}$ is multinomial with the above

Table 6.20. Cell Means for Three Lists on Heterogeneous Population

	L3—Yes		L3—No	
	L2—Yes	L2—No	L2—Yes	L2—No
L1—Yes	0.185	0.105	0.105	0.105
L1—No	0.105	0.105	0.105	0.185

means. (This is an extension of the result given in Section 4.4.1 for binomial variables with random probabilities.) Even when the p_{ih} are nonrandom, provided the heterogeneity is moderate the distribution is very close to multinomial, and use of this distribution will slightly overestimate standard errors [see Darroch et al. (1993)].

Under certain assumptions about the nature of the population heterogeneity, certain conclusions about the pattern of the cell means follow. We want these assumptions to be as weak and flexible as possible, while ensuring the implied pattern of dependence is tractable.

Rasch Model and Quasi-Symmetry We suppose that each individual h has a parameter s_h determining susceptibility to capture on any list. Each list i has a parameter u_i determined by the penetration of the list. Let us assume

$$\text{logit}(p_{ih}) = u_i + s_h. \tag{6.67}$$

This is the Rasch model, commonly used in educational testing theory for modeling heterogeneity of individuals in their response to questionnaires [see Rasch (1980)]. The essence of this model is that there is no tendency for particular (groups of) individuals to appear more or less on particular lists. Rather high penetration lists are uniformly more likely to capture all individuals. High susceptibility individuals are uniformly more likely to appear on all the lists. This is implausible if the lists are of different natures, for instance drug therapy lists and police lists in the previous example. When lists have different natures, we might group them into similar groups and make the Rasch assumption within, but not across, these groups.

As usual, let η_ω denote the log-mean of the R_ω. Substituting the Rasch assumption into Eq. (6.66) a little algebra leads us to

$$\begin{aligned} \eta_\omega &= m + \sum_{i=1}^{t} u_i x_i + \log\left(\sum_{h=1}^{N} e^{s_h x_\bullet} p_h(00..0)\right) \\ &= m + \sum_{i=1}^{t} u_i x_i + \gamma(x_\bullet) \end{aligned} \tag{6.68}$$

where the function γ is a logarithm moment with respect to the distribution $p_h(00..0)$ on s_h. Since the $\gamma(k)$ are a set of log-moments, they satisfy certain inequalities such as $\gamma(2) \geq 2\gamma(1)$, $\gamma(3) + \gamma(1) \geq 2\gamma(2)$. The important point however is that the term $\gamma()$, modeling departure from independence, depends on the capture history only through the total number of lists on which an individual appears. This is a so-called quasi-symmetry model of order t defined by the requirement that all interactions of the same order are equal. The quasi-symmetry model in Section 6.5.4 was of order 2. The additivity assumption underlying the Rasch model is equivalent to quasi-symmetry. The quasi-symmetry model Eq. (6.68) has been obtained by several authors including Darroch (1981), Fienberg (1981), Fienberg and Meyer (1983), and Darroch et al. (1993), under various models generating the heterogeneity.

No High-Order Interaction The Rasch/QS model does not impose any relationship between $\mu_{00..0}$ and the observable cell means and so N is not estimable without further assumptions. Following the discussion for homogeneous populations, we assume that the highest order interaction is zero. For instance, in the $t = 3$ list case, we assume that

$$u_{123} = \eta_{111} - \eta_{101} - \eta_{011} + \eta_{001} - \eta_{110} + \eta_{100} + \eta_{010} - \eta_{000} \tag{6.69}$$

equals zero. Substituting the expression (6.68) for the log-means η_ω we find that the u_i cancel and

$$\gamma(3) - 3\gamma(2) + 3\gamma(1) - \gamma(0) = 0.$$

This equates the third difference $\nabla^3\gamma$ of γ to zero, which is satisfied by any *quadratic* function γ. Since a linear term in $x_\bullet$ may be absorbed into the main effects $u_i x_i$, the linear predictor in Eq. (6.68) becomes

$$\eta_\omega = m + \sum_{i=1}^{3} u_i x_i + \gamma x_\bullet^2 \tag{6.70}$$

There are $2^t - 1 = 7$ observable counts and 5 parameters (m, $u1$, $u2$, $u3$, γ) leaving two degrees of freedom to assess the fit of this model.

It is instructive to show that this quadratic model is identical to the model with all two-way interactions equal, i.e., to the quasi-symmetry model. In the latter case, if the common value is γ then the linear predictor is

$$\eta_\omega = m + u_1x_1 + u_2x_2 + u_3x_3 + \gamma(x_{12} + x_{23} + x_{13}).$$

After expanding

$$\gamma x_\bullet^2 = \gamma(x_1^2 + x_2^2 + x_3^2 + 2x_1x_2 + 2x_1x_3 + 2x_2x_3)$$

we note that (i) $x_i^2 = x_i$, each design vector x_i comprising zeros and ones, (ii) the interaction design vectors $x_{ij} = x_i x_j$, and (iii) the linear terms are absorbed into the main effects. In other words, adding a quadratic term $\gamma x_\bullet^2$ is apparently equivalent to assuming equal two-way interactions. However, when the quadratic model is derived from the Rasch assumption, the values $\gamma(k)$ must satisfy certain inequalities listed earlier and in the case $t = 3$ it is sufficient that the parameter $\gamma > 0$. Heuristically, this is merely because we cannot have a negative amount of heterogeneity—under heterogeneity there will always be too many individuals caught very many and very few times and a plot of η_ω against $x_\bullet$ will be concave upwards.

For the general case of t lists, the assumption of no highest order interaction implies that the tth difference $\nabla^t\gamma$ of the function γ is zero, which implies that $\gamma()$ is a polynomial of degree $t - 1$. Since the linear term can be absorbed into the main effects the linear predictor becomes

$$\eta_\omega = m + \sum_{i=1}^{t} u_i x_i + \sum_{k=2}^{t-1} \gamma_k x_\bullet^k. \tag{6.71}$$

There are $1 + t + (t - 2) = 2t - 1$ parameters in this model and $2^t - 1$ observable counts leaving $2^t - 2t$ degrees of freedom for assessing fit. Except for inequality restrictions on the "heterogeneity parameters" $\gamma_1, \ldots, \gamma_{t-1}$, this is equivalent to the QS model with interactions of equal order assumed equal.

Partial Rasch Model The Rasch model assumes additivity of individual parameters s_h and list parameters u_i. A weaker assumption is that additivity only holds for a subset of the lists. For instance, with three lists we might assume that the pattern of heterogeneity is given by the parameters s_h for lists 1 and 2, and by a different pattern s'_h for list 3. Substituting this into Eq. (6.67) we find that

$$\eta_\omega = m + \sum_{i=1}^{t} u_i x_i + \gamma(x_1 + x_2, x_3)$$

and with the extra assumption of no three-way interaction this becomes

$$\begin{aligned} \eta_\omega &= m + \sum_{i=1}^{t} u_i x_i + \gamma_1(x_1 + x_2)^2 + \gamma_2(x_1 + x_2 + x_3)^2 \\ &= m + \sum_{i=1}^{3} u_i^* x_i + u_{12} x_{12} + u^*(x_{13} + x_{23}) \end{aligned} \tag{6.72}$$

where $u_i^* = u_i + \gamma_1 + \gamma_2$, $u_{12} = \gamma_1 + 2\gamma_2$ and $u^* = 2\gamma_2$. This is just the model with no three-way interaction, the interactions u_{13}, u_{23} having common value u^*.

More generally, for t lists, suppose that the Rasch model only holds for a subset of the lists indexed by $i \in G$. Then the log-linear model requires a term

$$\gamma \left(\sum_{i \in G} x_i \right)^2$$

that is equivalent to all interactions involving $i \in G$ being equal. Log-linear models with patterns of equality between interactions can be thought of as Rasch models on subsets of the lists.

The message of all of this seems to be that heterogeneity causes departure from independence and that log-linear models with equal interaction terms can be thought of as heterogeneity models. In practice, this means that we should just model the apparent dependencies between the lists, regardless of whether we believe the source of these dependencies is a genuine dependency or just a by-product of heterogeneity, or a combination of both.

Even when we find a model that fits the observed counts very well, the method described may give estimates that are grossly in error, simply because we are extrapolating observed frequency patterns to unobserved frequencies. As Fienberg (1972b) puts it

> This is analogous to, and has the same dangers as, fitting an arbitrary curve to a series of points (x, y), where $x > 0$, with the intention of estimating y at $x = 0$. Perhaps worse than fitting an arbitrary curve, however, is arbitrarily making the curve a straight line, in light of evidence to the contrary, and then estimating y from the intercept. This is analogous to the position of those who automatically assume that the ... (m lists) ... are independent.

6.6.4 Illustration: United States Census

The data in Table 6.21 relate to individuals in a certain area of St. Louis, Missouri. The first list (DR) is the 1988 dress rehearsal for the 1990 United States census. The second list is a follow up list (FU), collected after the main census. The third list (GR) is from government records. These lists should be independent because researchers compiling one list do not make reference to the other lists. However, there is clearly the possibility of considerable heterogeneity in individuals' probabilities of appearing on the lists.

One important characteristic of the Rasch model is that it should apply to sub-groups of the population as well as the population as a whole. This provides a check of model validity. If the model is correct then it should fit not only the data as a whole but sub-groups of the data, even when these sub-groups are chosen to be as different as possible. The data in Table 6.21 have been broken down into four groups depending on the age-group of the individual and whether or not they were renting the accommodation.

We consider three models. These are the partial Rasch/quasi-symmetry model Eq. (6.72) which says that the pattern of heterogeneity is the same for

Table 6.21. United States Census Data [from Darroch et al. (1993)]

DR	FU	GR	Group 1	Group 2	Group 3	Group 4	All
0	0	0	*	*	*	*	*
0	0	1	59	43	35	43	180
0	1	0	65	70	69	53	257
0	1	1	19	11	10	13	53
1	0	0	75	73	77	71	296
1	0	1	19	12	13	1	51
1	1	0	217	144	262	155	778
1	1	1	79	58	91	72	300
On list DR			390	287	443	305	1425
On list FU			380	283	432	293	1388
On list GR			176	124	149	135	584
On any list			533	411	557	414	1915

the DR and FU lists, but possibly different for the GR list. The design matrix for this model is

$$X_{PQS} = \begin{pmatrix} 1 & 0 & 0 & 0 & 0 & 0 \\ 1 & 0 & 0 & 1 & 0 & 0 \\ 1 & 0 & 1 & 0 & 1 & 0 \\ 1 & 0 & 1 & 1 & 1 & 4 \\ 1 & 1 & 0 & 0 & 1 & 1 \\ 1 & 1 & 0 & 1 & 1 & 4 \\ 1 & 1 & 1 & 0 & 4 & 4 \\ 1 & 1 & 1 & 1 & 4 & 9 \end{pmatrix}$$

with the parameter vector $(m, u_1, u_2, u_3, \gamma_1, \gamma_2)$ and one degree of freedom (note that the first line of data is not used). The second model is the full Rasch/quasi-symmetry model, which is the previous model with $\gamma_1 = 0$, or equivalently with the fifth column of the design matrix removed. There are two degrees of freedom for this model. The third model is the full independence (or equivalently homogeneity) model with $\gamma_1 = \gamma_2 = 0$ and three degrees of freedom.

Deviances are listed in Table 6.22 for the four groups separately and for the collective data. The independence model fits badly but much worse for the aggregated data. This is consistent with there being underlying heterogeneity that increases with more aggregation. The Rasch model does not fit any better. On examining estimates of the interactions from the full model, u_{12} seems to differ drastically from u_{13}, u_{23}, all of which should be equal under quasi-symmetry. The partial quasi-symmetry model only requires that $u_{13} = u_{23}$, and fits spectacularly well for three of the groups, and reasonably for the fourth. For the aggregated data, the deviance is again very small.

We have also listed, under the well-fitted partial quasi-symmetry model, esti-

Table 6.22. Models Fits to United States Census Data

Model	Group 1	Group 2	Group 3	Group 4	All
Independence	94.06	58.42	70.15	80.06	301.0
Full QS	94.04	52.99	67.89	73.01	291.8
Partial QS	0.15	0.01	0.11	3.45	0.71
$\hat{\mu}_{000}$	292	669	490	768	1962
$\text{var}(\log(\hat{\mu}_{000}))$	0.168	0.243	0.245	0.275	0.056

mates of the number of missed individuals $\hat{\mu}_{000}$ in each group and the variance of the logarithms of these estimates. These may be used to construct intervals for $\log(\hat{\mu}_{000})$ and then for N. It is worth noting that under the full quasi-symmetry model, or the independence model, the estimates of undercount are around 10 times smaller—fitting an incorrect model produces drastically misleading results. Luckily, the data themselves starkly point to the incorrectness of these models. If only two lists were on hand then the data could not provide such information and we would be forced to rely on the (completely wrong) assumption of independence.

FURTHER READING

6.4.5 *The article of McCullagh (1980) was influential in advancing the use of cumulative logit models. Further research on theory and applications may be found in Aranda–Ordaz (1983), Clayton (1974), Landis et al. (1987), McCullagh (1984), and Snapinn and Small (1986). Cumulative link models are discussed by McCullagh (1980) who also provided an algorithm for ML estimation based on multivariate generalized linear models. See also, Burridge (1981), Pratt (1981), Farewell (1982), and Genter and Farewell (1985).*

Section 6.5: Models for Incomplete Tables

6.5.2, 6.5.3 *The term "quasi-independence" was introduced into the statistical literature by Goodman (1968) but the method was used much earlier, especially by that author, in the analysis of social mobility data [see Goodman (1961, 1962, 1963, 1965)]. Clear discussions and examples of applications of quasi-independence models may be found in Fienberg (1978) and Bishop, Fienberg, and Holland (1975) who also describe in detail the calculation of degrees of freedom. The set S of nonstructural zeros has a bearing on whether or not the MLE is unique and whether it can be computed directly. Examples were direct computation is possible are given in Bishop, Fienberg, and Holland (1975, pp. 193–199). Most technical details on existence and uniqueness are available in Goodman (1968), Haberman (1974), Savage (1973), and Fienberg (1970, 1972a).*

6.5.4 *The term "quasi-symmetry" was introduced into the statistical literature by Caussinus (1965) who also showed that when combined with marginal homogene-*

ity it reduces to symmetry. Testing marginal homogeneity without assuming QS has been discussed by Madansky (1963) and Lipsitz (1988).

Generalization to multiway classification is given by Bishop, Fienberg, and Holland (1975, Section 8.3) who also showed how to fit the model using the "trick" described in this section. A model nested between QI and QS was given by Goodman (1979) which he calls the quasi-uniform association model. For further discussion of QS models see Darroch (1981, 1986) and Darroch and McCloud (1986).

6.7 EXERCISES

6.1. Consider a four-way cross-classification of factors W, X, Y, Z.

a. Show that the model symbolized by (WXZ, WYZ) assumes that, conditional on W and Z, X and Y are independent. Are three any other relations of (conditional) independence?

b. For what models, containing no three-factor interactions, is X and Y independent conditional on W and Z?

c. Are the marginal and partial associations of W and X the same for the model (WX, XYZ)?

d. Are the marginal and partial associations of W and X the same for the model (WX, WZ, XY, YZ)?

6.2. Consider a four-way cross-classification of factors R, X, Y, Z. Think of R as the response factor. In each case below, write down a Poisson log-linear model formula. If this model is equivalent to a multinomial logit model for the response R then describe the model.

a. R conditionally independent of X

b. Only X and Z affect R but their effects do not interact.

c. There are no three-factor interactions.

6.3. Try to think of a plausible context and a set of fictitious data illustrating the following mathematical situations.

a. Two factors are independent conditional on the value of a third factor but are marginally dependent.

b. Two factors are marginally independent but conditional on a third factor are dependent.

c. Give log-linear model formulas describing the dependence in your examples and relate these to the collapsibility conditions.

6.4. Ashford and Sowden (1970) collected the data shown in Table 6.23. The counts are from a survey of British miners between 20 and 64 years of age who were smokers but did not show signs of the lung disease pneumoconiosis. Besides their age-group, each miner was tested for the binary responses breathlessness and wheezing.

Table 6.23. Breathlessness and Wheezing of Miners by Age-Group

	Breathless		Not Breathless		
Age-Group	Wheeze	No Wheeze	Wheeze	No Wheeze	Total
20–24	9	7	95	1841	1952
25–29	23	9	105	1654	1791
30–34	54	19	177	1863	2113
35–39	121	48	257	2357	2783
40–44	169	54	273	1778	2274
45–49	269	88	324	1712	2393
50–54	404	117	245	1324	2090
55–59	406	152	225	967	1750
60–64	372	106	132	526	1136
Total	1827	600	1833	14022	18282

a. Fit the log-linear model treating age as a factor and containing only second-order interactions.

b. For each age-group i, estimate the log-odds ratio ψ_i for the association of breathlessness and wheezing. Plot these against the center x_i of the age-group.

c. Consider the following log-linear model for the cell mean μ_{ijk} for age-group i, breathlessness $j = 1, 2$, and wheezing $k = 1, 2$.

$$\log(\mu_{ijk}) = m + (ab)_{ij} + (bw)_{jk} + \beta_{jk}x_i.$$

Explain and fit this model. Find a better model.

6.5. Consider the occupational survey data in Table 6.4 and supplied as the *Splus* object `section6.3.4.dat`.

a. Consider aspiration as a binary response (high or low) and a model that says that IQ, economic status, and residence all contribute additively to the logit probability of high aspirations. Write down a `GLIM` model formula.

b. Consider economic status as a binary response (high or low) and a model that says that residence affects the logit probability of high status but that this effect differs depending on IQ and sex. Test the goodness-of-fit of this model.

c. Assuming as little as possible, test if IQ is affected by sex.

d. Assuming the sex effect (if it exists), is additive and unaffected by other factors, test whether or not IQ is affected by sex.

e. Give an approximate 90% confidence interval for the relative odds of having high economic status for males compared to females, ignoring all other factors.

6.6. The data in Table 6.24 was already given in the exercises after Chapter 4. The data is a cross-classification of survey results for 608 white women living in public housing projects by (i) their proximity to a black family, with P = + indicating close, (ii) frequency of contacts with blacks, with C = + indicating high frequency, and (iii) general local attitudes towards blacks, with A = + indicating a more positive attitude.

Table 6.24. Racial Sentiments of White Women by Proximity, Contacts, and Local Attitudes [from Wilner et al. (1955)]

Proximity (P)	Contact (C)	Attitudes (A)	S+	S−
+	+	+	77	32
+	+	−	30	36
+	−	+	14	19
+	−	−	15	27
−	+	+	43	20
−	+	−	36	37
−	−	+	27	36
−	−	−	41	118

a. Fit the full log-linear model and construct a normal probability plot of the standardized estimates (see Section 4.5.5).

b. Find a logistic model for sentiment in terms of the three explanatory variables P, C, and A. Under this model, estimate the proportional increase in odds of having positive sentiments for (i) close relative to far proximity, (ii) frequent relative to infrequent contacts, and (iii) positive, relative to negative local attitudes.

6.7. 1441 employees of a certain firm were surveyed by questionnaire. Employees were categorized by their age-group (<25, 25–35, >35), their sex, their education level (1 = low, 2 = medium, 3 = high), their color (1 = white, 2 = black) and their income (<$25000 per year = 1). The data are in the *Splus* object `employees.dat` at the Wiley website.

a. Are these data over-dispersed?

b. Consider high or low income as a binary response. Give a `GLIM` model formula which would be equivalent to a model which says that the logit probability of high income is an additive function of each of the other factors age, sex, color, and education.

c. Test the goodness-of-fit of the model which says that income does not depend on color. Note that all the fitted models are over-dispersed.

d. Test the hypothesis that income does not depend on color against the alternative that it does depend on color but that this dependence is the same regardless of age, sex, or education level.

e. Consider the model formula `SEX.AGE.COL.(INC+EDUC)`. Give two interpretations of this formula in terms of binomial logistic regressions.

6.8. Table 6.25 summarizes the results of a survey of 218 Baltimore mothers of children displaying behavioral problems at school, classified by (i) whether or not they have previously lost a baby and (ii) the birth order of the child. Children without siblings were excluded from the study. A set of 147 mothers whose children did not display behavioral problems were used as controls.

Table 6.25 Relationship of Behavioral Problems to Birth Order and Previous Loss of Siblings [from Cochran (1954)]

	Birth Order						
	2		3–4		5+		Total
Losses	Yes	No	Yes	No	Yes	No	
Problems	20	82	26	41	27	22	218
Controls	10	54	16	30	14	23	147
Total	30	136	42	71	41	45	365

a. Use a log-linear model to describe the response of being a problem child in terms of birth order and loss.

b. Use a log-linear model to describe the response loss in terms of birth order and having a problem child.

c. Find the simplest log-linear model with acceptable deviance.

d. Can the data be collapsed over birth order?

6.9. Cox (1970) describes an experiment where ingots are subjected to different heating and soaking times in their preparation and then the binary response variable, 'readiness' for rolling, measured. The data are given in Table 6.26. The numbers tabulated are the numbers 'ready' from the number in parentheses. These data are also given in exercise set 4.

a. Fit the log-linear model which says that the logit probability of response is an additive function of a term depending on heating time and a term depending on soaking time.

b. Fit the log-linear model which says that the logit probability of response is, for given soaking time, a linear function of heating time and, for given heating time, a linear function of soaking time.

c. Do both heating time and soaking time affect the probability of response?

Table 6.26. Readiness of Ingots for Rolling by Heating and Soaking Times [from Cox (1970)]

	Heating Time				
Soaking	7	14	27	51	Totals
1.0	10 (10)	31 (31)	55 (56)	10 (13)	106 (110)
1.7	17 (17)	43 (43)	40 (44)	1 (1)	101 (105)
2.2	7 (7)	31 (33)	21 (21)	1 (1)	60 (62)
2.8	12 (12)	31 (31)	21 (22)	0 (0)	64 (65)
4.0	9 (9)	19 (19)	15 (16)	1 (1)	44 (45)
Total	55 (55)	155 (157)	152 (159)	13 (16)	365 (377)

6.10. For 132 schizophrenic patients, Table 6.27 presents a cross-tabulation of the frequency of visits the patient received and the length of stay required in the hospital, taken from Wing (1962). One anticipates that less frequent visits would be associated with a poorer prognosis and longer stay in the hospital.

Table 6.27. Cross-Classification of 132 Schizophrenic Patients by Frequency of Visitors Received and Length of Stay in Hospital

	2–10 years	10–20 years	>20 years	Total
Regular visits	43	16	3	62
Irregular visits	6	11	10	27
Never visited	9	18	16	43
Total	58	45	29	132

a. Fit the independence model.

b. Fit the row-effects model by defining appropriate scores for the row factor.

c. Fit the bilinear model by defining appropriate scores for the column factor.

6.11. The data in Table 6.28 categorizes 3888 pregnant women by their marital status and daily caffeine consumption. These data were collected by Martin and Bracken (1987) as part of a study of the determinants of low birth weight infants and were actually used as explanatory variables. There is nothing to prevent us from investigating the associations of these two explanatory variables, however. Summarize the dependence, if any, of the caffeine consumption distribution on marital status.

Table 6.28. Caffeine Consumption

	Caffeine Consumption (mg/day)				
	0	1–150	150–300	>300	Total
Married	652	1537	598	242	3029
Separated/Widowed	36	46	38	21	141
Single	218	327	106	67	718

Note: Cross-classification of 3888 pregnant women by marital status and caffeine consumption [from Martin and Bracken (1987)].

6.12. The data in Table 6.29 are a classification of 2000 sixth-grade children by school performance and weight category.

Table 6.29. School Performance Data

	Underweight	Normal	Overweight	Obese
Poor at school	36	160	65	50
Satisfactory at school	180	840	300	185
Above average	34	100	35	15

Note: Cross-classification of 2000 sixth-grade students by weight and school performance [from Rasmussen (1992)].

a. Perform an overall test of independence.

b. Fit the row effects model and use it to summarize the difference in the school performance distribution for the different weights.

c. Fit a column effects model to describe how the weight distribution changes with differing school performance.

d. Fit the bilinear model and assess the fit compared to the row and column effects models. Summarize the association in one parameter estimate and interpret this estimate.

6.13. Researchers were interested in peoples' attitude to a national guaranteed income and its relationship to age. Table 6.30 lists survey results that relate to this issue. Opinions were strongly disagree (SD), disagree(D), undecided (U), favor (F), and strongly favor (SF).

a. Fit a row effects model to describe how the opinion distribution changes with age.

b. Fit a column effects model to describe how the age distribution changes with opinion.

c. Fit the bilinear model and assess the fit compared to the row and column effects models. Summarize the association in one parameter estimate and interpret this estimate.

Table 6.30. Opinion on National Guaranteed Income

Age	SD	D	U	F	SF
18–30	76	81	60	62	21
31–40	76	74	36	51	13
41–50	61	63	27	41	8
51–60	48	46	23	28	5
Over 60	39	36	4	18	3

Note: Cross-classification of opinions of surveyed individuals on national guaranteed income and age-group [from Milton et al. (1986)].

6.14. A group of 450 statistics students had their grade (one of A–F) recorded. In addition, it was recorded whether or not the student has done algebra and/or logic at the university. It is of interest to relate the statistic grade to previous enrollment in algebra and/or logic.

Table 6.31. Student Statistic Grades

		Grade					
Logic	Algebra	A	B	C	D	F	Total
Yes	Yes	10	20	25	5	0	60
Yes	No	10	15	70	25	5	120
No	Yes	15	20	40	5	0	80
No	No	15	15	75	30	55	190

Note: Cross-classification of students by grade in statistics, and enrollment in algebra and/or logic courses [from Byrkof (1992)].

a. Test independence of the three factors: logic, algebra, and statistics grade.

b. Treating the logic/algebra interaction as a four-level factor, fit a row effects model for statistics grade treated as an ordinal response.

c. Investigate whether the interaction of algebra/logic is necessary either as a main effect or in the row effect.

6.15. The data in Table 6.32 are a classification of 174 polio victims by their age, whether or not they had received the Salk vaccine and a binary classification, paralyzed or not paralyzed. It is of interest to measure the effect of Salk vaccine on the likelihood of paralysis, adjusting for age of the victim.

a. Fit the log-linear model which says that the odds-ratio for vaccine/paralysis is the same for each age-group.

b. Test whether the effect of age on the underlying paralysis rate can be modeled linearly and give a confidence interval for the proportional

Table 6.32. Paralysis of Polio Victims

Age	Vaccine	Paralysis	No Paralysis
0–4	Yes	14	20
	No	24	10
5–9	Yes	12	15
	No	15	3
10–14	Yes	2	3
	No	2	3
15–19	Yes	4	7
	No	6	1
20–39	Yes	3	12
	No	5	7
40+	Yes	0	1
	No	2	3

Note: Cross-tabulation of individuals by age, paralysis, and vaccination status [from Chin et al. (1961)].

increase in odds of paralysis for each year of age of the victim.

c. Fit equivalent logistic models to (a) and (b).

6.16. The data in Table 6.33 are a fuller version of the Framingham Heart Data given earlier on p. 311. The variable cholesterol was ignored in the earlier tabulation.

Table 6.33. Framingham Heart Study

Heart Disease	Serum Cholesterol	Blood Pressure <126	126–145	146–165	>165
Present	<200	2	3	3	4
	200–219	3	2	0	3
	220–259	8	11	6	6
	≥260	7	12	11	11
Absent	<200	117	121	47	22
	200–219	85	98	43	20
	220–259	119	209	68	43
	≥260	67	99	46	33

Note: Cross-tabulation of individuals by cholesterol levels (in mg/100 cc), blood pressure (systolic in mm Hg) and absence or presence of heart disease [from Cornfield (1962)].

a. Fit the log-linear model which says that the log-odds of heart disease depends additively on blood pressure and cholesterol, treated as categorical variables.

b. Fit the log-linear model which says that the log-odds of heart disease depends additively on blood pressure and cholesterol, treated as covariates.

c. Give confidence intervals for the effects of unit increases in blood pressure and cholesterol on the odds of heart disease.

6.17. The data in Table 6.34 refer to genital displays in a colony of six squirrel monkeys, labeled A–F. For each display one monkey is active and the other passive.

Table 6.34. Cross-Classification of Genital Displays by Active and Passive Squirrel Monkeys [from Ploog (1967)]

	Passive						
Active	A	B	C	D	E	F	Total
A	—	1	5	8	9	0	23
B	29	—	14	46	4	0	93
C	0	0	—	0	0	0	0
D	2	3	1	—	38	2	46
E	0	0	0	0	—	1	1
F	9	25	4	6	13	—	57

a. Fit the quasi-independence model to the data ignoring the main diagonal. What is the degrees of freedom?

b. Fit the quasi-symmetry model.

c. Compute a reliable P-value for testing the goodness-of-fit of the quasi-symmetry model, taking account of small expected values.

6.18. The data in Table 6.35 are a cross-classification of individuals by their social and professional status measured in 1954 and, for the same individuals, in 1962.

Table 6.35. Cross-Classification of Individuals by Socioprofessional Status in 1954 and 1962 [from Caussinus (1965)]

	Status in 1962					
1954	1	2	3	4	5	6
1	187	13	17	11	3	1
2	4	191	4	9	22	1
3	22	8	182	20	14	3
4	6	6	10	323	7	4
5	1	3	4	2	126	17
6	0	2	2	5	1	153

a. Why would you not expect the symmetry model to apply to these data? Fit the symmetry model and evaluate the fit.

b. Fit the QS model. Use it to test marginal homogeneity.

c. Fit the QI model to the data ignoring the main diagonal.

6.19. The data in Table 6.36 gives a cross-classification of Danish father–son pairs by the job status of the father and of the son. These data are comparable to the British Mobility Data.

Table 6.36. Cross-Classification of Socioprofessional Status of Danish Fathers and Sons [from Svalastoga (1959)]

	Son's Status				
Father	1	2	3	4	5
1	18	17	16	4	2
2	24	105	109	59	21
3	23	84	289	217	95
4	8	49	175	348	198
5	6	8	69	201	246

a. Why would you not expect the symmetry model to apply to these data? Fit the symmetry model and evaluate the fit.

b. Fit the QS model. Use it to test marginal homogeneity.

c. Fit the QI model to the data ignoring the main diagonal.

6.20. The data in Table 6.37 relate to the number of lambs produced by 227 Merino ewes in 1952 and 1953.

Table 6.37. Cross-Classification of 227 Merino Ewes by Number of Lambs in 1952 and 1953 [from Tallis (1962)]

	1952			
1953	1	2	3	Total
0	58	52	1	111
1	26	58	3	87
2	8	12	9	29
Total	92	122	13	227

a. Why would you not expect the symmetry model to apply to these data? Fit the symmetry model and evaluate the fit.

b. Fit the QS model. Use it to test marginal homogeneity.

c. Fit the QI model to the data ignoring the main diagonal.

6.21. Table 6.38 summarizes the results of a survey reported by Brunswick (1971). The three variables are age, gender, and health concerns. Respondents were asked to choose one of the four headings describing health concerns. Obviously, males do not worry about menstrual problems and these are structural zeros.

Table 6.38. Cross-Classification of Health Concerns of Teenagers by Age and Gender [from Brunswick (1971)]

	Males		Females	
Concern	12–15	16–17	12–15	16–17
Sex	4	2	9	7
Menstrual	—	—	4	8
General	42	7	19	10
None	57	20	71	31

a. Is there evidence for over-dispersion for these data?

b. Which is the most significant two-way dependency?

c. Find a reasonable and simple log-linear model for these data and, where possible, interpret parameter estimates.

6.22. Sewell and Shah (1968) carried out a survey of adolescents of college-entry age. Those surveyed had IQ and socioeconomic status measured on a four-point ordinal scale and parental encouragement on a three-point ordinal scale. Variable *plans* indicated plans to attend college. The data are in the *Splus* object `sewell.dat` at the Wiley website.

a. Find a simple and well-fitting logistic model for the binary response of college plans.

b. Find a multinomial logit model for the response IQ.

c. Fit a model that makes use of the ordinal nature of notional response IQ and the explanatory factor SES. Let the effects of gender and parental encouragement be modeled by simple row effect terms and the SES factor by a bilinear term.

CHAPTER 7

Conditional Inference

Many first-year statistics courses describe a method of analyzing 2×2 tables based on the hypergeometric distribution. There is no mention of empirical logits, of likelihood ratio statistics, or approximate chi-square distributions. This is because it is a conditional test and the null distribution can be given exactly—the hypergeometric.

Conditional methods are also called *exact* methods. This is because the size of tests and coverage of confidence intervals are not approximated by the central limit theorem but can be computed exactly. The term is misleading however as a so-called exact 95% conditional confidence interval will have coverage 95% *or more*. It is the lower bound on the coverage which is exact.

Conditional inference can be implemented in a wide variety of cases to be detailed in this chapter. For data sets of only moderate size and complexity, a truly incredible amount of computation is required and only recently have both hardware and software become available to make conditional methods a realistic alternative. Manuals for the packages *StatXacT* and *LogXacT* are supplied at the Wiley website and output from these packages are featured in the later sections of this chapter.

We will study conditional inference for logistic and log-linear models in this chapter. The methods described are alternatives to the unconditional methods detailed in Chapters 3, 4 and 6. We go through the calculations for the 2×2 and 2×3 tables in detail. While these computations are quite tedious, it is important to have an appreciation for the type of computations involved for more complicated models and this is only achieved by rolling up your sleeves and getting your hands dirty. After these relatively simple cases we consider the general linear model, in particular, applications to general two- and three-way contingency tables. In the final section, a method of approximating conditional inferences using the saddlepoint approximation is described.

7.1 CONDITIONAL INFERENCE FOR 2 × 2 TABLES

As in Section 3.1.2, imagine a sample of n cancer patients. We randomly assign n_1 to be given treatment 1 and the remaining $n_2 = n - n_1$ to be given treatment 2. After one year, the number of survivals under each treatment Y_1, Y_2 are recorded and on the basis of these two random variables we wish to compare the relative merits of the two treatments.

	Survive	Die	Total
Treatment 1	Y_1	$n_1 - Y_1$	n_1
Treatment 2	Y_2	$n_2 - Y_2$	n_2
Total	T	$n - T$	n

Several alternative sampling schemes for such data were listed in Section 3.1.2. While it makes no ultimate difference to the mathematics which scheme is used, to be concrete we will suppose the data come from a prospective clinical trial as described above. In this case we have two binomial variables Y_i with independent $\mathrm{Bi}(n_i, \pi_i)$ distributions, with n_1, n_2 fixed.

We respect the notation in Section 3.2.1. The logit parameters are denoted ψ_i and the parameter of interest is

$$\Delta = \psi_1 - \psi_2 = \log\left\{\frac{\pi_1(1-\pi_2)}{\pi_2(1-\pi_1)}\right\}.$$

The joint distribution of Y_1, Y_2 is the product of two binomial probabilities and using the parameters (Δ, ψ_2) instead of (π_1, π_2) this has the form

$$P_B(y_1, y_2; \Delta, \psi_2) = \binom{n_1}{y_1}\binom{n_2}{y_2}\frac{\exp\{\Delta y_1 + \psi_2(y_1 + y_2)\}}{\{1+\exp(\Delta+\psi_2)\}^{n_1}\{1+\exp(\psi_2)\}^{n_2}}$$

as given in Eq. (3.10). Taking the logarithm, and discarding additive terms that are not functions of the parameters, the log-likelihood function is

$$\ell(\Delta, \psi_2) = \Delta Y_1 + \psi_2 T - n_1\log(1+e^{\Delta+\psi_2}) - n_2\log(1+e^{\psi_2}), \qquad (7.1)$$

where $T = Y_1 + Y_2$ is the total number of successes. By inspection, T is sufficient for ψ_2 and so the distribution of the data conditional on T is distributed free of ψ_2. This conditional distribution can thus be used to perform exact inference on Δ, just as the distribution of a single binomial random variable can be used for exact inference on its probability parameter (see Section 2.6.1).

7.1.1 Conditional Distribution and Likelihood

The distribution of Y_1 conditional on $T = t$ is found by dividing the joint distribution of Y_1, T by the distribution of T. The joint distribution of Y_1, T is just the above joint distribution of (Y_1, Y_2) with $t = y_1 + y_2$ substituted and the distribution of T is this joint distribution summed over possible values of Y_1. This gives the conditional distribution

$$p_c(y_1|n, n_1, t, \Delta) = \binom{n_1}{y_1}\binom{n-n_1}{t-y_1} e^{\Delta y_1} \Bigg/ \sum_y \binom{n_1}{y}\binom{n-n_1}{t-y} e^{\Delta y} \quad (7.2)$$

where y_1 and the summation variable y both range over integer values between $\max(0, t - n_2)$ and $\min(n_1, t)$. This range is suppressed in subsequent formulas. We will also suppress the dependence on n, n_1, and t. The rather complicated sum is simply a normalizing constant to ensure that the probability function sums to one. There is one free variable in this probability function, y_1, because $T = t$ is fixed.

The distribution is called the *generalized* or *tilted* hypergeometric distribution and is denoted GHG(n, n_1, t, Δ). When $\Delta = 0$, which corresponds to no difference between the two treatments, this distribution is the ordinary hypergeometric distribution HG(n, n_1, t), whose probability function is

$$p_c(y_1) = \binom{n_1}{y_1}\binom{n-n_1}{t-y_1} \Bigg/ \binom{n}{t}.$$

This distribution involves no unknowns at all. It is symmetric when $n_1 = n_2$ or when $t = n - t$ but is asymmetric otherwise.

The most important aspect of Eq. (7.2) is that it depends on Δ only and not on ψ_2. We may use this distribution to estimate or test any value of Δ without the complicating issue of having an unknown value of ψ_2 to estimate. This property is unique to the log-odds (or monotonic functions of it) and if we decide to measure the difference between the two treatments in any other way, then the exact method to be described will not be possible. This is one of many good reasons for using log-odds rather than say $\pi_1 - \pi_2$ to measure the treatment differences.

The conditional probability function generates a conditional likelihood simply by considering it as a function of Δ rather than of y_1. Taking logarithms and discarding additive terms that do not depend on Δ gives the conditional log-likelihood

$$\ell_c(\Delta) = c + \Delta y_1 - \log\left\{\sum_y \binom{n_1}{y}\binom{n_2}{t-y} e^{\Delta y}\right\}. \tag{7.4}$$

We may plot this as a function of Δ to summarize our knowledge of Δ from the data y_1, t. This is an alternative to plotting the profile log-likelihood function in the unconditional approach. The conditional log-likelihood is not maximized at the same point as the profile log-likelihood. In other words, the conditional maximum likelihood estimator is not the same as the unconditional maximum likelihood estimator.

Example 7.1: Mice-Tumor Data (Revisited) We will use the mice-tumor data of Essenberg (1952) described in Example 3.1 (Table 7.1). This allows us to make direct comparisons of conditional and unconditional results.

The profile log-likelihood function was shown in Figure 3.1. Computing this curve involved solving the quadratic to find $\hat{\psi}_2(\Delta)$ and substituting this into the log-likelihood function $\ell(\Delta, \psi_2)$, see p. 130. Note that $\max(0, t - n_2) = 8$ and $\min(n_1, t) = 23$ and so the conditional log-likelihood function is

$$\ell_c(\Delta) = c + 21\Delta - \log\left\{\sum_8^{23} \binom{23}{y}\binom{32}{40-y} e^{\Delta y}\right\}.$$

The profile and conditional log-likelihood functions are compared in Figure 7.1, scaled so that each has maximum value zero. While the curves are extremely similar for this particular data set, they are not generally so close. Especially, when one of n_1 or n_2 are small there will be much more difference. The unconditional estimate is 1.972 and the conditional estimate is 1.939. The estimated unlogged odds ratios are 7.183 and 6.952, respectively.

Table 7.1. Tumor Prevalence Among Mice by Exposure to Tobacco Smoke

	Tumor	No Tumor	Total
Smoke treated	21	2	23
Control	19	13	32
Total	40	15	55

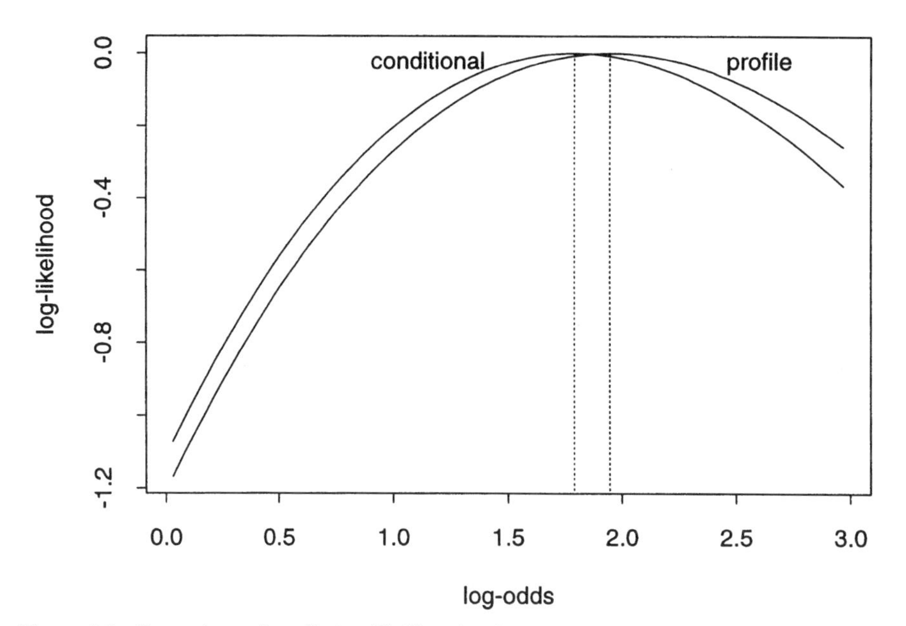

Figure 7.1. Comparison of profile log-likelihood and conditional log-likelihood against the log-odds parameter, for the mice-tumor data.

7.1.2 Conditional Tests and P-Values

Consider testing $\mathcal{H}_0 : \Delta = \delta_0$ against $\mathcal{H}_1 : \Delta > \delta_0$ on the basis of the conditional distribution [see Eq. (7.2)]. We would reject the null hypothesis if y_1 is too large (see Exercise 3.9). The P-value is given by

$$\Pr(Y_1 \geq y_1 | t, \delta_0) = \sum_{y \geq y_1} \binom{n_1}{y} \binom{n - n_1}{t - y} \phi_0^y \Big/ \sum_{y} \binom{n_1}{y} \binom{n - n_1}{t - y} \phi_0^y$$

where $\log(\phi_0) = \delta_0$. Conditional or *exact* tests of Δ require computation of tail probabilities of the tilted hypergeometric distribution with parameter δ_0. There is no approximation involved.

Testing no Treatment Effect (Δ = 0) When $\delta_0 = 0$ the null distribution is hypergeometric. This distribution traditionally arises in the context of sampling a labeled finite population without replacement. Where is the sampling without replacement? For prospective clinical trials, it is in the random choice of the n_1 individuals to be given treatment 1.

Think of the n patients in the clinical trial as a population, exactly t of whom survive. We have a population broken into two groups: t survivors and $n-t$ non-survivors. The n_1 individuals assigned to treatment group 1 are a random sample without replacement. If $\Delta = 0$ then each individual has the same chance of being in treatment group 1. How many of the these n_1 individuals will be among the survivor group?, i.e., what is the distribution of Y_1?—hypergeometric. This distribution is completely generated by the random assignment to treatment groups. Note especially that uniformity of treatment effects across individuals is not required for meaningful inference on Δ (see also Section 3.1.2).

Computing ordinary hypergeometric probabilities presents no special problems when n_1, n_2, t are moderate, but otherwise may involve factorials of high order. This can require careful programming. One simple approach is to (1) compute logarithmic terms in the expression

$$\log\{p_c(y)\} = c - \log y! - \log(t-y)! - \log(n_1 - y)! \\ - \log(n_2 - t + y)! \tag{7.5}$$

where c is not a function of y. The log-factorials can be well approximated by Stirling's formula, one form of which is

$$\log n! \approx \log\sqrt{2n} - \left(n + \frac{1}{6}\right) + \left(n + \frac{1}{2}\right)\log\left(n + \frac{1}{6}\right) \tag{7.6}$$

with error $O(n^{-2})$. (2) Take the value of c so that the collection of logged probabilities are centered at zero, which minimizes certain numerical inaccuracies. Then (3) exponentiate all these values and (4) normalize so that they add to one.

Testing a Specific Treatment Effect $(\Delta = \delta_0 \neq 0)$ Tilted probabilities are simply computed from untilted probabilities since from Eq. (7.2)

$$\log\{p_c(y; \delta_0)\} = c + \log\{p_c(y; 0)\} + y\delta_0$$

where c does not depend on y but relates to the normalizing constant. If we have computed the ordinary hypergeometric distribution then the filted probabilities are easily computed as follows: (1t) Take the logarithms of the ordinary probabilities and add $y\delta_0$, and (2t) choose c so that these are centered at zero. Then (3t) exponentiate and (4t) normalize so that they add to one.

Example 7.1: (*Continued*) Mice-Tumor Data. We have set out the computations for the (tilted) hypergeometric distribution in Table 7.2, which for the mice-tumor data is concentrated on the integers $\max(0, t - n_2) = 8$ up to $\min(n_1, t) = 23$. The first column is just the right-hand side of Eq. (7.5). Subtracting the mean of column (1) gives column (2). Exponentiating gives column (3) and normalizing gives the hypergeometric distribution in column (4). For testing

Table 7.2. Calculating the Tilted Hypergeometric Distribution

y	(1)	(2)	(3)	HG (4)	(2t)	(3t)	GHG (4t)
8	−120.06	−11.134	0.00	0.0000	−16.332	0.000	0.0000
9	−116.08	−7.158	0.00	0.0000	−11.663	0.000	0.0000
10	−113.01	−4.080	0.02	0.0000	−7.892	0.000	0.0000
11	−110.54	−1.611	0.20	0.0006	−4.730	0.009	0.0000
12	−108.56	0.370	1.45	0.0041	−2.056	0.128	0.0001
13	−107.00	1.926	6.86	0.0194	0.193	1.213	0.0008
14	−105.83	3.094	22.06	0.0622	2.054	7.799	0.0050
15	−105.03	3.895	49.15	0.1387	3.548	34.758	0.0223
16	−104.59	4.341	76.80	0.2167	4.688	108.613	0.0695
17	−104.49	4.435	84.33	0.2379	5.474	238.511	0.1527
18	−104.76	4.169	64.65	0.1824	5.902	365.699	0.2342
19	−105.40	3.527	34.03	0.0960	5.953	384.928	0.2465
20	−106.45	2.477	11.91	0.0336	5.596	269.437	0.1725
21	−107.97	0.962	2.62	0.0074	4.774	118.428	0.0758
22	−110.06	−1.130	0.32	0.0009	3.375	29.221	0.0187
23	−113.01	−4.083	0.02	0.0000	1.115	3.049	0.0020

$\mathcal{H}_0 : \Delta = 0$ against $\mathcal{H}_1 : \Delta > 0$ the P-value is

$$\Pr(Y_1 \geq 21 | t = 40) = 0.0074 + 0.0009 + 0.0000 = 0.0083.$$

Notice that there is no approximation in the figure 0.0083—hence the terminology *exact*. In Section 3.2 we obtained various unconditional and approximate P-values for testing the same hypotheses: 0.0028 using LR, 0.0044 using Pearson, 0.0083 using empirical logits, and 0.0067 using modified empirical logits. The fact that the conditional P-value agrees with the very simple empirical logit based P-value is an accident of this particular example. They need not be close in general.

It is instructive to consider how we might test $\mathcal{H}_0 : \Delta = 0.6931$ against $\mathcal{H}_1 : \Delta > 0.6391$. Note that $0.6931 = \log(2)$ and so the null hypothesis says that the odds of a tumor for treated mice is twice that for untreated mice. The alternative hypothesis says the odds ratio is even larger than 2. To compute the P-value, column (2t) was obtained by adding $0.6391y$ to column (1) and then subtracting the mean of the resulting column. Column (3t) is column (2t) exponentiated and the required tilted hypergeometric probabilities are obtained by dividing this by its total. The P-value is

$$\Pr(Y_1 \geq 21 | t = 40; \Delta = 0.6931) = 0.0758 + 0.0187 + 0.0020 = 0.0965.$$

There is moderate evidence that $\Delta > 0.6931$. This is consistent with the ML estimate 1.972 whose standard error is 0.823, computed in Section 3.2.3.

Difficulties with Testing Two-Sided Hypotheses Conditional tests for 2×2 tables are based on tail probabilities of the hypergeometric distribution. Computing one-sided P-values is straightforward in principle—we add up the probability of the observed value and all extreme values more extreme, either greater or smaller depending on the alternative hypothesis. For more complicated data structures, we need to calculate tail probabilities of more complicated, but still known, distributions.

For a two sided P-value we have difficulty in extending this idea, because both improbably high and improbably low values are extreme, regardless of whether the observed value happens to be higher or lower than expected. We thus compute not only the tail probability corresponding to the value observed but a tail probability on the opposite side of the distribution. How do we choose the opposite tail region? A common rule is to find the probability of all values y_1 as probable as, or less probable than, the observed value. This and other rules lead to annoying difficulties because of the discreteness of the distribution.

Example 7.2: Esophageal Cancer and Alcohol Consumption A retrospective study into the association of alcohol and Esophageal cancer was reported by Breslow and Day (1980). The data have been broken into five age-groups in Table 7.3. The last two columns give values of the Pearson chi-square statistic and the two-sided P-value for testing $\Delta = 0$, separately for each age-group. Particularly in the first two lines, there is considerable disagreement between the unconditional and exact approaches.

To see how the two-sided P-value becomes identical to the one-sided, consider the second line where four of the 30 alcohol exposed individuals have cancer, while only five of the 169 unexposed individuals have cancer. We have $n_1 = 30$, $t = 9$, and $n = 199$ and the null hypergeometric distribution is shown in Table 7.4. The one-sided P-value is $\Pr(Y_1 \geq 4) = 0.026 + 0.005 = 0.031$. To obtain the two-sided P-value, we consider all y-values on the opposite tail of the distribution less probable than the observed value of 4 whose probability

Table 7.3. Esophageal Cancer and Alcohol Consumption, by Age-Grouping

Age Group	Alcohol		No Alcohol		Exact		Pearson	
	Case	Control	Case	Control	One-Sided	Two-Sided	χ_1^2	P-Value
25–34	1	9	0	106	0.086	0.086[a]	10.69	0.001
35–44	4	26	5	164	0.031	0.031[a]	6.35	0.012
45–54	25	29	21	138	0.000	0.000	26.06	0.000
55–64	42	27	34	139	0.000	0.000	18.66	0.000
65–74	19	18	36	88	0.011	0.017	6.31	0.012
75+	5	0	8	31	0.001	0.001[a]	13.45	0.000

[a]For these cases, the exact two-sided P-value is identical to the one-sided.

Table 7.4. Hypergeometric Distribution for $n_1 = 30$, $t = 9$, and $n = 199$

y_1	0	1	2	3	4	5+
$\Pr(Y_1 = y_1)$	0.222	0.373	0.267	0.107	0.026	0.005

is 0.026. There is no such value and so the one- and two-sided P-value are identical!

This certainly seems unsatisfactory as a measure of evidence; if we dismiss the possibility that $\Delta < 0$ then we expect a given large value y_1 to provide more evidence against the null hypothesis. It has been argued by some authors that we should double the one-sided P-value in such situations. However, while doubling the one-sided value to 0.062 may give us a warm inner glow, using this P-value as the basis of a test will not give the correct size, except in an artificial sense described in Lloyd (1988).

We will certainly not be able to resolve this issue satisfactorily here as it speaks to some of the fundamental dilemmas in the frequentist approach to statistics. In practice however, since we often approximate distributions by a normal distribution that is symmetric, the two-sided P-value is automatically twice the one-sided to this level of approximation. This might be cold comfort to theorists, but perhaps explains why the kind of problems outlined here are given little emphasis in textbooks.

7.1.3 Conditional Intervals

By the duality of tests and confidence limits, a lower limit for Δ is the smallest value of δ_0 for which $\mathcal{H}_0 : \Delta = \delta_0$ is not rejected in favor of $\mathcal{H}_1 : \Delta > \delta_0$. This requires the solution of

$$P(y_1, \delta) = \Pr(Y_1 \geq y_1 | t, \delta) = \alpha/2.$$

The upper limit is found in a similar manner by computing lower tail probabilities. If the equation is to be solved by Newton–Raphson then we need the derivative of $P(y_1, \delta)$ with respect to δ which is given by

$$P(y_1, \delta)\{\mathrm{E}_c(Y_1 | Y_1 \geq y_1) - \mathrm{E}_c(Y_1)\}$$

where

$$\mathrm{E}_c(Y_1 | Y_1 \geq y_1, \delta) = \sum_{y \geq y_1} y \binom{n_1}{y} \binom{n_2}{t-y} \phi^{y_1} \Big/ \sum_{y \geq y_1} \binom{n_1}{y} \binom{n_2}{t-y} \phi^{y}.$$

An alternative and simple algorithm is bisection.

Example 7.1: (*Continued*) Mice-Tumor Data. Let us illustrate how to find an exact 2.5% upper limit $\hat{\Delta}_{.975}$ by bisection. We first need to find bounds for the solution of

$$Q(\delta) = \Pr(Y_1 \leq 21; \delta) = 0.025.$$

After some trial and error we find that $4 < \hat{\Delta}_{0.975} < 4.5$ since $Q(4) = 0.0402$ and $Q(4.5) = 0.0163$. We next compute the tilted hypergeometric tail probability $Q(4.25) = 0.0258$ and thus narrow the range to $4.25 < \hat{\Delta}_{0.975} < 4.5$. We eventually arrive at the upper limit 4.268. In a similar manner, computing upper tail probabilities $\Pr(Y_1 \geq 21;\, \delta)$ for various δ, we arrive at the lower limit 0.271.

The conditional interval (0.271, 4.268) is wider than the interval (0.359, 3.585) based on empirical logits. This is a general tendency, mostly attributable to the discreteness of the distribution, and conditional methods have been criticized on these grounds. On the other hand, the unconditional interval includes values that are rejected by the exact test of the previous section. A program `ghg.exe` for computing tilted hypergeometric tail probabilities is at the Wiley website.

Conditional ML Estimator and its Standard Error The conditional ML estimator, $\hat{\Delta}_c$, can be computed by numerically maximizing the conditional likelihood. A preferable algorithm is to find the value of Δ for which the derivative of the conditional log-likelihood is zero. From differentiating Eq. (7.4), the conditional score function is

$$U_c(\Delta) = y_1 - \sum_y y \binom{n_1}{y}\binom{n_2}{t-y} e^{\Delta y} \Big/ \sum_y \binom{n_1}{y}\binom{n_2}{t-y} e^{\Delta y}. \tag{7.7}$$

The rather complicated looking ratio of sums is actually the conditional mean of Y_1, i.e., the mean of the tilted hypergeometric distribution Eq. (7.2). The conditional ML estimator of Δ is that value which makes the conditional mean value of Y_1 equal to its observed value y_1. However, this mean is a complicated function of Δ and solution requires iterative methods. The Newton algorithm here has the very simple form

$$\hat{\Delta}_{\mathrm{NEW}} = \hat{\Delta}_{\mathrm{OLD}} + \frac{y_1 - \mathrm{E}_c(Y_1)}{\mathrm{Var}_c(Y_1)}. \tag{7.8}$$

The approximate variance of $\hat{\Delta}_c$ is obtained by applying the usual likelihood asymptotics to the conditional rather than the ordinary likelihood. The Fisher information I_c is the conditional variance of the above conditional score function. Treating t as fixed, the second term in U_c is not random and so

$$I_c = \text{Var}_c\{U_c(\Delta)\} = \text{Var}_c(Y_1).$$

This is automatically computed during the Newton–Raphson algorithm Eq. (7.8). The asymptotic standard error of $\hat{\Delta}_c$ is then

$$\text{se}(\hat{\Delta}_c) = \frac{1}{\sqrt{\text{Var}_c(Y_1)}}.$$

However, for 2 × 2 tables at least, it is not recommended to construct confidence intervals by approximating the distribution of $\hat{\Delta}_c$ by the normal distribution with this standard error. This violates the spirit of exact conditional calculations.

Example 7.1: (*Continued*) Mice-Tumor Data. The conditional estimate $\hat{\Delta}_c$ is the value of Δ solving

$$21 = \sum_{y=8}^{23} y \binom{23}{y} \binom{32}{40-y} e^{\Delta y} \Big/ \sum_{y} \binom{23}{y} \binom{32}{40-y} e^{\Delta y}.$$

The right-hand side is a ratio of functions, each being polynomials in $\phi = \exp(\Delta)$ of degree 23. We illustrate the Newton algorithm beginning with the unconditional ML estimate 1.972. For this value we find that $\text{E}_c(Y_1) = 21.049$ and $\text{Var}_c(Y_1) = 1.213$ and so our next guess is

$$1.972 + (21 - 21.049)/1.213 = 1.931.$$

This gives $\text{E}_c(Y_1) = 20.988$, etc., and we reach the value $\hat{\Delta}_c = 1.939$ after another three iterations, correct to three decimal places. For this value of Δ, $\text{Var}_c(Y_1) = 1.224$ and so

$$\text{se}(\hat{\Delta}_c) = 1/\sqrt{1.224} = 0.904.$$

The standard error of the unconditional ML estimator was 0.823 (Table 3.3).

7.1.4 Approximations to the Exact Method

There are many ways to approximate the exact confidence interval for Δ. For instance, one may approximate the distribution of $\hat{\Delta}_c$ by normal. However, if one has the computing power to find the estimate, one has the computing power to find the limits—both require similar iterative evaluation of tilted hypergeometric distributions. It is hard to see how the interval $(\hat{\Delta}_c \pm 1.96\text{se}(\hat{\Delta}_c))$ could ever be recommended.

There are two approximate methods that are worthy of mention, both involving approximations to the hypergeometric distribution of Y_1 itself.

Yates' Test of $\Delta = 0$ It can be shown that for large n, the tilted hypergeometric converges to normal so long as t/n or n_1/n do not converge to zero or one. In particular, the ordinary hypergeometric converges to normal with moments

$$\mathrm{E}_c(Y_1) = \frac{n_1 t}{n}, \quad \mathrm{Var}_c(Y_1) = \frac{n_1(n - n_1)t(n - t)}{n^2(n - 1)} \tag{7.9}$$

these being the usual hypergeometric formulas. Hence, the statistic

$$\frac{Y_1 - n_1 t/n}{\sqrt{n_1 n_2 t(n - t)/(n^2(n - 1))}}$$

is approximately standard normal. The square of this statistic in fact only differs from the Pearson statistic Eq. (2.26) by the factor $(n - 1)/n$ as is easily verified. A better test statistic is obtained by making a continuity correction. For testing against a two-sided alternative this gives

$$\frac{(|Y_1 - n_1 t/n| - \frac{1}{2})^2}{n_1 n_2 t(n - t)/(n^2(n - 1))} \tag{7.10}$$

proposed by Yates (1934) and often called the *Yates corrected* χ^2 statistic (if we ignore the factor $(n - 1)/n$). The motivation is to better approximate the hypergeometric distribution, not to produce a more chi-square statistic.

For moderate sample sizes there is little point in making any of these approximations and the exact hypergeometric distribution should be used. The package *StatXacT* will compute estimates, P-values, and confidence intervals based directly on Eq. (7.2) for any value of Δ, so there is no longer any impediment to doing the exact analysis as earlier described. For larger sample sizes the large factorials in Eq. (7.2) make Yates' method potentially useful when adequate computing power is not available.

For testing values other than zero, it is as computationally difficult to find the conditional moments as computing the entire distribution. Yates' method is only of practical use for testing $\Delta = 0$.

Saddlepoint Approximation Another more sophisticated method of approximating the exact analysis is based on a quite different class of distributional approximation formulas called saddlepoint approximations. Applying the saddlepoint formula to the joint distribution of Y_1, T given in Eq. (3.10), then to the distribution of T and taking the ratio gives an approximation to the conditional tilted hypergeometric distribution. It can be computed without great difficulty and is spectacularly accurate. For the case of 2 × 2 tables, it is actually equivalent to using Stirling's approximation to all the factorials involved.

A general account of this method and its implementation in GLIM are given in Section 7.9. A review of saddlepoint methods, their use in statistics, and an application to a simple 2 × 2 table may be found in Barndorff–Nielsen and Cox (1979).

7.2 CONDITIONAL INFERENCE FOR 2 × k TABLES

The notation used is shown in Table 7.5 below. A total of n individuals are classified into k groups, n_i being in group i. Within each group the number exhibiting a binary response, y_i is determined. The total number of responses is t. The question of interest is whether or not the probability of response varies from group to group and, if so, how.

As in the two sample problem, there are several alternative sampling schemes that ultimately lead to identical statistical estimators and tests. With prospective sampling the Y_i are independent binomial variables with the response probability for group i denoted π_i. The corresponding logits are denoted ψ_i and we let $\Delta_i = \psi_i - \psi_k$ measure the difference in logits for group i compared to group k. The null hypothesis is

$$\mathcal{H}_0 : \pi_1 = \pi_2 = \cdots = \pi_k = \pi \Leftrightarrow \Delta_1 = \Delta_2 = \cdots = \Delta_{k-1} = 0.$$

The association parameters $\Delta_1, \ldots, \Delta_{k-1}$ are collectively denoted Δ. The choice of group k as the baseline is arbitrary and has no effect on final results. You should reread Section 3.3.1 where we wrote down the log-likelihood function

$$\ell(\Delta, \psi_k; y) = c + \sum_{i=1}^{k-1} y_i \Delta_i + t\psi_k - \sum_{i=1}^{k} n_i \log\{1 + \exp(\psi_i)\}. \tag{7.11}$$

Since T is sufficient for ψ_i, inference about the association parameters Δ, in particular tests of $\mathcal{H}_0$ above, may be based on the distribution of the data $Y_1, \ldots,$

Table 7.5. Generic k-Sample Data Set

Group	1	2	·	·	k	Total
Response	y_1	y_2	·	·	y_k	t
No response	$n_1 - y_1$	$n_2 - y_2$	·	·	$n_k - y_k$	$n - t$
Total	n_1	n_2	·	·	n_k	n

Y_k conditional on T. Neither this distribution, nor the corresponding conditional likelihood, will depend on the nuisance parameter ψ_k.

7.2.1 Conditional Distribution and Likelihood

Inference on a *particular* log-odds parameter, say Δ_1, is based on the distribution of Y_1 conditional on $Y_2, \ldots, Y_{k-1}$ and T. This is equivalent to simply analyzing the 2×2 table comprising columns 1 and k and this was fully covered in the previous section.

Inference on the full vector Δ is based on the joint distribution of $Y_1, \ldots, Y_k$ given T. To derive this conditional distribution, we reexpress the joint binomial distribution of $Y_1, \ldots, Y_k$ in the form

$$P_B(y_1, \ldots, y_k) = \left\{ \prod_{i=1}^{k-1} \binom{n_i}{y_i} e^{\psi_i y_i}(1 + e^{\psi_i})^{-n_i} \right\} \left\{ \binom{n_k}{y_k} e^{\psi_k y_k}(1 + e^{\psi_k})^{-n_k} \right\}$$

$$= \binom{n_k}{y_k} \left\{ \prod_{i=1}^{k-1} \binom{n_i}{y_i} e^{\Delta_i y_i} \right\} e^{t\psi_k} f(\Delta, \psi_k)$$

where f is a function whose form is unimportant. The joint distribution of $Y_1, \ldots, Y_{k-1}, T$ is obtained by replacing y_k by $t - \sum_{i=1}^{k-1} y_i$ and the marginal distribution of T by summing over the y_i.

Taking the ratio of these two expressions we see that $e^{t\psi_k} f(\Delta, \psi_k)$ cancels and we obtained the conditional distribution

$$p_c(y|n, n_i, t, \Delta) = \binom{n_1}{y_1}\binom{n_2}{y_2} \cdots \binom{n_k}{y_k} e^{\Delta_1 y_1 + \ldots + \Delta_{k-1} y_{k-1}} / \kappa(\Delta) \tag{7.12}$$

where κ is a normalizing constant. The conditional log-likelihood is

$$\ell_c(\Delta) = c + \sum_{i=1}^{k-1} y_i \Delta_i - \log \left\{ \sum_{y(t)} \binom{n_1}{y_1}\binom{n_2}{y_2} \cdots \binom{n_k}{y_k} e^{\Delta_1 y_1 + \cdots + \Delta_{k-1} y_{k-1}} \right\}. \tag{7.13}$$

The distribution (7.12) is called the *multivariate* tilted hypergeometric distribution and reduces to the tilted hypergeometric when $k = 2$. Since the marginal distribution of each Y_i is tilted hypergeometric we may think of this distribution as that of k tilted hypergeometric variables constrained to add up to a fixed total t.

The conditional distribution is available for inference on the log-odds ratio parameters. In the special case that all the Δ_i are zero the joint distribution becomes

$$p_c(y) = \binom{n_1}{y_1}\binom{n_2}{y_2}\cdots\binom{n_k}{y_k}\bigg/\binom{n}{t} \qquad y \in \mathcal{Y}(t) \tag{7.14}$$

known as the multivariate hypergeometric distribution. Writing out the factorials in this probability function one finds terms for all the column and row totals on the numerator and terms for each of the $2k$ counts in the body of the table in the denominator. This pattern is worth noting as it extends directly to contingency tables of larger size and higher dimension.

Note that the arguments $y_1, \ldots, y_k$ in Eqs. (7.12) and (7.14) are constrained to add up to t and so these are really a $k-1$ dimensional distributions, involving $k-1$ parameters $\Delta_1, \ldots, \Delta_{k-1}$. However, there are extra constraints on the y_i to prevent any of the counts becoming negative similar to, but more complicated than, the condition $\max(0, n_2 - t) \le y \le \min(n_1, t)$ for the hypergeometric distribution. The set of allowable values of $y_1, \ldots, y_{k-1}$ will be denoted $\mathcal{Y}(t)$. As we move to more complex data structures, determining this set becomes more and more difficult. We will enumerate this set in the next example.

7.2.2 Conditional Tests and P-Values

For the 2 × 2 table there was only one free variable Y_1 and tests were based on tail probabilities of this variable. For $k > 2$, the hypothesis

$$\mathcal{H}_0 : \Delta_1 = \Delta_2 = \cdots = \Delta_{k-1} = 0$$

concerns more than one parameter and so there is no unique best test statistic. Consequently there are many possible conditional tests corresponding to different choices of test statistic. Using the conditional distribution Eq. (7.14), we can compute the tail probability of any statistic exactly, by enumerating all possible tables. However, there being many possible *exact* P-values, we see again that this name can be confusing.

Moreover, the procedure becomes computationally intensive for even moderately large k and marginal totals. There are two issues. First, one must enumerate all possible tables whose entries are consistent with the observed marginal totals of the observed table, i.e., enumerate the support set $\mathcal{Y}(t)$. This can become a formidable combinatoric problem. Second, the probabilities of each of these tables must be computed which often presents numerical problems. These computational difficulties have been solved in a wide range of cases.

Two very common and easily computed test statistics are the LR statistic and Pearson statistic. Both of these arise from consideration of the unconditional likelihood. One might expect to use the conditional LR statistic

$$2\{\ell_c(\hat{\Delta}_c) - \ell_c(\delta_0)\}$$

as a test statistic. The reason for not using this are computational, it being necessary to compute the test statistic for every possible data set. Except for cases where unconditional likelihood methods are inconsistent, using an unconditional statistic is quite defensible as long as the conditional distribution is used to compute the P-value.

Example 7.3: Worked Example of a 2 × 3 Table We will go through the necessary calculations in detail for a fictitious data set (Table 7.6). It is hard to find a real data set that would have counts small enough to make this a practical exercise. Imagine 14 patients, 8 of whom are given treatment 1 and 6 treatment 2. After a certain time each patient is classified as `CURED`, `ILL`, or `DECEASED`. We wish to test whether the treatments are different overall in terms of producing a different distribution of patients in these three categories. This view is more natural if the study were prospective. Alternatively, imagine the data came from a retrospective study. We have three groups of patients known to be `CURED`, `ILL`, or `DECEASED` and we wish to compare the proportions given treatment 1 across the three groups.

Whether we consider the data as two independent trinomials or three independent binomials, the results will be identical but for didactic purposes we take the latter view. We let π_i represent the probability of being given treatment 1 given that the patient has responded `CURED` ($i = 1$), `ILL` ($i = 2$), or `DECEASED` ($i = 3$). The log-odds parameter Δ_1 represents the logged ratio of the odds of being cured compared to dying when given treatment 1 rather than treatment 2. The log-odds parameter Δ_2 compares odds of being ill with deceased. Fitted values under the null hypothesis that treatment and outcome are unrelated are given in parentheses. Estimates of the log-odds parameters are $\hat{\Delta}_1 = 2.890$, $\hat{\Delta}_2 = 0.405$. The modified estimates are 2.314 and 0.336 being, as always, less extreme than the unmodified estimates.

Table 7.6. Fictitious Data for a 2 × 3 Table

	Cured	Ill	Deceased	Total
Treatment 1	6 (4.00)	1 (1.71)	1 (2.28)	8
Treatment 2	1 (3.00)	2 (1.29)	3 (1.71)	6
Total	7	3	4	14

Table 7.7. Exact Analysis of a 2 × 3 Table

	1	2	3	4	5	6	7	Total
0				35 7.00[a] 9.56[a]	84 5.10[a] 6.25[a]	42 6.42[a] 7.83[a]	4 10.93[a] 14.62[a]	165
1			105 4.28 5.74[a]	420 1.21 1.24	378 1.36 1.38	84 4.71[b] 5.06[b]	3 11.18[a] 15.30[a]	990
2		63 5.44[a] 6.93[a]	420 1.21 1.24	630 0.19 0.19	252 2.38 2.42	21 7.78[a] 9.56[a]		1386
3	7 10.50[a] 13.38[a]	84 5.10[a] 6.25[a]	210 2.92 4.01	140 3.94 5.06[a]	21 8.17[a] 10.74[a]			462
Total	7	147	735	1525	735	147	7	3003

[a]Statistic is as extreme as observed.
[b]Observed value of statistics.

If we were specifically interested in survival/death or cured/not cured then we could collapse the table into a 2 × 2 table and use the earlier exact method. We will see that this information comes out of the multivariate test. Table 7.7 enumerates all possible 2 × 3 tables with the same marginal row and column totals as our data. The free variables Y_1, Y_2 are the respective numbers of cured and ill patients given treatment 1. Columns refer to values of Y_1 and rows to Y_2. You should check that the empty cells of the table correspond to pairs (y_1, y_2) logically inconsistent with the fixed totals of Table 7.6. In each cell are listed three figures. The first is the combinatoric numerator in Eq. (7.14) which in this case is a product of three terms of the form

$$\binom{7}{y_1}\binom{3}{y_2}\binom{4}{8 - y_1 - y_2}.$$

The next numbers are computed values of the Pearson statistic Eq. (3.22) and the LR statistic Eq. (3.21) for the given table. The grand total of the combinatorics is 3003 which is 14 choose 8 and the probability of each table is found by dividing the cell combinatoric by this total. The observed table is marked with a 'b' and tables with LR or Pearson statistics as large as the observed by 'a'. Notice that in the cell $(y_1, y_2) = (3,4)$, the value of LR is actually identical to the observed value of 5.06203 for the observed table (6,1).

The exact P-value for the Pearson statistic is obtained by accumulating the probabilities of the 11 cells with Pearson statistics as large or larger than the observed value of 4.71. This gives

$$\text{P} - \text{value} = \frac{35 + 84 + 42 + 4 + 84 + 3 + 63 + 21 + 7 + 84 + 21}{3003} = 0.149.$$

The mean and variance of the exact distribution are 2.15 and 3.73—not drastically in conflict with the χ_2^2 distribution. The approximate P-value however is $\Pr(\chi_2^2 \geq 4.71) = 0.095$. For the LR statistic there are 13 cells with values as extreme as 5.06 and the total probability mass is $693/3003 = 0.231$. Notice that this is a larger P-value than the exact Pearson P-value even though the observed value of the Pearson statistic is *smaller* than the observed value of the LR. The mean and variance of the exact distribution is 2.55 and 6.36 and the approximate P-value based on the χ_2^2 distribution is $\Pr(\chi_2^2 \geq 5.06) = 0.080$. The Pearson statistic is closer in distribution to its limiting χ_2^2 as was claimed as a general trend in Section 2.5.5, although we have actually been investigating the conditional distribution here.

The normalized column totals in Table 7.7 give the $HG(7, 8, 14)$ distribution which could be used to perform inference on the 2×2 table with categories **ILL** and **DECEASED** combined, i.e., the table with entries (6, 2, 1, 5). The log-odds parameter Δ of this collapsed table measures the association of treatment with the cured/not-cured dichotomy. To test $\Delta > 0$ we compute the P-value

$$\Pr(Y_1 \geq 6 | n_1 = 8, t = 7, n = 14, \Delta = 0) = \frac{147 + 7}{3003} = 0.0513.$$

The normalized row totals give the $HG(3, 8, 14)$ distribution that could be used to perform inference on the (rather ridiculous) 2×2 table with categories **CURED** and **DECEASED** combined. The computations here are quite tedious and it will be appreciated that the complexity increases both with the size of the counts and with the dimension k. For example, with a 2×4 table there would be three free variables and correspondingly more tables consistent with the marginal totals to enumerate.

7.2.3 Approximations to the Exact Method

As with the 2×2 table the (multivariate) hypergeometric distribution involved may be normally approximated with continuity correction. Under $\mathcal{H}_0 : \Delta_1 = \cdots = \Delta_{k-1} = 0$, the random vector $Y = (Y_1, \ldots, Y_{k-1})$ has components with means

$$\text{E}(Y_j | T = t) = m_j = \frac{n_j t}{n}, \quad j = 1, \ldots, k = 1 \tag{7.15}$$

which we collect together into a vector m. The variance matrix is

$$\mathrm{Cov}(Y_i, Y_j | T = t) = V_{ij} = \frac{t(n-t)}{n-1} \frac{n_i}{n} \left(\delta_{ij} - \frac{n_j}{n} \right) \tag{7.16}$$

where $\delta_{ij} = 1$ if $i = j$ and so the standardized random vector

$$Z = V^{-1/2}(Y - m)$$

has conditional mean zero and conditional dispersion the identity. Further, the conditional distribution is approximately multivariate standard normal provided that the largest eigenvalue of V^{-1} is large. This requires that n be large and that none of the n_j or t take extreme values. Under these circumstances, using the quadratic form

$$Q = Z^T Z = (Y - m)^T V^{-1} (Y - m)$$

as a test statistic, and comparing it to the χ^2_{k-1} distribution, will be an approximately correct conditional test.

It should be understood that even were the exact conditional distribution of Q used, the results would be different to those obtained using the LR or Pearson statistics—because a different test statistic, Q, is being used. Second, the normal approximation is typically poorer for multivariate discrete distributions and continuity correction is even more important. In this context, we reduce all the components of $Y - m$ by 0.5 in absolute value.

Example 7.1: (*Continued*) Mice-Tumor Data. From Eqs. (7.15 and 7.16), the random vector (Y_1, Y_2) has null mean (4.00, 1.71) and null variance matrix

$$V = (2548)^{-1} \begin{pmatrix} 2352 & -1008 \\ -1008 & 1584 \end{pmatrix} \Rightarrow V^{-1} = \begin{pmatrix} 1.489 & 0.948 \\ 0.948 & 2.212 \end{pmatrix}.$$

The quadratic form $Q = 4.378$ giving the P-value 0.112, with continuity correction $Q^* = 2.852$ giving the P-value 0.240. Of course, we do not expect that the normal approximation will be very accurate since the distribution being approximated in Table 7.7 is so discrete. The *exact* P-values corresponding to Q and Q^* can be computed by calculating the values of these statistics for every cell in that table and gives the results 0.178 and 0.204, respectively. The chi-square approximation is better for Q^* than for Q, as expected.

7.3 CONDITIONAL INFERENCE ON LINEAR MODELS

Having looked at conditional inference on the association parameters of 2×2 and $2 \times k$ tables, the following generalization should not come as too much of shock. Generalized linear models were described in detail in Section 4.6. The data $Y_1, \ldots, Y_k$ are assumed to have a distribution from the exponential family [see Eq. (4.41)]. For binomial and Poisson models, the scale parameter $a_i(\phi) = 1$ and the log-likelihood is of the form

$$\log \ell(\theta; y) = \sum_{i=1}^{k} y_i\theta_i - b(\theta_i)$$

where θ_i is called the canonical parameter and b a differentiable function. For binomial models $\theta_i = \text{logit}(\pi_i)$ and for Poisson models $\theta_i = \log \mu_i$. This parameter is modeled linearly in terms of a set of parameters β via $\theta_i = \sum_{j=1}^{p} x_{ij}\beta_j$ and so the log-likelihood becomes

$$\log \ell(\beta; y) = \sum_{j=1}^{p} \beta_j \sum_{i=1}^{k} y_i x_{ij} - b(\theta_i(\beta)) \tag{7.17}$$

Hence, the statistics $S_j = \sum_{i=1}^{n} x_{ij}y_i$ are sufficient for β_j. In matrix form $S = X^T y$ is sufficient for β. Suppose we wanted to test β_1, the other regression parameters $(\beta_2, \ldots, \beta_p)$ being considered nuisance parameters. Then the distribution of S_1 conditional on $S_2, \ldots, S_p$ depends on β_1 only, by the definition of sufficiency.

More generally, let $\mathcal{H}_0$ be a linear sub-model specified by the restriction $\omega(\beta) = \omega_0$. Without loss of generality we may write $\beta = (\omega, \lambda)$ where the nuisance parameter λ is left free under $\mathcal{H}_0$. Denote the sufficient statistic for λ_0 by $\mathcal{S}_0$ and the sufficient statistic for (ω, λ) by $\mathcal{S}_1$ (just called S above).

Inference about ω may be based on the distribution of $\mathcal{S}_1$ conditional on $\mathcal{S}_0$ since, by definition of sufficiency of S_0, this distribution does not depend on λ. Tests based on the conditional distribution will have guaranteed type 1 error conditionally, and therefore unconditionally. Similarly, confidence intervals can be constructed with coverage guaranteed to exceed 95%, regardless of the values of λ.

7.3.1 Computing the Conditional Distribution

The joint distribution of $(\mathcal{S}_0, \mathcal{S}_1)$ may be factorized into two terms, the conditional distribution and the marginal distribution of $\mathcal{S}_0$. Since $\mathcal{S}_0$ is contained in $\mathcal{S}_1$ we have

$$f_1(s_1; \omega, \lambda) = f_c(s_1; s_0, \omega) f_0(s_0; \omega, \lambda) \tag{7.18}$$

in obvious notation. By using the conditional likelihood $L_c(\omega; s_1) = f_c(s_1; \omega)$ we are discarding the last factor.

Let $\mathcal{Y}(s_0)$ be the conditional sample space of $\mathcal{S}_1$, i.e., the set of possible values s_1 of $\mathcal{S}_1$ consistent with $\mathcal{S}_0 = s_0$. Rearranging Eq. (7.18) we find that

$$f_c(s_1; s_0; \omega) = \frac{f_1(s_1; \omega, \lambda)}{\kappa(\omega, \lambda)} \qquad s_1 \in \mathcal{Y}(s_0) \tag{7.19}$$

where $\kappa(\omega, \lambda)$ is a normalizing constant, equal to either the sum or integral of the numerator over the set $\mathcal{Y}(s_0)$. Since we know that the conditional distribution is free of λ, we are at liberty to choose any convenient value for λ to make the calculation easy. We may also discard any factors of $f_1(s_1)$ not depending on s_1. A general algorithm for finding the conditional distribution is therefore as follows: Write down the distribution f_1 of the sufficient statistic $\mathcal{S}_1$. Discard any factors free of s_1 and replace λ by any convenient value (often zero's or one's). Then normalize the distribution with respect to the conditional sample space $\mathcal{Y}(s_0)$.

This algorithm sounds easy, however, the conditional sample space $\mathcal{Y}(s_0)$ can be extremely hard to determine, particularly for discrete models where its exact shape depends on the vagaries of integer arithmetic. The general algorithm can be avoided in some cases by appealing to sampling theory and other statistical notions.

7.3.2 Problems with Conditioning

There are some problems associated with using the conditional distribution of $\mathcal{S}_1$ given $\mathcal{S}_0$. Discarding the second factor in Eq. (7.18) is potentially wasteful of the information S_0 may contain about ω, and conditional methods are sometimes inefficient and/or conservative, especially for small samples. This may manifest itself in several ways. For instance, the asymptotic variance may be greater for the conditional than the unconditional estimator. On the other hand, the bias of the conditional estimator may be less than the unconditional estimator, and the latter can even be inconsistent.

For discrete data, the conditional distribution has a smaller sample space and is therefore more discrete than the unconditional distribution. Tests of prespecified size are not usually achievable for discrete data, and the more discrete the distribution the more "conservative" the test must be made if its size is not to fall below nominal. Conditional tests are more subject to this conservatism than unconditional ones. Indeed, in extreme cases the conditional sample space contains just one point, the conditional distribution is degenerate, and the conditional method completely fails.

On a more practical level, the unconditional distribution is typically very simple (independent binomial or Poisson variables). However, the conditional distribution often requires a huge amount of computation. This becomes less problematic with recent advances in algorithms and with each new generation of computers. However, the computational difficulty increases so quickly with data size that there will always be reasonable data sets for which exact conditional computations are not practical. In this case, there are Monte Carlo as well as analytic approximations available.

7.3.3 Conditional Tests and P-Values

Testing ω requires a test statistic $T(\mathcal{S}_1)$. The role of the test statistic is to rank all possible data sets s_1 as more or less hostile to the null hypothesis. The P-value is then

$$\Pr\{T(\mathcal{S}_1) \geq T(s_1)|s_0; \omega_0\}$$

but in most cases there will not be an analytic expression for the conditional distribution of $T(\mathcal{S}_1)$. Instead, we must integrate or sum the conditional probabilities $f_c(s_1; s_0, \omega_0)$ over those data sets for which T exceeds the observed value $T(s_1)$. In the worst case, this means going through every element of the possibly huge set $\mathcal{Y}(s_0)$, computing the test statistic T, and adding the conditional probability into the P-value if the test statistic exceeds the observed value. In some cases, it is possible to discard many possible data sets from the calculation as we proceed using the so-called network algorithm of Mehta and Patel (1983).

Common choices of test statistics are the Wald, Score, or LR statistic. Freeman and Halton (1951) suggested using the test statistic

$$T(\mathcal{S}_1) = f_c(\mathcal{S}_1; s_0, \omega_0),$$

i.e., the null conditional probability of the observed table. Superficially, this seems reasonable however data that is improbable under $\mathcal{H}_0$ need not be hostile to $\mathcal{H}_0$ if it is almost as improbable under $\mathcal{H}_1$. At least in general, the conditional probability itself is not a good test statistic. On the other hand, there are computational advantages over Wald, Score, or LR statistics since these require more computation and their value must be computed for every possible data set s_1 in $\mathcal{Y}(s_0)$.

Confidence intervals for ω may be obtained with conditionally correct coverage by inverting the conditional P-value i.e. by finding values of ω for which the conditional probability $\Pr(T(S_1 \geq t|s_0)$ is at least α. This is typically only done when ω is scalar. Confidence regions for vector parameters are harder to digest and in any case their shape is often sensitive to the choice of test statistic.

7.3.4 Conditional Estimation and Confidence Limits

For a generalized linear model with canonical link, the score function $U_j = \partial f_1(s_1; \beta)/\partial\beta_j$ has the form $S_j - E(S_j)$ where S_j is the jth component of $\mathcal{S}_1$. From Eq. (7.19) we find the conditional score is

$$\frac{\partial \log f_c}{\partial \beta_j} = \frac{\partial f_1(s_1; \beta)}{\partial \beta_j} - \frac{\partial \log \kappa}{\partial \beta_j} = S_j - \mathrm{E}(S_j; \beta) - \frac{\partial \log \kappa}{\partial \beta_j}.$$

Since conditional scores must have conditional mean zero it follows that

$$\frac{\partial \log f_c}{\partial \beta_j} = S_j - E(S_j | s_0; \omega). \tag{7.20}$$

The conditional score function equals S_j minus its conditional, rather than unconditional, mean. Whereas the unconditional ML estimators are obtained by equating the sufficient statistics to their means, the conditional ML estimaors are obtained by equating the sufficient statistics to their conditional means.

The conditional estimator takes a lot of computing. This is because the conditional means are *much* more complicated functions than the unconditional means. In finding the root of Eq. (7.20), each iteration requires a conditional expectation that requires us to compute a full conditional distribution. The test is not so bad as it only requires a single conditional distribution, the null distribution.

The Newton–Raphson algorithm for solving Eq. (7.20) requires the information matrix which, from the above score, is the derivative matrix of the conditional mean. Since this is nonrandom, it must coincide with the expected information which is the variance matrix of the score function. An algorithm for finding the conditional MLE $\hat{\Omega}_c$ of ω is therefore,

$$\hat{\Omega}_{\text{NEW}} = \hat{\Omega}_{\text{OLD}} + \text{Var}_c(\mathcal{S}_1)^{-1}\{s_1 - \text{E}_c(\mathcal{S}_1)\}.$$

For instance, in the previous section we studied a particular 2×3 table. The parameter there was $\omega = (\Delta_1, \Delta_2)^T$ and the sufficient statistic $\mathcal{S}_1 = (Y_1, Y_2)^T$ with observed value $s_1 = (6, 1)$. An algorithm for estimating (Δ_1, Δ_2) iteratively solves

$$\begin{pmatrix} \Delta_1 \\ \Delta_2 \end{pmatrix} = \begin{pmatrix} \Delta_1 \\ \Delta_2 \end{pmatrix} + \begin{pmatrix} \text{Var}_c(Y_1) & \text{Cov}_c(Y_1, Y_2) \\ \text{Cov}_c(Y_1, Y_2) & \text{Var}_c(Y_2) \end{pmatrix} \begin{pmatrix} 6 - \text{E}_c(Y_1) \\ 1 - \text{E}_c(Y_2) \end{pmatrix}$$

where the conditional means, variances, and covariances are computed from the multivariate tilted hypergeometric distribution. For the unconditional ML estimator, the same equations are solved but the moments are computed from the much simpler joint binomial distribution.

The conditional ML estimator maximizes the conditional likelihood and will typically differ from the unconditional ML estimator. Under asymptotic conditions where the dimension of the nuisance parameter increases, the conditional ML estimator is, but the unconditional ML typically is not, consistent for ω [see Andersen (1970) and Section 3.6.4 for an example].

Confidence limits are computed by inverting the test of the previous section, and results will depend on the tests statistic chosen. Assuming that large values of $T(\mathcal{S}_1)$ are supportive of large values of ω, a lower limit is found by solving

$$\Pr\{T(\mathcal{S}_1) \geq T(s_1)|s_0; \omega\} = \alpha/2$$

for ω. An upper limit is found by solving this equation with $\leq$ replacing $\geq$. Again, iterative methods are required and the full conditional distribution is computed with each iteration. This is about as computationally intensive as finding the estimator.

7.4 CONDITIONAL INFERENCE ON $r \times c$ TABLES

We have just finished a general account of conditional inference. We will apply the ideas there to the particular case of two-way tables. This includes the 2×2 table as a special case. The counts in the table are assumed to be Poisson, as in Chapter 6, because this makes the theory easier. Identical results are obtained from a multinomial formulation.

As in Section 6.3 imagine two categorical classifications labeled A and B. We think of these as a row factor on r levels and a column factor on c levels. Individuals are classified by the row and column factor and the counts arranged in a two-way table with Y_{ij} being the number with $A = i$, $B = j$. The distributions of the Y_{ij} are independent Poisson with means μ_{ij} and the saturated model is of the form

$$\log(\mu_{ij}) = m + a_i + b_j + (ab)_{ij} \tag{7.21}$$

the usual hypothesis of first interest being that the interaction parameters $(ab)_{ij}$ are all zero, i.e., independence of the A–B classification. Under the saturated model the sufficient statistic $\mathcal{S}_1$ is the two-way table of counts $\{y_{ij}\}$. Under the null hypothesis of independence, the sufficient statistic $\mathcal{S}_0$ comprises the marginal row totals $\{y_{i\bullet}\}$ and column totals $\{y_{\bullet j}\}$.

It is convenient to summarize the association of factors A and B through the parameters

$$\Theta_{ij} = \frac{\mu_{ij}\mu_{\bullet}}{\mu_{i\bullet}\mu_{\bullet j}}.$$

This is the odds-ratio for the 2×2 table with i and "not i" defining rows and j and "not j" defining columns. We take the interest parameter ω to be these odds ratios, only $(r-1)(c-1)$ of which are functionally independent. The unconditional distribution of $\{y_{ij}\}$ is that of independent Poisson variables with means μ_{ij}. Hence the marginal distribution of $\mathcal{S}_1$ is

$$
\begin{aligned}
f_1(s_1;\omega) &= f(\{y_{ij}\};\{\Theta_{ij}\}) \\
&\propto \prod_{i=1}^{r}\prod_{j=1}^{c} \mu_{ij}^{y_{ij}}/y_{ij}! \\
&\propto \prod_{i=1}^{r}\prod_{j=1}^{c} \left(\frac{\Theta_{ij}\mu_{i\bullet}\mu_{\bullet j}}{\mu_{\bullet\bullet}}\right)^{y_{ij}} \Big/ y_{ij}! \\
&\propto \prod_{i=1}^{r}\prod_{j=1}^{c} \Theta_{ij}^{y_{ij}}/y_{ij}!
\end{aligned}
$$

where in the last line we set the nuisance parameters $\lambda = (\mu_{\bullet\bullet}, \mu_{i\bullet}, \mu_{\bullet j})$ all equal to 1. We can do this because we are trying to find the conditional distribution which will be free of the nuisance parameters [see Eq. (7.18)] and the discussion following. Denoting $\mathcal{Y}(s_0)$ the set of all tables $\{y_{ij}\}$ with marginal row and column totals the same as those of the observed table we have

$$
f_c(\{y_{ij}\};\{\Theta_{ij}\}) = \frac{\displaystyle\prod_{i=1}^{r}\prod_{j=1}^{c} \Theta_{ij}^{y_{ij}}/y_{ij}!}{\kappa(\{\Theta_{ij}\}, s_0)} \qquad \{y_{ij}\} \in \mathcal{Y}(s_0) \tag{7.22}
$$

where κ is the sum of the numerator over the set $\mathcal{Y}(s_0)$.

Under the null hypothesis of independence all the Θ_{ij} equal 1.0, and the conditional distribution simplifies to

$$
\frac{\displaystyle\prod_{i=1}^{r} y_{i\bullet}! \prod_{j=1}^{c} y_{\bullet j}!}{\displaystyle y_{\bullet\bullet}! \prod_{ij} y_{ij}!} \qquad \{y_{ij}\} \in \mathcal{Y}(s_0). \tag{7.23}
$$

This can be derived from Eq. (7.22) using a generalization of the binomial theorem. However, it is easier to imagine simple random sampling without replacement from a finite population of size $y_{\bullet\bullet}$ whose members have characteristics $A = i$ with frequencies $\{y_{i\bullet}\}$. As individuals are sampled they are labeled with characteristic $B = j$ with frequencies $\{y_{\bullet j}\}$. The name of the distribution is generalized multivariate hypergeometric.

The explicit formula for the conditional probability is misleadingly simple; computing a P-value requires enumeration of the set $\mathcal{Y}(s_0)$ of all possible tables consistent with the given marginal row and column totals. Not only is this set often large, but it can be hard to determine. We worked through an explicit example for a 2×3 table on p. 411, and it is worthwhile to look back at that comparatively simple example.

The most commonly used test statistics for testing independence of A and B

are the Pearson and LR statistics, respectively, given by

$$\sum_{i=1}^{r}\sum_{j=1}^{c}\frac{(y_{ij}-\hat{e}_{ij})^2}{\hat{e}_{ij}}, \qquad \sum_{i=1}^{r}\sum_{j=1}^{c} 2y_{ij}\log\left(\frac{y_{ij}}{\hat{e}_{ij}}\right)$$

where in each case $\hat{e}_{ij} = y_{i\bullet}y_{\bullet j}/y_{\bullet\bullet}$ is the fitted value under the null hypothesis of independence. The conditional P-value is the sum of conditional probabilities of all those tables $\{y_{ij}\}$ in $\mathcal{Y}(s_0)$ for which the value of the test statistic is as large as the observed value.

Conditional ML estimates of the association parameters Θ_{ij} are obtained by equating any $(r-1)(c-1)$ counts in the table $\{y_{ij}\}$ to their conditional mean. Computing expected values requires the full conditional distribution, and therefore enumeration of the set $\mathcal{Y}(s_0)$.

For the special case of the 2×2 table there is really only one association parameter Θ_{11} under test and, conditional on the marginal totals, most reasonable test statistics including the Pearson and LR are functionally related. They then lead to identical results as long as their exact distribution is used. For higher dimensional tables, different test statistics are not functionally related and can give quite different results, particularly when the tables have both very high and very low counts, as in the example below.

Example 7.4: Congenital Malformation and Alcohol The data set below is a good one for illustrating the differences between different statistical procedures. The salient feature of this data set is that there are both very large and very small null expected values in the table. Table 7.8 gives the frequency distribution of drinking habits of the mothers of 32,481 normal babies and of 93 babies with congenital malformation.

The value of the Pearson statistic is 12.07 and of the LR statistic 6.20, and using the approximating χ_4^2 distribution the corresponding P-values are 0.017 and 0.185, respectively, i.e., the Pearson statistic suggests strongly that drinking habits do affect the likelihood of malformation while the LR statistic does not. Since these statistics are supposed to be asymptotically equivalent and yet differ

Table 7.8. Congenital Malformations and Alcohol Consumption of the Mother [from Graubard and Korn (1987)]

	Alcohol Consumption					
Malformation	0	<1	1–2	3–5	≥6	Total
Absent	17,066	14,464	788	126	37	32.481
Present	48	38	5	1	1	93
Total	17,114	14,502	793	127	38	32,574

by so much, it comes as no surprise that the null distributions are not close to χ^2_4. The above P-values then do not accurately measure how hostile the data are to the independence hypothesis.

The exact method enumerates all possible tables with the same row and column totals as the observed data table. There is a huge number of such tables. By computing the Pearson or LR statistic for each of these tables and accumulating the probabilities of those for which the test statistic is more extreme than the value for the observed table, we obtain the exact conditional P-value 0.034 for the Pearson statistic and 0.133 for the LR statistic. The two statistics still give quite different measures of evidence, and using the conditional distribution has not, of course, removed this dilemma. Some slightly edited output from *StatXacT* is given below, see the supplied manual for details.

```
PEARSON CHI-SQUARED TEST
Statistic based on the observed 2 by 5 table:
    CH(x) = Pearson Chi-squared Statistic = 12.07
Asymptotic p-values: (based on X2 distribution with 4 df)
    Pr{CH(X) .GE. 12.07}=0.0168
Exact p-value:
    Pr{CH(X) .GE. 12.07}=0.0341
LIKELIHOOD RATIO TEST
Statistic bsed on the observed 2 by 5 table:
    LI(x) = Likelihood ratio statistic = 6.200
Asymptotic p-value: (based on X2 distribution with 4 df)
    Pr{LI(X) .GE. 6.200} =0.1847
Exact p-value:
    Pr{LI(X) .GE. 6.200}=0.1333
```

A better statistic than either of these would make use of (i) the ordinal nature of the drinking factor and (ii) the one-sided nature of the suspected alternative hypothesis, i.e., we expect that drinking more increases the probability of malformation. The row-effects model of Section 6.4.2 suggests the test statistic

$$T = \sum_{i=1}^{r} \sum_{j=1}^{c} u_i v_j y_{ij}$$

where the u_i, v_j are scores assigned to rows and columns. Simple expressions for the null mean and variance are available (see p. 5–11 of Cytel's *StatXacT* manual). For the present data, there are only two rows and the u-scores actually make no difference to the final result; we choose scores $u = (1, 2)$. For the columns we choose scores $v = (0, 0.5, 1.5, 4, 6)$. This gives $t = 9213$ leading to the standardized statistic $z = 2.442$. Using the asymptotic normal distribution gives the one-sided P-value 0.007, while using the exact distribution gives the one-sided P-value 0.019.

7.5 CONDITIONAL INFERENCE ON 2 × 2 × k TABLES

We now return to the $2 \times 2 \times k$ table for which unconditional methods were given in Section 3.4. We will respect the notation established there. The data are a set of $2k$ independent counts having distributions

$$Y_{i(j)} \stackrel{d}{=} Bi(n_{i(j)}, \pi_{i(j)}) \quad i = 1, 2 \qquad j = 1, \ldots, k,$$

where $\pi_{i(j)}$ denotes the probability of a randomly chosen individual from group j responding when given treatment i. The logits of these parameters are denoted $\psi_{i(j)}$ and the log-odds ratio for the jth groups is

$$\Delta_{(j)} = \psi_{1(j)} - \psi_{2(j)} = \log\left\{\frac{\pi_{1(j)}(1 - \pi_{2(j)})}{\pi_{2(j)}(1 - \pi_{1(j)})}\right\}.$$

This measures the relative effectiveness of treatment 1 compared to treatment 2 on the log-odds scale for group/table j. There are thus $2k$ parameters in total that we choose to be $\Delta_{(1)}, \ldots, \Delta_{(k)}$ and $\psi_{2(1)}, \ldots, \psi_{2(k)}$, the first set being of primary interest in assessing the treatment. Where there is no confusion we drop subscripts to denote a vector. Thus Δ refers to the association parameters collectively and Y_1 is the vector with jth component $Y_{1(j)}$.

The uniform treatments hypothesis is that the $\Delta_{(j)}$ are equal. It is useful to further reparametrize the association parameters in terms of

$$(\overline{\Delta}, \eta_1, \eta_2, \ldots, \eta_{k-1})$$

where $\overline{\Delta}$ is the average of the $\Delta_{(j)}$ and $\eta_j = \Delta_{(j)} - \overline{\Delta}$ is the difference of $\Delta_{(j)}$ from the average. The uniform treatment hypothesis is equivalent to all the η_j equaling zero.

7.5.1 Conditional Likelihoods and Sufficient Statistics

As derived in Section 3.4, the log-likelihood function is

$$\begin{aligned}\ell(\Delta, \psi_2) &= c + \sum_{j=1}^{k} \{\Delta_{(j)} y_{1(y)} + \psi_{2(j)}(y_{1(j)} + y_{2(j)}) + \kappa_j(\Delta_{(j)}, \psi_{2(j)})\} \\ &= c + \overline{\Delta} Y_{1+} + \sum_{j=1}^{k-1} \eta_j (y_{1(j)} - y_{1(k)}) + \psi_{2(j)}(y_{1(j)} + y_{2(j)}) \\ &\quad + \sum_{j=1}^{k} \kappa_j(\eta_j, \psi_{2(j)}) \end{aligned} \tag{7.24}$$

where c does not depend on the parameters and κ_j does not depend on the data. It follows that each $Y_{1(j)}$ is sufficient for each $\Delta_{(j)}$, $T_{(j)} = Y_{1(j)} + Y_{2(j)}$ is sufficient for $\psi_{2(j)}$, and $Y_{1(j)} - Y_{1(k)}$ is sufficient for η_j. Conditional tests of any and all the parameters can be constructed.

Inference on the $\boldsymbol{\eta}_j$ Hypotheses concerning the η_j are tested using the distribution of the $k - 1$ sufficient statistics $Y_{1(j)} - Y_{1(k)}$ conditional on the totals $T_{(j)}$ as well as Y_{1+}. This distribution depends on the η_i only. The computational difficulty of implementing tests based on this conditional distribution is in enumerating all the possible $2 \times 2 \times k$ tables consistent with all these marginal totals.

Testing the uniform treatments hypothesis is a test that all the η_j are zero. The alternative hypothesis is the full model so this is really a test of goodness of fit. Within this more general framework, details of this test are given in Section 7.8 and will not be covered in this section.

Inference on the $\boldsymbol{\Delta}_{(j)}$ The joint distribution of the sufficient statistic vector Y_1 conditional on the totals $T_{(j)}$ depends only on Δ. What is the joint conditional distribution of vector Y_1? Each $Y_{1(j)}$ has tilted hypergeometric distribution with parameters $(n_{1(j)}, T_{(j)}, n_{(j)}, \Delta_{(j)})$. Since the tables are assumed independent the joint distribution will be that of k independent nonidentical tilted hypergeometrics.

Any simple hypothesis about Δ, for instance the hypothesis that all the $\Delta_{(j)}$ are zero, is conceptually straightforward. We choose a test statistics such as LR and then compute a tail probability using the conditional distribution with the hypothesized value of Δ substituted.

Tests of the average treatment effect $\overline{\Delta}$ use the distribution further conditioned on the $k-1$ statistics $Y_{1(j)} - Y_{1(k)}$. This is a one-dimensional distribution depending on $\overline{\Delta}$ only.

Inference on $\overline{\Delta}$ Assuming Uniform Treatments Assuming the uniform treatments hypothesis, the log-likelihood Eq. (7.24) becomes

$$\ell(\Delta, \psi_{2(j)}) = c + \overline{\Delta} Y_{1+} + \sum_{j=1}^{k} T_{(j)}\psi_{2(j)} + \kappa(\overline{\Delta}, \psi_{2(j)}). \tag{7.25}$$

and so the sufficient statistic for $\overline{\Delta}$ is Y_{1+}, the total number of responses under treatment 1. This does *not* mean we can add the tables and proceed as if we only had a single 2×2 table. Collapsing tables can be very misleading, and this was discussed under the name Simpson's Paradox in Section 3.6.1. The cautions there apply to conditional as well as to unconditional inference.

Inference on $\overline{\Delta}$ should be based on the distribution of Y_{1+} given *all* the $t_{(j)}$.

This distribution is that of the sum of k independent tilted hypergeometric variables which is algebraically extremely complex to write down in all but trivial cases. However, in principle, this distribution could be used to give exact P-values and confidence intervals. A slight complication is that when Y_{1+} takes extreme values of its distribution only one-sided exact intervals are possible. This is the case for confidence intervals based on any bounded discrete distribution.

7.5.2 Conditional Estimation of a Common Treatment Effect

Using the general theory in Section 7.3.4, the conditional ML estimator is obtained by equating the sufficient statistic, here Y_{1+}, to its conditional mean. The Newton–Raphson algorithm also requires the conditional variance. Because the $Y_{1(j)}$ are independent, the conditional moments are

$$\mathrm{E}_c\{Y_{1+}\} = \sum_{j=1}^{k} \mathrm{E}_c\{Y_{1(j)}\}, \qquad \mathrm{Var}_c\{Y_{1+}\} = \sum_{j=1}^{k} \mathrm{Var}_c\{Y_{1(j)}\}$$

and the terms in these sums are the means and variances of tilted hypergeometric distributions. The conditional ML estimator for $2 \times 2 \times k$ tables was proposed by Birch (1964).

Example 3.4: (*Continued*) Smoking in Three Cities. Consider the data used for illustrative purposes in Section 3.4 which is tabulated again. For variety, think of healthy lungs rather than impaired lungs as defining a success, being a nonsmoker as treatment 1 and a smoker as treatment 2. This does not alter the log-odds ratios $\Delta_{(j)}$ at all but we have rearranged the data (Table 7.9) to reflect the new definition.

There are a total of 10 + 14 + 6 = 30 responses (healthy) under treatment 1 (nonsmokers). Identical results will be obtained from the original formulation of the data where there are 12 + 9 + 6 = 27 responses (impaired) under treatment 1 (smokers). Assume the uniform treatment hypothesis, i.e., that the effect of smoking is the same in each city. We want to estimate the common log-odds ratio $\bar{\Delta}$.

Table 7.9. Lung Health for Smokers/Nonsmokers in Three Cities

	Los Angeles		New York		Washington, DC	
	Healthy	Impaired	Healthy	Impaired	Healthy	Impaired
Nonsmoker	10	6	14	12	6	12
Smoker	3	12	5	9	2	6

			$\Delta = 0$		$\Delta = 0.1$	
City	$Y_{1(j)}$	Distribution	m_j	v_j	m_j	v_j
Los Angeles	10	$GHG(16, 13, 31, \overline{\Delta})$	6.710	1.948	6.904	1.946
New York	14	$GHG(26, 19, 40, \overline{\Delta})$	12.350	2.328	12.583	2.322
Washington, DC	6	$GHG(18, 8, 26, \overline{\Delta})$	5.538	1.227	5.660	1.206
Total	30	hard	24.598	5.503	25.147	5.474

The null distributions, means, and variances of $Y_{1(j)}$ for $j = 1, 2, 3$ are tabulated. We begin with the (very poor) guess $\overline{\Delta} = 0$ for which Y_{1+} has mean 24.598 and so our next guess is

$$0 + (30 - 24.598)/5.503 = 0.981$$

For this value of $\overline{\Delta}$ we find that $\mathrm{E}(Y_{1+}) = 29.744$ and $\mathrm{Var}(Y_{1+}) = 4.857$ and so our next guess for $\hat{\Delta}_c$ is $0.981 + (30 - 29.744)/4.857 = 1.034$. For this value of $\overline{\Delta}$ the conditional mean of Y_{1+} is 30.000 to three decimal places. The standard error is

$$\mathrm{se}(\hat{\Delta}_c) = \frac{1}{\sqrt{\mathrm{Var}_c\{Y_{1+}\}}} = \frac{1}{\sqrt{4.857}} = 0.454.$$

The unconditional ML estimate was $\hat{\Delta} = 1.067$ with standard error 0.463.

Approximate Conditional MLE Computation of the conditional estimator requires special software such as *StatXacT*. If this is unavailable, a very similar estimator can be obtained by bias correction. For the 2 × 2 × k table with k large, Nam (1993) has shown that an approximately unbiased estimator is $c\hat{\Delta}_{ML}$ where the factor

$$c = 1 - \sum_{j=1}^{k} n_{1(j)}n_{2(j)}p_j(1 - p_j)/n_j^2 \Big/ \sum_{j=1}^{k} n_{1(j)}n_{2(j)}p_j(1 - p_j)/n_j \qquad (7.26)$$

and $p_j = \mathrm{E}(T_{(j)}/n_j)$ is estimated by $T_{(j)}/n_j$. The formula is correct for k diverging, regardless of how small the sample sizes $n_{i(j)}$. Notice that $0 < c < 1$ so that the unconditional estimator always exaggerates the value of Δ.

If the probabilities p_j are not too variable then they may be removed from the formula. Also if the $n_{i(j)}$ are the same for each j, then $c = 1 - 1/n_j$ which, for $n_j = 2$, recovers the correction factor $1/2$ for the case of binary pairs (see Section 3.6.4). The formula is extremely accurate in practice and an approximate standard error is obtained by multiplying the standard error of the unconditional ML estimator by the same constant c. An *Splus* function `cmle.nam`, which computes the factor c, is at the Wiley website. Other approximations to the conditional estimator are available for instance, based on saddlepoint approximation (see Section 7.9).

Example 3.4: (*Continued*) Smoking in Three Cities. For these data the value of c is calculated to be 0.9694 so there is not much difference between the conditional and unconditional estimators. The unconditional estimate was 1.067. Multiplying this by c gives 1.034, which is identical to the conditional ML estimator to this order of accuracy.

7.5.3 Conditional Tests of a Common Treatment Effect

For testing $H_0 : \overline{\Delta} = \delta_0$ against $\mathcal{H}_1 : \overline{\Delta} > \delta_0$ we compute the conditional tail probability

$$\Pr(Y_{1+} \geq y_{1+} | t_{(1)}, \ldots, t_{(k)}; \delta_0).$$

Even for the simplest case $\delta_0 = 0$, which is a test of no (average) treatment effect, the distribution on the right-hand side is a sum of ordinary hypergeometrics and is hard to work with.

A very close approximation can be obtained by approximating each of these hypergeometrics by a normal distribution with continuity correction. The mean and variance of Y_{1+} will be the sum of the means and varianes of the $Y_{1(j)}$. Let m_j denote the conditional mean of $Y_{1(j)}$ and v_j the conditional variance given by Eq. (7.9). Then referring the continuity corrected test statistic

$$\frac{\left[\left|Y_{1+} - \sum_{j=1}^{k} m_j\right| - \frac{1}{2}\right]^2}{\sum_{j=1}^{k} v_j} \tag{7.27}$$

to the χ_1^2 distribution gives a close approximation to the exact test of uniform treatment differences.

The same idea can be used for testing values of δ_0 other than zero, but m_j, v_j are not simply computed. When used to test $\overline{\Delta} = 0$, Eq. (7.27) is known as the *Mantel–Haenszel* test statistic. The normal approximation to Y_{1+} above is extremely accurate; first each of the summed terms $Y_{1(j)}$ is close to normal provided $n_{(j)}$ is large. Second, Y_{1+} is itself a sum of k such terms and so will be approximately normal when k is large.

Example 3.4: (*Continued*) Smoking in Three Cities. Assuming that the effect of smoking, if existent, is the same for all three cities, suppose we desire to test whether or not smoking has a deleterious effect on lung condition at all, i.e.,

$$\mathcal{H}_0 : \overline{\Delta} = 0 \text{ against } \mathcal{H}_1 : \overline{\Delta} > 0.$$

The exact P-value for testing these hypotheses is 0.0170, obtained using *StatXact*. Let us see how closely we can approximate this value.

Under the null hypothesis the conditional mean and variance of Y_{1+} is 24.598 and 5.503. The observed value $y_{1+} = 10 + 14 + 6 = 30$ which is higher than the mean value of 24.598 and so the P-value is

$$\Pr(Y_{1+} \geq 30) \approx 1 - \Phi\left(\frac{29.5 - 24.598}{\sqrt{5.503}}\right)$$

$$= 1 - \Phi(2.089) = 0.0183,$$

with continuity correction. Thus, there is considerable evidence that smoking increases the chance of impaired lungs.

We could test nonzero values of $\overline{\Delta}$ by calculating the means and variances of the $Y_{1(j)}$ from their tilted hypergeometric distributions. For instance, to test $\overline{\Delta} = 0.1$ the mean of Y_{1+} (given earlier) is 25.147 and variance is 5.474 and the normal approximation to $\Pr(Y_{1+} \geq 30)$ is 0.0316. This computation is not trivial but is much easier than finding the exact distribution of Y_{1+}.

We could continue finding values of $\overline{\Delta}$ to test which are just accepted at the level $\alpha/2$ to build up a confidence interval for $\overline{\Delta}$. Using the approximation to the conditional distribution, this can be computed using the supplied Fortran program `ghi.exe` or any module that computes the required means and variances. This calculation is really no harder than for a single 2×2 table, but it only approximates the conditional interval. Using *StatXacT* we find that the exact 95% confidence interval for the unlogged odds ratio ϕ is (1.071, 7.873). The conditional estimate is $\hat{\phi}_c = \exp(1.034) = 2.813$.

7.6 CONDITIONAL INFERENCE FOR THREE-WAY TABLES

We consider models for three-way contingency tables, as in Section 6.3, the factors being labeled **A**, **B**, and **C** on r, c and s levels, respectively. Individuals are classified by these three factors and the counts arranged as $r \times c$ tables stratified by the s levels of the factor **C**. The distributions of the Y_{ijk} are independent Poisson with means μ_{ijk} and the saturated model is of the form

$$\log(\mu_{ijk}) = m + a_i + b_j + c_j + (ab)_{ij} + (ac)_{ik} + (bc)_{jk} + (abc)_{ijk}. \tag{7.28}$$

Hierarchical models are denoted by their highest order interactions. Thus (**AB**, **BC**, **AC**) is the model with no three-way interaction and the above saturated model is ABC. The sufficient statistic for a model simply comprises the marginal tables corresponding to each term in the model. For instance, the marginal table corresponding to AC is $\{y_{i\bullet k}\}$, etc., and so the sufficient statistic for the no three-way model (AB, BC, AC) is

$$\{y_{ij\bullet}, y_{i\bullet k}, y_{\bullet jk}\}.$$

Consider the hierarchy of log-linear models listed below, together with their minimal sufficient statistics. We will consider testing (1), (2), and (3) against the next higher model. Tests against the full model are goodness-of-fit tests and are described in Section 7.8.

1. A+B+C: A, B, C mutually independent,

$$S = \{y_{i\bullet\bullet}, y_{\bullet j\bullet}, y_{\bullet\bullet y}\}$$

2. AB, C: BC jointly independent of A,

$$S = \{y_{ij\bullet}, y_{\bullet\bullet k}\}$$

3. AB, BC: A and C independent given B,

$$S = \{y_{ij\bullet}, y_{\bullet jk}\}$$

4. AB, BC, AC: no three-way dependence.

$$S = \{y_{ij\bullet}, y_{\bullet yk}, y_{i\bullet k}\}$$

Model (1: A + B + C) against (2: AB, C). This is a test of independence of the factors A and B, in the presence of C. By collapsibility conditions given in Section 6.3.3, we may collapse over factor C. Tests of independence of A and B were covered in the previous section.

Model (2: AB, C) against (3: AB, BC). We are testing the association of B and C, assuming an association of A and B. Exact inference is based on the distribution of $\{y_{\bullet jk}\}$ conditional on $\{y_{ij\bullet}, y_{\bullet\bullet k}\}$. Now conditional on B, A, and C are independent, i.e., conditioning on AB gives the same result as just conditioning on B. It follows that conditioning on $y_{ij\bullet}$ gives the same distribution as conditioning on $y_{\bullet j\bullet}$ and so inference is based on the distribution of

$$\{y_{\bullet jk}\}|y_{\bullet j\bullet}, y_{\bullet\bullet k}.$$

This is the distribution of the marginal B–C table conditional on its marginal row and column totals, already described fully in the previous section.

An alternative route to the same answer is to use the collapsibility conditions on p. 333. These tell us that we may collapse over A without changing the association of B and C, leading to the exact test of independence of B and C, in the marginal B–C table.

Model (3: AB,BC) against (4: AB,BC,AC). The null model has A and C independent conditional on B. The alternative model has A and C dependent but their association parameter $(ac)_{ik}$ not dependent on B. We are thus testing conditional independence of A and C under the assumption that their association is the same for all levels of B. Testing a common odds ratio in the stratified $2 \times 2 \times k$ table is a special case, see Section 3.4.2.

Conditional Distribution Exact inference is based on the distribution of $\{y_{i\bullet k}\}$ conditional on $\{y_{ij\bullet}\}$ and $\{y_{\bullet jk}\}$. Write

$$\{y_{i\bullet k}\} = M_1(ik) + M_2(ik) + \cdots + M_c(ik)$$

where $M_j(ik)$ is the two-way table for A–C, for the stratum $B = j$. These tables are independent under Poisson sampling. The conditioning statistics are the row and column totals of these tables. The conditional distribution of each table is the generalized multivariate hypergeometric with null probability function given by Eq. (7.23). Thus, the required distribution for inference is the convolution of these distributions as B ranges from 1 to c.

Computing just one distribution of the form [see Eq. (7.23)] can be a large computational task unless the dimensions of the A–C table is small, and this must be repeated for each B-stratum of the data. On the other hand, the counts will often be smaller in each of these tables making (i) computation a little easier and (ii) the accuracy of asymptotic procedures more suspect. Of course, while the conditional distribution is hard to enumerate, it is very easy to simulate. Take a population of size $y_{\bullet\bullet\bullet}$ and randomly assign them labels B satisfying the distribution $\{y_{\bullet j\bullet}\}$. For the subset of the population with $B = j$, (i) randomly assign labels A in the distribution $\{y_{ij\bullet}\}$ and (ii) independently assign labels C in the distribution $\{y_{\bullet jk}\}$. Repeat this for all B-strata and add the resulting tables to obtain the table $\{y^*_{i\bullet k}\}$. The distribution of this randomly generated table follows the distribution Eq. (7.23).

Conditional Estimation Conditional ML estimators of the A–C association parameters are obtained by equating the rs counts in the marginal table $\{y_{i\bullet k}\}$ to their conditional expected values. For each (i, k), imagine reducing the factor A to measure $A = i$ or $A \neq i$ and similarly the factor C to measure $C = k$ or $C \neq k$. The stratified A–C tables then collapse into c 2×2 tables. The distribution of y_{ijk} conditional on the marginal totals of this 2×2 table is the simple titled hypergeometric distribution of Section 7.1, and so the distribution of $\{y_{i\bullet k}\}$ is the c-fold convolution of these distributions. From this distribution can be computed $\text{E}_c(\{y_{i\bullet k}\})$. This is repeated for all cells (i, k) and we attempt to solve the equations

$$\{y_{i\bullet k}\} = \mathrm{E}_c(\{y_{i\bullet k}\}) \qquad i = 1, \ldots, r-1 \qquad k = 1, \ldots, s-1.$$

Iteration is required to solve these nonlinear equations and so the entire computation must be repeated several times. Clearly, a formidable amount of computing is required.

Example 3.4: (*Continued*) Smoking in Three Cities. Currently software is available only for estimating or testing the common odds ratio in a 2×2 table, stratified by some other factor. The association parameter is a single odds ratio, assumed constant across the strata. The data were given on p. 425.

For factors `A`, `B`, `C` refer to lung function, smoking habits, and city, respectively. Lung function has two levels, healthy (H) or impaired (I). We are interested in the association of lung function and smoking habits, assuming this association is the same in each city. This is the log-linear model (`AB`, `AC`, `BC`) and the null model of no association is (`AC`, `BC`). To test the A–B association, we consider the distribution of the marginal table $\{y_{ij\bullet}\}$ conditional on the six row and six column totals. The A–B association can be described by one log-odds ratio Δ, which we take to be the log ratio of odds of healthy lungs for nonsmokers compared to smokers. Similarly, in the marginal table with counts $\{y_{ij\bullet}\} = (30, 40, 10, 29)$ there is only one free statistic; we arbitrarily choose the statistic $T = y_{11\bullet}$ with observed value 30 whose conditional distribution is to be used.

```
ESTIMATION AND TESTING OF COMMON ODDS RATIO

Summary of Exact Distribution of T :

           Min  Max   Mean  Std-dev  Observed  Standardized
             5   40  24.60    2.346        30         2.303

Exact p-values for testing that the Common odds Ratio is 1:
    One-sided: Pr{S .Ge. 30}=0.0170
    Two-sided: Method 1: 2*One-sided                 = 0.0339
              Method 2: (Sum of Probs .LE. 0.0118   = 0.0315
              Method 3: Pr{|S-Mean| .GE. |s-Mean|}  = 0.0315

Exact Estimation of Common Odds Ratio:
  Conditional maximum likelihood estimate: 2.813
  95.00% Confidence Intervals: (1.071, 7.873)
```

The output of *StatXact* from the command `OD` with subcommand `EX` is listed above. The conditional mean and variance of $T = y_{11\bullet}$ as well as the maximum and minimum values are given. The standardized value

$$\frac{30 - 24.6}{2.346} = 2.303$$

may be used to approximate the conditional tail probability by a standard normal tail probability. This gives the one-sided P-value 0.011 for testing $\Delta > 0$. However, it does not include a continuity correction. With continuity correction (replacing 30 by 29.5) the z-value is 2.094 giving one-sided P-value 0.018. It is hard to understand why the uncorrected z-statistic is given by *StatXacT* since it gives a much poorer approximation to the exact one-sided P-value, which in this case is 0.017.

Two-Sided Tests in StatXacT Two-sided test results, using three different methods, are also listed. The first simply doubles the one-sided value, regardless of how asymmetric the distribution of T may be. This seems hard to justify when we are claiming the methods are exact! The second ranks tables by their null probability. It was noted earlier that this is not justifiable in general, and is presumably done to reduce computations. The third ranks tables by how far T is from its expected value, 5.4 for the observed table. The P-value is the null probability of all tables for which T differs from its mean by 5.4 or more.

It would be preferable to rank the tables using a more sensible measure of hostility to the null hypothesis, such as the likelihood ratio, either conditional or unconditional, or the Pearson chi-square. However, looking at the closeness of the P-values for the above three methods, the result would probably not be much different for this data set.

7.7 CONDITIONAL LOGISTIC REGRESSION

Logistic regression is useful for modeling one binary factor in terms of other explanatory variables. Since Logistic linear models may be represented as log-linear models (Section 6.2.3), exact inference may be performed using either formulation with identical results.

Let Y_i be independent binomial random variables with denominator n_i and log-odds probability $\psi_i(\beta)$ a linear function of the parameter β with design matrix X. Then the log-likelihood function is

$$\ell(\beta) = \sum_{j=1}^{p} \beta_j S_j - \sum_{i=1}^{k} n_i \log(1 + \exp\{\psi_i(\beta)\}) \tag{7.29}$$

where $S_j = \sum_{i=1}^{k} x_{ij} Y_i$ is sufficient for β_j. Inference on any parameter can be performed exactly by conditioning on the sufficient statistics for all the other parameters.

Table 7.10. Cancer Remission Data

x	8	10	12	14	16	18	20
Y	0	0	0	0	0	1	2
n	2	2	3	3	3	1	3
Y/n	0.00	0.00	0.00	0.00	0.00	1.00	0.67
x	22	24	26	28	32	34	38
Y	1	0	1	1	0	1	2
n	2	1	1	1	1	1	3
Y/n	0.50	0.00	1.00	1.00	0.00	1.00	0.67

Example 4.3: (*Continued*) Cancer Remission. Below is the cancer remission data (Table 7.10) that we looked at in Chapter 4. There were y remissions out of n treated patients. The covariate x measures level of cell activity after treatment and is a possible predictor of remission. Consider the logistic regression model

$$\psi(x) = \beta_0 + \beta_1 x.$$

The interest parameter is β_1. The results for unconditional inference were given in Section 4.1.3.

The sufficient statistics for (β_0, β_1) are $S_0 = \sum_{i=1}^{k} Y_i$ and $S_1 = \sum_{i=1}^{k} x_i Y_i$ with observed values $s_0 = 9$, $s_1 = 244$. Conditional inference is based on the distribution of S_1 given $s_0 = 9$, and some output from the package *LogXacT* is given in the *LogXacT* manual at the Wiley Website.

We first compare estimates of $r = 100(\exp(\beta) - 1)$ giving the percentage increase in odds of remission for each unit increase in x. The conditional ML estimate of r is $100(\{\exp(0.138) - 1\} = 14.8\%$ with 95% confidence interval (3.5, 31.2)%. The unconditional ML estimator is $100\{\exp(0.145) - 1\} = 15.6\%$ and 95% interval (2.9, 28.9)%.

Some of the results of testing $\beta = 0$ against both one and two-sided alternatives are summarized in Table 7.11. The unconditional test statistics W, S, and LR when referred to their asymptotic χ_1^2 distribution give approximate two-sided P-values. Since the Wald test disagrees with the Score and LR tests, we suspect that the asymptotics may not be accurate. There are two conditional tests listed. The first uses the conditional score statistic that reduces to

$$S_c = \frac{\{S_1 - \mathrm{E}(S_1 | s_0)\}^2}{\mathrm{Var}(S_1 | s_0)} = 7.637$$

where formulas for the conditional mean and variance of S_1 are given on p. 435. The P-value 0.0041 is based on the conditional distribution of S_c. The second conditional test is based on the test statistic S_1, and the one-sided P-value 0.0033, is the conditional probability that $S_1 \geq 244$. All P-values are

Table 7.11. Conditional and Unconditional P-Values for Testing $\boldsymbol{\beta} = 0$

Test Statistic	Mode	Sided	Value	P-Value
Wald	Asymptotic	2	5.959	0.0146
Score	Asymptotic	2	7.931	0.0049
LR	Asymptotic	2	8.299	0.0040
S_c	Exact	2	7.637	0.0041
T_1	Exact	1	244	0.0033

very small, and we expect asymptotics to work poorly in the extreme tails of the distribution. At a practical level, the differences between unconditional and conditional inference are modest in this example.

The conditional and unconditional results become less similar as the amount of conditioning is increased. This can be illustrated by considering the quadratic regression model

$$\psi(x) = \beta_0 + \beta_1 x + \beta_2 x^2.$$

The sufficient statistics for $(\beta_0, \beta_1, \beta_2)$ are

$$S_0 = \sum_{i=1}^{n} Y_i, \qquad S_1 = \sum_{i=1}^{n} x_i Y_i, \qquad S_2 = \sum_{i=1}^{n} x_i^2 Y_i$$

with observed values $s_0 = 9$, $s_1 = 244$, and $s_2 = 7112$. Exact inference on β_1 uses the distribution of S_1 given $(S_0, S_2) = (9, 7112)$ while exact inference on β_2 uses the distribution of S_2 given $(S_0, S_1) = (9, 244)$. Table 7.12 gives estimates and 95% confidence intervals for β_1 and β_2 using the unconditional and conditional distributions.

For the linear parameter β_1 the conditional estimate is around half the magnitude of the unconditional estimate. However, the standard errors of these estimates are quite large judging by the widths of the confidence intervals. Note, however, that the conditional upper limit of β_1 is infinite. This is because con-

Table 7.12. Conditional Versus Unconditional Estimates

Parameter	Mode	Estimate	95% CI	One-Sided P-Value
β_1	Unconditional	0.962	(−0.049, 1.973)	0.0310
	Conditional	0.493	(−0.069, ∞)	0.0445
β_2	Unconditional	−0.016	(−0.035, 0.003)	0.0463
	Conditional	−0.03	(−0.031, 0.001)	0.0406

ditional on $(S_0, S_2) = (9, 7112)$ the possible values of S_1 are greatly reduced, so much so that the observed value of $s_1 = 244$ is now the largest possible value. Why is the sample space so reduced? Because conditionally we only consider tables that have $S_0 = 9$ successes in total, and that also satisfy

$$64y_1 + 100y_2 + 144y_2 + 196y_4 + 256y_5 + 324y_6 + 400y_7 + 484y_8+$$
$$576y_9 + 676y_{10} + 784y_{11} + 1024y_{12} + 1156y_{13} + 1444y_{14} = 7112$$

It is very hard indeed to find *integer* vectors $(y_1, \ldots, y_{14})$ that satisfy the second condition, on top of the first. It is not nearly so difficult to find solutions of $(s_0, s_1) = (9, 224)$ which explains why conditional and unconditional inferences on the quadratic parameter β_2 are much more similar.

We can make the differences even more acute by replacing the x-values by real numbers very close to them, say 8 by 8.001, 10 by 10.002, etc. We find that $(s_0, s_1, s_2) = (9, 244.09, 7117.3)$, the above linear combination of the y_i's has real numbers for coefficients, and the *only* way to get the answer 7117.3 is with the observed data table. The conditional distribution is then degenerate and no nontrivial inference is possible from it. The same occurs for inference on β_0 or β_1.

This is quite a serious problem for conditional logistic regression. Reasonable results are only obtained after rounding the predictors and why this should have such a large effect on results is hard to understand or justify. For this reason many researchers have great misgivings about using conditional methods when the predictors are continuous.

Approximation for Testing* $\boldsymbol{\beta}$ *= 0 For testing whether or not $\beta = 0$ in the logistic regression model for N binary observations Y_i with covariates x_i, the conditional distribution of $S_1 = \sum x_i Y_i$ given $S_0 = \sum Y_i$ is required.

Imagine taking a sample without replacement of size s_0 from a population of size N with labels x_i on the individuals. The distribution of S_1 is the distribution of the total of the labels of the sampled individuals. Let

$$m_1 = \sum_{i=1}^{k} x_i n_i \Big/ \sum_{i=1}^{k} n_i, \qquad m_2 = \sum_{i=1}^{k} x_i^2 n_i \Big/ \sum_{i=1}^{k} n_i$$

and let $n = \sum n_i$. Then equations for the mean and variance of this conditional distribution are

$$\mathrm{E}(S_1 | s_0) = s_0 m_1, \quad \mathrm{Var}(S_1 | s_0) = \frac{s_0(n - s_0)(m_2 - m_1^2)}{n - 1},$$

as may be found in any textbook on sampling theory. The conditional P-value can be approximated by a normal tail probability provided the distribution is not too discrete. Further details are given in Cox and Snell (1989, Section 2.2). An *Splus* function `cox.test` is at the Wiley website. As is always the case when approximating a discrete distribution by a continuous one, the continuity correction is desirable. However, for irregular covariate values, the continuity correction cannot be easily implemented.

Example 4.3: (*Continued*) Cancer Remission. For the simple linear regression on the cancer remission data $s_0 = 9$, $s_1 = 244$, and $n = 27$. By calculation $m_1 = 20.094$ and $m_2 = 487.259$ and so we find

$$\mathrm{E}(S_1 | s_0 = 9) = 180.666, \quad \mathrm{Var}(S_1 | s_0 = 9) = 525.197.$$

The exact one-sided P-value we just computed was 0.003. The approximate one-sided P-value for testing $\beta = 0$ against $\beta > 0$ is

$$\begin{aligned}\Pr(S_1 \geq 244 | s_0 = 9) &\approx \Pr(\mathcal{N}(180.666, 525.197) \geq 243.5) \\ &= 1 - \Phi(2.7417) \\ &= 0.0031.\end{aligned}$$

7.8 CONDITIONAL GOODNESS-OF-FIT TESTS

Goodness-of-fit tests are tests against the full model. In the context of log-linear models with three factors, this is the model (`ABC`) with sufficient statistic the full three-way table, i.e., $S_1 = \{y_{ijk}\}$. To construct an exact goodness-of-fit test of one of the models (1)–(4), first choose a test statistic, such as the Pearson or LR goodness-of-fit statistic. Conditional inference requires enumeration of all those three-way tables $\{y_{ijk}\}$ consistent with the observed value of $\mathcal{S}_0$, the sufficient statistic under the null model. The conditional P-value is the total of conditional probabilities of all such tables.

The conditional distributions can be thought of as arising from certain simple sampling experiments without replacement. When the counts and/or dimensions of the tables are large, then simulating from the conditional distribution via such a sampling experiment is a practical alternative to enumeration of the conditional distribution.

Model (1: `A, B, C`). For testing the goodness-of-fit of the independence model (`A, B, C`) we need the distribution of $\{y_{ijk}\}$ conditional on the marginal tables $\{y_{i\bullet\bullet}\}$, $\{y_{\bullet j\bullet}\}$, and $\{y_{\bullet\bullet k}\}$. The null-conditional distribution is

$$p_c(y_{ijk}) = \frac{\prod_{i=1}^{r} y_{i\bullet\bullet}! \prod_{j=1}^{c} y_{\bullet j\bullet}! \prod_{k=1}^{s} y_{\bullet\bullet k}!}{(y_{\bullet\bullet\bullet}!)^2 \prod_{i,j,k} y_{ijk}!} \qquad \{y_{ijk}\} \in \mathcal{Y}(s_0) \qquad (7.30)$$

for those tables compatible with the marginal tables. Again, the computational difficulty is in determining the support of this conditional distribution.

For tables with large counts and dimensions, it is more computationally feasible to simulate from the conditional distribution than to enumerate it. This is easily done. Imagine a population of size $y_{\bullet\bullet\bullet}$. We randomly assign labels $A = i$ so that the distribution of these labels across the population is $\{y_{i\bullet\bullet}\}$. We similarly assign labels $B = j$ according to the distribution $\{y_{\bullet j\bullet}\}$, and finally labels $C = k$ according to the distribution $\{y_{\bullet\bullet k}\}$. If y^*_{ijk} is the resulting number of individuals in the population with factor labels $A = i$, $B = j$, and $C = k$ then the distribution of $\{y^*_{ijk}\}$ follows Eq. (7.30).

Model (3: AB, BC). Testing the goodness-of-fit of this model is a test of association of factors A and C, without the assumption that the association is the same for each level of B. We need the distribution of $\{y_{ijk}\}$ conditional on the marginal tables $\{y_{ij\bullet}\}$ and $\{y_{\bullet jk}\}$. The conditional support set $\mathrm{Y}(s_0)$ includes all those table $\{y_{ijk}\}$ that are identical to the observed tables, when collapsed over C and when collapsed over A. Determining all such tables is not easy. An algorithm for simulating the conditional distribution was given earlier on p. 429, where we were testing the model (AB, AB, BC) against (AB, BC).

Model (4: AB, BC, AC). Testing goodness-of-fit of this model is an exact test of the uniform treatments hypothesis. We might think of this as a test of the uniformity of the AB association across different levels of C. We need the distribution of $\{y_{ijk}\}$ conditional on the marginal tables $\{y_{ij\bullet}\}$, $\{y_{i\bullet k}\}$, and $\{y_{\bullet jk}\}$. The conditional support set $\mathcal{Y}(s_0)$ includes all those tables $\{y_{ijk}\}$ that are identical to the observed tables, when collapsed over A, when collapsed over B, and when collapsed over C.

For the case of the $2 \times 2 \times s$ table, this test was originated by Zelen (1971) and is implemented in *StatXacT*. Despite the theoretical shortcomings outlined on p. 416, the test statistic used is $p_c(y_{ijk})$, presumably to reduce computations.

Example 3.4: (*Continued*) Smoking in Three Cities. For the fictitious data on smoking and lung impairment stratified by three cities, we looked earlier at exact inference for the common odds ratio. The output below is from *StatXacT* and can be used to test whether or not the odds ratios for the three cities are the same. The asymptotic P-value is based on a statistic [see Eq. (3.34)] that is a simple quadratic form with approximate unconditional χ^2_{s-1} distribution. For the exact test, the conditional probability itself is used as the test statistic, with value 0.0733 here. One does not know the effect of using the exact conditional distribution as opposed to the unconditional distribution here since different test statistics are used. However, both tests clearly indicate that there is no evidence whatsoever against the uniform treatments hypothesis.

```
TEST FOR HOMOGENEITY OF ODDS RATIOS
[3  2×2 informative tables ]

Observed Statistics:
    BD: Breslow and Day Statistic=1.734

Asymptotic p-value: (based on X2 with 2 df)
    Pr{BD.GE.1.734}=0.4202
Exact p-value: Pr{ZE.LE.0.07333}= 0.4902
```

7.9 SADDLEPOINT APPROXIMATION*

The saddlepoint method can be used to approximate distributions of sums of random variables, with much more accuracy than typically obtained from the central limit theorem. In this section, a nontechnical overview is given as well as formulas for implementing the approximation in logistic or log-linear models.

Let S be a statistic with cumulant generating function $K_S(t)$, where t may now be complex. When it is real this is the logarithm of the moment generating function and when it is purely complex it is the logarithm of the characteristic function. The generating function $K_S(t)$ determines the distribution $f_S(s)$ through the inversion equation

$$f_S(s) = \frac{1}{2\pi i} \int e^{nK(t) - ts} dt, \tag{7.31}$$

the integral being taken along a strip in the complex plane where $K(t)$ exists.

We suppose that $K_S(t)$ is of order n and for definiteness write $K_S(t) = nK(t)$ where $K(t)$ is $O(1)$. A typical example would be an identical, independent sum $S = \sum_{i=1}^{n} Y_i$, in which case n is the sample size and $K(t)$ the common generating function of the Y_i. For most, though not all applications, n is indeed a sample size. However, mathematically n is just a parameter and we are interested in approximating the distribution of S as n increases. For large n, $K_S(t)$ is often locally quadratic (the central limit theorem). More accurate approximations to $f_S(s)$ may be obtained by approximating $K_S(t)$ about $K_S(0) = 0$ by a cubic or quartic function, so-called Edgeworth expansions. The key to the saddlepoint method however is to approximate the function

$$G_{ns}(t) = nK(t) - ts,$$

*See p. 15 in Chapter 1 for an explanation.

in the exponent of Eq. (7.31) rather than $K_S(t)$ and to approximate it not near $t = 0$ but near its maximum value. Because of the exponential in the integrand of Eq. (7.31), the behavior of $G_{ns}(t)$ near its maximum should largely determine the value of the integral. We need the maximizing value $\tilde{t}(s)$, determined by the *saddlepoint* equation

$$nK'(t) - s = 0.$$

Actually, $\tilde{t}$ is not a maximum—it is a saddlepoint of the surface $G_{ns}(t)$ in the complex plane, that will be a maximum when approached along the appropriate contour. The second derivative of $G_{ns}(t)$ at $\tilde{t}$, giving the curvature of this maximum, is $K_S''(\tilde{t})$. The so-called saddlepoint approximation to $f_S(s)$ is given by

$$\tilde{f}_S(s) = \frac{1}{\sqrt{2\pi K_S''(\tilde{t})}} \exp\{K_S(\tilde{t}) - \tilde{t}_s\}. \tag{7.32}$$

If fact, this is just the leading term of a more general expansion, but use of the leading term alone is so phenomenally accurate that higher order terms are seldom required. The *relative* error of the saddlepoint approximation is $O(n^{-1})$ uniformly in s. Compare this to the central limit theorem whose *absolute* error is $O(n^{-1/2})$, nonuniformly in s. This makes the central limit theorem of limited use for approximating smallish tail probabilities for moderate sample sizes. And it is precisely tail probabilities that are of interest in statistics!

Approximate Tail Probabilities An approximation formula for the cumulative distribution function may be obtained by integrating the saddlepoint approximation to the density, and rearranging so that it involves a normal density term. Details are omitted here, but the incredibly simple result is

$$\tilde{F}_S(s) = \Phi(r) + \phi(r)\left(\frac{1}{r} - \frac{1}{\omega}\right) \tag{7.33}$$

where Φ, ϕ are standard normal cdf and pdf, respectively, and where

$$r(s) = \text{sign}(\tilde{t})\sqrt{2\{s\tilde{t} - K_S(\tilde{t})\}}, \quad \omega(s) = \tilde{t}\sqrt{K_S''(\tilde{t})}. \tag{7.34}$$

Note that $\tilde{t}$ is a function of s and $K_S(\tilde{t}) = nK(\tilde{t})$.

The approximation $\tilde{F}_S(s)$ has relative error $O(n^{-3/2})$, for fixed $s \neq 0$. In the extreme tails (mathematically when r is $O(\sqrt{n})$), the relative error is $O(n^{-1})$. When $s = 0$ we find that $r = 0$ and the formula cannot be used. In this case another very accurate approximate is available but we are seldom interested in approximating central probabilities of a distribution.

Multivariate Distributional Approximation When S is a vector random variable of dimension p we let $K_S(t) = nK(t)$ be the joint cumulant generating function (cgf) with argument $t \in \Re^p$. The saddlepoint approximation to the joint density function of S is

$$\tilde{f}_S(s) = (2\pi)^{-p/2}|K''_S(\tilde{t})|^{-1/2} \exp\{K_S(\tilde{t}) - s^T\tilde{t}\}. \tag{7.35}$$

where now the saddlepoint $\tilde{t}(s)$ is the solution of the system of saddlepoint equations

$$\frac{\partial K_S}{\partial t_i}(t_1, \ldots, t_p) = s_i, \qquad i = 1, \ldots, p,$$

K''_S is the Jacobian matrix of second partial derivatives of $K_S(t)$ and $|.|$ denotes determinant.

Conditional Distributional Approximation When S is multivariate, denote the first component by S_1 and the other $p-1$ components by S_2. Write the cumulant generating function as $K_S(t) = nK(t_1, t_2)$. The distribution of S_1 given S_2 is the ratio of the joint distribution of $S = (S_1, S_2)$ to the distribution of S_2. Both these may be approximated by the multivariate saddlepoint approximation Eq. (7.35). This ends up giving

$$\tilde{f}_{S_1|S_2}(s) = \left(\frac{|K''_S(\tilde{t}_1, \tilde{t}_2)|}{2\pi |K^{22}_S(0, \tilde{t}_{2(1)})|} \right)^{1/2} \exp\{K_S(\tilde{t}) - K_S(0, \tilde{t}_{2(1)}) + s_s^T \tilde{t}_{2(1)} - s^T\tilde{t}\}, \tag{7.36}$$

where $\tilde{t} = (\tilde{t}_1, \tilde{t}_2)$ is the saddlepoint for the joint distribution of S, and $\tilde{t}_{2(1)}$ is the saddlepoint for the distribution of S_2, obtained from the equations

$$\frac{\partial K_S}{\partial t_2}(0, t_2) = s_2.$$

The matrix K^{22}_S is the $(p-1) \times (p-1)$ Jacobian matrix of K_S with respect to t_2 only, and with $t_1 = 0$ substituted. The conditional distribution function of S_1 given S_2 is approximated by Eq. (7.33), but with

$$r(s) = \text{sign}(\tilde{t}_1)\{2(K_S(0, \tilde{t}_{2(1)}) - K_S(\tilde{t}_1, \tilde{t}_2) - s_2^T\tilde{t}_{2(1)} + s^T\tilde{t})\}^{1/2} \tag{7.37}$$

and

$$\omega(s) = \tilde{t}_1 \left(\frac{|K_S^{22}(0, \tilde{t}_{2(1)})|}{|K_S''(\tilde{t}_1, \tilde{t}_2)|} \right)^{1/2}. \qquad (7.38)$$

Adjustment for Discrete Variables The expansion $\tilde{F}_S(s)$ came from approximating the integrated density. If S_1 is discrete then the integral becomes a sum. It turns out that tail probabilities of S_1 conditional on S_2 can still be approximated as above but better results are obtained using one of the following two adjustments. The first adjustment is the simplest. Simply replace $\tilde{t}_1$ by $1 - \exp(-\tilde{t}_1)$ in Eq. (7.38). The second adjustment requires us to replace s_1 by $s_1 - 1/2$ in the saddlepoint equations defining $\tilde{t}_{2(1)}$ and $\tilde{t} = (\tilde{t}_1, \tilde{t}_2)$ and also to replace $\tilde{t}_1$ in Eq. (7.38) by

$$e^{\tilde{t}_1/2} - e^{-\tilde{t}_1/2} = 2\sinh(\tilde{t}_1).$$

Application to Logistic and Log-Linear Models Consider independent variables $Y_1, \ldots, Y_n$ with binomial or Poisson distributions. The density of Y_i then has the form

$$f_i(y; \theta) = \exp\{y\theta_i - b(\theta_i) + c(y)\} \qquad (7.39)$$

where θ_i is the canonical parameter, namely $\text{logit}(p_i)$ for the binomial distribution and $\log \mu_i$ for the Poisson distribution. Simple algebra shows that the cumulant generating function of Y_i is

$$K_i(t) = b(\theta_i + t) - b(\theta_i).$$

Assume a linear predictor $\theta_i = x_i^T \beta$ for the canonical parameter, where x_i^T is the ith row of the design matrix X. Then the log-likelihood function is

$$\ell(\beta) = s^T \beta - \sum_{i=1}^{n} b(x_i^T \beta) \qquad (7.40)$$

where the sufficient statistic S for β has jth component $S_j = \sum x_{ij} Y_i$. By elementary properties of cumulant generating functions, the joint cgf of S at the point $t = (t_1, \ldots, t_p)$ is

$$K_S(t) = \sum_{i=1}^{n} K_i(x_i^T t) = \sum_{i=1}^{n} b\{x_i^T(\beta + t)\} - b(x_i^T \beta).$$

We now show that all the quantities required for the approximation to the conditional density and distribution function may be obtained from solving ordinary likelihood equations. First, the saddlepoint $(\tilde{t}_1, \ldots, \tilde{t}_p)$ in Eqs. (7.36) and (7.37) satisfy the saddlepoint equations which here reduce to

$$s_j = \frac{\partial K_S}{\partial t_j}(t_1, \ldots, t_p) = \sum_{i=1}^{n} x_{ij} b'\{x_i^T(\beta + t)\} \qquad j = 1, \ldots, p. \tag{7.41}$$

Comparing these with the likelihood equations

$$s_j = \sum_{i=1}^{n} x_{ij} b'(x_i^T \beta)$$

for $\hat{\beta}$, we see that the saddlepoints $\tilde{t}_j = \hat{\beta}_j - \beta_j$. Similarly, the saddlepoints $\tilde{t}_{2(1)}$ for approximating the distribution of $(S_2, \ldots, S_p)$ satisfy Eq. (7.41) except that $t_1 = 0$, or equivalently $\hat{\beta}_1 = \beta_1$, is substituted. In other words, the saddlepoints $(\tilde{t}_2, \ldots, \tilde{t}_p)$ are just the ML estimators $\hat{\beta}_j - \beta_j$ with the value of β_1 assumed known. Substituting $\tilde{t}_j = \hat{\beta}_j - \beta_j$ in the log-likelihood Eq. (7.40) we find

$$\begin{aligned} K_S(\tilde{t}) &= \sum_{i=1}^{n} b(\hat{\theta}_i) - s^T\hat{\beta} - \sum_{i=1}^{n} b(\theta_i) + s^T\beta \\ &= \ell(\beta) - \ell(\hat{\beta}) \end{aligned}$$

and similarly

$$K_S(0, \tilde{t}_{2(1)}) = \ell(\beta) - \ell\{\beta_1, \hat{\beta}_2(\beta_1)\}$$

in obvious notation. The quantity in braces in Eq. (7.36) and also in Eq. (7.37) is, therefore,

$$\ell\{\beta_1, \hat{\beta}_2(\beta_1)\} - \ell(\hat{\beta}_1, \hat{\beta}_2)$$

or *half* the difference between the deviance $D(\hat{\beta}_1)$ of the model with β_1 held fixed and the unrestricted deviance $D(\hat{\beta})$. Finally, we interpret $K_S''(\tilde{t})$ as the inverse of the estimate variance matrix $\hat{V}$ for $\hat{\beta}$ and $K_S^{22}(0, \tilde{t}_{2(1)})$ as the inverse of the estimated variance matrix $\hat{V}(\beta_1)$ of $(\hat{\beta}_2, \ldots, \hat{\beta}_p)$ for β_1 known. All these quantities can be obtained in almost any package, for instance using the `OFFSET` facility in `GLIM`.

We are now in a position to express the earlier approximation formulas in terms of simple and easily computed statistical quantities. Treating the approximate conditional density Eq. (7.36) as a likelihood, we obtain the approximate conditional log-likelihood function for β_1 as

$$\tilde{\ell}_c(\beta_1) = \{\log|V(\beta_1)| - D(\beta_1)\}/2. \tag{7.42}$$

Computing the determinant is not always convenient (for instance in `GLIM`) and is subject to numerical instability for large dimension parameters. The conditional likelihood may be used to obtain the conditional ML estimator, and its approximate conditional variance is the reciprocal of the second derivative of this function at the maximum, easily computed numerically.

The conditional P-value is computed from $F_{S_1|S_2}(s_1)$ at the observed value s_1 of S_1. Equation (7.33) is very easily implemented for this purpose, where

$$r(\beta_1) = \text{sign}(\hat{\beta}_1 - \beta_1)\sqrt{D(\beta_1) - D(\hat{\beta})}$$

is just the deviance test statistic for one-sided testing of β_1, and

$$\omega(\beta_1) = (\hat{\beta}_1 - \beta_1)\sqrt{|V(\beta_1)|/|\hat{V}|}.$$

An *Splus* function `cl.saddle` has been supplied; it computes the conditional log-likelihood for any parameter in a generalized linear model. The supplied function `cpvalue.saddle` further computes the saddlepoint approximation to the conditional P-value.

Example 7.5: Crying Babies Data On each of 18 successive days, those babies not crying at a specified time in a hospital ward were selected as subjects (Table 7.13). On each day, one of these babies was randomly chosen to be given a treatment and after a specified time, the number of babies not crying were recorded. The underlying probability of a baby crying will differ from day to day due to hospital conditions as well as the fact that different babies are chosen. We thus have 18 2×2 contingency tables and the data are collectively a $2 \times 2 \times 18$ table. Assuming that the odds ratio θ measuring the effect of the treatment is the same for all days, we desire inference on θ in the presence of 18 nuisance parameters. Under such conditions we expect significant bias in the unconditional inference, and conditional ML estimation is indicated (see Section 3.6.4).

Unconditional inference on θ yields the estimate $\hat{\theta} = 1.432$ with standard error 0.734. The associated one-sided P-value is 0.025 or 0.015 using the change in deviance 4.664. The unconditional and conditional likelihoods are compared in Figure 7.2.

The sufficient statistic S_1 for θ is the total number of successes for the treated babies with observed value $s_1 = 15$. The conditioning statistics S_2 are the marginal totals of the eighteen 2×2 tables. The conditional likelihood was approximated using the supplied *Splus* function `cl.saddle`, based on Eq. (7.42). To within drawing accuracy this cannot be distinguished from the exact con-

Table 7.13. Crying Babies Data [from Cox and Snell (1994, p. 4)]

	Untreated		Treated	
Day	Crying	Not Crying	Crying	Not Crying
1	5	3	0	1
2	4	2	0	1
3	4	1	0	1
4	5	1	1	0
5	1	4	0	1
6	5	4	0	1
7	3	5	0	1
8	4	4	0	1
9	2	3	0	1
10	1	8	1	0
11	1	5	0	1
12	1	8	0	1
13	3	5	0	1
14	1	4	0	1
15	2	4	0	1
16	1	7	0	1
17	2	4	1	0
18	3	5	0	1

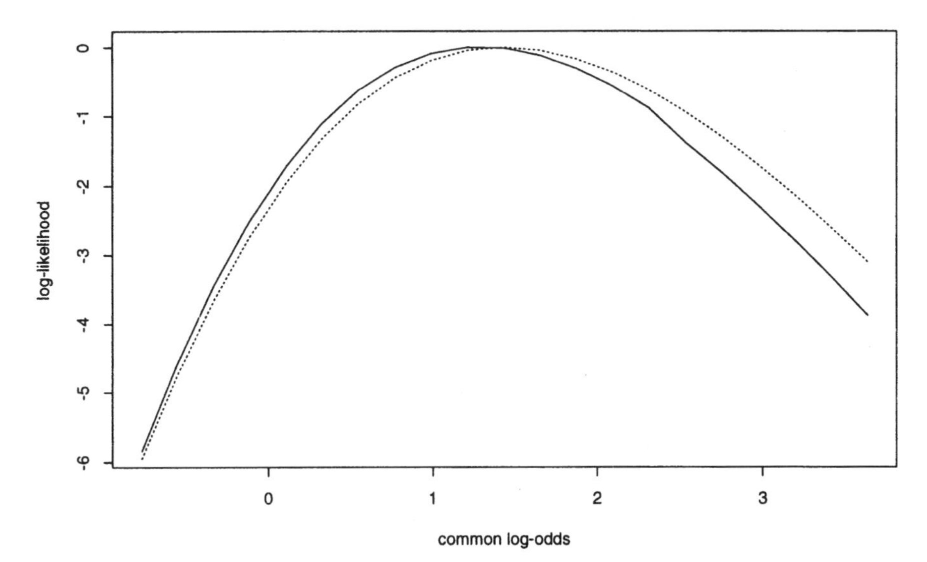

Figure 7.2. Profile and conditional log-likelihoods for crying babies data. Curves are profile log-likelihood function (dotted), and saddlepoint approximation to the conditional log-likelihood (solid). The error in this approximation cannot be distinguished by eye on this scale.

ditional likelihood. The conditional estimate is $\hat{\theta} = 1.256$ with standard error 0.686.

The conditional P-value is computed using `cpvalue.saddle` and the result is 0.048 whereas the exact conditional P-value is 0.045, using *StatXacT*. Implementing the continuity correction is easily done. Change the first row of data to 4.5, 3.5, 0.5, 0.5. Then $s_1 = 14.5$ rather than 14 and all marginal totals are unchanged. This gives the answer 0.044, very close to the exact answer 0.045. We did not have to choose the first day for the data adjusment—any day could be chosen provided that adding $-1/2$, $1/2$, $1/2$, $-1/2$ does not produce any negative counts. The data can be adjusted in this way for virtually any data set and model, but this is hard to program in any generality; I have not included this continuity adjustment in the function `cpvalue.saddle`.

FURTHER READING

There is an enormous literature on conditional inference. Proponents of unconditional methods claim that the "exact" test is too conservative when used with a fixed significance level and that less conservative and correspondingly more powerful tests can be obtained from an unconditional analysis. This conservatism can be moderated by taking half the probability of the observed in calculating the P-value [see Lancaster (1961), Plackett (1984), Barnard (1989, 1990), Mehta and Walsh (1992), and Vollset, Hirji, and Afifi (1991). However, this is no longer a P-value in the usual sense.

Proponents of conditional methods argue that it is illogical and inappropriate to average P-values over tables with other marginal totals than the observed and that the perceived advantages of unconditional tests are based on their not being correctly evaluated conditionally. An excellent review article on exact methods for contingency tables is Agresti (1992).

Section 7.1: Conditional Inference for 2 × 2 Tables

7.1.1 *The "exact" test was originally proposed by Fisher (1934, 1935) and independently by Irwin (1935) and Yates (1934) and is known as the Fisher–Yates–Irwin test. An unconditional test was proposed by Barnard (1945, 1947) but later retracted in Barnard (1949). Discussion of the issues involved has continued in a short chronological bibliography which follows: McDonald, Davis, and Millikin (1977); Berkson (1978a,b); Kempthorne (1979); Basu (1979); Barnard (1979, 1982); Upton (1982); Yates (1984); Suissa and Shuster (1984, 1985); Bhapkar (1986); Haber (1987, 1989); Lloyd (1988a); Greenland (1991); and Lloyd (1992).*

7.1.2 *Dupont (1988) has pointed out another serious problem with exact two-sided P-values. Adding a single extra case to a contingency table with high counts (eg. $n_1 = n_2 = 100$) may sometimes alter the P-value by a huge amount. He thus recommends doubling the one-sided P-value and suggests that the notion of the P-value being the probability of anything be abandoned. This is an extreme position and has been argued against by Lloyd (1988b) where an alternative to doubling the one-sided P-value is suggested.*

7.1.3 *Iterative methods for calculating the conditional ML estimate of the log-odds ratio were first given by Cornfield (1956) with further details given in Plackett (1981). Algorithms for computing confidence limits were given by Thomas (1971). A Fortran program is given by Batista and Pike (1977).*

Section 7.5: Conditional Inference for 2 × 2 × k Tables

7.5.2 *Exact inference on an assumed common treatment effect was first described by Birch (1964) who also demonstrated that the conditional test was uniformly most powerful unbiased using the theory of Lehmann (1959). Birch also gave the conditional estimator and the simple algorithm [see Eq. (7.8)] for its calculation.*

The exact test of the uniform treatments hypothesis was given by Zelen (1971). Efficient algorithms for its implementation were given by Thomas (1975) and Pagano and Tritchler (1983b). One problem with this and other conditional tests is their lack of power for extreme values of the conditioning statistics. For instance, if $Y_{1+} = 0$ *then each of* $Y_{1(j)}$ *must also be zero and so the distribution is degenerate. Thus, the P-value is 1.0 and the power of the (conditional) test is uniformly zero. For any data set where there are no responses under treatment 1 in every table it will not be possible to test for equal treatments. While it is claimed that this demonstrates the "conservativism" of exact methods, it can also be argued that no sensible test is possible in this circumstance and that the power ought to be zero.*

7.5.3 *Efficient algorithms for computing P-values and confidence intervals using the exact convolution of tilted hypergeometric distributions were given by Thomas (1975), Pagano and Tritchler (1983a) and by Mehta et al. (1985). The analogue of the Yates/Pearson chi-square statistic for the* $2 \times 2 \times k$ *table was given by Cochran (1954). Five years later Mantel and Haenszel (1959) proposed a very similar statistic but with hypergeometric variance terms and the continuity correction included, namely Eq. (7.27). It is perhaps surprising that these tests significantly predate proper development of the conditional test they approximate.*

Exact inference on any linear combination of the $\Delta_{(j)}$ *is actually possible in principle under arbitrary linear assumptions about these* $\Delta_{(j)}$*. Thus, inference on* $\bar{\Delta}$ *is possible either with or without assuming uniform treatments but computational difficulties and possible power loss are more severe in the latter case. The exact procedures are obtained by reorganizing the likelihood appropriately to identify sufficient statistics, and are implemented in LogXacT.*

Section 7.9: Saddlepoint Approximation*

7.9 *Accessible review papers on saddlepoint approximations and their use in statistics are Barndorff-Nielsen and Cox (1979) and Reid (1988). The interested reader is also referred to Barndorff-Nielsen and Cox (1989), as well as Jensen (1995) for a fully technical account. Davison (1988b) describes the practical implementation of saddlepoint approximations for generalized linear models. The material of this section is a slight specialization and simplification of his work.*

*See p. 15 in Chapter 1 for an explanation.

Bibliography

Agresti, A. (1980). Generalised odds ratios for ordinal data. *Biometrics* **36,** 59–67.

Agresti, A. (1984). *Analysis of Ordinal Categorical Data*, Wiley, New York.

Agresti, A. (1990). *Categorical Data Analysis*, Wiley, New York.

Agresti, A. (1992). A survey of exact inference for contingency tables, *Statistical Science* **7,** 131–177.

Aitkin, M., Anderson, D. A., Francis, B., and Hinde, J. (1989). *Statistical Modelling in GLIM*, Clarendon Press, Oxford.

Aitkin, M. (1990). Model choice in contingency table analysis using the posterior Bayes factor. *Comput. Statist. Data Anal.* **13,** 245–251.

Aitkin, M. (1991). Posterior Bayes factors (with discussion). *J. Roy. Statist. Soc. B* **53,** 111–142.

Akaike, H. (1973). Information theory and an extension of the maximum likelihood principle, in *Second International Symposium on Information Theory*. B. N. Petrov and F. Csaki, Eds. Akademia Kiado, Budapest, 267–281.

Allen, D. M. (1974). The relationship between variable selection and data augmentation and a method of prediction. *Technometrics* **16,** 125–127.

Amari, S. I. (1985). *Differential Geometrical Methods in Statistics*, Lecture Notes in Statistics, New York, Springer–Verlag.

Amemiya, T. (1981). Qualititative response models: A survey. *J. Econom. Lit.* **19,** 1483–1536.

Andersen, E. B. (1970). Asymptotic properties of conditional maximum likelihood estimators. *J. Roy. Statist. Soc. B*, **32,** 283–301.

Anscombe, F. J. (1964). Normal likelihood functions. *Ann. Inst. Statis. Math* **26,** 1–19.

Aranda-Ordaz, F. J. (1981). On two families of transformations to additivity for binary response data. *Biometrika* **68,** 357–363.

Aranda-Ordaz, F. J. (1983). An extension of the proportional hazards model for grouped data. *Biometrics* **39,** 109–117.

Armitage, P. (1955). Tests for linear trends in proportions and frequencies. *Biometrics* **11,** 375–386.

Armitage, P. and Berry G. (1987). *Statistical Methods in Medical Research*, 2nd ed., Blackwell Scientific Publications, Oxford.

Ashford, J. R. and Sowden, R. R. (1970). Multivariate probit analysis. *Biometrics* **26,** 535–546.

Atkinson, A. C. (1969). Test for discriminating between models. *Biometrika* **56,** 337–347.

Atkinson, A. C. (1970). A method for discriminating between models (with discussion). *J. Roy. Statist. Soc. B* **32,** 323–353.

Atkinson, A. C. (1981). Robustness, transformations and two graphical displays for outlying and influential observations in regression. *Biometrika* **68,** 13–20.

Atkinson, A. C. (1985). *Plots, Transformations and Regression. An Introduction to Diagnostic Regression Analysis*, Clarendon Press, Oxford.

Ayesh, R., Mitchell, S. C. and Waring, R. H. (1987). Sodium aurothiomalate toxicity and sulphoxidation capacity in rheumatoid arthritis patients. *Brit. J. Rheumatology* **26,** 197–201.

Azzalini, A. (1996). *Statistical Inference: Based on the Likelihood.* Chapman and Hall, London.

Bahadur, R. R. (1958). Examples of inconsistency of maximum likelihood estimates. *Sankhya* **20,** 207–210.

Basu, D. (1979). In dispraise of the exact test: Reactions. *J. Statist. Plann. Inf.* **3,** 189–192.

Barnard, G. A. (1945). A new test for 2 × 2 tables. *Nature* **156,** 177–182.

Barnard, G. A. (1947). Significance tests for 2 × 2 tables. *Biometrika* **34,** 123–138.

Barnard, G. A. (1949). Statistical inference. *J. Roy. Statist. Soc. B* **11,** 115–139.

Barnard, G. A. (1979). In contradiction to J. Berkson's 'In dispraise of the exact test.' *J. Statist. Plann. Inf.* **3,** 181–188.

Barnard, G. A. (1982). Conditionality versus similarity in the analysis of 2 × 2 tables, in *Essays in Honour of C. R. Rao*, North-Holland, Amsterdam.

Barndorff-Nielsen, O. E. (1983). On a formula for the distribution of the maximum likelihood estimator. *Biometrika*, **70,** 343–365.

Barndorff-Nielsen, O. E. and Cox, D. R. (1979). Edgeworth and saddle-point approximations with statistical applications (with discussion), *J. Roy. Statist. Soc. B* **41,** 279–312.

Barndorff-Nielsen, O. E. and Cox, D. R. (1989). *Asymptotic Techniques for Use in Statistics.* Chapman and Hall. London.

Bartlett, M. (1953a). Approximate confidence intervals, *Biometrika* **40,** 12–19.

Bartlett, M. (1953b). Approximate confidence intervals II: More than one unknown parameter. *Biometrika* **40,** 306–317.

Bartlett, M. (1955). Approximate confidence intervals III: A bias correction, *Biometrika* **42,** 201–204.

Basu, D. (1959). The family of ancillary statistics. *Sankhya* **21,** 247–256.

Basu, D. (1964). Recovery of ancillary information. *Sankhya* **A 26,** 3–16.

Batista, J. and Pike, M. C. (1977). Algorithm AS115: Exact two sided confidence limits for the odds ratio in a 2×2 table. *Appl. Statist.* **26,** 214–220.

Berger, J. O. and Sellke, T. (1987). Testing a point null hypothesis; the irreconcilability of P values and evidence. *J. Amer. Statist. Assoc.* **82,** 112–122.

Berkson, J. (1988). Some difficulties of interpretation encountered in the application of the chi-square test. *J. Amer. Statist. Assoc.* **33,** 526–536.

Berkson, J. (1978). In dispraise of the exact test. *J. Statist. Planning Inf.* **2,** 27–42.

Berkson, J. (1978). Do the marginal totals of the 2×2 table contain relevent information concerning the table proportions? *J. Statist. Planning Inf.* **2,** 43–44.

Bhapkar, V. P. (1986). On conditionality and likelihood with nuisance parameters in models for contingency tables, Technical Report 253, Department of Statistics, University of Kentucky.

Bhattacharya, A. (1946). On some analogues of the amount of information and their use in statistical estimation. *Sankhya* **8,** 1–14.

Birch, M. W. (1963). Maximum likelihood in three way contingency tables. *J. Roy. Statist. Soc. B*, **25,** 220–223.

Birnbaum, A. (1969). Concepts of statistical evidence, in *Philosophy Science and Method: Essays in Honour of E. Nagel*, St. Martin's Press, New York, 112–143.

Bishop, Y. V. V. (1971). Effects of collapsing multidimensional contingency tables. *Biometrics* **27,** 545–562.

Bishop, Y. V. V. and Fienberg, S. E. (1969). Incomplete two-dimensional contingency tables. *Biometrics* **25,** 119–128.

Bishop, Y. V. V., Fienberg, S. E., and Holland, P. W. (1975). *Discrete Multivariate Analysis: Theory and Practice*, M.I.T. Press, Cambridge.

Blyth, C. R. (1986). Approximate binomial confidence intervals, *J. Amer. Statist. Assoc.* **81,** 468–474.

Blyth, C. R. and Still, H. A. (1983). Binomial confidence intervals, *J. Amer. Statist. Assoc.* **78,** 108–116.

Blyth, C. R. and Staudte, R. G. (1995). Estimating statistical hypotheses. *Stat. Prob. Letters* **23,** 45–52.

Box, G. E. P. and Tidwell, P. W. (1962). Transformations of the independent variable. *Technometrics*, **4,** 531–550.

Bradley, R. A. and Terry, M. E. (1952). Rank analysis of incomplete block designs I. The method of paired comparisons. *Biometrika* **39,** 324–345.

Breslow, N. E. (1980). Odds ratio estimates when the data are sparse. *Biometrika* **68,** 73–84.

Breslow, N. E. and Day, N. E. (1980). *The Analysis of Case-Control Studies*, IARC Scientific Publications No. 32, Lyon, France.

Brown, C. C. (1982). On a goodness of fit test for the logistic model based on score statistics. *Comm. Statist. A* **11,** 1087–1105.

Brunswick, A. F. (1971). Adolescent health, sex and fertility. *Amer. J. Public Health* **61,** 711–720.

Burridge, J. (1981). A note on maximum likelihood estimation for regression models using grouped data. *J. Roy. Statist. Soc. B* **43,** 41–45.

Byrkof, D. R. (1992). *Statistics Today: A Comprehensive Introduction.* Benjamin-Cummings. New York.

Carmichael, C. L., Rugg-Gunn, P. J., and Ferrel, R. S. (1989). The relationship between fluoridation, social class and caries experience in five year old children in Newcastle and Northumberland in 1987. *Brit. Dent. J.* **167,** 57–61.

Carroll, R. J. and Ruppert, D. (1988). *Transformations and Weighting in Regression.* Chapman and Hall. London.

Carroll, R. J., Wu, C. J. F., and Ruppert, D. (1987). The effect of estimating weights in generalised least squares. *J. Amer. Statist. Assoc.* **83,** 1045–1054.

Carter-Saltzman, L. (1980). Biological and socio-cultural effects on handedness: comparison between biological and adoptive families. *Science* **209,** 1263–1265.

Casella, G. (1986). Refining binomial confidence intervals. *Can. J. Statist.* **14,** 113–129.

Casella, G. and Berger, R. L. (1987). Reconciling Bayesian and frequentist evidence in the one-sided testing problem. *J. Amer. Statist. Assoc.* **82,** 106–111.

Caussinus, H. (1965). Contribution a l'analyse statistique des tableau de correlation. *Ann. Fac. Sci. Univ. Toulouse* **29,** 77–182.

Centerwell, B. S., Armstrong, C. W., Funkhowser, L. S., and Elzay, R. P. (1986). Erosion of dental enamel among competitive swimmers at a gas-chlorinated swimming pool. *Amer. J. Epid.* **123,** 641–647.

Chambers, E. A. and Cox, D. R. (1967). Discrimination between alternative binary response models. *Biometrika* **54,** 573–578.

Chao, A. (1989). Estimating population size from sparse data in capture-recapture experiments. *Biometrics* **45,** 427–438.

Chapman, D. G. and Robbins, H. E. (1951). Minimum variance estimation without regularity assumptions. *Ann. Math. Statist.* **22,** 581–586.

Chapman, J. W. (1976). A comparison of χ^2, $-2\log R$ and multinomial probability criteria for significance tests when the expected frequencies are small. *J. Amer. Statist. Assoc.* **71,** 854–863.

Chatterjee, S. and Hadi, A. S. (1986). Influential observations, high leverage points and outliers in regression. *Statist. Sci.* **1,** 379–416.

Chin, T., Marine W., Hall, E., Gravelle, C., and Speers, J. (1961). The influence of Salk vaccination on the epidemic patterns and the spread of the virus in the community. *Amer. J. Hyg.* **73,** 67–94.

Clayton, D. G. (1974). Some odds ratio statistics for the analysis of ordered categorical data. *Biometrika* **61,** 525–531.

Cleveland, W. S. (1979). Robust locally weighted regression and smoothing scatter plots. *J. Amer. Statist. Assoc.* **74,** 829–836.

Clogg, C. C. and Goodman, L. A. (1984). Latent structure analysis of a set of multidimensional contingency tables. *J. Amer. Statist. Assoc.* **79,** 762–771.

Clogg, C. C. and Goodman, L. A. (1985). Simultaneous latent structure analysis in several groups. *Sociol. Method.* **15,** 81–110.

Clogg, C. C. and Shockey, J. W. (1988). Multivariate analysis of discrete data, in *Handbook of Multivariate Experimental Psychology*, J. R. Nesselroade and R. B. Cattell, (Eds.), Plenum Press, New York.

Cochran, W. G. (1954). Some methods of strengthening the common χ^2 test, *Biometrics* **10,** 417–451.

Cochran, W. G. (1955). A test of a linear function of the deviations between observed and expected numbers. *J. Amer. Statist. Assoc.* **50,** 377–397.

Cohen, J. (1960). A coefficient of agreement for nominal scales. *Educ. Psychol. Meas.* **20,** 37–46.

Collett, D. (1991). *Modelling Binary Data*, Chapman and Hall. London.

Copas, J. B. (1983). Plotting p against x. *Appl. Statist.* **32,** 25–31.

Cormack, R. M. (1992). Interval estimation for mark-recapture studies of closed populations. *Biometrics* **48,** 567–576.

Cornfield, J. (1956). A statistical problem arising from retrospective studies, *Proceedings of 3rd Berkeley Symposium* **4,** 135–148.

Cornfield, J. (1962). Joint dependence of risk of coronary heart disease on serum cholesterol and systolic blood pressure: a discriminant function analysis, *Fed. Proc.* **21,** Supplement 11, 58–61.

Cox, D. R. (1958). Some problems connected with statistical inference. *Ann. Math. Statist.*, **29,** 357–372.

Cox, D. R. (1961). Further results on tests of separate families of hypotheses. *J. Roy. Statist. Soc. B* **24,** 406–424.

Cox, D. R. (1970). *The Analysis of Binary Data*, Chapman and Hall, London (2nd ed., 1989, with E. J. Snell).

Cox, D. R. (1971). On the choice between alternative ancillary statistics. *R. Roy. Statist. Soc., B* **33,** 251–255.

Cox, D. R. and Hinkley, D. V. (1974). *Theoretical Statistics*, Chapman and Hall, London.

Cox, D. R. and Snell, E. J. (1989). *Analysis of Binary Data*, 2nd ed., Chapman and Hall, London.

Cramer, H. (1946). *Mathematical Methods of Statistics*, Princeton University Press, New Jersey.

Cressie, N. and Read, T. R. C. (1984). Multinomial goodness of fit tests, *J. Roy. Statist. Soc. B* **46,** 440–464.

Cressie, N. and Read, T. R. C. (1989). Pearson X^2 and the loglikelihood ratio statistic G^2: a comparative review. *Int. Statist. Rev.* **57,** 19–43.

Darroch, J. N. (1981). The Mantel-Haenszel test and tests of marginal symmetry: fixed effects and mixed models for a categorical response. *Int. Statist. Rev.* **49,** 285–307.

Darroch, J. N. (1986). Quasi-symmetry. *Encyclopaedia of Statistical Sciences* **7.** Wiley-Interscience New York.

Darroch, J. N. and McCloud, P. I. (1986). Category distinguishability and observer agreement. *Austral. J. Statist.* **28,** 371–388.

Darroch, J. N., Fienberg, S. E., Glonek, G. F. V., and Junker, B. W. (1993). A three-sample multiple-recapture approach to census population estimation with heterogeneous catchability. *J. Amer. Statist. Assoc.* **88,** 1137–1148.

Davison, A. C. (1988a). Contribution to discussion of Copas (1988), *J. Roy. Statist. Soc. B* **50,** 258–259.

Davison, A. C. (1988b). Approximate conditional inference in generalised linear models. *J. Roy. Statist. Soc. B*, **50,** 445–461.

Davison, A. C. and Tsai, C. L. (1992). Regression model diagnostics, *Int. Statist. Rev.* **60,** 337–354.

De Bruijn, N. D. (1970). *Asymptotic Methods in Analysis*, 3rd ed., North-Holland, Amsterdam.

Deming, W. E. and Stephan, F. F. (1940). On a least squares adjustment of a sampled frequency table when the expected marginal totals are known. *Ann. Math. Statist.* **11,** 427–444.

Devore, J. (1982a). Linkage studies of the tomato. *Trans. Roy. Canada Inst.* **1931,** 1–19.

Devore, J. (1982b). A genetic and biochemical study on pericarp pigments in a cross between two cultivars of grain sorghum. *Sorghum Bicolor. Heredity* **1976,** 413–416.

Draper, C. C., Voller, A., and Carpenter, R. G. (1972). The epidemiological interpretation of seriological data in malaria. *Amer. J. Trop. Med. Hygiene* **21,** 696–703.

Drost, F. C., Kallenberg, W. C. M., Moore, D. S., and Oosterhoff, J. (1989). Power approximations to multinomial tests of fit. *J. Amer. Statist. Assoc.* **84,** 130–141.

Dupont, W. D. (1988). Sensitivity of Fisher's exact test to minor perturbations in 2 × 2 contingency tables. *Stat. Med.* **5,** 629–635.

Edwards, A. W. F. (1972). *Likelihood*, Cambridge University Press, Cambridge.

Efron, B. (1975). Defining the curvature of a statistical problem (with applications to second order efficiency) (with discussion). *Ann. Statist.* **3,** 1189–1242.

Efron, B. and Hinkley, D. V. (1978). Assessing the accuracy of the maximum likelihood estimator; observed versus expected information. *Biometrika* **65,** 457–487.

Essenberg, J. M. (1952). Cigarette smoke and the incidence of primary neoplasm of the lung in albino mice. *Science* **116,** 561–562.

Farewell, V. T. (1982). A note on regression analysis of ordinal data with variability of classification. *Biometrika* **69,** 533–538.

Fienberg, S. E. (1970). Quasi-independence and maximum likelihood estimation in incomplete contingency tables. *J. Amer. Statist. Assoc.* **65,** 1610–1616.

Fienberg, S. E. (1972a). The analysis of incomplete multi-way contingency tables. *Biometrics* **28,** 177–202.

Fienberg, S. E. (1972b). The multiple recapture census for closed populations and incomplete contingency tables. *Biometrika* **59,** 591–603.

Fienberg, S. E. (1978). *The Analysis of Cross-Classified Categorical Data*. M.I.T. Press, Cambridge.

Fienberg, S. E. (1981). Recent advances in theory and methods for the analysis of categorical data: making the link to statistical practice. *Bull. Int. Statist. Inst.* **49,** 763–791.

Fienberg, S. E. and Larntz, K. (1976). Loglinear representation for paired and multiple comparison models. *Biometrika* **63,** 245–254.

Fienberg, S. E. and Meyer, M. M. (1983). Log-linear models and categorical data analysis with psychometric and econometric applications. *J. Econometrics* **22,** 191–214.

Finney, D. J. (1971). *Probit Analysis*, 3rd ed., Cambridge University Press, Cambridge.

Firth, D., Glosup, J., and Hinkley, D. V. (1991). Model checking with non-parametric curves. *Biometrika* **78,** 245–252.

Fisher, R. A. (1922). On the mathematical foundations of theoretical statistics, *Phil. Trans. Roy. Soc. London A* **222,** 309–368.

Fisher, R. A. (1925). The theory of statistical estimation. *Proc. Camb. Phil. Soc.* **22,** 700–725.

Fisher, R. A. (1934). Two new properties of the mathematical likelihood, *Proc. R. Soc. A* **144,** 285–307.

Fisher, R. A. (1935). *The Design of Experiments*, 8th ed., Oliver and Boyd, Edinburgh.

Fisher, R. A. (1956). *Statistical Methods and Scientific Inference*, 2nd ed., Oliver and Boyd, Edinburgh.

Fowlkes, E. B. (1987). Some diagnostics for binary logistic regression via smoothing. *Biometrika* **74,** 503–515.

Frame, S., Moore, J., Peters, A., and Hall, D. (1985). Maternal height and shoe size as predictors of pelvic disproportion: an assessment. *Brit. J. Obstet. Gynaecol.* **92,** 1239–1245.

Freedman, D., Pisani, R., and Purves, R. (1978). *Statistics.* Norton: New York.

Freeman, G. H. and Halton, J. H. (1951). Note on an exact treatment of contingency, goodness-of-fit and other problems of significance. *Biometrika* **38,** 141–149.

Frischer, M., Leyland, A., Cormack, R. M., Goldberg, D. J., Bloor, M., Green, S. T., Taylor, A., Covell, R., McKeganey, N., and Platt, S. (1993). Estimating the population prevalence of injecting drug use and infection with human immunodeficiency virus among injecting drug users in Glasgow, Scotland. *Amer. J. Epidem.* **138,** 170–181.

Gallant, A. R. (1987). *Non-linear Statistical Models*, Wiley, New York.

Gart, J. J. and Zweiful, J. R. (1967). On the bias of various estimators of the logit and its variance with applications to quantal bioassay, *Biometrika* **54,** 181–187.

Genter, F. C. and Farewell, V. T. (1985). Goodness-of-link testing in ordinal regression models. *Canad. J. Statist.* **3,** 37–44.

Gilula, Z. (1983). Latent conditional independence in two-way contingency tables: a diagnostic approach. *Brit. J. Math. Statist. Psychol.* **36,** 114–122.

Gilula, Z. (1984). On some similarities between canonical correlation models and latent class models for two-way contingency tables. *Biometrika* **71,** 523–539.

Glass, D. V. (Ed.) (1954). *Social Mobility in Britain.* The Free Press. Glencoe, IL.

Godambe, V. P. (1960). An optimum property of regular maximum likelihood estimation. *Ann. Math. Statist.* **31,** 1208–1212.

Godambe, V. P. and Heyde, C. C. (1987). Quasi-likelihood and optimal estimating functions. *Int. Statist. Rev.* **55,** 231–244.

Good, I. J. (1957). Saddlepoint methods for multinomial distributions, *Ann. Math. Statist.* **28,** 861–881.

Good, I. J. (1987). Discussion of 'Testing a point null hypothesis; the irreconcilability of P values and evidence,' by Berger J. and Sellke T., *J. Amer. Statist. Assoc.* **82,** 123–125.

Good, I. J. and Crook, J. F. (1974). The Bayes/non-Bayes compromise and the multinomial distribution. *J. Amer. Statist. Assoc.* **69,** 711–720.

Goodman, L. A. (1961). Statistical methods for the mover-stayer model. *J. Amer. Statist. Assoc.* **56,** 841–868.

Goodman, L. A. (1962). Statistical methods for analysing processes of change. *Amer. J. Sociol.* **68,** 57–78.

Goodman, L. A. (1963). Statistical methods for the preliminary analysis of transaction flows. *Econometrica* **31,** 197–208.

Goodman, L. A. (1965). On the statistical analysis of mobility tables. *Amer. J. Sociol.* **70,** 546–585.

Goodman, L. A. (1968). The analysis of cross-classified data: independence, quasi-independence and interaction in contingency tables with or without missing cells. *J. Amer. Statist. Assoc.* **63,** 1091–1131.

Goodman, L. A. (1974). Exploratory latent structure analysis using both identifiable and unidentifiable models. *Biometrika* **61,** 215–231.

Goodman, L. A. (1978). *Analysing Qualitative/Categorical Data.* Addison-Wesley. London.

Goodman, L. A. (1979). Simple models for the analysis of association in cross-classifications having ordered categories. *J. Amer. Statist. Assoc.* **74,** 537–552.

Goodman, L. A. (1983). The analysis of dependence in cross-classifications having ordered categories, using log-linear models for frequencies and log-linear models for odds. *Biometrics* **39,** 149–160.

Goodman, L. A. and Kruskall, W. H. (1979). *Measures of Association for Cross-Classifications.* Springer-Verlag, New York.

Graubard, B. I. and Korn, E. L. (1987). Choice of column scores for testing independence in ordered $2 \times k$ tables. *Biometrics* **43,** 471–476.

Green, P. J. (1984). Iterative reweighted least squares for maximum likelihood estimation and some robust and resistant alternatives. *J. Roy. Statist. Soc. B* **46,** 149–192.

Greenland, S. (1991). On the logical justification of conditional tests for two-by-two contingency tables. *Amer. Statist.* **45,** 248–251.

Guerrero, M. and Johnson, R. A. (1982). Use of the Box-Cox transformations with binary response models. *Biometrika* **65,** 309–314.

Haber, M. (1987). A comparison of some conditional and unconditional exact tests for 2 by 2 contingency tables. *Comm. Statist. B* **16,** 999–1013.

Haber, M. (1989). Do the marginal totals of a 2×2 contingency table contain information regarding the table proportions? *Comm. Statist. A* **18,** 147–156.

Haberman, S. J. (1974). *The Analysis of Frequency Data*, University of Chicago Press, Chicago.

Haberman, S. J. (1977). Log-linear models and frequency tables with small expected cell counts. *Ann. Statist.* **5,** 1148–1169.

Haldane, J. B. S. (1955). The estimation and significance of the logarithm of a ratio of frequencies. *Ann. Human Genet.* **20,** 309–311.

Harris, P. and Peers, H. W. (1980). The local power of the efficient scores test statistic. *Biometrika* **67,** 525–529.

Hastie, T. and Tibshirani, R. (1990). *Generalised Additive Models*, Chapman and Hall, London.

Hastie, T. and Loader, C. (1993). Local regression: Automatic kernel carpentry. *Stat. Sci.*, **8,** 120–143.

Henry, N. W. (1983). Latent structure analysis. *Encyclopedia of Statistical Sciences* **4,** 497–504.

Hoeffding, W. (1956). On the distribution of the number of successes in independent trials. *Ann. Math. Statist.*, **27,** 713–721.

Holst, L. (1972). Asymptotic normality and efficiency for certain goodness of fit tests. *Biometrika* **59,** 137–145.

Horn, S. D. (1977). Goodness of fit tests for discrete data; a review and an application to a health impairment scale. *Biometrics* **33,** 237–248.

Hosmer, D. W. and Lemeshow, S. (1980). A goodness-of-fit test for the multiple logistic regression model. *Comm. Statist.* **A10,** 1043–69.

Hosmer, D. W. and Lemeshow, S. (1989). *Applied Logistic Regression*, Wiley, New York.

Hotelling, H. (1931). The generalisation of student's ratio. *Ann. Math. Statist.* **2,** 360–378.

Irwin, J. O. (1935). Tests of significance for differences between percentages based on small numbers, *Metron* **12,** 83–94.

Jackson, O. A. Y. (1968). Some results on tests of separate families of hypotheses. *Biometrika* **55,** 355–363.

Jensen, J. L. (1995). *Saddlepoint Approximations.* Clarendon Press, Oxford.

Jobson, J. D. and Fuller, W. A. (1980). Least squares estimation when the covariance matrix and parameter vector are functionally related. *J. Amer. Statist. Assoc.*, **75,** 176–181.

Kabaila, P. V. and Lloyd, C. J. (1997). Tight upper confidence limits from discrete data. *Austral. J. Statist.* **39,** 193–204.

Kabaila, P. V. and Lloyd, C. J. (1998). Profile upper confidence limits from discrete data. Res. Rep. No. 187, Department of Statistics, University of Hong Kong.

Kay, R. and Little, S. (1987). Transformations of the explanatory variables in the logistic regression model for binary data. *Biometrika* **74,** 495–501.

Kempthorne, O. (1979). In dispraise of the exact test: reactions. *J. Statist. Plann. Inf.* **3,** 199–213.

Kiefer, J. and Wolfowitz, J. (1956). Consistency of the maximum likelihood estimator in the presence of infinitely many nuisance parameters, *Ann. Math. Statist.* **27,** 887–906.

Koehler, K. (1986). Goodness-of-fit tests for log-linear models in sparse contingency tables, *J. Amer. Statist. Assoc.* **81,** 483–493.

Koehler, K. and Larntz, K. (1980). An empirical investigation of goodness-of-fit statistics for sparse multinomials. *J. Amer. Statist. Assoc.* **75,** 336–334.

Kornguth, M. L. and Miller, M. H. (1985). Comparison of inbreeding in schools of nursing by size of faculty. *J. Nursing Educ.* **24,** 23–32.

Kullback, S. (1959). *Information Theory and Statistics*, Wiley, New York.

Lancaster, H. O. (1961). Significance tests in discrete distributions. *J. Amer. Statist. Assoc.* **56,** 223–234.

Landau, E. (1927). Vorlesungen uber Zahlentheorie, **2,** p. 3–5.

Landis, J. R., Lepkowski, J. M., Davis, C. S., and Miller, M. (1987). Cumulative logit models for weighted data from complex sample surveys. *Proc. Soc. Statist. Amer. Statist. Assoc.*, 165–170.

Landwehr, J. M., Pregibon, D., and Shoemaker, A. C. (1984). Graphical methods for assessing logistic regression models (with discussion). *J. Roy. Statist. Soc. B* **79,** 61–83.

Lang, J. B. (1996). On the comparison of multinomial and Poisson log-linear models, *J. Roy. Statist. Soc. B* **58,** 253–266.

Larntz, K. (1978). Small-sample comparison of exact levels for chi-squared goodness-of-fit statistics. *J. Amer. Statist. Assoc.* **73,** 253–263.

Lazarfeld, P. F. and Henry, N. W. (1968). *Latent Structure Analysis*. Houghton-Mifflin, Boston.

Lee, E. T. (1974). A computer program for linear logistic regression analysis, *Comp. Prog. Biomed.* **4,** 80–92.

Lehmann, E. L. (1959). *Testing Statistical Hypotheses*, Wiley, New York (2nd ed., 1986).

Lipsitz, S. (1988). Methods for analysing repeated categorical outcomes. Unpublished Ph.D. Dissertation, Department of Biostatistics, Harvard University.

Lister, W. R. (1984). Bass singers have a more masculine sex ratio in their siblings than tenors. *I. R. C. Soc. Med. Sci.* **12,** 234–239.

Lloyd, C. J. (1988a). Some issues arising from the analysis of 2 × 2 contingency tables. *Austral. J. Statist.* **30,** 35–46.

Lloyd, C. J. (1988b). Doubling the one sided P-value in testing independence in 2 × 2 tables against a two-sided alternative. *Stat. Med.* **7,** 1297–1306.

Lloyd, C. J. (1990). Confidence intervals from the difference between two correlated proportions. *J. Amer. Statist. Assoc.* **85,** 1154.

Lloyd, C. J. (1991). Estimating the effect of alcohol on the risk of a fatal road accident, *J. Roy. Statist. Soc. A* **153,** 29–52.

Lloyd, C. J. (1992). Effective conditioning. *Austral. J. Statist.* **34,** 241–260.

Lloyd, C. J. (1996). Estimating the response attributable to a continuous risk factor. *Biometrika* **83,** 563–573.

MacGregor, I. D. M. and Balding, J. W. (1988). Bedtimes and family size in English school children. *Ann. Hum. Biol.* **15,** 435–441.

Madansky, A. (1963). Tests of homogeneity for correlated samples. *J. Amer. Statist. Assoc.* **58,** 97–119.

Mantel, N. and Haenszel, W. (1959). Statistical aspects of the analysis of data from retrospective disease. *J. Natl. Cancer. Inst.* **22,** 719–748.

Marsden, J. (1987). An analysis of serological data by the method of spline approximation. M.Sc. Thesis, University of Reading, Statistics Department.

Martin, T. R. and Bracken, M. B. (1987). The association between low birth weight and caffeine consumption during pregnancy. *Amer. J. Epid* **126,** 813–821.

McCullagh, P. (1980). Regression models for ordinal data (with discussion) *J. Roy. Statist. Soc. B* **42,** 109–142.

McCullagh, P. (1984). Generalised linear models. *Europ. J. Oper. Res.* **16,** 285–292.

McCullagh, P. (1986). The conditional distribution of goodness of fit statistics for discrete data. *J. Amer. Statist. Assoc.* **81,** 104–107.

McCullagh, P. and Nelder, J. A. (1989). *Generalized Linear Models*, Chapman and Hall, London.

McCutcheon, A. L. (1987). *Latent Class Analysis*, Sage Publications, Beverly Hills.

McDonald, L. L., Davies, B. M., and Milliken, C. A. (1977). A non-randomised unconditional test for comparing proportions in a 2×2 table. *Technometrics* **19,** 145–150.

McFadden, D. (1974). Conditional logit analysis of qualitative choice behaviour, pp. 105–142, in *Frontiers in Econometrics*, P. Zarembka (Ed.). Academic Press, New York.

McFadden, D. (1981). Economic models of probabilistic choice, in *Structural Analysis of Discrete Data with Econometric Applications*, pp. 198–272, M.I.T. Press, Cambridge.

McFadden, D. (1982). Qualitative response models, pp. 1–37, in *Advances in Econometrics*, Cambridge University Press, Cambridge.

McFadden, D. (1984). Econometric analysis of qualitative response models. *Handbook of Econometrics* **2,** 1395–1457.

Mehta, C. R. and Walsh, S. J. (1992). Comparison of exact mid-p and Mantel-Haenszel confidence intervals for the common odds ratio across several 2×2 contingency tables. *Amer. Statist.* **46,** 146–150.

Mehta, C. R. and Patel, N. R. (1983). A network algorithm for performing Fisher's exact test in $r \times c$ contingency tables. *J. Amer. Statist. Assoc.* **78,** 427–434.

Mehta, C. R., Patel, N. R., and Gray R. (1985). Computing an exact confidence interval for the common odds ratio in several 2×2 contingency tables, *J. Amer. Statist. Assoc.* **80,** 967–973.

Milton, C. and McTeen (1986). *Introduction to Statistics*. D. C. Heath and Company.

Mitra, S. K. (1958). On the limiting power function of the frequency chi-square test. *Ann. Statist.* **29,** 1221–1233.

Montgometry, D. C. and Peck, E. A. (1982). *Introduction to Linear Regression Analysis*, Wiley, New York.

Moore, D. S. and Spruill, M. C. (1975). Unified large sample theory of generalised chi-square statistics for tests of fit. *Ann. Statist.* **3,** 599–616.

Morgan, B. J. T. (1992). *Analysis of Quantal Response Data*. Chapman and Hall.

Morris, C. (1975). Central limit theorems for multivariate sums. *Ann. Statist.* **3,** 165–188.

Mukerjee, R. (1989). Third-order comparison of unbiased tests: A simple formula for the power difference in the one-parameter case. *Sankhya A* **51,** 212–232.

Nadaraya, E. A. (1964). On estimating regression. *Theor. Prob. Appl.*, **9,** 141–142.

Nam, Jun-Mo (1993). Bias corrected maximum likelihood estimator of the log common odds ratio. *Biometrika* **80,** 688–694.

Nelder, J. A. and Wedderbrun, R. W. M. (1972). Generalized linear models, *J. Roy. Statist. Soc. B* **135,** 370–384.

Neyman, J. (1949). Contributions to the theory of the X^2 test, *Proceedings of the 4th Berkeley Symposium* **1,** 239–273.

Neyman, J. (1959). Optimal asymptotic tests of composite statistical hypotheses, pp. 213–234, in *Probability and Statistics*, Almqvist and Wiksell.

Neyman, J. and Pearson, E. S. (1928). On the use and interpretation of certain test criteria for purposes of statistical inference. *Biometrika* **20A,** 175–240.

Neyman, J. and Pearson, E. S. (1933). On the problem of the most efficient test of statistical hypotheses. *Phil. Trans. R. Soc. A* **231,** 289–337.

Neyman, J. and Scott, E. L. (1948). Consistent estimates based on partially consistent observations. *Econometrica* **16,** 1–32.

Norton, P. G. and Dunn, V. V. (1985). Snoring as a risk factor for disease: an epidemiological study. *Brit. Med. J.* **291,** 630–632.

O'Gorman, T. W. and Woolson, R. F. (1993). The effect of category choice on the odds ratio and several measures of association in case-control studies. *Comm. Statist. A* **22,** 1157–1171.

Pagano, M. and Tritchler, D. (1983a). On obtaining permutation distributions in polynomial time. *J. Amer. Statist. Assoc.* **78,** 435–440.

Pagano, M. and Tritchler, D. (1983b). Algorithms for the analysis of several 2×2 contingency tables. *SIAM J. Sci. Statist. Comput.* **4,** 302–309.

Palmgren, J. and Ekholm A. (1987). Exponential family non-linear models for categorical data with errors of observation. *Appl. Stochastic Models Data Anal.* **3,** 111–124.

Pearson, K. (1900). On a criterion that a given system of deviations from the probable in the case of a correlated system of variables is such that it can reasonably be supposed to have arisen from random sampling. *Philos. Mag. Series 5* **50,** 157–175. Also reprinted in (1948), *Karl Pearson's Early Statistical Papers*, Cambridge University Press, Cambridge.

Peers, H. W. (1971). Likelihood ratio and associated test criteria. *Biometrika* **58,** 577–587.

Pitman, E. J. G. (1938). The estimation of the location and scale parameters of a continuous population of any given form. *Biometrika* **29,** 322–335.

Plackett, R. L. (1977). The marginal totals of a 2×2 table. *Biometrika* **64,** 37–42.

Plackett, R. L. (1984). Discussion of paper by Yates. *J. Roy. Statist. Soc. A* **147,** 426–463.

Ploog, D. W. (1967). The behaviour of squirrel monkeys (*Saimiri sciureus*) as revealed by sociometry, bio-acoustics and brain stimulation, pp. 149–184, in *Social Communication Among Primates*, S. Altmann (Ed.) University of Chicago Press, Chicago.

Poisson, S. D. (1837). *Recherches sur la Probabilite des jugements en matiere criminelle et en matiere civile.* Bachelier, Paris.

Pratt, J. W. (1981) Concavity of the log-likelihood. *J. Amer. Statist. Assoc.* **76,** 103–106.

Pratt, J. W. (1987). Discussion of Berger and Sellke's paper 'Testing a point null hypothesis; the irreconcilability of P values and evidence.' *J. Amer. Statist. Assoc.* **82,** 123–125.

Pregibon, D. (1980). Goodness of link tests for generalised linear models, *Appl. Statist.* **29,** 15–24.

Prentice, R. L. (1976a). Use of the logistic model in retrospective studies, *Biometrics* **32,** 1–22.

Prentice, R. L. (1976b). A generalisation of the probit and logit methods for dose-response curves. *Biometrics* **32,** 761–768.

Rao, C. R. (1946). On the linear combination of observations and the general theory of least squares. *Sankhya* **7,** 237–256.

Rao, C. R. (1947). Large sample tests of statistical hypotheses concerning several parameters with applications to problems of estimation. *Proc. Cam. Phil. Soc.* **44,** 50–57.

Rao, C. R. (1957). Maximum likelihood estimation for the multinomial distribution, *Sankhya* **18,** 139–148.

Rao, C. R. (1958). Maximum likelihood estimation for the multinomial distribution with an infinite number of cells. *Sankhya* **20,** 211–218.

Rao, C. R. (1961). Apparent anomalies and irregularities in MLE. *Sankhya A* **24,** 73–101.

Rasch, G. (1980). *Probabilistic Models for some Intelligence and Attainment Tests.* University of Chicago Press, Chicago.

Rasmussen, S. (1992). *Introduction to Statistics with Data Analysis.* Brookes-Cole. London.

Reading, V. M. and Weale, R. A. (1986). Eye strain and visual display units. *Lancet* **i,** 905–906.

Reid, N. (1988). Saddlepoint expansions and statistical inference (with discussion). *Statistical Science* **3,** 213–227.

Rosenberg, L., Palmer, J. R., Kelly, J. P., Kaufman, D. W., and Shapiro, S. (1988). Coffee drinking and non-fatal myocardial infarction in men under 55 years of age. *Amer. J. Epidem.* **128,** 570–578.

Rosner, B. (1986). *Fundamentals of Biostatistics*, Holden and Day, San Francisco.

Ross, G. J. S. (1990). *Non-linear Estimation.* Springer-Verlag. New York.

Roth, J. A., Eilber, F. R., Nizze, J. A., and Morton, D. L. (1975). Lack of correlation between skin reactivity to dinitochlorobenzene and croton oil in patients with cancer. *New Engl. J. Med.* **295,** 386–389.

Sandland, R. L. and Cormack, R. M. (1984). Statistical inference for Poisson and multinomial models for capture-recapture experiments. *Biometrika* **71,** 27–33.

Santner, T. J. and Snell, M. K. (1980). Small sample confidence intervals for $p_1 - p_2$ and p_1/p_2 in 2×2 contingency tables. *J. Amer. Statist. Assoc.* **75,** 386–394.

Savage, I. R. (1973). Incomplete contingency tables; conditions for the existence of unique MLE, pp. 87–90, in *Mathematics and Statistics. Essays in honour of Harold Bergstrom*, Chalmers Institute of Technology, Goteberg, Sweden.

Schafer, G. (1982). Lindley's paradox. *J. Amer. Statist. Assoc.* **77,** 325–351.

Schiff, S. E., Peleg, E., and Goldenberg, M. (1989). The use of aspirin to prevent pregnancy induced hypertension and lower the ratio of thromboxane Az to prostacyclin in relatively high risk patients. *New Eng. J. Med.* **321,** 351–356.

Seber, G. A. F. (1982). *The Estimation of Animal Abundance and Related Parameters*, 2nd ed. Griffin, London.

Seber, G. A. F. and Wild, C. J. (1989). *Non-linear Regression*, Wiley, New York.

Serfling, R. J. (1980). *Approximation Theorems of Mathematical Statistics*, Wiley, New York.

Sewell, W. H. and Orenstein, A. M. (1965). Community of residence and occupational choice. *Amer. J. Sociol.* **70,** 551–563.

Silvapulle, M. J. (1981). On the existence of maximum likelihood estimators for the binomial response model. *J. Roy. Statist. Soc. B* **43,** 310–313.

Small, K. A. (1987). A discrete choice model for ordered alternatives. *Econometrica* **55,** 409–424.

Small, K. A. (1988). Discrete choice econometrics. *J. Math. Psychol.* **32,** 80–87.

Snapinn, S. M. and Small, R. D. (1986). Tests of significance using regression models for ordered categorical data. *Biometrics* **42,** 583–592.

Sprott, D. A. (1973). Practical uses of the likelihood function, in *Inference and Indecision*, S. Portnoy (Ed.), Halstead Press, New York.

Srole, L., Langner, T. S., Michael, S. T., Kirkpatrick, P., Opler, M. K., and Rennie, T. A. C. (1978). *Mental Health in the Metropolis: The Midtown Manhattan Study*, rev. ed. NYU Press, New York.

Stevens, W. L. (1951). Asymptotic regression. *Biometrics* **7,** 247–267.

Stone, M. (1974). Cross-validatory choice and assessment of statistical predictions (with discussion). *J. Roy. Statist. Soc. B* **36,** 111–147.

Suissa, S. and Shuster, J. J. (1984). Are uniformly most powerful unbiased tests really best? *Amer. Statist.* **38,** 204–206.

Suissa, S. and Shuster, J. J. (1985). Exact unconditional sample sizes for the 2 by 2 binomial trial. *J. Roy. Statist. Soc. A* **148,** 317–327.

Svalastoga, K. (1959). *Prestige, Class and Mobility.* Heinemann: London.

Tallis, G. M. (1962). The maximum likelihood estimation of correlation from contingency tables. *Biometrics* **18,** 342–353.

Theil, H. (1969). A multinomial extension of the linear logit model. *Int. Econ. Rev.* **10,** 251–259.

Theil, H. (1970). On the estimation of relationships involving qualititative variables. *Amer. J. Sociol.* **76,** 103–154.

Thomas, D. G. (1971). Exact confidence intervals for the odds ratio in a 2 × 2 table. *Appl. Statist.* **20,** 105–110.

Thomas, D. G. (1975). Exact and asymptotic methods for the combination of 2 × 2 tables. *Comput. Biomed. Res.* **8,** 423–446.

Upton, G. J. G. (1982). A comparison of alternative tests for the 2 × 2 comparative trial. *J. Roy. Statist. Soc. A* **145,** 86–105.

Vollset, S. E., Hirji, K. F., and Afifi, A. A. (1991). Evaluation of exact and asymptotic interval estimators in logistic analysis of matched case-control studies. *Biometrics* **47,** 1311–1325.

Wald, A. (1941). Asymptotically most powerful tests of statistical hypotheses, *Ann. Math. Statist*, **12,** 1–19.

Wald, A. (1943). Tests of statistical hypotheses concerning several parameters when the number of observations is large. *Trans. Amer. Math. Soc.* **54,** 426–482.

Wald, A. (1949). Note on the consistency of the maximum likelihood estimate, *Ann. Math. Statist.* **20,** 595–601.

Wand, M. P. and Jones, M. C. (1995). *Kernel Smoothing.* Chapman and Hall. London.

Wang, D. Q. (1990). *Computing algorithms for testing and fitting some categorical data models.* Unpublished Ph.D. Dissertation, La Trobe University, Australia.

Wang, P. C. (1985). Adding a variable in generalised linear models. *Technometrics* **27,** 273–276.

Wang, P. C. (1987). Residual plots for detecting non-linearity in generalised linear models. *Technometrics* **29,** 435–438.

Wang, Y. H. (1993). On the number of successes in independent trials. *Statistica Sinica,* **3,** 295–312.

Watson, G. S. (1964). Smooth regression analysis. *Sankhya A*, **26,** 359–372.

Wilks, S. S. (1938). The large sample distribution of the likelihood ratio for testing composite hypotheses. *Ann. Math. Statist.* **9,** 60–62.

Williams, D. A. (1982a). Extra-binomial variation in logistic linear models, *Appl. Statist.* **31,** 144–148.

Williams, D. A. (1982b). Contribution to the discussion of Atkinson (1982), *J. Roy. Statist. Soc. B* **44,** 33.

Wilner, D. M., Walkley, R. P., and Cook, S. W. (1955). *Human Relations in Interracial Housing: A Study of the Contact Hypothesis.* University of Minnesota Press, Minneapolis.

Wing, J. K. (1962). Institutionalism in mental hospitals. *Brit. J. Soc. Clin. Psychol.* **1,** 38–51.

Yates, F. (1934). Contingency tables involving small numbers and the χ^2 test. *J. Roy. Statist. Soc. Suppl.* **1,** 217–235.

Yates, F. (1984). Tests of significance for 2 × 2 tables (with discussion), *J. Roy. Statist. Soc. A* **147,** 426–463.

Zacks, S. (1971). *The Theory of Statistical Inference*, Wiley, New York.

Zelen, M. (1971). The analysis of several 2 × 2 contingency tables, *Biometrika* **58,** 129–137.

Index

Added variable plots, 209, 218
Algorithms
  EM, 56
  Fisher scoring, 16
  IRLS, 137, 181, 250
  iterative proportional fitting, 56
  Newton–Raphson, 16, 417
  for nonlinear models, 191
Ancillary statistic, 29
Anscombe residuals, 199
Aranda–Ordaz link test, 219
Association of categorical variables, 71
Asymptotics, 19–31
  of alternative test statistics, 45
  of ML estimator, 22
  of Poisson process, 26
  of test power, 45
Attributable response, 284–287

Bachman–Landau notation, 50
Beta-binomial distribution, 231
Bilinear model, 345
Bias
  correction of kernel smoother, 274
  from hidden factors, 153
Breslow–Day statistic, 152, 436

Canonical link, 249
Capture–recapture, 367
Categorical variables, 68
  association measures, 70
  spread measures, 70
Central limit theorem
  illustration of accuracy, 20, 28, 37, 53
  ML estimator, 22
  for score function, 22
  transformation of, 51
Clinical trials, 121, 155
  binomial model for, 121
Collapsibility, 155, 329, 333
Complementary log–log, 185, 219
Concentration coefficient, 71
Conditional independence, 328
Conditional inference
  and nuisance parameters, 36
  approximation to, 405, 406, 412
  goodness-of-fit, 100
  for linear models, 413
  for log-linear models, 440
  for logistic regression, 440
  on log-odds ratio, 398
  for $r \times c$ tables, 418
  saddlepoint approximation to, 439
  for sparse tables, 425
  for three-way tables, 427
  for 2×2 tables, 396
  for $2 \times 2 \times k$ tables, 422
  for $2 \times k$ tables, 407
  worked example, 411
  Yates correction, 406, 413
Conditional likelihood
  saddlepoint approximation, 441
Confidence intervals, 48–49
  definition of, 48
  from inverting a P-value, 48
  for nonlinear parameters, 193
Consistency
  breakdown for sparse tables, 46, 160, 161
  of ML estimator, 23
Constructed variables, 210
Continuation ratio, 349
Contrasts, 140
Cook's distance, 212
Correlated binary trials, 231

Cramer–Rao bound, 23
Cross-sectional studies, 122
Cumulative logits, 352

Deletion residuals, 201
Deviance, 89–91
  closeness to Pearson statistic, 92
  for log-linear model, 300
  residuals, 199
Diagnostics, 198–219
  added variables, 209
  constructed variables, 210
  Cook's distance, 212
  leverage and influence, 211–214
  normal plots, 205–208
  residuals, 198–211
Dose response models, 184–190
  effective dose, 189
  tolerance distributions, 185

Effective dose, 189
Entropy, 70
Estimation
  of attributable response, 284
  of effective dose, 189
  IRLS, 137, 250
  maximum likelihood, 10
  of P-value, 33
  of relative risk, 220
Exact inference
  approximation to, 405, 406, 412
  on binomial probability, 101
  goodness-of-fit, 100
  for linear models, 413
  for log-linear models, 439
  for logistic regression, 439
  for Poisson mean, 101
  for $r \times c$ tables, 418
  saddlepoint approximation to, 438
  on a single parameter, 100
  for three-way tables, 427
  for 2×2 tables, 396
  for $2 \times 2 \times k$ tables, 422
  for $2 \times k$ tables, 407
  worked example, 411
  Yates correction, 406, 412

Generalized additive models
  application to seed germination, 279
  GAM command, 278–279
Generalized linear models, 175, 248–251
  for 2×2 tables, 136
  weighted LS approximation, 236
GLIM
  output, 137, 147, 149, 215, 243, 304, 309, 313, 322, 323, 325, 344, 347, 351, 359, 367, 376
Goodness-of-fit, 133
  against a smooth, 294
  for binary data, 182
  conditional tests, 435–437
  illustrations, 37, 39, 42, 93
  null and nonnull distributions, 45, 96
  power comparison, 45, 96
  power divergence statistic, 95
  unconditional tests, 86–100
Goodness-of-link tests, 218–219
Graphical models, 334

Half normal plots, 206
Hat matrix, 200, 212
Heterogeneity
  beta-binomial distribution, 231
  effects of, 123, 227
  Rasch model for, 376–381
  of response probability, 123, 227
Hidden factors, 153
Hypergeometric distribution
  computation of, 400
  mean and variance, 406
  multivariate, 408
  ordinary, 406
  tilted, 397, 408

Incomplete tables, 352–367
  from multiple record systems, 368
Independence
  log-linear model formula, 331
  quasi, 354–357
  types of, 327
  types of models, 332
Influence, 211–214
  Cook's distance, 212
Information
  definition, 21
  for generalized linear model, 250
  for logistic regression, 181
  for multinomial models, 75
  with nuisance parameters, 25
  observed or expected, 22
  orthogonality, 24
  for Poisson models, 81
  transformation of, 22, 52
  and variance, 23
Interpretation of model formulas, 329

Invariance
of logistic models, 157
to parameterization, 6
Iterative reweighted LS, 137, 250

Kernel smoother, 267

Latent variables, 310
Least squares
iterative reweighted, 137, 250
Leverage, 201, 211–214
Likelihood
for binomial models, 120
definition of, 10
estimating equations, 21
general, 21
for log-linear models, 299
for logistic regression, 180
for multinomial models, 74
for Poisson models, 81
profiled, 43
residuals, 202
Linear smoothers, 268
Link function, 137, 184
canonical, 249
symmetry of, 185, 219
testing goodness of, 218–219
Link residuals, 198
Local smoothing, 266–268
Locally weighted regression, 268
Loess, 286
Log-linear models
conditional inference for, 440
exact inference for, 440
for multinomial logit models, 319
for multiple record systems, 367–383
for ordinal row effects, 343
for quasi-independence, 354–357
for Rasch model, 378
unconditional inference, 299–311
with interaction as response, 324
Log-odds
advantages of, 155–158
definition of, 120
inconsistent estimation of, 161
problems with, 158
under retrospective sampling, 155
Log-odds ratio, 125
Logistic regression
conditional inference, 431–435
exact inference for, 440
model for relative risk, 223
unconditional inference, 176–180
Logits
central limit theorem for, 127
compared to probit, 188
cumulative, 352
definition of, 120
empirical, 127
modified empirical, 128
LR statistic
for binomial models, 133
breakdown for sparse tables, 161
definition of, 42
for goodness-of-fit, 87
for multinomial models, 88
partition of, 142
for Poisson models, 88
LR test
advantages of, 45
nonnull distribution, 100
relation to Wald and score, 43
Lugganini–Rice formula, 438

Mantel–Haenszel test, 426
Marginal independence, 328
Matched pairs, 160
Maximum likelihood
definition of, 10
existence for log-linear models, 353
existence for logistic models, 251
inconsistency of, 161
simulation of null distribution, 37
MINITAB
output, 240, 243, 245
Mobility models, 357–362
Model selection, 248
Models
for association data, 121–123
bilinear, 345
Bradley–Terry, 366
dose-response, 184–190
elements of, 2–3
generalized additive, 278–279
generalized linear, 248–251
for incomplete tables, 352–367
for independence, 326–338
latent variable, 310
log-linear, 299–311
logistic regression, 176–180
for mobility, 357–362
multinomial logit, 311–326
nested and nonnested, 8

Models (continued)
nonlinear, 190–194
for ordinal data, 339–352
for over-dispersion, 226–236
parallel odds, 346
pictures of, 334
for quasi-symmetry, 362–365
for ranking preferences, 365
for relative risk, 219–226
row effects, 341
saturated, 4
Monte Carlo
conditional inference, 429, 434
envelopes for normal plots, 206
estimation of variance matrix, 29
null distribution of MLE, 37
Multinomial
connection with Poisson, 83–85, 118, 137, 313
distribution, 73
likelihood, 75
LR statistic, 88
Multinomial conversion formula, 320
Multinomial logit models, 311–326
definition of, 316
generalized, 318
as log-linear models, 319
Multiple record systems, 367–383
as incomplete table, 368
Rasch model, 376

Nested and nonnested models, 8
Noncentral chi-square, 45, 100
Nonlinear models, 190–194
Nonorthogonality, 214–217
definition of, 25
as diagnostic, 27, 215
effects of, 25, 37, 215
in parallel lines model, 217
Normal approximation
illustration of accuracy, 20, 27, 37, 53
to P-values, 35
for score function, 22
Normal probability plots, 205
Nuisance parameters, 34
and conditioning, 36
problems with, 33
sufficient statistics for, 36

O_p notation, 51
Ordinal data, 339–352
bilinear model, 345
continuation ratio, 349
cumulative logits, 352
local logits, 339
local odds ratios, 340
parallel odds model, 346
row effects model, 341
scores, 340, 346
Over-dispersion, 226–236
from correlated trials, 231
detection of, 231
of log-linear models, 307–311
and weights, 239
Williams procedure, 233

P-values
approximate, 35
conditional, 36
distribution of, 47
as evidence, 32, 46
inversion to confidence intervals, 48
with nuisance parameters, 33
problems with, 46, 402
saddlepoint approximation, 442
two-sided, 38, 402
Parallel odds model, 346
Partial Rasch model, 380
Partitioning of chi-square, 142
Pearson residuals, 199
Pearson statistic, 91–94, 133, 134
closeness to deviance, 92
generalized, 94
nonnull distribution, 100
as score statistic, 91
Poisson
connection with multinomial, 83–85, 118, 137, 313
distribution, 80
likelihood, 81
LR statistic, 88
model for 2×2 tables, 122
model for $2 \times k$ tables, 135
Power
of alternative test statistics, 45, 96
divergence statistic, 95
Pregibon link test, 219
Probits, 185
compared to logits, 188
Profile likelihood, 43, 102
for binomial probability, 102, 104, 105
for Poisson mean, 103
for odds ratio, 130, 398
Prospective studies, 121, 155, 399

Quasi log-linear models, 355
Quasi-independence, 354–357
Quasi-symmetry models, 362–365, 378

Ranking preference models, 365
Rasch model
  for heterogeneity, 376
  as log-linear model, 378
  partial, 380
  and quasi-symmetry, 378
Raw residuals, 198
Relative risk, 219–226
  point and interval estimates, 220
  logistic regression model for, 223
  and Simpson's paradox, 222
  smoothing of, 283
Residuals, 198–211
  Anscombe, 199
  deletion, 201
  deviance, 199
  likelihood, 201
  for log-linear models, 306
  multiple comparisons of, 203
  normal plots, 205–208
  ordinary or raw, 198
  Pearson, 199
  scaled, 199
  smoothing of, 287–292
  standardized, 199
Retrospective studies, 122, 410
  estimability of odds ratios, 155
  estimability of relative risk, 221
Row effects model, 341

Saddlepoint approximation, 436–443
  to conditional distribution, 438
  continuity correction, 439
  for generalized linear models, 439–441
  multivariate, 438
  Splus function for, 441
Scaled residuals, 199
Score function
  asymptotics, 22
  expansion of, 22
  general, 21
  for generalized linear model, 250
  for logistic regression, 180
  for multinomial models, 75
  for Poisson models, 81
Score test
  definition of, 41
  and Pearson statistic, 91
  relation to Wald and LR, 43
Shannon index of diversity, 70
Simpson's paradox, 152
  for relative risk, 222
Simulation envelope, 206
Smoothing
  attributable response, 286
  bias correction, 274
  of binary data, 272–282
  choice of constant, 271
  effective degrees of freedom, 271
  effective sample size, 273
  goodness-of-fit tests, 294
  kernel, 267, 277
  linear, 268
  loess, 286
  multiple binary regressions, 277
  reasons for, 266
  relative risk, 283
  residuals, 287–292
  spline, 278
  standard error, 273
  for testing constant scale, 292
Splus
  functions, 277, 366
  output, 280
Standardized residuals, 199
StatXacT, 406
Stratified tables, 143
Structural zeros, 354
Sufficient statistics
  for log-linear models, 299

Testing
  Bayes factors, 46
  constant scale via smoothing, 292
  goodness-of-link, 218–219
  LR statistic, 42
  Mantel–Haenszel, 426
  nonnested models, 8
  for over-dispersed models, 235
  particular contrasts, 140
  posterior Bayes, 46
  Score statistic, 40
  several parameters, 39
  a single parameter, 32
  trend, 140
  Wald statistic, 39
Three-way tables
  conditional inference, 427–431
  log-linear models for, 326–333
Tilted hypergeometric distribution, 397, 408

Tolerance distribution, 184
Transformation, 50–54
  of asymptotic variance, 52
  Box–Cox, 196
  of central limit theorem, 51
  complementary log–log, 185
  of confidence intervals, 104
  of covariates, 194
  logit, 120
  probit, 185
Twins data, 160
2×2 tables
  conditional inference, 396–407
  unconditional inference, 124–131
$2 \times 2 \times k$ tables
  conditional inference, 422–427
  unconditional inference, 143–152
$2 \times k$ tables
  conditional inference, 407–413
  unconditional inference, 131–143
Two-way tables
  conditional inference, 418–421
  log-linear models for, 326–333

Unscaled residuals, 198

Wald test
  definition of, 39
  relation to score and LR, 43
Weighted least squares, 238
  approximation to GLM, 236–239
Williams procedure, 233

Yates correction, 92, 406, 413

WILEY SERIES IN PROBABILITY AND STATISTICS
ESTABLISHED BY WALTER A. SHEWHART AND SAMUEL S. WILKS

Probability and Statistics Section

*ANDERSON · The Statistical Analysis of Time Series
ARNOLD, BALAKRISHNAN, and NAGARAJA · A First Course in Order Statistics
ARNOLD, BALAKRISHNAN, and NAGARAJA · Records
BACCELLI, COHEN, OLSDER, and QUADRAT · Synchronization and Linearity: An Algebra for Discrete Event Systems
BASILEVSKY · Statistical Factor Analysis and Related Methods: Theory and Applications
BERNARDO and SMITH · Bayesian Statistical Concepts and Theory
BILLINGSLEY · Convergence of Probability Measures
BOROVKOV · Asymptotic Methods in Queuing Theory
BOROVKOV · Ergodicity and Stability of Stochastic Processes
BRANDT, FRANKEN, and LISEK · Stationary Stochastic Models
CAINES · Linear Stochastic Systems
CAIROLI and DALANG · Sequential Stochastic Optimization
CONSTANTINE · Combinatorial Theory and Statistical Design
COOK · Regression Graphics
COVER and THOMAS · Elements of Information Theory
CSÖRGŐ and HORVÁTH · Weighted Approximations in Probability Statistics
CSÖRGŐ and HORVÁTH · Limit Theorems in Change Point Analysis
DETTE and STUDDEN · The Theory of Canonical Moments with Applications in Statistics, Probability, and Analysis
*DOOB · Stochastic Processes
DRYDEN and MARDIA · Statistical Analysis of Shape
DUPUIS and ELLIS · A Weak Convergence Approach to the Theory of Large Deviations
ETHIER and KURTZ · Markov Processes: Characterization and Convergence
FELLER · An Introduction to Probability Theory and Its Applications, Volume 1, *Third Edition,* Revised; Volume II, *Second Edition*
FULLER · Introduction to Statistical Time Series, *Second Edition*
FULLER · Measurement Error Models
GHOSH, MUKHOPADHYAY, and SEN · Sequential Estimation
GIFI · Nonlinear Multivariate Analysis
GUTTORP · Statistical Inference for Branching Processes
HALL · Introduction to the Theory of Coverage Processes
HAMPEL · Robust Statistics: The Approach Based on Influence Functions
HANNAN and DEISTLER · The Statistical Theory of Linear Systems
HUBER · Robust Statistics
IMAN and CONOVER · A Modern Approach to Statistics
JUREK and MASON · Operator-Limit Distributions in Probability Theory
KASS and VOS · Geometrical Foundations of Asymptotic Inference
KAUFMAN and ROUSSEEUW · Finding Groups in Data: An Introduction to Cluster Analysis

*Now available in a lower priced paperback edition in the Wiley Classics Library.

Probability and Statistics (Continued)

KELLY · Probability, Statistics, and Optimization
LINDVALL · Lectures on the Coupling Method
McFADDEN · Management of Data in Clinical Trials
MANTON, WOODBURY, and TOLLEY · Statistical Applications Using Fuzzy Sets
MORGENTHALER and TUKEY · Configural Polysampling: A Route to Practical Robustness
MUIRHEAD · Aspects of Multivariate Statistical Theory
OLIVER and SMITH · Influence Diagrams, Belief Nets and Decision Analysis
*PARZEN · Modern Probability Theory and Its Applications
PRESS · Bayesian Statistics: Principles, Models, and Applications
PUKELSHEIM · Optimal Experimental Design
RAO · Asymptotic Theory of Statistical Inference
RAO · Linear Statistical Inference and Its Applications, *Second Edition*
RAO and SHANBHAG · Choquet-Deny Type Functional Equations with Applications to Stochastic Models
ROBERTSON, WRIGHT, and DYKSTRA · Order Restricted Statistical Inference
ROGERS and WILLIAMS · Diffusions, Markov Processes, and Martingales, Volume I: Foundations, *Second Edition;* Volume II: Îto Calculus
RUBINSTEIN and SHAPIRO · Discrete Event Systems: Sensitivity Analysis and Stochastic Optimization by the Score Function Method
RUZSA and SZEKELY · Algebraic Probability Theory
SCHEFFE · The Analysis of Variance
SEBER · Linear Regression Analysis
SEBER · Multivariate Observations
SEBER and WILD · Nonlinear Regression
SERFLING · Approximation Theorems of Mathematical Statistics
SHORACK and WELLNER · Empirical Processes with Applications to Statistics
SMALL and McLEISH · Hilbert Space Methods in Probability and Statistical Inference
STAPLETON · Linear Statistical Models
STAUDTE and SHEATHER · Robust Estimation and Testing
STOYANOV · Counterexamples in Probability
TANAKA · Time Series Analysis: Nonstationary and Noninvertible Distribution Theory
THOMPSON and SEBER · Adaptive Sampling
WELSH · Aspects of Statistical Inference
WHITTAKER · Graphical Models in Applied Multivariate Statistics
YANG · The Construction Theory of Denumerable Markov Processes

Applied Probability and Statistics Section

ABRAHAM and LEDOLTER · Statistical Methods for Forecasting
AGRESTI · Analysis of Ordinal Categorical Data
AGRESTI · Categorical Data Analysis
ANDERSON, AUQUIER, HAUCK, OAKES, VANDAELE, and WEISBERG · Statistical Methods for Comparative Studies
ARMITAGE and DAVID (editors) · Advances in Biometry
*ARTHANARI and DODGE · Mathematical Programming in Statistics
ASMUSSEN · Applied Probability and Queues
*BAILEY · The Elements of Stochastic Processes with Applications to the Natural Sciences
BARNETT and LEWIS · Outliers in Statistical Data, *Third Edition*
BARTHOLOMEW, FORBES, and McLEAN · Statistical Techniques for Manpower Planning, *Second Edition*

*Now available in a lower priced paperback edition in the Wiley Classics Library.

Applied Probability and Statistics (Continued)

BATES and WATTS · Nonlinear Regression Analysis and Its Applications
BECHHOFER, SANTNER, and GOLDSMAN · Design and Analysis of Experiments for Statistical Selection, Screening, and Multiple Comparisons
BELSLEY · Conditioning Diagnostics: Collinearity and Weak Data in Regression
BELSLEY, KUH, and WELSCH · Regression Diagnostics: Identifying Influential Data and Sources of Collinearity
BHAT · Elements of Applied Stochastic Processes, *Second Edition*
BHATTACHARYA and WAYMIRE · Stochastic Processes with Applications
BIRKES and DODGE · Alternative Methods of Regression
BLOOMFIELD · Fourier Analysis of Time Series: An Introduction
BOLLEN · Structural Equations with Latent Variables
BOULEAU · Numerical Methods for Stochastic Processes
BOX · Bayesian Inference in Statistical Analysis
BOX and DRAPER · Empirical Model-Building and Response Surfaces
BOX and DRAPER · Evolutionary Operation: A Statistical Method for Process Improvement
BUCKLEW · Large Deviation Techniques in Decision, Simulation, and Estimation
BUNKE and BUNKE · Nonlinear Regression, Functional Relations and Robust Methods: Statistical Methods of Model Building
CHATTERJEE and HADI · Sensitivity Analysis in Linear Regression
CHILÈS and DELFINER · Geostatistics: Modeling Spatial Uncertainty
CHOW and LIU · Design and Analysis of Clinical Trials: Concepts and Methodologies
CLARKE and DISNEY · Probability and Random Processes: A First Course with Applications, *Second Edition*
*COCHRAN and COX · Experimental Designs, *Second Edition*
CONOVER · Practical Nonparametric Statistics, *Second Edition*
CORNELL · Experiments with Mixtures, Designs, Models, and the Analysis of Mixture Data, *Second Edition*
*COX · Planning of Experiments
CRESSIE · Statistics for Spatial Data, *Revised Edition*
DANIEL · Applications of Statistics to Industrial Experimentation
DANIEL · Biostatistics: A Foundation for Analysis in the Health Sciences, *Sixth Edition*
DAVID · Order Statistics, *Second Edition*
*DEGROOT, FIENBERG, and KADANE · Statistics and the Law
DODGE · Alternative Methods of Regression
DOWDY and WEARDEN · Statistics for Research, *Second Edition*
DRYDEN and MARDIA · Statistical Shape Analysis
DUNN and CLARK · Applied Statistics: Analysis of Variance and Regression, *Second Edition*
ELANDT-JOHNSON and JOHNSON · Survival Models and Data Analysis
EVANS, PEACOCK, and HASTINGS · Statistical Distributions, *Second Edition*
FLEISS · The Design and Analysis of Clinical Experiments
FLEISS · Statistical Methods for Rates and Proportions, *Second Edition*
FLEMING and HARRINGTON · Counting Processes and Survival Analysis
GALLANT · Nonlinear Statistical Models
GLASSERMAN and YAO · Monotone Structure in Discrete-Event Systems
GNANADESIKAN · Methods for Statistical Data Analysis of Multivariate Observations, *Second Edition*
GOLDSTEIN and LEWIS · Assessment: Problems, Development, and Statistical Issues
GREENWOOD and NIKULIN · A Guide to Chi-Squared Testing
*HAHN · Statistical Models in Engineering
HAHN and MEEKER · Statistical Intervals: A Guide for Practitioners
HAND · Construction and Assessment of Classification Rules

*Now available in a lower priced paperback edition in the Wiley Classics Library.

Applied Probability and Statistics (Continued)

HAND · Discrimination and Classification
HEIBERGER · Computation for the Analysis of Designed Experiments
HINKELMAN and KEMPTHORNE: · Design and Analysis of Experiments, Volume 1: Introduction to Experimental Design
HOAGLIN, MOSTELLER, and TUKEY · Exploratory Approach to Analysis of Variance
HOAGLIN, MOSTELLER, and TUKEY · Exploring Data Tables, Trends and Shapes
HOAGLIN, MOSTELLER, and TUKEY · Understanding Robust and Exploratory Data Analysis
HOCHBERG and TAMHANE · Multiple Comparison Procedures
HOCKING · Methods and Applications of Linear Models: Regression and the Analysis of Variables
HOGG and KLUGMAN · Loss Distributions
HOSMER and LEMESHOW · Applied Logistic Regression
HØYLAND and RAUSAND · System Reliability Theory: Models and Statistical Methods
HUBERTY · Applied Discriminant Analysis
JACKSON · A User's Guide to Principle Components
JOHN · Statistical Methods in Engineering and Quality Assurance
JOHNSON · Multivariate Statistical Simulation
JOHNSON and KOTZ · Distributions in Statistics
Continuous Multivariate Distributions
JOHNSON, KOTZ, and BALAKRISHNAN · Continuous Univariate Distributions, Volume 1, *Second Edition*
JOHNSON, KOTZ, and BALAKRISHNAN · Continuous Univariate Distributions, Volume 2, *Second Edition*
JOHNSON, KOTZ, and BALAKRISHNAN · Discrete Multivariate Distributions
JOHNSON, KOTZ, and KEMP · Univariate Discrete Distributions, *Second Edition*
JUREČKOVÁ and SEN · Robust Statistical Procedures: Aymptotics and Interrelations
KADANE · Bayesian Methods and Ethics in a Clinical Trial Design
KADANE AND SCHUM · A Probabilistic Analysis of the Sacco and Vanzetti Evidence
KALBFLEISCH and PRENTICE · The Statistical Analysis of Failure Time Data
KELLY · Reversability and Stochastic Networks
KHURI, MATHEW, and SINHA · Statistical Tests for Mixed Linear Models
KLUGMAN, PANJER, and WILLMOT · Loss Models: From Data to Decisions
KLUGMAN, PANJER, and WILLMOT · Solutions Manual to Accompany Loss Models: From Data to Decisions
KOVALENKO, KUZNETZOV, and PEGG · Mathematical Theory of Reliability of Time-Dependent Systems with Practical Applications
LAD · Operational Subjective Statistical Methods: A Mathematical, Philosophical, and Historical Introduction
LANGE, RYAN, BILLARD, BRILLINGER, CONQUEST, and GREENHOUSE · Case Studies in Biometry
LAWLESS · Statistical Models and Methods for Lifetime Data
LEE · Statistical Methods for Survival Data Analysis, *Second Edition*
LEPAGE and BILLARD · Exploring the Limits of Bootstrap
LINHART and ZUCCHINI · Model Selection
LITTLE and RUBIN · Statistical Analysis with Missing Data
LLOYD · The Statistical Analysis of Categorical Data
MAGNUS and NEUDECKER · Matrix Differential Calculus with Applications in Statistics and Econometrics
MALLER and ZHOU · Survival Analysis with Long Term Survivors
MANN, SCHAFER, and SINGPURWALLA · Methods for Statistical Analysis of Reliability and Life Data
McLACHLAN and KRISHNAN · The EM Algorithm and Extensions

*Now available in a lower priced paperback edition in the Wiley Classics Library.

Applied Probability and Statistics (Continued)

McLACHLAN · Discriminant Analysis and Statistical Pattern Recognition
McNEIL · Epidemiological Research Methods
MEEKER and ESCOBAR · Statistical Methods for Reliability Data
MILLER · Survival Analysis
MONTGOMERY and PECK · Introduction to Linear Regression Analysis, *Second Edition*
MYERS and MONTGOMERY · Response Surface Methodology: Process and Product in Optimization Using Designed Experiments
NELSON · Accelerated Testing, Statistical Models, Test Plans, and Data Analyses
NELSON · Applied Life Data Analysis
OCHI · Applied Probability and Stochastic Processes in Engineering and Physical Sciences
OKABE, BOOTS, and SUGIHARA · Spatial Tesselations: Concepts and Applications of Voronoi Diagrams
PANKRATZ · Forecasting with Dynamic Regression Models
PANKRATZ · Forecasting with Univariate Box-Jenkins Models: Concepts and Cases
PIANTADOSI · Clinical Trials: A Methodologic Perspective
PORT · Theoretical Probability for Applications
PUTERMAN · Markov Decision Processes: Discrete Stochastic Dynamic Programming
RACHEV · Probability Metrics and the Stability of Stochastic Models
RÉNYI · A Diary on Information Theory
RIPLEY · Spatial Statistics
RIPLEY · Stochastic Simulation
ROUSSEEUW and LEROY · Robust Regression and Outlier Detection
RUBIN · Multiple Imputation for Nonresponse in Surveys
RUBINSTEIN · Simulation and the Monte Carlo Method
RUBINSTEIN and MELAMED · Modern Simulation and Modeling
RYAN · Statistical Methods for Quality Improvement
SCHUSS · Theory and Applications of Stochastic Differential Equations
SCOTT · Multivariate Density Estimation: Theory, Practice, and Visualization
*SEARLE · Linear Models
SEARLE · Linear Models for Unbalanced Data
SEARLE, CASELLA, and McCULLOCH · Variance Components
SENNOTT · Stochastic Dynamic Programming and the Control of Queueing Systems
STOYAN, KENDALL, and MECKE · Stochastic Geometry and Its Applications, *Second Edition*
STOYAN and STOYAN · Fractals, Random Shapes and Point Fields: Methods of Geometrical Statistics
THOMPSON · Empirical Model Building
THOMPSON · Sampling
TIJMS · Stochastic Modeling and Analysis: A Computational Approach
TIJMS · Stochastic Models: An Algorithmic Approach
TITTERINGTON, SMITH, and MAKOV · Statistical Analysis of Finite Mixture Distributions
UPTON and FINGLETON · Spatial Data Analysis by Example, Volume 1: Point Pattern and Quantitative Data
UPTON and FINGLETON · Spatial Data Analysis by Example, Volume II: Categorical and Directional Data
VAN RIJCKEVORSEL and DE LEEUW · Component and Correspondence Analysis
WEISBERG · Applied Linear Regression, *Second Edition*
WESTFALL and YOUNG · Resampling-Based Multiple Testing: Examples and Methods for p-Value Adjustment
WHITTLE · Systems in Stochastic Equilibrium
WOODING · Planning Pharmaceutical Clinical Trials: Basic Statistical Principles

*Now available in a lower priced paperback edition in the Wiley Classics Library.

Applied Probability and Statistics (Continued)

WOOLSON · Statistical Methods for the Analysis of Biomedical Data
*ZELLNER · An Introduction to Bayesian Inference in Econometrics

Texts and References Section

AGRESTI · An Introduction to Categorical Data Analysis
ANDERSON · An Introduction to Multivariate Statistical Analysis, *Second Edition*
ANDERSON and LOYNES · The Teaching of Practical Statistics
ARMITAGE and COLTON · Encyclopedia of Biostatistics: Volumes 1 to 6 with Index
BARTOSZYNSKI and NIEWIADOMSKA-BUGAJ · Probability and Statistical Inference
BERRY, CHALONER, and GEWEKE · Bayesian Analysis in Statistics and Econometrics: Essays in Honor of Arnold Zellner
BHATTACHARYA and JOHNSON · Statistical Concepts and Methods
BILLINGSLEY · Probability and Measure, *Second Edition*
BOX · R. A. Fisher, the Life of a Scientist
BOX, HUNTER, and HUNTER · Statistics for Experimenters: An Introduction to Design, Data Analysis, and Model Building
BOX and LUCEÑO · Statistical Control by Monitoring and Feedback Adjustment
BROWN and HOLLANDER · Statistics: A Biomedical Introduction
CHATTERJEE and PRICE · Regression Analysis by Example, *Second Edition*
COOK and WEISBERG · An Introduction to Regression Graphics
COX · A Handbook of Introductory Statistical Methods
DILLON and GOLDSTEIN · Multivariate Analysis: Methods and Applications
DODGE and ROMIG · Sampling Inspection Tables, *Second Edition*
DRAPER and SMITH · Applied Regression Analysis, *Third Edition*
DUDEWICZ and MISHRA · Modern Mathematical Statistics
DUNN · Basic Statistics: A Primer for the Biomedical Sciences, *Second Edition*
FISHER and VAN BELLE · Biostatistics: A Methodology for the Health Sciences
FREEMAN and SMITH · Aspects of Uncertainty: A Tribute to D. V. Lindley
GROSS and HARRIS · Fundamentals of Queueing Theory, *Third Edition*
HALD · A History of Probability and Statistics and their Applications Before 1750
HALD · A History of Mathematical Statistics from 1750 to 1930
HELLER · MACSYMA for Statisticians
HOEL · Introduction to Mathematical Statistics, *Fifth Edition*
HOLLANDER and WOLFE · Nonparametric Statistical Methods, *Second Edition*
HOSMER and LEMESHOW · Applied Survival Analysis: Regression Modeling of Time to Event Data
JOHNSON and BALAKRISHNAN · Advances in the Theory and Practice of Statistics: A Volume in Honor of Samuel Kotz
JOHNSON and KOTZ (editors) · Leading Personalities in Statistical Sciences: From the Seventeenth Century to the Present
JUDGE, GRIFFITHS, HILL, LÜTKEPOHL, and LEE · The Theory and Practice of Econometrics, *Second Edition*
KHURI · Advanced Calculus with Applications in Statistics
KOTZ and JOHNSON (editors) · Encyclopedia of Statistical Sciences: Volumes 1 to 9 wtih Index
KOTZ and JOHNSON (editors) · Encyclopedia of Statistical Sciences: Supplement Volume
KOTZ, REED, and BANKS (editors) · Encyclopedia of Statistical Sciences: Update Volume 1
KOTZ, REED, and BANKS (editors) · Encyclopedia of Statistical Sciences: Update Volume 2
LAMPERTI · Probability: A Survey of the Mathematical Theory, *Second Edition*

*Now available in a lower priced paperback edition in the Wiley Classics Library.

Texts and References (Continued)

LARSON · Introduction to Probability Theory and Statistical Inference, *Third Edition*
LE · Applied Categorical Data Analysis
LE · Applied Survival Analysis
MALLOWS · Design, Data, and Analysis by Some Friends of Cuthbert Daniel
MARDIA · The Art of Statistical Science: A Tribute to G. S. Watson
MASON, GUNST, and HESS · Statistical Design and Analysis of Experiments with Applications to Engineering and Science
MURRAY · X-STAT 2.0 Statistical Experimentation, Design Data Analysis, and Nonlinear Optimization
PURI, VILAPLANA, and WERTZ · New Perspectives in Theoretical and Applied Statistics
RENCHER · Methods of Multivariate Analysis
RENCHER · Multivariate Statistical Inference with Applications
ROSS · Introduction to Probability and Statistics for Engineers and Scientists
ROHATGI · An Introduction to Probability Theory and Mathematical Statistics
RYAN · Modern Regression Methods
SCHOTT · Matrix Analysis for Statistics
SEARLE · Matrix Algebra Useful for Statistics
STYAN · The Collected Papers of T. W. Anderson: 1943–1985
TIERNEY · LISP-STAT: An Object-Oriented Environment for Statistical Computing and Dynamic Graphics
WONNACOTT and WONNACOTT · Econometrics, *Second Edition*

WILEY SERIES IN PROBABILITY AND STATISTICS

ESTABLISHED BY WALTER A. SHEWHART AND SAMUEL S. WILKS

Editors
Robert M. Groves, Graham Kalton, J. N. K. Rao, Norbert Schwarz, Christopher Skinner

Survey Methodology Section

BIEMER, GROVES, LYBERG, MATHIOWETZ, and SUDMAN · Measurement Errors in Surveys
COCHRAN · Sampling Techniques, *Third Edition*
COUPER, BAKER, BETHLEHEM, CLARK, MARTIN, NICHOLLS, and O'REILLY (editors) · Computer Assisted Survey Information Collection
COX, BINDER, CHINNAPPA, CHRISTIANSON, COLLEDGE, and KOTT (editors) · Business Survey Methods
*DEMING · Sample Design in Business Research
DILLMAN · Mail and Telephone Surveys: The Total Design Method
GROVES and COUPER · Nonresponse in Household Interview Surveys
GROVES · Survey Errors and Survey Costs
GROVES, BIEMER, LYBERG, MASSEY, NICHOLLS, and WAKSBERG · Telephone Survey Methodology
*HANSEN, HURWITZ, and MADOW · Sample Survey Methods and Theory, Volume I: Methods and Applications
*HANSEN, HURWITZ, and MADOW · Sample Survey Methods and Theory, Volume II: Theory

*Now available in a lower priced paperback edition in the Wiley Classics Library.

Survey Methodology (Continued)

KISH · Statistical Design for Research

*KISH · Survey Sampling

LESSLER and KALSBEEK · Nonsampling Error in Surveys

LEVY and LEMESHOW · Sampling of Populations: Methods and Applications, *Third Edition*

LYBERG, BIEMER, COLLINS, de LEEUW, DIPPO, SCHWARZ, TREWIN (editors) · Survey Measurement and Process Quality

SKINNER, HOLT, and SMITH · Analysis of Complex Surveys

*Now available in a lower priced paperback edition in the Wiley Classics Library.